Structure, Activity, and Function of Protein Methyltransferases

Structure, Activity, and Function of Protein Methyltransferases

Editors

Albert Jeltsch
Arunkumar Dhayalan

MDPI • Basel • Beijing • Wuhan • Barcelona • Belgrade • Manchester • Tokyo • Cluj • Tianjin

Editors
Albert Jeltsch
Institute of Biochemistry and
Technical Biochemistry
Univesity of Stuttgart
Stuttgart
Germany

Arunkumar Dhayalan
Department of Biotechnology
Pondicherry University
Puducherry
India

Editorial Office
MDPI
St. Alban-Anlage 66
4052 Basel, Switzerland

This is a reprint of articles from the Special Issue published online in the open access journal *Life* (ISSN 2075-1729) (available at: www.mdpi.com/journal/life/special_issues/protein_methyltransferases).

For citation purposes, cite each article independently as indicated on the article page online and as indicated below:

LastName, A.A.; LastName, B.B.; LastName, C.C. Article Title. *Journal Name* **Year**, *Volume Number*, Page Range.

ISBN 978-3-0365-4139-6 (Hbk)
ISBN 978-3-0365-4140-2 (PDF)

© 2022 by the authors. Articles in this book are Open Access and distributed under the Creative Commons Attribution (CC BY) license, which allows users to download, copy and build upon published articles, as long as the author and publisher are properly credited, which ensures maximum dissemination and a wider impact of our publications.
The book as a whole is distributed by MDPI under the terms and conditions of the Creative Commons license CC BY-NC-ND.

Contents

About the Editors vii

Preface to "Structure, Activity, and Function of Protein Methyltransferases" ix

Arunkumar Dhayalan and Albert Jeltsch
Special Issue "Structure, Activity, and Function of Protein Methyltransferases"
Reprinted from: *Life* **2022**, *12*, 405, doi:10.3390/life12030405 1

Alexia Klonou, Sarantis Chlamydas and Christina Piperi
Structure, Activity and Function of the MLL2 (KMT2B) Protein Lysine Methyltransferase
Reprinted from: *Life* **2021**, *11*, 823, doi:10.3390/life11080823 5

Sara Weirich, Mina S. Khella and Albert Jeltsch
Structure, Activity and Function of the Suv39h1 and Suv39h2 Protein Lysine Methyltransferases
Reprinted from: *Life* **2021**, *11*, 703, doi:10.3390/life11070703 19

Coralie Poulard, Lara M. Noureddine, Ludivine Pruvost and Muriel Le Romancer
Structure, Activity, and Function of the Protein Lysine Methyltransferase G9a
Reprinted from: *Life* **2021**, *11*, 1082, doi:10.3390/life11101082 35

Mariam Markouli, Dimitrios Strepkos and Christina Piperi
Structure, Activity and Function of the SETDB1 Protein Methyltransferase
Reprinted from: *Life* **2021**, *11*, 817, doi:10.3390/life11080817 61

Samantha Tauchmann and Juerg Schwaller
NSD1: A Lysine Methyltransferase between Developmental Disorders and Cancer
Reprinted from: *Life* **2021**, *11*, 877, doi:10.3390/life11090877 83

Philipp Rathert
Structure, Activity and Function of the NSD3 Protein Lysine Methyltransferase
Reprinted from: *Life* **2021**, *11*, 726, doi:10.3390/life11080726 99

Alexandra Daks, Elena Vasileva, Olga Fedorova, Oleg Shuvalov and Nickolai A. Barlev
The Role of Lysine Methyltransferase SET7/9 in Proliferation and Cell Stress Response
Reprinted from: *Life* **2022**, *12*, 362, doi:10.3390/life12030362 113

Magnus E. Jakobsson
Structure, Activity and Function of the Dual Protein Lysine and Protein N-Terminal Methyltransferase METTL13
Reprinted from: *Life* **2021**, *11*, 1121, doi:10.3390/life11111121 127

Michael Tellier
Structure, Activity, and Function of SETMAR Protein Lysine Methyltransferase
Reprinted from: *Life* **2021**, *11*, 1342, doi:10.3390/life11121342 139

Charlène Thiebaut, Louisane Eve, Coralie Poulard and Muriel Le Romancer
Structure, Activity, and Function of PRMT1
Reprinted from: *Life* **2021**, *11*, 1147, doi:10.3390/life11111147 157

Vincent Cura and Jean Cavarelli
Structure, Activity and Function of the PRMT2 Protein Arginine Methyltransferase
Reprinted from: *Life* **2021**, *11*, 1263, doi:10.3390/life11111263 181

Aishat Motolani, Matthew Martin, Mengyao Sun and Tao Lu
The Structure and Functions of PRMT5 in Human Diseases
Reprinted from: *Life* **2021**, *11*, 1074, doi:10.3390/life11101074 . . . . **195**

Somlee Gupta, Rajashekar Varma Kadumuri, Anjali Kumari Singh, Sreenivas Chavali and Arunkumar Dhayalan
Structure, Activity and Function of the Protein Arginine Methyltransferase 6
Reprinted from: *Life* **2021**, *11*, 951, doi:10.3390/life11090951 . . . . **211**

Levon Halabelian and Dalia Barsyte-Lovejoy
Structure and Function of Protein Arginine Methyltransferase PRMT7
Reprinted from: *Life* **2021**, *11*, 768, doi:10.3390/life11080768 . . . . **233**

Rui Dong, Xuejun Li and Kwok-On Lai
Activity and Function of the PRMT8 Protein Arginine Methyltransferase in Neurons
Reprinted from: *Life* **2021**, *11*, 1132, doi:10.3390/life11111132 . . . . **247**

Apolonia Witecka, Sebastian Kwiatkowski, Takao Ishikawa and Jakub Drozak
The Structure, Activity, and Function of the SETD3 Protein Histidine Methyltransferase
Reprinted from: *Life* **2021**, *11*, 1040, doi:10.3390/life11101040 . . . . **259**

About the Editors

Albert Jeltsch

Prof. Albert Jeltsch studied Biochemistry in Hannover, where he received his PhD in 1994. After completing his PostDoc and Assistant Professorship (2003) at Justus-Liebig University in Giessen, he was appointed as an Associate Professor of Biochemistry at Jacobs University Bremen in 2003, where he became a Full Professor in 2006. Prof. Jeltsch moved to University of Stuttgart in 2011, where he heads the Department of Biochemistry at the Institute of Biochemistry and Technical Biochemistry. The group of Prof. Jeltsch investigates the structure, mechanism and function of bacterial and mammalian DNA methyltransferases. They have long standing expertise in the field of rational and evolutionary protein design of DNA-interacting enzymes and in the design of chimeric methylation enzymes for epigenome editing and gene regulation in eukaryotic cells. In addition, they study the specificity and activity of protein lysine methyltransferases and methyllysine-reading domains and the biological role of protein methylation in cells.

Arunkumar Dhayalan

Dr. Arunkumar Dhayalan completed his PhD in Biochemistry in the year 2009 at the Jacobs University, Germany. After his post-doctoral studies at the same University, he joined Pondicherry University, India, as an Assistant Professor in the year 2010. He became an Associate Professor in the year 2021. At Pondicherry University, Dr. Dhayalan is working in two major areas, viz., (i) protein arginine methyltransferases and readers of histone arginine modifications and (ii) the functional characterization of the DEAD box family of RNA helicases.

Preface to "Structure, Activity, and Function of Protein Methyltransferases"

Protein methylation is an essential post-translational modification of histone and non-histone proteins involved in numerous important biological processes. This book presents a collection of review articles on individual protein methyltransferase enzymes written by leading experts in the field. It includes review articles on protein lysine methyltransferases and protein arginine methyltransferases, as well as the less abundant protein histidine methyltransferases and protein N-terminal end methyltransferases. The topics covered in the individual reviews include structural aspects (domain architecture, homologs and paralogs, and structure), biochemical properties (mechanism, sequence specificity, product specificity, regulation, and histone and non-histone substrates), cellular features (subcellular localization, expression patterns, cellular roles and function, biological effects of substrate protein methylation, connection to cell signaling pathways, and connection to chromatin regulation) and the role of protein methyltransferases in diseases. The reviews also provide an outlook, open questions, and directions for future research. We are optimistic that this review book is a useful resource for scientists working on protein methylation and protein methyltransferases and those interested in joining this emerging research field.

Albert Jeltsch and Arunkumar Dhayalan
Editors

Editorial

Special Issue "Structure, Activity, and Function of Protein Methyltransferases"

Arunkumar Dhayalan [1,*] and Albert Jeltsch [2,*]

1. Department of Biotechnology, Pondicherry University, Puducherry 605014, India
2. Institute of Biochemistry and Technical Biochemistry, University of Stuttgart, Allmandring 31, 70569 Stuttgart, Germany
* Correspondence: arun.dbt@pondiuni.edu.in (A.D.); albert.jeltsch@ibtb.uni-stuttgart.de (A.J.)

Post-translational modifications (PTMs) largely expand the functional diversity of the proteome [1]. Protein methylation is an essential PTM, which regulates numerous cellular events by altering the functionality of proteins [2]. The methylation of histones regulates the chromatin structure and participates in the epigenetic regulation of gene expression in diverse biological processes, including development and differentiation [3]. Methylation also controls the activity of numerous non-histone proteins [2,4], where it often plays key roles in the regulation of their (i) stability, (ii) enzymatic activity, (iii) sub-cellular distribution, and (iv) interactions with other proteins. Aberrant protein methylation is implicated in various pathologies, including cancers [5,6].

Protein methylation mainly occurs at lysine and arginine residues and is catalyzed by protein lysine methyltransferases (PKMTs) and protein arginine methyltransferases (PRMTs), respectively. Protein methylation also occurs at the N-terminal α-amino group of proteins and other amino acids such as histidine and glutamine [4] (Figure 1). PKMTs methylate the lysine residues of proteins at three different levels and generate (i) monomethyllysine, (ii) dimethyllysine, and (iii) trimethyllysine (Figure 1). PKMT activity is exhibited by two protein domain families: (i) the SET domain-containing enzymes and (ii) the seven β strand domain-containing enzymes (7BS) [7,8]. The 7BS methyltransferase family is larger and contains several enzymes that methylate a wide range of substrates including arginine residues, DNA and RNA, in addition to the lysine residues [4,9]. It was predicted that the human genome encodes more than 100 PKMTs [10]. In contrast, PRMTs are only found in the 7BS family. The human genome contains at least nine different PRMTs, which are grouped into three types based on the nature of the methylarginine produced upon their enzymatic activity on arginine residues. Type I, II, and III PRMT enzymes catalyze the formation of asymmetric dimethylarginine, symmetric dimethylarginine, and monomethylarginine, respectively [11,12] (Figure 1).

A PubMed search for the terms "methylation" AND "lysine", "methylation" AND "arginine", or "methylation" AND "histidine" resulted in 622, 202 and 20 publications, respectively, for the year 2021 alone, suggesting that protein methylation research is a very exciting and active area of research. The goal of this thematic Special Issue is to collect and compile focused reviews about individual PMTs written by specialists in the field, which, according to our literature research, is currently not available for the majority of PMT enzymes. All review articles in this Special Issue cover central topics such as structure, biochemistry, cellular functions, and association with diseases, if any, and provide future perspectives. This Special Issue addresses an urgent demand in the field; currently, reviews on PMTs often analyze several unrelated enzymes, so the details and peculiarities of each one are not explored. The collection in this Special Issue, hopefully, will become a useful resource for researchers in the entire protein methylation and protein methyltransferase field.

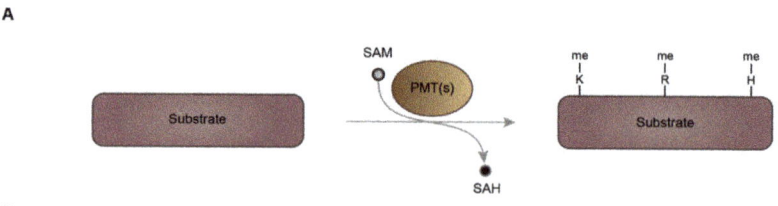

Figure 1. Methylation of proteins by PMTs. (**A**) Schema depicting the methylation of proteins by various PMTs at lysine (K), arginine (R), and histidine (H) residues. (**B**) The possible types of methylation modifications at lysine, arginine, and histidine residues. SAM, S-adenosyl-L-methionine; SAH, S-adenosyl-L-homocysteine.

Altogether, this Special Issue contains nine articles on individual PKMTs, six articles on individual PRMTs, and one article on the SETD3 histidine methyltransferase. In the context of PKMTs, Weirich et al. [13] reviewed the PKMTs SUV39H1 and SUV39H2 thoroughly, with a special focus on their substrate specificity profiles and non-histone protein substrates. Tauchmann and Schwaller [14] provided a review on the NSD1 PKMT, with detailed coverage on the role of NSD1 in developmental disorders and cancers. Rathert [15] presented the current literature status of the NSD3 PKMT, including structural aspects and the role of NSD3 in cancers. Klonou et al. [16] detailed the subunit composition of the MLL2 protein complex and discussed the role of the MLL2 complex in transcription and cellular functions. Poulard et al. [17] provided an elaborate overview of the G9a PKMT, covering various aspects, including structural features, non-histone substrates of G9a, the role of G9a on chromatin regulation, and its function in development, DNA repair, and diverse types of cancers. Markouli et al. [18] reviewed the structural, biochemical, and functional aspects of SETDB1 PKMT and discussed the role of SETDB1 in the physiological processes such as cell division, the formation of Promyelocytic leukemia nuclear bodies (PML-NBs), and the development of the nervous system and pathological conditions such as various types of cancers, neuropsychiatric diseases, genetic diseases, cardiovascular, and gastrointestinal diseases. Tellier [19] provided a review on various aspects of the SETMAR PKMT and discussed the role of SETMAR in non-homologous end joining (NHEJ) DNA repair pathway, restarting the collapsed replication fork, and chromosome decatenation in a detailed manner. Daks et al. [20] described the structure, substrate specificity, and cellular functions of SET7/9 PKMT, with particular focus on non-histone substrates and its role in cell proliferation and stress response. Finally, Jakobsson [21] covered the structural and biochemical features of the dual lysine methyltransferase METTL13, which is capable of catalyzing both N-terminal and lysine methylation, and highlighted its role in the regulation of global translation dynamics.

In addition to the above-mentioned PKMTs, this Special Issue also contains review articles about different PRMTs. Thiebaut et al. [22] reviewed different aspects of the major type I protein arginine methyltransferase PRMT1 and discussed the non-histone substrates of PRMT1, the role of PRMT1 in various cell signaling pathways, and DNA repair extensively. Cura and Cavarelli [23] provided a review on PRMT2, focusing on its structural features and its role in splicing. Motolani et al. [24] reviewed the functionally versatile major type II protein arginine methyltransferase PRMT5, focusing on its role in various human diseases such as various types of cancers, diabetes, cardiovascular, and neurodegenerative diseases. Gupta et al. [25] presented the current literature status of PRMT6 and covered topics such as structural features, kinetic mechanism, epigenetic functions, and non-histone substrates of PRMT6. Halebelian and Barsyte-Lovejoy [26] provided a comprehensive overview on various aspects of type III protein arginine methyltransferase PRMT7 and elaborated its role on gene expression, genome maintenance, pluripotency, differentiation, senescence, and stress response. Dong et al. [27] wrote a detailed review on the neuronal functions of the PRMT8, which is expressed exclusively in the brain, and discussed its potential role in neurological diseases.

This Special Issue also contains a review on the histidine methyltransferase SETD3. Witecka et al. [28] describe and discuss structural, biochemical, and functional aspects of SETD3 and highlight its role in actin polymerization and pathological conditions such as cancers.

We envisage that this Special Issue will be of interest to researchers of the protein methylation field and will promote further research on the writers of protein methylome and the functional outcomes of these chemical modifications.

Author Contributions: Conceptualization, A.D. and A.J. writing, A.D. and A.J. All authors have read and agreed to the published version of the manuscript.

Conflicts of Interest: The authors declare no conflict of interest.

References

1. Walsh, C.T.; Garneau-Tsodikova, S.; Gatto, G.J. Protein posttranslational modifications: The chemistry of proteome diversifications. *Angew. Chem. Int. Ed.* **2005**, *44*, 7342–7372. [CrossRef]
2. Cornett, E.M.; Ferry, L.; Defossez, P.A.; Rothbart, S.B. Lysine Methylation Regulators Moonlighting outside the Epigenome. *Mol. Cell* **2019**, *75*, 1092–1101. [CrossRef] [PubMed]
3. Jambhekar, A.; Dhall, A.; Shi, Y. Roles and regulation of histone methylation in animal development. *Nat. Rev. Mol. Cell Biol.* **2019**, *20*, 625–641. [CrossRef] [PubMed]
4. Clarke, S.G. Protein methylation at the surface and buried deep: Thinking outside the histone box. *Trends Biochem. Sci.* **2013**, *38*, 243–252. [CrossRef]
5. Biggar, K.K.; Li, S.S.C. Non-histone protein methylation as a regulator of cellular signalling and function. *Nat. Rev. Mol. Cell Biol.* **2015**, *16*, 5–17. [CrossRef] [PubMed]
6. Zhao, S.; Allis, C.D.; Wang, G.G. The language of chromatin modification in human cancers. *Nat. Rev. Cancer* **2021**, *21*, 413–430. [CrossRef] [PubMed]

7. Boriack-Sjodin, P.A.; Swinger, K.K. Protein Methyltransferases: A Distinct, Diverse, and Dynamic Family of Enzymes. *Biochemistry* **2016**, *55*, 1557–1569. [CrossRef]
8. Falnes, P.; Jakobsson, M.E.; Davydova, E.; Ho, A.; Malecki, J. Protein lysine methylation by seven-β-strand methyltransferases. *Biochem. J.* **2016**, *473*, 1995–2009. [CrossRef]
9. Petrossian, T.C.; Clarke, S.G. Uncovering the human methyltransferasome. *Mol. Cell. Proteomics* **2011**, *10*, M110.000976. [CrossRef]
10. Husmann, D.; Gozani, O. Histone lysine methyltransferases in biology and disease. *Nat. Struct. Mol. Biol.* **2019**, *26*, 880–889. [CrossRef]
11. Blanc, R.S.; Richard, S. Arginine Methylation: The Coming of Age. *Mol. Cell* **2017**, *65*, 8–24. [CrossRef]
12. Lorton, B.M.; Shechter, D. Cellular consequences of arginine methylation. *Cell. Mol. Life Sci.* **2019**, *76*, 2933–2956. [CrossRef]
13. Weirich, S.; Khella, M.S.; Jeltsch, A. Structure, activity and function of the suv39h1 and suv39h2 protein lysine methyltransferases. *Life* **2021**, *11*, 703. [CrossRef]
14. Tauchmann, S.; Schwaller, J. Nsd1: A lysine methyltransferase between developmental disorders and cancer. *Life* **2021**, *11*, 877. [CrossRef]
15. Rathert, P. Structure, activity and function of the NSD3 protein lysine methyltransferase. *Life* **2021**, *11*, 726. [CrossRef]
16. Klonou, A.; Chlamydas, S.; Piperi, C. Structure, activity and function of the mll2 (Kmt2b) protein lysine methyltransferase. *Life* **2021**, *11*, 823. [CrossRef]
17. Poulard, C.; Noureddine, L.M.; Pruvost, L.; Le Romancer, M. Structure, activity, and function of the protein lysine methyltransferase g9a. *Life* **2021**, *11*, 1082. [CrossRef]
18. Markouli, M.; Strepkos, D.; Piperi, C. Structure, Activity and Function of the SETDB1 Protein Methyltransferase. *Life* **2021**, *11*, 817. [CrossRef]
19. Tellier, M. Structure, Activity, and Function of SETMAR Protein Lysine Methyltransferase. *Life* **2021**, *11*, 1342. [CrossRef]
20. Daks, A.; Vasileva, E.; Fedorova, O.; Shuvalov, O.; Barlev, N.A. The Role of Lysine Methyltransferase SET7/9 in Proliferation and Cell Stress Response. *Life* **2022**, *12*, 362. [CrossRef]
21. Jakobsson, M.E. Structure, activity and function of the dual protein lysine and protein n-terminal methyltransferase mettl13. *Life* **2021**, *11*, 1121. [CrossRef]
22. Thiebaut, C.; Eve, L.; Poulard, C.; Le Romancer, M. Structure, activity, and function of prmt1. *Life* **2021**, *11*, 1147. [CrossRef]
23. Cura, V.; Cavarelli, J. Structure, activity and function of the prmt2 protein arginine methyltransferase. *Life* **2021**, *11*, 1263. [CrossRef]
24. Motolani, A.; Martin, M.; Sun, M.; Lu, T. The structure and functions of prmt5 in human diseases. *Life* **2021**, *11*, 1074. [CrossRef] [PubMed]
25. Gupta, S.; Kadumuri, R.V.; Singh, A.K.; Chavali, S.; Dhayalan, A. Structure, activity and function of the protein arginine methyltransferase 6. *Life* **2021**, *11*, 951. [CrossRef] [PubMed]
26. Halabelian, L.; Barsyte-Lovejoy, D. Structure and function of protein arginine methyltransferase prmt7. *Life* **2021**, *11*, 768. [CrossRef] [PubMed]
27. Dong, R.; Li, X.; Lai, K.O. Activity and function of the prmt8 protein arginine methyltransferase in neurons. *Life* **2021**, *11*, 1132. [CrossRef] [PubMed]
28. Witecka, A.; Kwiatkowski, S.; Ishikawa, T.; Drozak, J. The structure, activity, and function of the setd3 protein histidine methyltransferase. *Life* **2021**, *11*, 1040. [CrossRef] [PubMed]

Review

Structure, Activity and Function of the MLL2 (KMT2B) Protein Lysine Methyltransferase

Alexia Klonou [1], Sarantis Chlamydas [1,2] and Christina Piperi [1,*]

1. Department of Biological Chemistry, Medical School, National and Kapodistrian University of Athens, 11527 Athens, Greece; alexiakl@med.uoa.gr (A.K.); schlamydas@med.uoa.gr (S.C.)
2. Research and Development Department, Active Motif, Inc., Carlsbad, CA 92008, USA
* Correspondence: cpiperi@med.uoa.gr; Tel.: +30-210-7462610

Abstract: The Mixed Lineage Leukemia 2 (MLL2) protein, also known as KMT2B, belongs to the family of mammalian histone H3 lysine 4 (H3K4) methyltransferases. It is a large protein of 2715 amino acids, widely expressed in adult human tissues and a paralog of the MLL1 protein. MLL2 contains a characteristic C-terminal SET domain responsible for methyltransferase activity and forms a protein complex with WRAD (WDR5, RbBP5, ASH2L and DPY30), host cell factors 1/2 (HCF 1/2) and Menin. The MLL2 complex is responsible for H3K4 trimethylation (H3K4me3) on specific gene promoters and nearby *cis*-regulatory sites, regulating bivalent developmental genes as well as stem cell and germinal cell differentiation gene sets. Moreover, MLL2 plays a critical role in development and germ line deletions of *Mll2* have been associated with early growth retardation, neural tube defects and apoptosis that leads to embryonic death. It has also been involved in the control of voluntary movement and the pathogenesis of early stage childhood dystonia. Additionally, tumor-promoting functions of MLL2 have been detected in several cancer types, including colorectal, hepatocellular, follicular cancer and gliomas. In this review, we discuss the main structural and functional aspects of the MLL2 methyltransferase with particular emphasis on transcriptional mechanisms, gene regulation and association with diseases.

Keywords: MLL2; structure; H3K4me3; chromatin regulation; disease; dystonia; cancer

1. Introduction

Chromatin remodeling is a key feature of gene regulation and activity, with histone modifications playing a primary role in the modulation of the chromatin landscape and gene expression. Among the most prominent histone modifications is the methylation of histone 3 (H3) lysine (K) residues, detected on gene enhancers and specific gene promoter regions. Mono- and di-methylation of H3K4 (H3K4me1/me2) is mainly observed in enhancers whereas H3K4me3 is present on active gene promoters. Several protein lysine methyltransferases (PKMTs), including Mixed Lineage Leukemia 1-5 (MLL1-5/KMT2A-E), SET Domain-Containing 7 (SET7), SET and MYND Domain-Containing 3 (SMYD3), SET9 and PR/SET Domain 9 (PRDM9), are responsible for the transfer of methyl groups onto H3K4. The largest group of human lysine 4 (K4) HMKTs is the Mixed Lineage Leukemia (MLL/KMT2) protein family, named after the association with a subset of incurable acute leukemias of its founding member. All family members are characterized by a highly conserved catalytically active Su(var)3-9, Enhancer of zeste and Trithorax (SET) domain [1].

In yeast, there is a single MLL homolog comprised of a SET domain (SET1) which catalyzes mono-, di- and tri-methylation of histone H3K4, whereas in *Drosophila melanogaster* there are three homologs, namely, Set1, Trithorax-related (Trr) and Trithorax (Trx), responsible for H3K4 methyltransferase activity [2]. The Trithorax group of proteins has been identified as regulators of Homeotic (Homeobox) genes in *Drosophila* and are essential for body patterning in multicellular organisms. Their activity is antagonized by the Polycomb group of proteins (PcP) which exerts the repressive role in Homeobox genes expression.

Mammalian cells possess six SET1-like H3K4 methyltransferases, including the four MLL1-4 family proteins and Set1A and Set1B (KMT2F, KMT2G). Sequence homology has shown that two human homologs exist for each of the H3K4 methyltransferase proteins in *Drosophila*. More specifically, MLL1/KMT2A and MLL2 (4)/KMT2B have a similar domain structure to Trx, MLL3/KMT2C and MLL4 (2)/KMT2D are homologous to Trr while SET1A and SET1B are homologous to Set1/dSet1 [2]. Although MLL5 (KMT2E) was originally considered as an MLL family member, its divergent SET domain from the other family members as well as the lack of lysine methyltransferase activity, have re-classified it to a different subgroup of SET domain proteins.

Of importance, the MLL family members deposit distinct H3K4 methylation states and target different genomic regions. SET1A/B enzymes establish global H3K4me3 levels through a crosstalk with the monoubiquitination of the H2B process, whereas MLL1 and MLL2 catalyze the H3K4me3 modifications at specific gene promoters. MLL2 further implements H3K4me3 at bivalently marked gene promoters, while MLL3/4 enzymes mediate H3K4me1 at transcription enhancers throughout the human genome [3,4].

Although a direct functional role of H3K4 in transcription is still under investigation, the aberrant transcription mediated by MLL family members has a significant impact in gene regulation and normal cell physiology with an ultimate connection to developmental disorders and cancer [5].

Herein, we discuss the major structural and biochemical characteristics of the MLL2 (KMT2B) methyltransferase with emphasis on its cellular and molecular functions as well as its connection to diseases.

2. The MLL2 Protein

Genome duplication during mammalian evolution resulted in two paralogs in each MLL subgroup (MLL1/KMT2A and MLL2(4)/KMT2B) which are analogous proteins within the Trx-related subgroup, referred to as the MLX family (MLL-TRX) [6].

The *MLL2 (KMT2B)* gene (OMIM 606834) is located on chromosome 19q13.12 and consists of an 8.5–9 kb transcript, spanning 20 kb of genomic DNA. It is expressed in most human tissues [7] and has a similar genomic structure with *MLL1*, present in chromosome 11q23.

The MLL2 protein is 2715 amino acids in length and its structural organization includes the catalytically active C-terminal SET domain, an AT hook, a CXXC domain and several plant homeotic domains (PHD) in the N-terminal region (Figure 1) [6,8]. The SET domain forms a pocket that binds to methyltransferase cofactor S-adenosylmethionine and the N-terminal tail of histone H3 catalyzing the methylation reaction [9]. Prior to the C-terminal SET domain, the MLL2 protein displays additional structurally distinctive characteristics which determine its non-redundant role and the intrinsic biochemical and molecular functions.

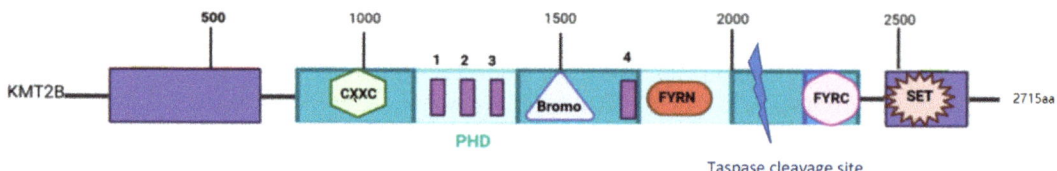

Figure 1. Domain structure of MLL2 (KMT2B) protein. MLL2 protein contains a CXXC domain composed of two zinc ions and four cysteine residues (CXXC), 4 plant homeotic domains (PHD) in the N-terminal region, a FY-rich N-terminal (FYRN) as well as a FY-rich C-terminal (FYRC) domain and a catalytically active C-terminal SET domain.

The CXXC domain composed of two zinc ions and four cysteine residues (Cys4), also known as the Zinc-finger (ZF)-CXXC domain, recognizes and binds to non-methylated CpG DNA, being critical for the association of MLL2 to chromatin [10]. Both MLL1 and

MLL2 contain a CXXC domain serving as a localization mechanism through the recognition of CpG islands present in most active promoters. However, MLL3/4 do not possess a CXXC domain as well as SETD1A/B which, however, are located at a complex with the CXXC domain-containing protein CFP1 which stabilizes them at promoters.

Next to the ZF-CXXC domain, MLL2 protein contains multiple PHD fingers, PHD1 to PHD4 [11] which possess a Cys4-His-Cys3 motif, coordinated by two zinc ions and mediating binding to methylated histone H3 [12]. Although all MLL family members contain PHD fingers, they exhibit different interaction specificities with the PHD3 of MLL2 being mostly involved in binding to H3K4me3 tails. Between PHD3 and PHD4, there is a bromodomain (BRD) which does not serve as a reader of acetylated lysine as commonly observed, but rather supports the PHD3 function [13]. Following the BRD, there is another PHD and a FY-rich N-terminal (FYRN) as well as a FY-rich C-terminal (FYRC) domain which allows the non-covalent dimerization of the N- and C-terminal fragments upon proteolytic cleavage [14–16].

MLL2 and MLL1 can be cleaved by threonine aspartase 1 (Taspase 1) [17]. Upon cleavage, the two fragments associate via a FYRN and FYRC domain interaction and form at the junction a new FYR domain [18] which has proved essential for methyltransferase activity [14–16]. Mice deficient in *Taspase 1* exhibit defects in cell proliferation and at the progression of the cell cycle, indicating the functional significance of the MLL2 cleavage [16].

Additional DNA-binding motifs have been detected in MLL1 and MLL2 in the form of multiple HMG-like N-terminal AT hooks which enable binding to AT-rich DNA, discriminating their binding activities from the other MLL family members [18].

In addition to the characteristic SET domain, MLL2 contains a CXXC domain followed by 4 PHD (PHD1-4), a single bromo domain (Bromo, BRD) as well as a FYRN and FYRC domain (created using BioRender).

3. The MLL2 Protein Complex

All MLL family proteins contain a highly conserved SET domain at their C-terminus. They all form multi-protein complexes known as COMPASS (complex of proteins associated with Set1) and COMPASS-like complexes based on their homology with Drosophila Trr, Trx and dSet1 [19,20]. These complexes share four common subunits, the so-called WRAD module.

The WRAD module regulates the enzymatic active form of the complex, confers stability and enables recruitment to chromatin. It is composed of WDR5 (WD repeat domain 5, homolog of Swd3), RbBP5 (retinoblastoma-binding protein 5, homolog of Swd1), ASH2L (absent, small or homeotic-2 like, homolog of Bre2) and DPY30 (homolog of Sdc1) subunits which are critical for H3K4 methylation activity [9,21–23].

Each COMPASS complex contains additional unique subunits on top of the main interacting proteins that enable their functional diversity (Figure 2). For MLL1/MLL2 COMPASS complexes, these proteins are Menin and host cell factors 1/2 (HCF1/2) [24] as well as the lens epithelium-derived growth factor (LEDGF), also named PSIP1/p75, which is capable of interacting indirectly with the complex through Menin.

Menin has been shown to be necessary for MLL1 target gene expression such as *Meis 1*, *Hoxa9*, *CDKN1B* and *CDKN2C*, which are required for MLL fusion protein-mediated leukemogenesis. The interaction of MLL1 and Menin forms a binding pocket for LEDGF which promotes transcriptional activation and is necessary for leukemogenesis. MLL2 shares the same interaction with Menin but not with LEDGF which is considered a unique coactivator of MLL1 complex activity. Apart of the nuclear member of the A-kinase anchoring protein family, AKAP95, MLL2 interacts only with very few interacting partners [25].

For the MLL3 and MLL4 COMPASS, the unique subunits are PTIP-associated 1 (PA1), PAX transactivation domain-interacting protein (PTIP), nuclear receptor coactivator 6 (NcoA6) and the H3K27me3 demethylase ubiquitously transcribed tetratricopeptide repeat X chromosome (UTX) [26–28], while the SET1A/B COMPASS complexes contain the WD repeat domain 82 (WDR82), CXXC finger protein 1 (CFP1) and HCF1 [29].

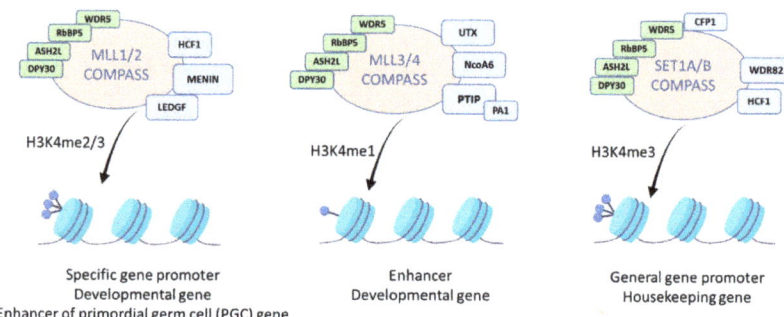

Figure 2. Subunit composition of the different COMPASS protein complexes showing their H3K4 methylation activities and distinctive genomic target regions.

The COMPASS complexes are responsible for mono-, di- and trimethylation of H3K4 [30] and according to their unique subunits, as well as the interaction with transcriptional regulators, the COMPASS complexes exhibit a differential specificity and genome localization. Specific reader proteins can recognize H3K4 methylation and connect the relevant information underlying this modification with the basal transcription machinery; thus, enhancing transcription. Currently, a range of writers, readers and erasers that bind to methylated H3K4 via different domains [such as PHD, ZF-CXXC, tandem Tudor domain (TTD) or double chromodomains (DCD)], has been detected and is summarized in Table 1 [31–33].

Table 1. H3K4 writers, readers and erasers.

Writers	Readers	Erasers
MLL1 MLL2 MLL3 MLL4 SET1A SET1B	BPTF (Bromodomain PHD Finger Transcription Factor) INGs (inhibitor of growth) RAG2 (Recombination Activating 2) TAF3 (TATA-Box Binding Protein Associated Factor 3) CHD1 (Chromodomain Helicase DNA Binding Protein 1)	JARID1A-D (Lysine-specific demethylase 5A, KDM5A) LSD1 (Lysine-specific histone demethylase 1A)/KDM1A (me1/2) JMJD2A (Jumonji domain-containing 2A)/KDM4A (Lysine Demethylase 4A) LSD1/KDM1B (me1/2)

Complex recruitment is mediated by several mechanisms to specific gene loci, either through binding to histone modifications, specific transcription factors, cofactors and long noncoding RNAs (lncRNAs). Of interest, both the core subunits as well as the complex-specific ones can interact with transcription factors to recruit the complexes to specific gene loci. Among them, Menin has been demonstrated to interact with estrogen receptor-α (ERα) and enables MLL2 recruitment to the gene locus [34]. Other transcription factors that associate with MLL complexes include the AP2δ (activating protein 2δ) [35], MYC [36], NF-E2 (nuclear factor, erythroid 2) [37], NF-Y (nuclear transcription factor Y) [38], USF1 (upstream transcription factor 1) [39], E2Fs [40], NANOG [41], PAX7 (Paired Box 7) [42] and p53 [43].

MLL1 has been shown to interact with transcription cofactors, including the lysine acetyltransferases MOZ, CBP and MOF, which mediate gene expression through H4K16 acetylation. Furthermore, MLL1, but not MLL2, has been demonstrated to bind to the PAF1 complex, serving as a bridge for RNA pol II, indicating a unique function of MLL1 [44]. Another distinct property of MLL1 is the interaction with repressive factors that results in negative regulation, including the PcG proteins HPC2, BMI-1 and HDAC1, c-terminal binding protein (CtBP) corepressors [45].

4. Structural Nucleosome Recognition by MLL Complexes

Recently, elegant single-particle cryo-electron microscopy (cryo-EM) studies have shed some light on the way MLL complexes recognize H3K4 within the nucleosome core particles (NCP).

The MLL1 complex was shown to dock on the NCP through the RbBP5 and ASH2L proteins which interact both with the nucleosomal DNA and the H4 tail. This configuration enabled the catalytic SET domain of MLL1 to align at the nucleosomal dyad and facilitate the symmetric access of H3K4 substrate to the NCP [46].

Additionally, there is evidence that the methylation of H3K4 can be induced by mono-ubiquitination of the histone H2B. A specific H2B mark on lysine 120 (H2BK120ub1) has been shown to disrupt chromatin compaction and allow the open chromatin structure. Cryo-EM studies of MLL1 and MLL3 have demonstrated their association with NCPs that contain H2BK120ub1 or unmodified H2BK120. The RBBP5 of MLL1 or MLL3 binds directly to H2B-conjugated ubiquitin. This interaction enables access to the H3 tail, which is required for H3K4 methylation. The differential organization of WDR5 and RBBP5 in MLL1 and MLL3 complexes accounts for their distinct enzymatic activities [23,47].

Another recent structural study demonstrated that the activity of the MLL family members on the NCP requires DPY30 [46]. It was shown that DPY30 interacts with the ASH2L intrinsically disordered regions (IDRs) to control MLL1 binding to NCP and regulate the complex activity. Of note, this interaction of DPY30-ASH2L-IDRs was shown to regulate all MLL family members regardless of their respective intrinsic activities. DPY30 was shown to affect global H3K4me3 levels by MLL1/MLL2, but also to mediate H3K4me1 by MLL3 in vitro. Moreover, DPY30 was essential for establishing *de novo* H3K4me3 in ESCs since its knockdown caused a global reduction in H3K4me3 [48].

Altogether, the core subunits of each MLL complex, as well as the proteins containing IDRs, exert important biophysical properties in MLL complexes and modulate their activity in chromatin.

5. MLL2 Role in Transcription Regulation

Several studies using Chromatin Immunoprecipitation (ChIP) followed by NGS sequencing (ChIP-seq) have shown that MLL2 can establish narrow H3K4me3 peaks at regions proximal to active gene promoters as well as co-exist with H3K27me3 in bivalent genes of embryonic stem cells (ESC). The purification of a minimal catalytical MLL2 complex (MLL2C) has demonstrated a specific methyltransferase activity for H3K4 methylation (H3K4me1/me2/me3) on recombinant histone octamers and recombinant chromatin. The stimulatory effect of the MLL2-related H3K4me on transcription has been validated using a well-established chromatin-templated in vitro transcription system [49]. Specifically, H3K4me3 at bivalent genes has been demonstrated to be mediated by the MLL2 COMPASS complex indicating an important role of MLL2 during development [50]. MLL2, but not MLL1, was shown to establish H3K4me3 on bivalent promoters in mouse ESCs (mESCs), thereby activating genes critical for the differentiation of stem cells [51–54]. Bivalent genes commonly harbor both H3K4me3 and H3K27me3 marks at the promoters in mESCs and are typically expressed only at very low levels. However, upon differentiation, they become either predominantly H3K27me3- or H3K4me3-marked and, subsequently, silenced or activated, respectively [55–57].

By using specific antibodies recognizing two different epitopes in the C-terminal portion of MLL2 (ab CT1 and more C-terminal ab CT2) and the ChIP-seq technique, a large number of MLL2-binding regions were identified with 70% localized to promoters, 14% to gene bodies and 16% to intergenic regions. The high occupancy of MLL2 in promoters was consistent with previous studies indicating its activity in bivalent genes. A further analysis with the ab CT1 revealed that more than 6000 MLL2 binding sites were outside of TSS, located in the intergenic and gene body regions. Moreover, the same study reported that around 39% of them shared marks with active enhancers, including p300 and H3K27ac [50]. Interestingly, a direct causal role for MLL2C-mediated H3K4 methylation was demonstrated in transcription activation and a combinatorial, synergistic effect of the p300 acetyltransferase responsible for H3K27Ac at the enhancer regions, indicating a role of MLL2 in the enhancer function in combination with histone acetylation marks [25]. Subsequently, MLL2 depletion in mESC infected with lentiviral shRNA and a ChIP-seq analysis demonstrated that MLL2 was responsible for establishing H3K4me3 marks at the non-TSS MLL2-associated genes [50].

In order to reveal the molecular mechanisms that underly MLL2 targeting to chromatin, the CRISP/Cas9 technology was used to generate MLL2 knockout mESC and was used for future rescue experiments. MLL2 was shown to bind unmethylated CpG-containing DNA indicating a possible involvement of the CXXC domain. Using several CXXC mutants, it was observed that MLL2 depends on its CXXC domain for recruitment to chromatin [50]. Through structural studies analyzing different CXXC domains in mammalian proteins, it was shown that the MLL2 CXXC domain is specific for unmethylated CpG regions in dsDNA, indicating a gene regulation role of MLL2 and a potential crosstalk with the methylation of DNA at promoters. Interestingly, CXXC domain swap experiments between MLL1 and MLL2 revealed different subnuclear localization and genomic binding patterns and, thus, a differential gene regulation [53,54]. Furthermore, it was shown that the small amino acid differences which are present around the CXXC domain of MLL2 guide it to target genes different from those of MLL1 [51]. Moreover, the CXXC domain of MLL1, but not of MLL2, associates to the Paf1/RNA Polymerase II (pol II) Complex Component (PAF1) transcription elongation complex [44,52].

Studies on MLL2 knockout mESC have demonstrated that MLL2 is not required for the expression of pluripotency genes (such as *Klf4, Oct4*), but is rather necessary for the expression of master regulators required for the primordial germ cell (PGC) specification, such as *Prdm1, 14, DDx4 and Lin28b*, during ESC differentiation [50]. Gene expression profiling and cell differentiation studies in parental MLL2 knockout and CXXC mutant mESC, demonstrated that the H3K4 methyltransferase activity of MLL2 as well as its CXXC domain are required for PGC induction [50]. Moreover, it was shown that the MLL2 COMPASS regulates the activity of enhancers and promoters of PGC gene regulators through H3K4 trimethylation, further indicating an essential role of H3K4me3 in the establishment of the differentiation of embryonic cells.

A functional interplay between H3K27me3, H3K4me3 and methylation of DNA has been detected to fine tune the expression of MLL2 gene targets in mammalian ES cells. Mechanistic studies have revealed a significant role of the MLL2 and SET1A/B complex in counteracting H3K27me3 and the methylation of DNA [49,58,59].

Of importance, a study showed that a large set of genes which exhibited increased levels of H3K27me3 upon MLL2 depletion, could be rescued by the removal of DNA methylation or the depletion of members of the PRC2 complex [58]. This overlap of the repressive mechanisms could be attributed to the potential of the H3K27me3 mark to recruit and regulate DNA methylation deposition. This hypothesis was further confirmed by demethylation with a 5-Aza-2-deoxycytidine (5dAza) treatment in mESCs (both WT and *Mll2* KO), which revealed the concomitant interplay of H3K27me3, MLL2-dependent transcription regulation and DNA methylation. Genome-scale screening indicated that the depletion of CXXC1 (component of SET1A/B complexes) in MLL2 knockout mESCs was sufficient to rescue the loss of expression of the ~1200 MLL2-dependent genes. Interestingly, the rescue of these genes expression was not correlated with the re-appearance of the H3K4me3, showing that MLL2 and H3K4me3 may have a more instructional role in gene regulation for certain types of genes, as has been reported in previous studies [60,61].

Additionally, in the vast majority of MLL2 targeted genes, the deletion of MLL2 led to the reduction in gene expression, while this repression was shown to be restored by either the removal of H3K27me3 or restoring the DNA demethylation. It was observed that there is a big overlap among genes at least partially rescued that could be explained from the fact that DNA methylation impacts H3K27me3 deposition [58]. The removal of DNMTs in triple knockout mice was further connected with global alterations of H3K27me3 levels and the dilution of the repressive effect on Polycomb-targeted genes [62].

Transcriptional kinetic studies in *Mll2*-conditional knockout mammalian ESCs have revealed the order of the events that lead to gene silencing and the crosstalk between the action of MLL2, RNA pol II processing and DNA methylation. Focusing at the *MagohB* gene, they have uncovered the mechanistic role of MLL2 in gene expression. The presence of MLL2 maintained an open chromatin state at the promoter of the target gene, regulated RNA pol II association and was correlated with active chromatin marks and high levels of mRNA. The depletion of Mll2 led to a rapid decrease in active marks (H3K4me3 and H3K9Ac) and an increase in DNA methylation at the *MagohB* gene promoter. Interestingly, DNA methylation seemed to be a secondary event to gene silencing [63].

To further evaluate the role of DNA methylation in antagonizing MLL2 in d25 GV oocytes, a study assessed the distribution of H3K4me3 in the absence of DNA methylation by using conditional knockout mice for MLL2 and double knockout for DNMT3a/b genes. It was revealed that there are two complementary, independent mechanisms of H3K4m3 trimethylation. One mechanism was transcription-dependent and was not connected to MLL2 activity, while the other relied on the specific targeting of MLL2 in unmethylated CpG-rich regions, mainly at distal elements and intergenic regions. Interestingly, these regions are protected by DNA methylation during oogenesis. The non-canonical role of MLL2 was not connected to gene expression but rather marked the bivalent chromatin state at repressed H3K27me3-marked promoters [55,64].

Additional studies on the epigenetic and expression profiling of target genes were performed to detect pathways that are regulated by MLL2 enzymatic activity and revealed mechanistic insights into the functional role of MLL2 [65]. Upon ChIP-seq and RNA-seq profiling in both wild-type and MLL2 null mammalian cells (HCT116 cells), MLL2 was found to participate in retinoic acid receptor signaling by promoting retinoic acid-responsive gene transcription. Among the genes associated with MLL2-enriched loci was the Ankyrin repeat and SOCS box protein 2 (*ASB2*) which in previous studies had been demonstrated to be induced by retinoic acid in leukemia cells. ASB2 expression in myeloid leukemia cells has been shown to induce the inhibition of proliferation and chromatin condensation. However, $MLL2^{-/-}$ cell lines have shown a reduction in ASB2 expression due to an effect on H3K4me3 levels [65].

The same genome-wide study demonstrated the involvement of MLL2 in different cellular pathways. Among the transcription factors that were regulated by MLL2 was the Nuclear Receptor Subfamily 3 Group C Member 1 (NR3C1) and p53. It has been demonstrated that p53 contains a sequence similar to an autoinhibitory N-terminal loop of the MLL2-SET domain. The N-terminal loop of MLL2 adopts a similar conformation as the H3 tail and, thus, enters the substrate-binding pocket of another MLL2-SET enzyme. This specific sequence and conformation of p53 makes it a perfect candidate substrate of the complex [66]. Further biochemical experiments and a mass spectrometry analysis have shown that p53 could be methylated by MLL complexes, at the K503 site, identifying a non-histone substrate for the MLL family. Preliminary experiments suggest that this newly found methylation site of p53 may affect its transcription activity and be implicated in human pathologies, including cancer [66].

6. MLL2 Role in Human Physiology

The *MLL2* gene was originally identified by its homology to *MLL1* and was further detected to be broadly expressed in human tissues. The ability of both paralogs to bind Menin/LEDGF, has proved critical for their normal functions [7,67].

Of importance, *Mll2* germ-line deletions have been associated with early growth retardation, neural tube defects and increased apoptosis that leads to embryonic death before E11.5 [68]. *Mll2* was involved in the preservation of the mesodermal marker Mox1 and Hoxb1 as well as in the deregulation of HoxB cluster genes. However, after E11.5, *Mll2* loss was not associated with notable pathologies indicating that it is not required for the late development and homeostasis of somatic or stem cells [68].

However, MLL2 is also implicated in germinal cell differentiation and contributes to enriched H3K4me3 marks observed in the active genes of spermatogonial stem cells. Spermatogenesis was lost upon its deletion [69]. Moreover, in oocytes global H3K4me3 mediated by MLL2 has been observed and deletions in *Mll2* resulted in anovulation and death. The elevated transcription of apoptotic factors and p53 as well as the loss of global H3K4me2/3 was also detected [70]. Additionally, it was shown that MLL2 is autonomously required for fertility and participates in epigenetic reprogramming during fertilization. However, in mid-gestation, *Mll2* deletion did not affect the global methylation of H3K4 and hematopoiesis, as observed with *Mll1* [69].

Although the majority of hematopoietic cell types do not depend on MLL2 for their function, macrophages have been demonstrated to require *MLL2* for proper cytokine signaling. Upon stimulation by lipopolysaccharide (LPS), a Rosa-CreERT2 model of $Mll2^{-/-}$ macrophages from bone marrow, displayed attenuated intracellular NF-κB signaling due to reduced Toll receptor 4 (TLR4) activation. This was attributed to the loss of Phosphatidylinositol Glycan Anchor Biosynthesis Class P protein (Pigp) which adds glycophosphatidylinositol to transmembrane proteins. In turn, this induced the loss of CD14 anchoring at the cellular membrane which co-operates with TLR4 in response to LPS. Apart of the *Pigp* gene promoter, several other *Mll2* targets exhibited reduced H3K4me3 peaks in TSS and a respective increase in H3K27me3 mark which relate to repressed or bivalent genes. Therefore, MLL2 possibly maintains the expression of the target genes through H3K4me3 promoter enrichment and the resistance of invading repression complexes. However, H3K4 hypomethylated genes in $Mll2^{-/-}$ macrophages exhibited no change in expression levels, indicating a higher sensitivity of some genes to H3K4me3 promoter depletion than others [71,72].

Furthermore, MLL2 is involved in cell growth control by regulating the activity of the *MYC* oncogene. MLL2 is attracted to the *MYC* enhancer by a process that involves β-catenin and promotes the transcription of *MYC* via H3K4me3 methylation [73].

7. MLL2 Implication in Diseases

The regulation of transcription by MLL family members is very important for human health, and mutations in *MLL* genes have been detected in several developmental disorders as well as in hematological and non-hematological cancers.

An important physiological role of MLL2 has been demonstrated in the control of voluntary movement. Specifically, MLL2 haploinsufficiency has been linked to the most severe type of a hyperkinetic movement disorder, the early onset-generalized children dystonia, which is defined by involuntary twisting postures due to sustained or intermittent contractions of agonist and antagonist muscles [5,74,75]. The patients present heterozygous mutations in the *MLL2* gene and characteristic brain magnetic resonance imaging findings with a typical facial appearance and possible progress to cranial and laryngeal dystonia over time [76].

Gene expression profiling in patients harboring *MLL2* mutations has shown that certain proteins associated with dystonia, such as torsin family 1 member A (TOR1A), THAP domain-containing, apoptosis-associated protein 1 (THAP1) and dopamine receptor D2 (D2R) are decreased in cerebrospinal fluid and fibroblasts, indicating MLL2 implication in disease pathogenesis that needs further investigation [5].

Moreover, in adult mice, conditional *MLL2* deletion in excitatory forebrain neurons resulted in learning impairment due to increased activity of genes involved in hippocampal plasticity via H3K4me2/3 [50,76].

Another disease-promoting role has been attributed to MLL2 in respect to cell proliferation enhancement and carcinogenesis [31]. As originally identified, somatic mutations of *MLL1* have been associated with cancer onset. The *MLL1* gene exhibits a considerable number of rearrangements with several other translocation partner genes, possibly attributed to the inability of developing hematopoietic cells to repair the frequent chromosomal double-strand DNA breaks [31]. The MLL1-fusion proteins are coded by exons 8-13 forming the C-terminal part of the hybrid protein and a variable number of fusion partner exons coding the N-terminal. Upon the translocation of *MLL1* to its fusion partners, the H3K4 methyltransferase activity is lost due to the loss of the SET domain. More than 135 *MLL1* rearrangements have been identified up to date, being mostly in frame translocations that lead to the generation of gain-of-function oncoproteins with altered activities [77]. The fusion of translocation gene partners results in the formation of complexes which may interact with other methyltransferases such as the disruptor of telomeric silencing 1-like (DOT1L) to induce H3K79 methylation and alter gene expression in favor of a leukemic transformation.

It is interesting to note that although MLL2 exhibits a structural similarity to MLL1, it is not related to chromosomal translocations and exhibits a lower affinity for DNA binding at unmethylated CpG sequences, being unable to replace MLL1 in leukemic oncoproteins [68,77,78]. Several common genes can be fused with *MLL1*, including MLL-ENL, MLL-ELL, MLL-AF4, MLL-AF9, MLL-AF10 and MLL-PTD, accounting for 80–90% of MLLs, whereas MLL1-rearranged leukemias account for 10% of all leukemias [31,77].

In MLL1-rearranged leukemias (MLL-AF9), deletion of the *MLL2* gene (wt) was shown capable of decreasing the leukemic cell survival, but WT-*MLL1* deletion had no impact on leukemia cell function, since targeting the N-terminal part, that is shared in the MLL1-fusion protein, did ablate leukemia cells [79]; thus, indicating that the activities of the two genes are not redundant, as previously suggested [31,80].

Of interest, conditional or germline *Mll2* mutations in mice were not capable of inducing carcinogenesis [67,68]. However, *MLL2* mutations detected in cancers are mostly nonsense, missense or frameshift, and mainly involve the PHD and SET domains [6]. Mutation rates are higher in uterine corpus endometrial carcinoma (UCEC), esophageal sarcomatoid carcinoma and in gastric cancer [81–83]. Additionally, somatic mutations of *MLL2* have been detected in neurofibromatosis 1-glioblastoma (NF1-GBM), leading to the truncation of the MLL2 protein and have been associated with early steps of gliomagenesis [84].

The overexpression of MLL2 has also been detected in pancreatic cancer cells and additional translocations have been observed in glioblastomas [85].

In colorectal cancer, MLL2 has been reported to promote cell proliferation through physical interaction with β-catenin which allows the recruitment of MLL2 to the enhancer element of *c-MYC*, inducing its transcription [73]. MLL2 target genes, profiling in both wild-type and MLL2 null mammalian colon cancer cells (HCT116 cells), revealed that MLL2 promotes retinoic acid-responsive gene transcription such as *ASB2* which was previously induced in leukemia cells. Other transcription factors that were regulated by MLL2 include NR3C1 and p53, explaining the potential mechanistic implication of MLL2 in cancer progression [66].

Additionally, MLL2 has been revealed in genomic studies as a recurrent target for the integration on oncogenic viruses (hepatitis B virus and adeno-associated virus type 2) of hepatocellular carcinoma (HCC) tissues [86,87], indicating a potential relationship of elevated MLL2 expression with liver cancer progression that needs to be further investigated.

Furthermore, in follicular lymphoma (FL), *MLL2* mutations were frequently detected at a similar rate to t(14;18) translocation which is the molecular hallmark of the disease, indicating a central role of MLL2 in tumorigenesis [88].

Finally, in squamous-cell cancer of the head and neck (SCCHN), somatic mutations of *MLL2* were frequently detected at a 17.9% mutation rate [89]. Since these mutations were inactivating, it is suggested that MLL2 has a tumor-suppressor role in head and neck cancer, potentially changing the expression of global gene sets.

8. Conclusions

Taken together, all the significant progress that has occurred in recent years in understanding chromatin accessibility mechanisms and their role in gene regulation,

H3K4me1/me2/me3-enriched genomic regions were demonstrated to be of primary importance. Furthermore, structural and functional studies of the MLL methyltransferase family in mediating these histone marks in specific tissues have revealed unique, non-redundant functions despite the similarity between paralogs.

MLL2 is particularly significant in mediating H3K4me3 in specific promoters of development-related genes, but is also required for H3K4me3 accumulation on bivalent promoters in ES cells. Moreover, an extensive range of H3K4 methylation-reader domains has been detected in many transcriptional coactivators demonstrating the direct stimulatory effects of the MLL complex-mediated H3K4 methylation on transcription. Therefore, studies determining the factors that enable the recruitment of MLL1/2 complexes to specific loci in the genome are highly demanded. MLL2 plays multiple and significant roles in the regulation of physiological voluntary movement; it is involved in childhood dystonia and in the pathogenesis of several malignancies. It is, thus, important to determine how to target MLL2 with small molecule inhibitors in different settings. Current efforts are directed to the development of inhibitors that target H3K4 methyltransferase activities or MLL1/2-associated subunit interactions in controlling the H3K4 methyltransferase function of MLL complexes. Major efforts were focused on the identification of chemicals that treat leukemias caused by MLL1 rearrangements. Two molecules are currently in phase I/II clinical trials (NCT04065399, NCT04067336) for Menin–MLL inhibition (SNDX-5613 from Syndax Pharmaceuticals and KO-539 from Kura Oncology) for MLL-rearranged leukemias which show promising results. Selected inhibitors can either act on proteins recruited to the MLL1 complex that are required to maintain the leukemic state or block the methyltransferase activity of MLL1 by interrupting its interaction with WDR5, Menin or LEDGF [90]. Other approaches include the direct inhibition of MLL1 activity, associated metabolic pathways and protein degradation or, alternatively, the inhibitory targeting of the BRD4 domain recruited to the *MYC* gene, switching-off *MYC*-dependent leukemia [90,91]. Importantly, the core subunits of MLL complexes are frequently amplified in different cancer types, exhibiting an oncogenic role and, therefore, present potential targets for cancer patients that need to be further explored [31].

Furthermore, recent experimental evidence suggests that MLL-associated transcriptional regulatory mechanisms, independent of the H3K4 methyltransferase activities of the complexes, are also involved in gene regulation and need to be taken into consideration as well as further investigated in functional studies.

Author Contributions: Conceptualization, C.P.; methodology, A.K., S.C. and C.P.; software, A.K. and S.C.; validation, S.C. and C.P.; formal analysis, C.P.; investigation, A.K., S.C. and C.P.; resources, C.P.; data curation, A.K., S.C. and C.P.; writing—original draft preparation, A.K., S.C. and C.P.; writing—review and editing, S.C. and C.P.; visualization, C.P.; supervision, C.P.; project administration, C.P.; funding acquisition, C.P. All authors have read and agreed to the published version of the manuscript.

Funding: This research received no external funding.

Institutional Review Board Statement: Not applicable.

Informed Consent Statement: Not applicable.

Data Availability Statement: Not applicable.

Conflicts of Interest: The authors declare no conflict of interest.

References

1. Gu, B.; Lee, M.G. Histone H3 lysine 4 methyltransferases and demethylases in self-renewal and differentiation of stem cells. *Cell Biosci.* **2013**, *3*, 1–14. [CrossRef]
2. Herz, H.M.; Garruss, A.; Shilatifard, A. SET for life: Biochemical activities and biological functions of SET domain-containing proteins. *Trends Biochem. Sci.* **2013**, *38*, 621–639. [CrossRef] [PubMed]
3. Hu, D.; Gao, X.; Morgan, M.A.; Herz, H.-M.; Smith, E.R.; Shilatifard, A. The MLL3/MLL4 Branches of the COMPASS Family Function as Major Histone H3K4 Monomethylases at Enhancers. *Mol. Cell. Biol.* **2013**, *33*, 4745–4754. [CrossRef] [PubMed]

4. Piunti, A.; Shilatifard, A. Epigenetic balance of gene expression by polycomb and compass families. *Science* **2016**, *352*, 6290. [CrossRef] [PubMed]
5. Park, K.; Kim, J.A.; Kim, J. Transcriptional regulation by the KMT2 histone H3K4 methyltransferases. *Biochim. Biophys. Acta Gene Regul. Mech.* **2020**, *1863*, 194545. [CrossRef]
6. Rao, R.C.; Dou, Y. Hijacked in cancer: The KMT2 (MLL) family of methyltransferases. *Nat. Rev. Cancer* **2015**, *15*, 334–346. [CrossRef]
7. Fitzgerald, K.T.; Diaz, M.O. MLL2: A new mammalian member of the trx/MLL family of genes. *Genomics* **1999**, *59*, 187–192. [CrossRef]
8. Zhang, J.; Walsh, M.F.; Wu, G.; Edmonson, M.N.; Gruber, T.A.; Easton, J.; Hedges, D.; Ma, X.; Zhou, X.; Yergeau, D.A.; et al. Germline Mutations in Predisposition Genes in Pediatric Cancer. *N. Engl. J. Med.* **2015**, *373*, 2336–2346. [CrossRef]
9. Li, Y.; Han, J.; Zhang, Y.; Cao, F.; Liu, Z.; Li, S.; Wu, J.; Hu, C.; Wang, Y.; Shuai, J.; et al. Structural basis for activity regulation of MLL family methyltransferases. *Nature* **2016**, *530*, 447–452. [CrossRef] [PubMed]
10. Allen, M.D.; Grummitt, C.G.; Hilcenko, C.; Min, S.Y.; Tonkin, L.M.; Johnson, C.M.; Freund, S.M.; Bycroft, M.; Warren, A.J. Solution structure of the nonmethyl-CpG-binding CXXC domain of the leukaemia-associated MLL histone methyltransferase. *EMBO J.* **2006**, *25*, 4503–4512. [CrossRef] [PubMed]
11. Ali, M.; Hom, R.A.; Blakeslee, W.; Ikenouye, L.; Kutateladze, T.G. Diverse functions of PHD fingers of the MLL/KMT2 subfamily. *Biochim. Biophys. Acta Mol. Cell Res.* **2014**, *1843*, 366–371. [CrossRef]
12. Sanchez, R.; Zhou, M.M. The PHD finger: A versatile epigenome reader. *Trends Biochem. Sci.* **2011**, *36*, 364–372. [CrossRef] [PubMed]
13. Wang, Z.; Song, J.; Milne, T.A.; Wang, G.G.; Li, H.; Allis, C.D.; Patel, D.J. Pro isomerization in MLL1 PHD3-Bromo cassette connects H3K4me readout to CyP33 and HDAC-mediated repression. *Cell* **2010**, *141*, 1183–1194. [CrossRef] [PubMed]
14. Hsieh, J.J.D.; Cheng, E.H.Y.; Korsmeyer, S.J. Taspase1: A threonine aspartase required for cleavage of MLL and proper HOX gene expression. *Cell* **2003**, *115*, 293–303. [CrossRef]
15. Hsieh, J.J.-D.; Ernst, P.; Erdjument-Bromage, H.; Tempst, P.; Korsmeyer, S.J. Proteolytic Cleavage of MLL Generates a Complex of N- and C-Terminal Fragments That Confers Protein Stability and Subnuclear Localization. *Mol. Cell. Biol.* **2003**, *23*, 186–194. [CrossRef]
16. Takeda, S.; Chen, D.Y.; Westergard, T.D.; Fisher, J.K.; Rubens, J.A.; Sasagawa, S.; Kan, J.T.; Korsmeyer, S.J.; Cheng, E.H.Y.; Hsieh, J.J.D. Proteolysis of MLL family proteins is essential for Taspase1-orchestrated cell cycle progression. *Genes Dev.* **2006**, *20*, 2397–2409. [CrossRef]
17. Yokoyama, A.; Kitabayashi, I.; Ayton, P.M.; Cleary, M.L.; Ohki, M. Leukemia proto-oncoprotein MLL is proteolytically processed into 2 fragments with opposite transcriptional properties. *Blood* **2002**, *100*, 3710–3718. [CrossRef]
18. Zeleznik-Le, N.J.; Harden, A.M.; Rowley, J.D. 11q23 translocations split the "AT-hook" cruciform DNA-binding region and the transcriptional repression domain from the activation domain of the mixed-lineage leukemia (MLL) gene. *Proc. Natl. Acad. Sci. USA* **1994**, *91*, 10610–10614. [CrossRef]
19. Shilatifard, A. The COMPASS family of histone H3K4 methylases: Mechanisms of regulation in development and disease pathogenesis. *Annu. Rev. Biochem.* **2012**, *81*, 65–95. [CrossRef]
20. Ford, D.J.; Dingwall, A.K. The cancer COMPASS: Navigating the functions of MLL complexes in cancer. *Cancer Genet.* **2015**, *208*, 178–191. [CrossRef] [PubMed]
21. Patel, A.; Dharmarajan, V.; Vought, V.E.; Cosgrove, M.S. On the mechanism of multiple lysine methylation by the human mixed lineage leukemia protein-1 (MLL1) core complex. *J. Biol. Chem.* **2009**, *284*, 24242–24256. [CrossRef]
22. Cao, F.; Chen, Y.; Cierpicki, T.; Liu, Y.; Basrur, V.; Lei, M.; Dou, Y. An Ash2L/RbBP5 heterodimer stimulates the MLL1 methyltransferase activity through coordinated substrate interactions with the MLL1 SET domain. *PLoS ONE* **2010**, *5*, e14102. [CrossRef] [PubMed]
23. Xue, H.; Yao, T.; Cao, M.; Zhu, G.; Li, Y.; Yuan, G.; Chen, Y.; Lei, M.; Huang, J. Structural basis of nucleosome recognition and modification by MLL methyltransferases. *Nature* **2019**, *573*, 445–449. [CrossRef]
24. Hughes, C.M.; Rozenblatt-Rosen, O.; Milne, T.A.; Copeland, T.D.; Levine, S.S.; Lee, J.C.; Hayes, D.N.; Shanmugam, K.S.; Bhattacharjee, A.; Biondi, C.A.; et al. Menin associates with a trithorax family histone methyltransferase complex and with the Hoxc8 locus. *Mol. Cell* **2004**, *13*, 587–597. [CrossRef]
25. Jiang, H.; Lu, X.; Shimada, M.; Dou, Y.; Tang, Z.; Roeder, R.G. Regulation of transcription by the MLL2 complex and MLL complex-associated AKAP95. *Nat. Struct. Mol. Biol.* **2013**, *20*, 1156–1163. [CrossRef] [PubMed]
26. Cho, Y.W.; Hong, T.; Hong, S.H.; Guo, H.; Yu, H.; Kim, D.; Guszczynski, T.; Dressler, G.R.; Copeland, T.D.; Kalkum, M.; et al. PTIP associates with MLL3- and MLL4-containing histone H3 lysine 4 methyltransferase complex. *J. Biol. Chem.* **2007**, *282*, 20395–20406. [CrossRef]
27. Goo, Y.-H.; Sohn, Y.C.; Kim, D.-H.; Kim, S.-W.; Kang, M.-J.; Jung, D.-J.; Kwak, E.; Barlev, N.A.; Berger, S.L.; Chow, V.T.; et al. Activating Signal Cointegrator 2 Belongs to a Novel Steady-State Complex That Contains a Subset of Trithorax Group Proteins. *Mol. Cell. Biol.* **2003**, *23*, 140–149. [CrossRef]
28. Patel, S.R.; Kim, D.; Levitan, I.; Dressler, G.R. The BRCT-Domain Containing Protein PTIP Links PAX2 to a Histone H3, Lysine 4 Methyltransferase Complex. *Dev. Cell* **2007**, *13*, 580–592. [CrossRef]

29. Lee, J.H.; Tate, C.M.; You, J.S.; Skalnik, D.G. Identification and characterization of the human Set1B histone H3-Lys 4 methyltransferase complex. *J. Biol. Chem.* **2007**, *282*, 13419–13428. [CrossRef]
30. Mohan, M.; Herz, H.-M.; Smith, E.R.; Zhang, Y.; Jackson, J.; Washburn, M.P.; Florens, L.; Eisenberg, J.C.; Shilatifard, A. The COMPASS Family of H3K4 Methylases in Drosophila. *Mol. Cell. Biol.* **2011**, *31*, 4310–4318. [CrossRef]
31. Poreba, E.; Lesniewicz, K.; Durzynska, J. Aberrant activity of histone–lysine n-methyltransferase 2 (Kmt2) complexes in oncogenesis. *Int. J. Mol. Sci.* **2020**, *21*, 9340. [CrossRef]
32. Kim, J.; Daniel, J.; Espejo, A.; Lake, A.; Krishna, M.; Xia, L.; Zhang, Y.; Bedford, M.T. Tudor, MBT and chromo domains gauge the degree of lysine methylation. *EMBO Rep.* **2006**, *7*, 397–403. [CrossRef]
33. Musselman, C.A.; Khorasanizadeh, S.; Kutateladze, T.G. Towards understanding methyllysine readout. *Biochim. Biophys. Acta Gene Regul. Mech.* **2014**, *1839*, 686–693. [CrossRef]
34. Dreijerink, K.M.A.; Mulder, K.W.; Winkler, G.S.; Höppener, J.W.M.; Lips, C.J.M.; Timmers, H.T.M. Menin links estrogen receptor activation to histone H3K4 trimethylation. *Cancer Res.* **2006**, *66*, 4929–4935. [CrossRef] [PubMed]
35. Tan, C.C.; Sindhu, K.V.; Li, S.; Nishio, H.; Stoller, J.Z.; Oishi, K.; Puttreddy, S.; Lee, T.J.; Epstein, J.A.; Walsh, M.J.; et al. Transcription factor Ap2δ associates with Ash2l and ALR, a trithorax family histone methyltransferase, to activate Hoxc8 transcription. *Proc. Natl. Acad. Sci. USA* **2008**, *105*, 7472–7477. [CrossRef]
36. Ullius, A.; Lüscher-Firzlaff, J.; Costa, I.G.; Walsemann, G.; Forst, A.H.; Gusmao, E.G.; Kapelle, K.; Kleine, H.; Kremmer, E.; Vervoorts, J.; et al. The interaction of MYC with the trithorax protein ASH2L promotes gene transcription by regulating H3K27 modification. *Nucleic Acids Res.* **2014**, *42*, 6901–6920. [CrossRef] [PubMed]
37. Demers, C.; Chaturvedi, C.P.; Ranish, J.A.; Juban, G.; Lai, P.; Morle, F.; Aebersold, R.; Dilworth, F.J.; Groudine, M.; Brand, M. Activator-Mediated Recruitment of the MLL2 Methyltransferase Complex to the β-Globin Locus. *Mol. Cell* **2007**, *27*, 573–584. [CrossRef]
38. Deng, C.; Li, Y.; Liang, S.; Cui, K.; Salz, T.; Yang, H.; Tang, Z.; Gallagher, P.G.; Qiu, Y.; Roeder, R.; et al. USF1 and hSET1A Mediated Epigenetic Modifications Regulate Lineage Differentiation and HoxB4 Transcription. *PLoS Genet.* **2013**, *9*, e1003524. [CrossRef] [PubMed]
39. Fossati, A.; Dolfini, D.; Donati, G.; Mantovani, R. NF-Y recruits Ash2L to impart H3K4 trimethylation on CCAAT promoters. *PLoS ONE* **2011**, *6*, e17220. [CrossRef]
40. Tyagi, S.; Chabes, A.L.; Wysocka, J.; Herr, W. E2F Activation of S Phase Promoters via Association with HCF-1 and the MLL Family of Histone H3K4 Methyltransferases. *Mol. Cell* **2007**, *27*, 107–119. [CrossRef] [PubMed]
41. Bertero, A.; Madrigal, P.; Galli, A.; Hubner, N.C.; Moreno, I.; Burks, D.; Brown, S.; Pedersen, R.A.; Gaffney, D.; Mendjan, S.; et al. Activin/Nodal signaling and NANOG orchestrate human embryonic stem cell fate decisions by controlling the H3K4me3 chromatin mark. *Genes Dev.* **2015**, *29*, 702–717. [CrossRef]
42. Kawabe, Y.I.; Wang, Y.X.; McKinnell, I.W.; Bedford, M.T.; Rudnicki, M.A. Carm1 regulates Pax7 transcriptional activity through MLL1/2 recruitment during asymmetric satellite stem cell divisions. *Cell Stem Cell* **2012**, *11*, 333–345. [CrossRef] [PubMed]
43. Tang, Z.; Chen, W.Y.; Shimada, M.; Nguyen, U.T.T.; Kim, J.; Sun, X.J.; Sengoku, T.; McGinty, R.K.; Fernandez, J.P.; Muir, T.W.; et al. SET1 and p300 act synergistically, through coupled histone modifications, in transcriptional activation by p53. *Cell* **2013**, *154*, 297. [CrossRef]
44. Muntean, A.G.; Tan, J.; Sitwala, K.; Huang, Y.; Bronstein, J.; Connelly, J.A.; Basrur, V.; Elenitoba-Johnson, K.S.; Hess, J.L. The PAF complex synergizes with MLL fusion proteins at HOX loci to promote leukemogenesis. *Cancer Cell* **2010**, *17*, 609–621. [CrossRef] [PubMed]
45. Xia, Z.B.; Anderson, M.; Diaz, M.O.; Zeleznik-Le, N.J. MLL repression domain interacts with histone deacetylases, the polycomb group proteins HPC2 and BMI-1, and the corepressor C-terminal-binding protein. *Proc. Natl. Acad. Sci. USA* **2003**, *100*, 8342–8347. [CrossRef] [PubMed]
46. Park, S.H.; Ayoub, A.; Lee, Y.T.; Xu, J.; Kim, H.; Zheng, W.; Zhang, B.; Sha, L.; An, S.; Zhang, Y.; et al. Cryo-EM structure of the human MLL1 core complex bound to the nucleosome. *Nat. Commun.* **2019**, *5*, 5540. [CrossRef]
47. Vedadi, M.; Blazer, L.; Eram, M.S.; Barsyte-Lovejoy, D.; Arrowsmith, C.H.; Hajian, T. Targeting human SET1/MLL family of proteins. *Protein Sci.* **2017**, *26*, 662–676. [CrossRef]
48. Lee, Y.T.; Ayoub, A.; Park, S.H.; Sha, L.; Xu, J.; Mao, F.; Zheng, W.; Zhang, Y.; Cho, U.S.; Dou, Y. Mechanism for DPY30 and ASH2L intrinsically disordered regions to modulate the MLL/SET1 activity on chromatin. *Nat. Commun.* **2021**, *19*, 2953. [CrossRef]
49. An, W.; Roeder, R.G. Reconstitution and Transcriptional Analysis of Chromatin In vitro. *Methods Enzymol.* **2003**, *377*, 460–474. [CrossRef]
50. Hu, D.; Gao, X.; Cao, K.; Morgan, M.A.; Mas, G.; Smith, E.R.; Volk, A.G.; Bartom, E.T.; Crispino, J.D.; Di Croce, L.; et al. Not All H3K4 Methylations Are Created Equal: Mll2/COMPASS Dependency in Primordial Germ Cell Specification. *Mol. Cell* **2017**, *65*, 460–475.e6. [CrossRef]
51. Bach, C.; Mueller, D.; Buhl, S.; Garcia-Cuellar, M.P.; Slany, R.K. Alterations of the CxxC domain preclude oncogenic activation of mixed-lineage leukemia 2. *Oncogene* **2009**, *28*, 815–823. [CrossRef]
52. Milne, T.A.; Kim, J.; Wang, G.G.; Stadler, S.C.; Basrur, V.; Whitcomb, S.J.; Wang, Z.; Ruthenburg, A.J.; Elenitoba-Johnson, K.S.J.; Roeder, R.G.; et al. Multiple Interactions Recruit MLL1 and MLL1 Fusion Proteins to the HOXA9 Locus in Leukemogenesis. *Mol. Cell* **2010**, *38*, 853–863. [CrossRef]

53. Xu, C.; Liu, K.; Lei, M.; Yang, A.; Li, Y.; Hughes, T.R.; Min, J. DNA Sequence Recognition of Human CXXC Domains and Their Structural Determinants. *Structure* **2018**, *26*, 85–95.e3. [CrossRef]
54. Tomizawa, S.I.; Kobayashi, Y.; Shirakawa, T.; Watanabe, K.; Mizoguchi, K.; Hoshi, I.; Nakajima, K.; Nakabayashi, J.; Singh, S.; Dahl, A.; et al. Kmt2b conveys monovalent and bivalent H3K4me3 in mouse spermatogonial stem cells at germline and embryonic promoters. *Development* **2018**, *145*, dev169102. [CrossRef]
55. Denissov, S.; Hofemeister, H.; Marks, H.; Kranz, A.; Ciotta, G.; Singh, S.; Anastassiadis, K.; Stunnenberg, H.G.; Stewart, A.F. Mll2 is required for H3K4 trimethylation on bivalent promoters in embryonic stem cells, whereas Mll1 is redundant. *Development* **2014**, *141*, 526–537. [CrossRef] [PubMed]
56. Sze, C.C.; Cao, K.; Collings, C.K.; Marshall, S.A.; Rendleman, E.J.; Ozark, P.A.; Chen, F.X.; Morgan, M.A.; Wang, L.; Shilatifard, A. Histone H3K4 methylation-dependent and -independent functions of set1A/COMPASS in embryonic stem cell self-renewal and differentiation. *Genes Dev.* **2017**, *31*, 1732–1737. [CrossRef]
57. Bernstein, B.E.; Mikkelsen, T.S.; Xie, X.; Kamal, M.; Huebert, D.J.; Cuff, J.; Fry, B.; Meissner, A.; Wernig, M.; Plath, K.; et al. A bivalent chromatin structure marks key developmental genes in embryonic stem cells. *Cell* **2006**, *125*, 315–326. [CrossRef]
58. Douillet, D.; Sze, C.C.; Ryan, C.; Piunti, A.; Shah, A.P.; Ugarenko, M.; Marshall, S.A.; Rendleman, E.J.; Zha, D.; Helmin, K.A.; et al. Uncoupling histone H3K4 trimethylation from developmental gene expression via an equilibrium of COMPASS, Polycomb and DNA methylation. *Nat. Genet.* **2020**, *52*, 615–625. [CrossRef]
59. Sze, C.C.; Ozark, P.A.; Cao, K.; Ugarenko, M.; Das, S.; Wang, L.; Marshall, S.A.; Rendleman, E.J.; Ryan, C.A.; Zha, D.; et al. Coordinated regulation of cellular identity–associated H3K4me3 breadth by the COMPASS family. *Sci. Adv.* **2020**, *6*, eaaz4764. [CrossRef]
60. Margaritis, T.; Oreal, V.; Brabers, N.; Maestroni, L.; Vitaliano-Prunier, A.; Benschop, J.J.; van Hooff, S.; van Leenen, D.; Dargemont, C.; Géli, V.; et al. Two distinct repressive mechanisms for histone 3 lysine 4 methylation through promoting 3′-end antisense transcription. *PLoS Genet.* **2012**, *8*, e1002952. [CrossRef] [PubMed]
61. Clouaire, T.; Webb, S.; Bird, A. Cfp1 is required for gene expression-dependent H3K4 trimethylation and H3K9 acetylation in embryonic stem cells. *Genome Biol.* **2014**, *15*, 451. [CrossRef] [PubMed]
62. Brinkman, A.B.; Gu, H.; Bartels, S.J.; Zhang, Y.; Matarese, F.; Simmer, F.; Marks, H.; Bock, C.; Gnirke, A.; Meissner, A.; et al. Sequential ChIP-bisulfite sequencing enables direct genome-scale investigation of chromatin and DNA methylation cross-talk. *Genome Res.* **2012**, *22*, 1128–1138. [CrossRef] [PubMed]
63. Ladopoulos, V.; Hofemeister, H.; Hoogenkamp, M.; Riggs, A.D.; Stewart, A.F.; Bonifer, C. The Histone Methyltransferase KMT2B Is Required for RNA Polymerase II Association and Protection from DNA Methylation at the MagohB CpG Island Promoter. *Mol. Cell. Biol.* **2013**, *33*, 1383–1393. [CrossRef]
64. Hanna, C.W.; Taudt, A.; Huang, J.; Gahurova, L.; Kranz, A.; Andrews, S.; Dean, W.; Stewart, A.F.; Colomé-Tatché, M.; Kelsey, G. MLL2 conveys transcription-independent H3K4 trimethylation in oocytes. *Nat. Struct Mol. Biol.* **2018**, *25*, 73–82. [CrossRef]
65. Guo, C.; Chang, C.C.; Wortham, M.; Chen, L.H.; Kernagis, D.N.; Qin, X.; Cho, Y.W.; Chi, J.T.; Grant, G.A.; McLendon, R.E.; et al. Global identification of MLL2-targeted loci reveals MLL2's role in diverse signaling pathways. *Proc. Natl. Acad. Sci. USA* **2012**, *109*, 17603–17608. [CrossRef] [PubMed]
66. Li, Y.; Zhao, L.; Tian, X.; Peng, C.; Gong, F.; Chen, Y. Crystal Structure of MLL2 Complex Guides the Identification of a Methylation Site on P53 Catalyzed by KMT2 Family Methyltransferases. *Structure* **2020**, *28*, 1141–1148.e4. [CrossRef]
67. Crump, N.T.; Milne, T.A. Why are so many MLL lysine methyltransferases required for normal mammalian development? *Cell. Mol. Life Sci.* **2019**, *76*, 2885–2898. [CrossRef]
68. Glaser, S.; Schaft, J.; Lubitz, S.; Vintersten, K.; van der Hoeven, F.; Tufteland, K.R.; Aasland, R.; Anastassiadis, K.; Ang, S.L.; Stewart, A.F. Multiple epigenetic maintenance factors implicated by the loss of Mll2 in mouse development. *Development* **2006**, *133*, 1423–1432. [CrossRef]
69. Glaser, S.; Lubitz, S.; Loveland, K.L.; Ohbo, K.; Robb, L.; Schwenk, F.; Seibler, J.; Roellig, D.; Kranz, A.; Anastassiadis, K.; et al. The histone 3 lysine 4 methyltransferase, Mll2, is only required briefly in development and spermatogenesis. *Epigenetics Chromatin* **2009**, *2*, 1–16. [CrossRef] [PubMed]
70. Andreu-Vieyra, C.V.; Chen, R.; Agno, J.E.; Glaser, S.; Anastassiadis, K.; Stewart Francis, A.; Matzuk, M.M. MLL2 is required in oocytes for bulk histone 3 lysine 4 trimethylation and transcriptional silencing. *PLoS Biol.* **2010**, *8*, 53–54. [CrossRef]
71. Antunes, E.T.B.; Ottersbach, K. The MLL/SET family and haematopoiesis. *Biochim. Biophys. Acta Gene Regul. Mech.* **2020**, *1863*, 194579. [CrossRef]
72. Yang, W.; Ernst, P. Distinct functions of histone H3, lysine 4 methyltransferases in normal and malignant hematopoiesis. *Curr. Opin. Hematol.* **2017**, *24*, 322–328. [CrossRef]
73. Sierra, J.; Yoshida, T.; Joazeiro, C.A.; Jones, K.A. The APC tumor suppressor counteracts β-catenin activation and H3K4 methylation at Wnt target genes. *Genes Dev.* **2006**, *20*, 586–600. [CrossRef]
74. Zech, M.; Boesch, S.; Maier, E.M.; Borggraefe, I.; Vill, K.; Laccone, F.; Pilshofer, V.; Ceballos-Baumann, A.; Alhaddad, B.; Berutti, R.; et al. Haploinsufficiency of KMT2B, Encoding the Lysine-Specific Histone Methyltransferase 2B, Results in Early-Onset Generalized Dystonia. *Am. J. Hum. Genet.* **2016**, *99*, 1377–1387. [CrossRef]
75. Meyer, E.; Carss, K.J.; Rankin, J.; Nichols, J.M.E.; Grozeva, D.; Joseph, A.P.; Mencacci, N.E.; Papandreou, A.; Ng, J.; Barral, S.; et al. Mutations in the histone methyltransferase gene KMT2B cause complex early-onset dystonia. *Nat. Genet.* **2017**, *49*, 223–237. [CrossRef] [PubMed]

76. Ng, A.; Ng, A.; Galosi, S.; Salz, L.; Wong, T.; Schwager, C.; Amudhavalli, S.; Gelineau-Morel, R.; Chowdhury, S.; Friedman, J.; et al. Failure to thrive—An overlooked manifestation of KMT2B-related dystonia: A case presentation. *BMC Neurol.* **2020**, *20*, 1–6. [CrossRef]
77. Takahashi, S.; Yokoyama, A. The molecular functions of common and atypical MLL fusion protein complexes. *Biochim. Biophys. Acta BBA Gene Regul. Mech.* **2020**, *1863*, 194548. [CrossRef]
78. Risner, L.E.; Kuntimaddi, A.; Lokken, A.A.; Achille, N.J.; Birch, N.W.; Schoenfelt, K.; Bushweller, J.H.; Zeleznik-Le, N.J. Functional specificity of CpG DNA-binding CXXC domains in mixed lineage leukemia. *J. Biol. Chem.* **2013**, *288*, 29901–29910. [CrossRef]
79. Chen, Y.; Anastassiadis, K.; Kranz, A.; Stewart, A.F.; Arndt, K.; Waskow, C.; Yokoyama, A.; Jones, K.; Neff, T.; Lee, Y.; et al. MLL2, Not MLL1, Plays a Major Role in Sustaining MLL-Rearranged Acute Myeloid Leukemia. *Cancer Cell* **2017**, *31*, 755–770.e6. [CrossRef]
80. Thiel, A.T.; Blessington, P.; Zou, T.; Feather, D.; Wu, X.; Yan, J.; Zhang, H.; Liu, Z.; Ernst, P.; Koretzky, G.A.; et al. MLL-AF9-Induced Leukemogenesis Requires Coexpression of the Wild-Type Mll Allele. *Cancer Cell* **2010**, *17*, 148–159. [CrossRef]
81. Lu, H.; Yang, S.; Zhu, H.; Tong, X.; Xie, F.; Qin, J.; Han, N.; Wu, X.; Fan, Y.; Shao, Y.W.; et al. Targeted next generation sequencing identified clinically actionable mutations in patients with esophageal sarcomatoid carcinoma. *BMC Cancer* **2018**, *18*, 1–7. [CrossRef]
82. Kandoth, C.; McLellan, M.D.; Vandin, F.; Ye, K.; Niu, B.; Lu, C.; Xie, M.; Zhang, Q.; McMichael, J.F.; Wyczalkowski, M.A.; et al. Mutational landscape and significance across 12 major cancer types. *Nature* **2013**, *502*, 333–339. [CrossRef]
83. Genomic Alterations in Advanced Gastric Cancer Endoscopic Biopsy Samples Using Targeted Next-Generation Sequencing—PubMed. Available online: https://pubmed.ncbi.nlm.nih.gov/28744403/ (accessed on 28 June 2021).
84. Wong, W.H.; Junck, L.; Druley, T.E.; Gutmann, D.H. NF1 glioblastoma clonal profiling reveals KMT2B mutations as potential somatic oncogenic events. *Neurology* **2019**, *93*, 1067–1069. [CrossRef] [PubMed]
85. Huntsman, D.G.; Chin, S.F.; Muleris, M.; Batley, S.J.; Collins, V.P.; Wiedemann, L.M.; Aparicio, S.; Caldas, C. MLL2, the second human homolog of the Drosophila trithorax gene, maps to 19q13.1 and is amplified in solid tumor cell lines. *Oncogene* **1999**, *18*, 7975–7984. [CrossRef] [PubMed]
86. Nault, J.C.; Datta, S.; Imbeaud, S.; Franconi, A.; Mallet, M.; Couchy, G.; Letouzé, E.; Pilati, C.; Verret, B.; Blanc, J.F.; et al. Recurrent AAV2-related insertional mutagenesis in human hepatocellular carcinomas. *Nat. Genet.* **2015**, *47*, 1187–1193. [CrossRef]
87. Cancer Genome Atlas Research Network; Wheeler, D.A.; Roberts, L.R. Comprehensive and Integrative Genomic Characterization of Hepatocellular Carcinoma. *Cell* **2017**, *169*, 1327–1341.e23. [CrossRef]
88. Kishimoto, W.; Nishikori, M. Molecular pathogenesis of follicular lymphoma. *J. Clin. Exp. Hematop.* **2014**, *54*, 23–30. [CrossRef]
89. Mountzios, G.; Rampias, T.; Psyrri, A. The mutational spectrum of squamous-cell carcinoma of the head and neck: Targetable genetic events and clinical impact. *Ann. Oncol.* **2014**, *25*, 1889–1900. [CrossRef]
90. Chan, A.K.N.; Chen, C.-W. Rewiring the Epigenetic Networks in MLL-Rearranged Leukemias: Epigenetic Dysregulation and Pharmacological Interventions. *Front. Cell Dev. Biol.* **2019**, *7*, 81. [CrossRef]
91. Zhu, S.; Cheng, X.; Wang, R.; Tan, Y.; Ge, M.; Li, D.; Xu, Q.; Sun, Y.; Zhao, C.; Chen, S.; et al. Restoration of microRNA function impairs MYC-dependent maintenance of MLL leukemia. *Leukemia* **2020**, *34*, 2484–2488. [CrossRef]

Review

Structure, Activity and Function of the Suv39h1 and Suv39h2 Protein Lysine Methyltransferases

Sara Weirich [1], Mina S. Khella [1,2] and Albert Jeltsch [1,*]

[1] Institute of Biochemistry and Technical Biochemistry, University of Stuttgart, Allmandring 31, 70569 Stuttgart, Germany; sara.weirich@ibtb.uni-stuttgart.de (S.W.); mina.saad@ibtb.uni-stuttgart.de (M.S.K.)
[2] Biochemistry Department, Faculty of Pharmacy, Ain Shams University, African Union Organization Street, Abbassia, Cairo 11566, Egypt
* Correspondence: albert.jeltsch@ibtb.uni-stuttgart.de

Abstract: SUV39H1 and SUV39H2 were the first protein lysine methyltransferases that were identified more than 20 years ago. Both enzymes introduce di- and trimethylation at histone H3 lysine 9 (H3K9) and have important roles in the maintenance of heterochromatin and gene repression. They consist of a catalytically active SET domain and a chromodomain, which binds H3K9me2/3 and has roles in enzyme targeting and regulation. The heterochromatic targeting of SUV39H enzymes is further enhanced by the interaction with HP1 proteins and repeat-associated RNA. SUV39H1 and SUV39H2 recognize an RKST motif with additional residues on both sides, mainly K4 in the case of SUV39H1 and G12 in the case of SUV39H2. Both SUV39H enzymes methylate different non-histone proteins including RAG2, DOT1L, SET8 and HupB in the case of SUV39H1 and LSD1 in the case of SUV39H2. Both enzymes are expressed in embryonic cells and have broad expression profiles in the adult body. SUV39H1 shows little tissue preference except thymus, while SUV39H2 is more highly expressed in the brain, testis and thymus. Both enzymes are connected to cancer, having oncogenic or tumor-suppressive roles depending on the tumor type. In addition, SUV39H2 has roles in the brain during early neurodevelopment.

Keywords: protein lysine methylation; H3K9 methylation; PKMT; enzyme specificity; enzyme regulation; heterochromatin; protein post-translational modification

1. Introduction

The unstructured N-terminal tails of the histone proteins protrude from the core nucleosome and contain complex patterns of post-translational modifications (PTMs), including the methylation of lysine and arginine residues, lysine acetylation and the phosphorylation of serine and threonine [1–4]. These PTMs regulate many features of chromatin biology, gene expression and play a central role in developmental processes of multicellular organisms. In addition, aberrant histone PTMs are implicated in many diseases, such as cancer [5,6]. Acting in concert with DNA methylation and H4K20me3, H3K9me3 is a hallmark of constitutive heterochromatin in eukaryotes [7–10] and it is also enriched in silenced genes [11]. The suppressor of the variegation 3–9 gene has been genetically identified in screens for suppressors of position effect variegation in D. melanogaster in 1994 [12]. In 2000, its human homolog 1 (SUV39H1, also known as KMT1A) was biochemically identified as the first human protein lysine methyltransferase (PKMT) [13]. It introduces H3K9me3 together with a second human paralog called SUV39H2 (KMT1B) [14], and through H3K9me3 generation both of these enzymes have essential roles in heterochromatin formation and gene silencing. In addition, SUV39H1 and SUV39H2 were shown to methylate different non-histone substrate proteins, with essential functions in regulating protein stability, activity and protein–protein interactions (see below). The SUV39 PKMTs and their function in heterochromatin formation are evolutionarily conserved and

orthologous proteins can be detected in most organisms from fission yeast to humans including plants [15,16].

2. Domain Architecture and Structure of SUV39 Enzymes

SUV39H1 and SUV39H2 consist of two-conserved domains, one SET- and one chromodomain (Figure 1). The amino acid sequences of SUV39H1 and SUV39H2 are highly conserved, with 56% amino acid identity over the entire protein alignment. The SET (Su(var)3–9, Enhancer-of-zeste, Trithorax) domain is the catalytic domain of one large group of PKMTs, called SET-domain PKMTs [17,18]. The structure of the SUV39H2 SET domain has been solved and it shows a high similarity to the known SET domain structures of other H3K9 PKMTs such as Dim-5 or G9a [19]. This domain binds the methyl group donor S-adenosyl-L-methionine (AdoMet) and brings it in close contact to the target lysine residue in its active site pocket. Chromodomains are methylated lysine binding modules [20,21]. While the SUV39H1 chromodomain was shown to recognize H3K9me2/3 [22], the function of the SUV39H2 chromodomain has not yet been confirmed.

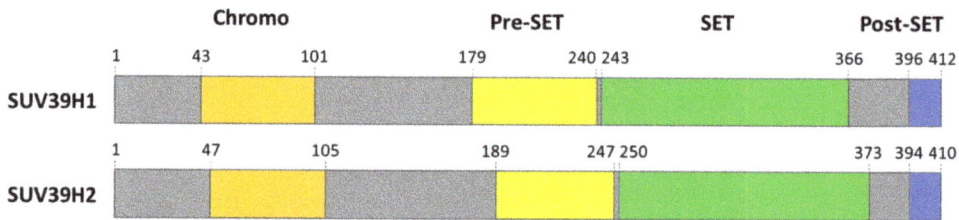

Figure 1. Scheme of the domain structure of human SUV39H1 and SUV39H2. The domain boundaries are indicated as listed in Uniprot.

2.1. Structure and Biochemical Properties of the SET Domain

One large group of PKMTs contains a SET domain as the catalytically active part, which consists of approximately 130 amino acids [17,18,23]. The SET domain comprises several small β-sheets that surround a knot-like structure in which the C-terminus of the protein is thread through an opening of a short loop in the preceding amino acid sequences. This structure brings together the two most-conserved motifs (NH(S/C)xxPN and ELx(F/Y)DY, where x denotes any amino acid residue) of the SET domain and forms the active site of the enzyme next to the AdoMet binding pocket and substrate peptide binding cleft. It is packed together with a Post-SET, Pre-SET or an additional I-SET domain that is inserted into the core SET domain.

The SET domain of SUV39H2 (Figure 2) has been structurally characterized and shown to contain an additional N-SET region, which is N-terminal to the Pre-SET regions and wraps around the core SET domain [19]. The H3K9 peptide binds in a groove formed by the I-SET and Post-SET domains, where it contacts the enzyme with backbone and side-chain interactions. Thus far, no structure has been solved for the SET domain of SUV39H1 but based on the amino acid sequence similarity, the overall folding and peptide interactions can be expected to be similar.

2.1.1. Biochemical Properties of the SUV39H1 SET Domain

SUV39H1 is able to introduce trimethylation at H3K9 in vitro, but the conversion of H3K9me2 into H3K9me3 is slow [13,24,25]. The SET domain of SUV39H1 introduces methyl groups on the H3 substrate in a non-processive manner [25]. Peptide SPOT array methylation experiments in the context of the H3K9 sequence revealed recognition of H3 residues between K4 and G12 with a highly specific readout of R8 (Figure 3A) [26]. Similar to G9a [27] and SUV39H2 (see below) [28], SUV39H1 shows a high specificity for an arginine at the −1 position (R8) (using K9 as reference position), replacing this R by any other amino acid completely abolished the catalytic activity. Apart from this, residues from the −5 to +3 positions are recognized with variable stringency. At the −5 site (K4), lysine and more weakly arginine were preferred. At the −3 position (T6), the enzyme prefers T, S, A and Y. At the −2 position, SUV39H1 accepts several residues, including polar (N, Q), small (A) and hydrophobic (L, P, W) ones. At the +1 position, the positively charged K and R are equally accepted as the native S10. At the +2 position, SUV39H1 tolerates only small amino acids such as A, G and S, in addition to the native amino acid T11 and at the +3 site, G, K and Q are preferred. In agreement with these findings, the catalytic activity of SUV39H1 has been shown to be influenced by the PTMs of this region of the H3 tail, for example, the trimethylation of K4 has been shown to reduce the activity of SUV39H1 [26,29,30].

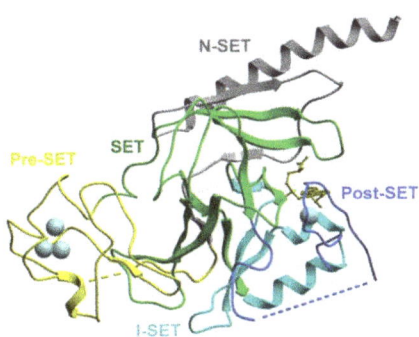

Figure 2. Structure of SUV39H2 in complex with AdoMet. The Pre-Set, SET, I-SET, Post-SET and N-SET domains are highlighted. The co-factor is shown as yellow sticks. Residues flanking un-resolved regions are connected by dotted lines. Taken from [19] with permission.

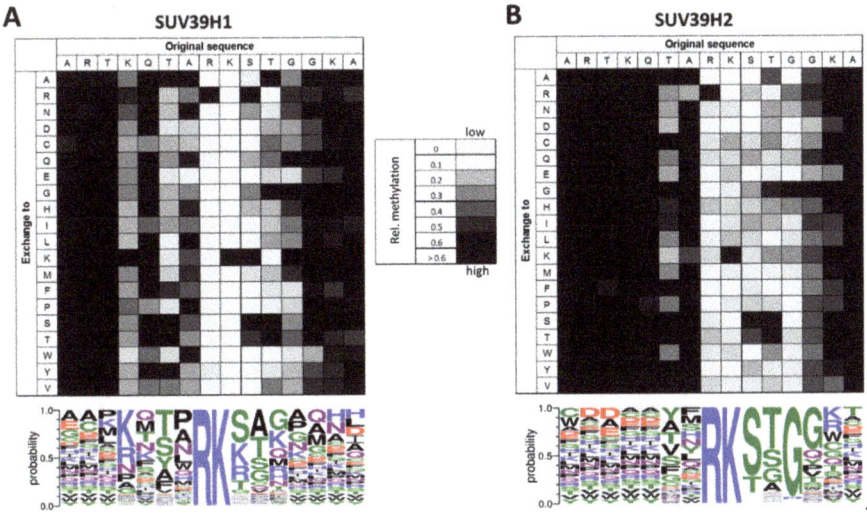

Figure 3. Specificity profiles of SUV39H1 (**A**) and SUV39H2 (**B**). Methylation of peptide substrates containing all possible single amino acid exchanges of the H3 sequence is shown. The horizontal axis represents the sequence of the peptide and in the vertical direction the amino acid that is altered in the corresponding peptide is indicated. Activity is encoded in a grayscale as indicated in the legend. The sequence logo describing the specificity has been prepared with Weblogo3 (http://weblogo.threeplusone.com/ (accessed on 30 April 2021)) [31] and is printed below. Activity data were taken from [26,28].

2.1.2. Biochemical Properties of the SUV39H2 SET Domain

Similar to SUV39H1, SUV39H2 introduces H3K9me3 in vitro [28,32]. It prefers the unmethylated H3 peptide as substrate [28,33] and the SUV39H2 catalytic SET domain introduces the first two methyl groups into H3K9me0 in a processive reaction [28], but similar to SUV39H1 (see above), the generation of H3K9me3 was slower than the generation of H3K9me2 [28,33]. The recognition of the H3K9 sequence by SUV39H2 has been investigated by peptide array methylation studies, which revealed accurate sequence recognition of the positions R8, S10, T11 and G12. In addition, the residues T6, A7, G13 and K14 were important for the enzyme activity (Figure 3B) [28]. Similar to SUV39H1 (see above) and G9a [27], SUV39H2 critically depends on the recognition of R8. This can be explained on the basis of the SUV39H2-SET domain structure, because D196 in SUV39H2 is ideally positioned to contact R8 with H-bonds. At the C-terminal side of the target lysine, S10 recognition could be mediated by D198 in SUV39H2, which is positioned identically as D209 in Dim-5, which takes over this role in this enzyme [34]. In agreement with the accurate readout of the R8, S10 and T11 positions, the modifications of R8 reduced the methylation activity of SUV39H2 and the phosphorylation of S10 or T11 completely blocked the enzyme [14,28].

Interestingly, the substrate specificity profiles of the two SUV39H enzymes differ from each other (Figure 3). Overall, SUV39H1 has stronger preferences for residues N-terminal to the target lysine, whereas SUV39H2 is more specific for residues C-terminal to the target lysine. One clear difference between SUV39H1 and SUV39H2 is the preference of SUV39H1 for R, K, S and T at the +1 site, where SUV39H2 is more specific and accepts mainly S and, more weakly, T. In general, SUV39H2 is more specific than SUV39H1, because it displays a high preference for the native H3 tail residues at six sequence positions (R8-G12), while SUV39H1 shows stringent readout of only one residue (R8). These differences indicate that the same methylation site on histone H3 is recognized in a different manner and both enzymes could have different non-histone substrate proteins, which may also be one reason explaining the emergence of different SUV39H paralogs in evolution. A similar observation was made for the paralogous SUV4-20H1 and SUV4-20H2 enzymes which also showed overlapping but distinct biological functions and properties [35].

In Clr4, the SUV39H homolog in *S. pombe*, automethylation was observed on K455 and K472, which are located in an autoregulatory loop (ARL) positioned between the SET and post-SET domain [36]. This ARL blocks the active center of the enzyme, but after automethylation, it undergoes a conformational change increasing the enzyme activity [36,37], which potentially connects the intracellular concentration of AdoMet to Clr4 activity [37]. Intriguingly, K392 in SUV39H2, which is analogous to Clr4 K472 and located in a similar flexible loop [19], has been shown to be automethylated as well and accordingly to change the enzyme activity and binding affinity to its substrate proteins [38], suggesting that automethylation might play a role in the regulation of SUV39H2 as well.

2.2. Structure and Biochemical Properties of the Chromodomain

Chromodomains are well-known methyllysine interaction domains [20,21]. Structural studies showed that the SUV39H1 chromodomain displays a generally conserved structure compared with other solved chromodomains [22]. The chromodomain fold comprises an N-terminal β-barrel consisting of three anti-parallel strands, which is followed by a long C-terminal α-helix that in the case of the SUV39H1 chromodomain is longer than typically observed with other chromodomains. Biochemical studies documented the specific binding of the SUV39H1 chromodomain to H3K9me3 and, more weakly, H3K9me2, but the overall binding affinities were lower than those observed with other chromodomains [22]. Modelling could identify a trimethyllysine binding cage that is structurally very similar to the one in HP1 proteins (Figure 4).

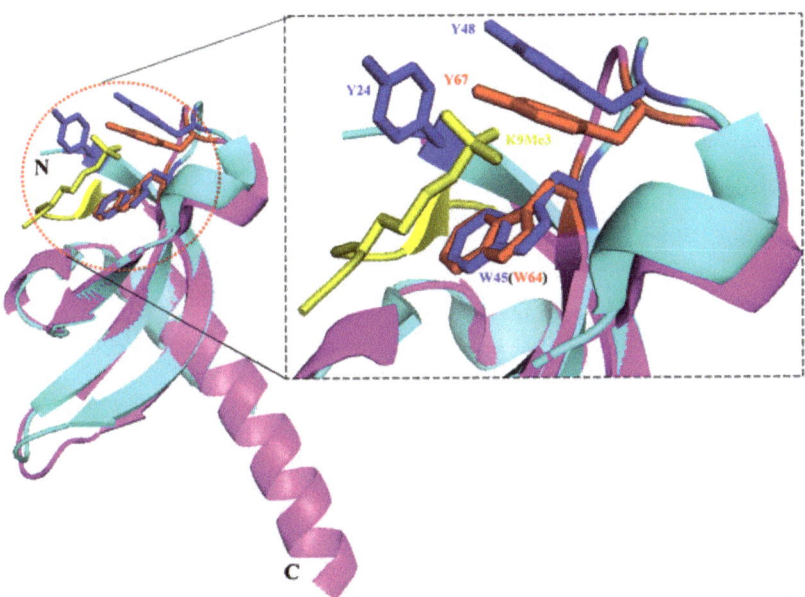

Figure 4. Model of H3K9me3 binding by human SUV39H1 chromodomain. The structures of human SUV39H1 and Drosophila melanogaster HP1 (PDB: 1KNE) chromodomains are aligned and shown in magenta and cyan, respectively [22]. The Y24, W45 and Y48 residues of Drosophila melanogaster HP1 chromodomain that are critical for H3K9me3 binding are shown as sticks in blue. The corresponding residues W64 and Y67 of human SUV39H1 chromodomain are shown as sticks in red. The H3K9me3 peptide present in the Drosophila melanogaster HP1 chromodomain is shown in yellow with trimethylated lysine 9 shown as sticks. Taken from [22] with permission.

Biochemical studies revealed that the chromodomain of SUV39H1 inhibits its methyltransferase activity, and this inhibition was relieved by H3K9me3 binding to the chromodomain [39]. Using designer chromatin templates for methylation kinetics, Müller et al. (2016) discovered a two-step activation switch acting in SUV39H1, where H3K9me3 recognition by the chromodomain firstly leads to the anchoring of the enzyme to chromatin. Secondly, the H3K9me3 interaction of the chromodomain led to an allosterically activation of the methylation activity of the SET domain. This process establishes a positive feedback loop for spreading of H3K9me2 and H3K9me3 over extended heterochromatic regions that was shown to be operational in cells as well [39].

In 2017, two additional papers shed more light on the targeting and regulatory role of the SUV39H1 chromodomain [40,41]. Collectively, these papers showed that the chromodomain of SUV39H1 binds to nucleic acids with basic surface residues that are distinct from the trimethyllysine binding cage. Binding was observed to the RNA associated with pericentric heterochromatin, which is retained in cis at its transcription sites. Binding to H3K9me3 and pericentromeric RNA was synergistic and both activities were required for the efficient targeting of SUV39H1 to heterochromatin, H3K9me3 deposition and heterochromatin silencing. The specificity of the nucleic acid binding was partially controversial; while one paper reported binding without sequence preference to ssRNA, ssDNA, dsRNA, dsDNA and RNA/DNA hybrids [40], the second one observed better binding of ssRNA than dsDNA [41]. Regarding the mechanism of the RNA-mediated regulation of SUV39H1, a two-step process similar to that suggested for the H3K9me3-dependent activation of SUV39H1 had been proposed [41]. In this model, the RNA interaction with the chromodomain targets the enzyme and it also leads to an allosteric activation of the catalytic activity of the SET domain by disrupting its inhibitory interaction with the chromodomain.

The targeting and regulation of SUV39H1 by RNA binding to its chromodomain is also consistent with the finding that the telomeric TERRA RNA associates with this domain and this interaction promotes the accumulation of H3K9me3 at damaged telomeres and end-to-end chromosome fusions [42]. Currently, it is not known if the chromodomain of SUV39H2 has similar roles.

2.3. Biochemical Properties of the N-Terminal Part of SUV39H1

The chromodomains of HP1 proteins are critical readers of pericentromeric H3K9me3 [43,44]. Strikingly, SUV39H1 binds directly to HP1 proteins [45] with its N-terminal part [46] and this interaction has been shown to recruit more SUV39H activity to existing H3K9me sites. This process constitutes a self-enforcing feedback loop necessary for the efficient deposition of pericentromeric H3K9me3 [46]. Similarly, in vitro and in vivo data indicated a role of the N-terminal extension to the chromodomain of SUV39H1 in RNA binding [40,41]. In addition, the regulation of SUV39H1 by its N-terminal part and chromodomain has been shown to be under regulation of post-translational modifications in the N-terminal part, because K105 and K123 in the N-terminal part of SUV39H1 were shown to be a target of lysine methylation by SET7/9 in response to DNA damage [47]. The methylation of SUV39H1 reduced its catalytic activity leading to a decreased pericentromeric H3K9me3 and an increased expression of satellite 2 and genome instability [47].

3. Biological Roles of SUV39H1 and SUV39H2

3.1. Expression Patterns of SUV39H1 and SUV39H2

The expression profiles of SUV39H1 and SUV39H2 in mice are overlapping during embryogenesis, but SUV39H2 remained expressed in adult testis where it is localized at meiotic heterochromatin [14]. In human, SUV39H1 shows little tissue specificity, despite some enrichment in thymus (Figure 5). SUV39H2 is ubiquitously expressed as well, but in adult tissues, the expression is enriched in cerebellum and testis (Figure 5). The expression of SUV39H1 has been observed to decline with age in hematopoietic stem cells [48]. This was shown to lead to a global decrease in H3K9me3 and perturbed heterochromatin function. SUV39H1 was found to be a target of microRNA miR-125b, the expression of which increases with age in human HSC [48]. Moreover, SUV39H1-mediated H3K9 trimethylation regulates the expression of several genes, and the dysregulation of SUV39H1 is observed in different cancers [49,50]. SUV39H2 is overexpressed in many cancer tissues, such as leukemia, lymphomas, lung cancer, breast cancer, colorectal cancer, gastric cancer and hepatocellular cancer [51]. It was found that SUV39H2 is degraded through the ubiquitin-proteasomal pathway and its half-life was reduced by interaction with the translationally controlled tumor protein (TCTP) [52].

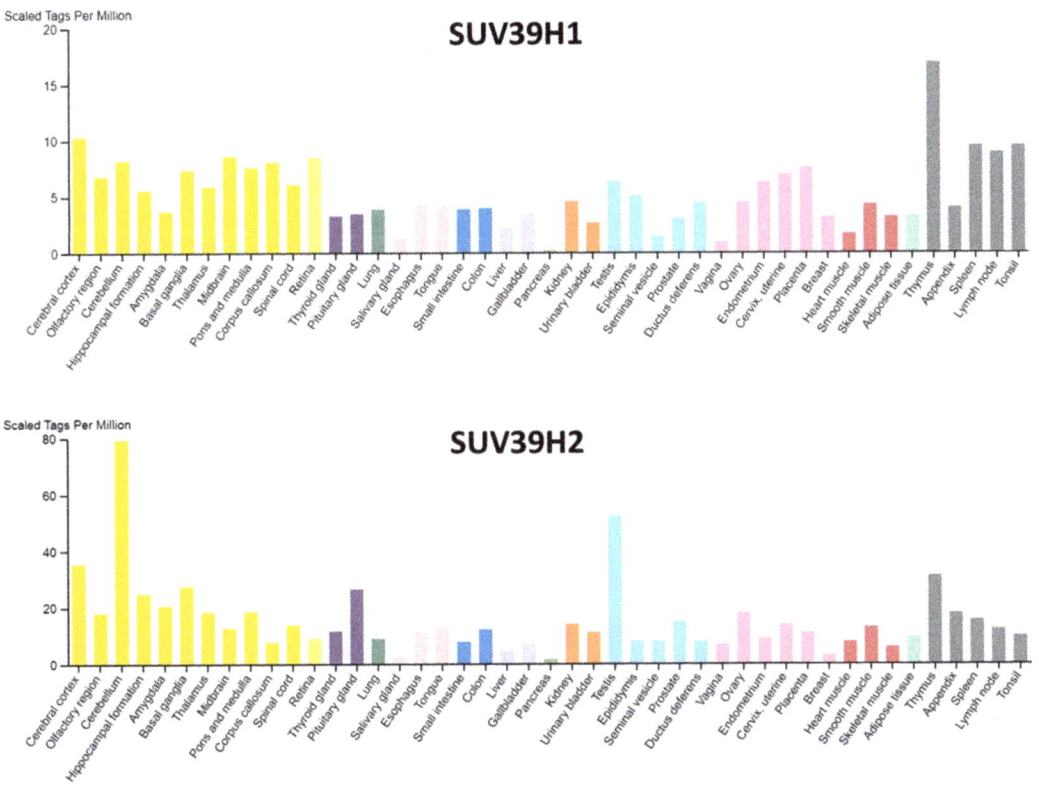

Figure 5. Expression of SUV39H1 and SUV39H2 in human tissues. Data were taken from https://www.proteinatlas.org/ (accessed on 30 April 2021) [53] using the FANTOM5 data set. Data are reported as Scaled Tags per million. Color-coding is based on tissue groups, each consisting of tissues with functional features in common.

3.2. Summary of the Functions of H3K9me2/3

As mentioned above, H3K9me3 is a key feature of constitutive heterochromatin in eukaryotes [7–10] and it also has roles in gene silencing in euchromatic regions [11]. While single SUV39H knock-out mice are viable, the deletion of both SUV39H1 and SUV39H2 is lethal, indicating that the roles of both enzymes are (at least partially) overlapping [13,54]. SUV39H1/2 double knock-out (SUV39H dn) resulted in a drastic loss of pericentric H3K9 trimethylation and also led to chromosomal instabilities [24,54,55]. SUV39H dn cells show severely diminished H3K9me3 levels over the pericentromere, resulting in a lack of accumulation of HP1 proteins and chromosomal instabilities [44,54,55]. In vivo, both SUV39H1 and SUV39H2 introduce H3K9me3 at pericentric heterochromatin, as shown by the finding that the reduction in heterochromatic H3K9 trimethylation in SUV39H dn cells was efficiently recovered by the ectopic expression of either SUV39H1 or SUV39H2 [32,44]. SUV39H1 and SUV39H2 introduced H3K9me3 in the pericentric regions plays a major role in silencing the expression of these regions, thereby repressing 'selfish' genetic elements and repetitive DNA and promoting genomic stability [24,54,56].

3.3. Chromatin Modification Network of SUV39H1 and SUV39H2

SUV39H1 and SUV39H2 also contribute to the chromatin modification network via different pathways, because HP1 proteins also recruit SUV4-20H enzymes to heterochromatic regions, where they generate H4K20me3 by using the H4K20me1 provided by SET8 as a substrate [57–59]. By this mechanism, the H3K9me3 introduced by SUV39H enzymes indirectly stimulates the generation of H4K20me3, another characteristic heterochromatic histone tail modification (Figure 6B). In fact, the knock-out of the SUV39H enzymes has also been shown to lead to decreased levels of heterochromatic H4K20me3 [58].

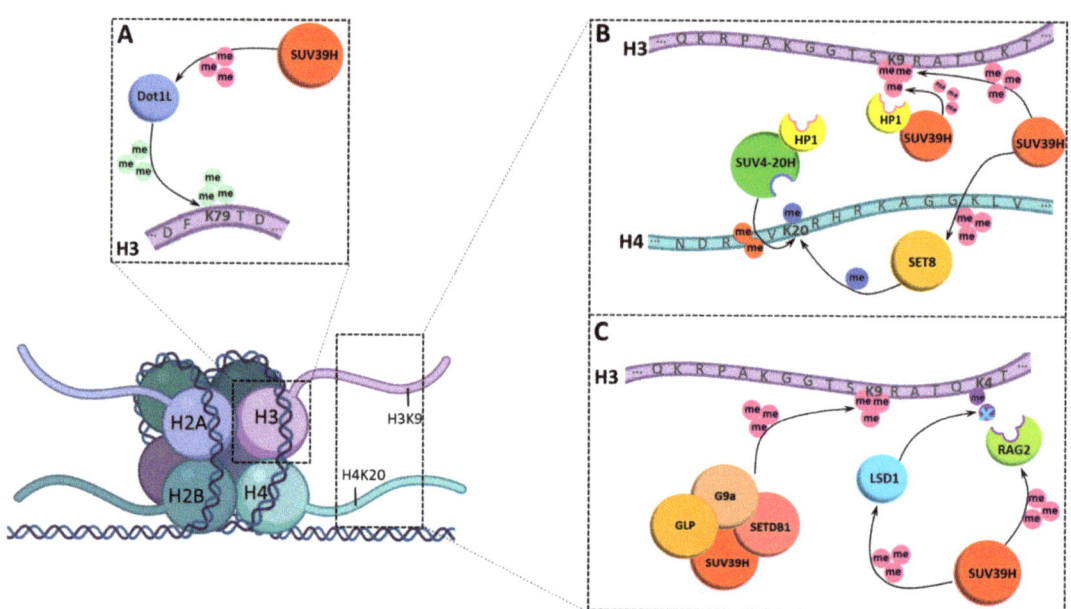

Figure 6. Compilation of the SUV39H centered chromatin network including interacting proteins and complex partners, histone methylation sites and non-histone substrates of SUV39H1 and SUV39H2. The upper left insert (**A**) illustrates SUV39H1 methylation of Dot1L, which itself methylates H3K79. The upper right insert (**B**) shows methylation of H3K9 by SUV39H. This is followed by recruitment of SUV39H and SUV4-20H by HP1 proteins to the H3K9me2/3 sites, leading to the spreading of H3K9me3 and introduction of H4K20me2/3 at H4K20me1 sites. Generation of H4K20me1 by SET8 is stimulated by SUV39H1 mediated methylation. The lower right insert (**C**) features the role of SUV39H in H3K9 methylation as members of the complex of SUV39H with G9a, GPL and SETDB1 PKMTs. SUV39H methylation of the LSD1 (SUV39H2) and RAG2 (SUV39H1) non-histone substrates is shown, which are creating a crosstalk with H3K4 methylation. LSD1 has a role in the removal of H3K4me1/2 and RAG2 is a reader of H3K4me3.

Another poorly understood observation is that SUV39H1 exists in multimeric complexes with the other H3K9 PKMTs such as G9a, GLP and SETDB1 (Figure 6C) and the deletion of SUV39H1 destabilizes the corresponding proteins and leads to a decrease in the H3K9 methylation signal at the global level [60]. Moreover, in SUV39H or G9a null cells, the remaining H3K9 PKMTs are destabilized at the protein level, indicating that the integrity of these PKMTs is interdependent. In this work, it was also shown that all four H3K9-specific PKMTs are recruited not only to major satellite repeats, a known SUV39H1 genomic target, but also to multiple G9a target genes [60]. Moreover, the functional cooperation between the four H3K9 PKMTs was demonstrated in the regulation of known G9a target genes.

3.4. Non-Histone Substrates of SUV39H1 and SUV39H2

As described above, the specificity profile of SUV39H1 differs from SUV39H2, suggesting that these paralogs could have non-redundant functions in the methylation of non-histone proteins. Based on the specificity profile, several SUV39H1 non-histone substrates were identified [26]. The methylation of RAG2, SET8 and DOT1L was confirmed in cells, which all have important roles in chromatin regulation (Figure 6A–C). The SUV39H1-mediated methylation of SET8 was shown to allosterically stimulate its activity [26]. The SET8 PKMT generates monomethylated H4K20 [61–63] that is the substrate used by the SUV4-20H enzymes for the generation of H4K20me3 [58,59,64]. This indicates that SUV39H1 controls heterochromatic H4K20 trimethylation through the following two processes: Firstly, SUV39H1-introduced H3K9me3 recruits HP1 proteins that recruit SUV4-20 enzymes, and secondly, it stimulates SET8 to generate more H4K20me1, which is used by SUV4-20H as a substrate (Figure 6B).

Other non-histone substrates of SUV39H1 also have chromatin associated roles: The methylation of RAG2 by SUV39H1 occurs within its NLS and it was shown to alter its sub-nuclear localization [26]. This observation suggests that SUV39H1 could have a direct influence on VDJ recombination catalyzed by RAG2, which is in agreement with data showing that SUV39H1 regulates class switch recombination in B cells [65] and H3K9me3 is associated with this process [66]. This process also contributes to the crosstalk of SUV39H1 with H3K4 methylation because RAG2 is a reader of H3K4me3 (Figure 6C) and H3K4me3 inhibits SUV39H1 activity.

DOT1L is an evolutionarily conserved 7-beta-strand histone PKMT specific for lysine 79 of H3 (H3K79), which has important roles in development and cancer [67]. DOT1L-deficient mouse embryos show reduced levels of heterochromatic H3K9me3 and H4K20me3 marks at centromeres and telomeres indicating that DOT1L plays an important role in heterochromatin formation as well [68]. Conversely, SUV39H1 can also methylate DOT1L, but the biological effects of this methylation event need further investigation [26].

Furthermore, SUV39H1 has been connected to immune function and bacterial infections, because the mycobacterial histone-like HupB protein has been shown to be methylated by SUV39H1 and this process participates in host defense [69]. The SUV39H1 methylation of HupB reduced the survival of mycobacteria inside host cells and it reduced the ability of mycobacteria to form biofilms.

SUV39H2 was found to trimethylate LSD1 at K322 [70] creating a crosstalk of SUV39H2 with H3K4 methylation, because LSD1 has a role in the removal of H3K4me1/2 (Figure 6C). SUV39H2-induced LSD1 methylation suppresses LSD1 polyubiquitination and subsequent degradation, revealing a novel regulatory mechanism of LSD1 in human cancer cells (Figure 6C). SUV39H2 was also reported to methylate K134 of H2AX and stimulate H2AX phosphorylation during DNA damage response [71]. However, the sequence context of H2AX-K134 differs from the specificity of SUV39H2 [28] and in vitro methylation of H2AX could not be confirmed in an independent study [72].

3.5. Connections to Diseases

As described above, SUV39H1 and H3K9me3 are predominately associated with the generation and maintenance of constitutive heterochromatin. In mammals, defective pericentric heterochromatin and aberrant transcription of pericentric repeats are associated with genomic instability and cancer [73,74]. These defects in constitutive heterochromatin are most evident in SUV39H1 and SUV39H2 double knockout mice, which exhibit reduced embryonic viability, small stature, chromosome instability, an increased risk of tumor formation and male infertility owing to defective spermatogenesis [54]. Both SUV39H1 and SUV39H2 are prognostic markers for different cancers, but dependent on the tumor type, either a high or low expression of SUV39H1 constitutes a risk, while a high expression of SUV39H2 is unfavorable in most cases (Figure 7).

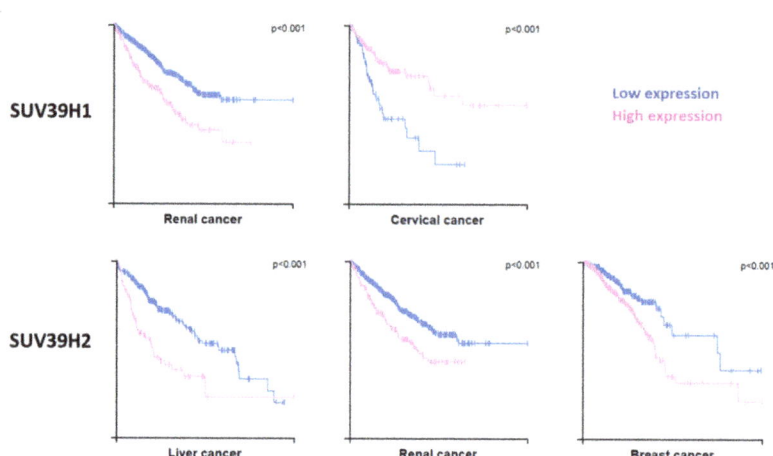

Figure 7. Cancer connection of SUV39H1 and SUV39H2. Kaplan–Meier plots are shown for cancers where high expression of SUV39H1 or SUV39H2 has significant ($p < 0.001$) association with patient survival. Data were retrieved at https://www.proteinatlas.org (accessed on 30 April 2021) [53].

Human SUV39H1 has been implicated in a variety of complex biological processes such as DNA damage repair [75–77], telomere maintenance [42,78], cell differentiation [79,80] and aging [81]. Several lines of evidence documented a tumor suppressive role of SUV39H1 through its stabilization and silencing of heterochromatin. It has been found that SUV39H-deficient mice develop B-cell lymphomas with increased frequencies [54], and SUV39H1 was observed to be downregulated in many leukemias [82]. The protective effect of SUV39H1 in leukemia was validated in mouse models using SUV39H1 overexpression or knockdown and the data provided a direct link between SUV39H1 and AML via the silencing of HOXB13 and SIC1 [82]. Similarly, tumorigenesis driven by Ras or Myc is accelerated by the loss of SUV39H1 [83,84]. Moreover, SUV39H1 was shown to reduce Cyclin D1 expression and, by this, trigger cell cycle arrest [85].

Accumulating evidence indicates that SUV39H2 acts mainly as an oncogene that contributes to the initiation and progression of cancers including invasion and metastasis [51]. As mentioned above, SUV39H2 is overexpressed in many cancers [51], including lung cancer [71,86], acute lymphoblastic leukemia [87], osteosarcoma [88] and glioma [89], and SUV39H2 knockdown resulted in the inhibition of glioma cell growth [89]. The important role of SUV39H2 in cancer is further illustrated by the finding that somatic mutations in this (and other PKMTs) are observed in tumor tissues [6,90]. Moreover, SUV39H2 has been connected with cancer through its regulatory effect on LSD1 [70] and several cancer relevant genes which mainly act as tumor suppressor genes have been shown to be repressed by SUV39H2 overexpression including FAS, P16, P21 and Twist1, while oncogenes like PSA and C-myc are overexpressed [51]. Another study showed that SUV39H2 promotes colorectal cancer proliferation and metastasis via tri-methylation of the SLIT1 promoter and suppression of SLIT1 transcription [91]. In addition, SUV39H2 downregulates the hedgehog interacting protein in glioma cells, thereby promoting hedgehog signaling [89]. Another study identified SUV39H2 as a tumor suppressor and showed that SUV39H2 overexpression in a non-small cell lung cancer (NSCLC) cell line leads to the inhibition of cell growth and proliferation by inducing G_1 cell cycle arrest [52].

In concordance with its high expression in cerebellum, SUV39H2 has also been connected with neuronal effects. It has been shown that stress-induced H3K9 methylation in the hippocampus was correlated with an upregulation of SUV39H2, suggesting that the enzyme plays a functional role in this process [92]. Recently, the A211S loss of function variant of SUV39H2 has been identified in autism spectrum disorder patients, where it causes altered H3K9-trimethylation and the dysregulation of protocadherin β cluster (Pcdhb cluster) genes in the developing brain [93]. This paper provided direct evidence of the role of SUV39H2 in autism spectrum disorder, and it discovered a molecular pathway of SUV39H2 dysfunction leading to H3K9me3 deficiency, followed by an elevated expression of Pcdhb cluster genes during early neurodevelopment.

Moreover, the SUV39H2 N324K loss of function mutation [28] has been identified to cause hereditary nasal parakeratosis in Labrador Retriever dogs [94], which is a monogenic, inherited, autosomal recessive disorder. Defects in the differentiation of the specialized nasal epidermis cells in affected dogs lead to the formation of crusts and fissures in the nasal planum already in young age, while the animals are otherwise healthy. As differentiation of the nasal epidermis involves selective activation of specific olfactory receptors and silencing of all others, SUV39H2 appears to have a function in this process. This process has been observed to be under the control of the G9a and GLP H3K9 methyltransferases as well as LSD1 [95], which all are connected to SUV39H2, as described above.

4. Perspectives and Outlook

The biological function of a PKMT is intimately connected to its substrates and the changes of the substrates' properties associated with target lysine methylation. For SUV39H1 and SUV39H2, H3K9 is a main substrate. It needs to be studied systematically, how intrinsic and external signals lead to changes in H3K9me3 levels in the heterochromatin and at defined genomic target loci. Moreover, both enzymes also methylate non-histone proteins. The identification of more non-histone proteins and deeper knowledge about their roles in the cellular signaling network will be important for the complete understanding of the biological role of SUV39H1 and SUV39H2. Combined achievements in both directions will help to understand the roles of SUV39H enzymes in diseases better and they may result in novel and specific therapeutic strategies.

Author Contributions: S.W., M.S.K. and A.J. wrote the text and prepared the figures. All authors have read and agreed to the published version of the manuscript.

Funding: This research was funded by DFG grant JE 252/7-4 (AJ) and the GERLS scholarship program (MSK) funding program number 57311832 by the German Academic Exchange Service (DAAD) and the Egyptian Ministry of Higher Education.

Institutional Review Board Statement: Not applicable.

Informed Consent Statement: Not applicable.

Data Availability Statement: Not applicable.

Conflicts of Interest: The authors declare no conflict of interest.

References

1. Kouzarides, T. Chromatin Modifications and Their Function. *Cell* **2007**, *128*, 693–705. [CrossRef] [PubMed]
2. Tan, M.; Luo, H.; Lee, S.; Jin, F.; Yang, J.S.; Montellier, E.; Buchou, T.; Cheng, Z.; Rousseaux, S.; Rajagopal, N.; et al. Identification of 67 Histone Marks and Histone Lysine Crotonylation as a New Type of Histone Modification. *Cell* **2011**, *146*, 1016–1028. [CrossRef]
3. Margueron, R.; Reinberg, D. Chromatin structure and the inheritance of epigenetic information. *Nat. Rev. Genet.* **2010**, *11*, 285–296. [CrossRef] [PubMed]
4. Bannister, A.; Kouzarides, T. Regulation of chromatin by histone modifications. *Cell Res.* **2011**, *21*, 381–395. [CrossRef]
5. Suvà, M.L.; Riggi, N.; Bernstein, B.E. Epigenetic Reprogramming in Cancer. *Science* **2013**, *339*, 1567–1570. [CrossRef] [PubMed]
6. Kudithipudi, S.; Jeltsch, A. Role of somatic cancer mutations in human protein lysine methyltransferases. *Biochim. Biophys. Acta (BBA) Bioenerg.* **2014**, *1846*, 366–379. [CrossRef] [PubMed]
7. Martin, C.; Zhang, Y. The diverse functions of histone lysine methylation. *Nat. Rev. Mol. Cell Biol.* **2005**, *6*, 838–849. [CrossRef]
8. Krishnan, S.; Horowitz, S.; Trievel, R.C. Structure and Function of Histone H3 Lysine 9 Methyltransferases and Demethylases. *ChemBioChem* **2011**, *12*, 254–263. [CrossRef]
9. Becker, J.; Nicetto, D.; Zaret, K.S. H3K9me3-Dependent Heterochromatin: Barrier to Cell Fate Changes. *Trends Genet.* **2016**, *32*, 29–41. [CrossRef]
10. Mozzetta, C.; Boyarchuk, E.; Pontis, J.; Ait-Si-Ali, S. Sound of silence: The properties and functions of repressive Lys methyltransferases. *Nat. Rev. Mol. Cell Biol.* **2015**, *16*, 499–513. [CrossRef]

11. Barski, A.; Cuddapah, S.; Cui, K.; Roh, T.-Y.; Schones, D.E.; Wang, Z.; Wei, G.; Chepelev, I.; Zhao, K. High-Resolution Profiling of Histone Methylations in the Human Genome. *Cell* **2007**, *129*, 823–837. [CrossRef]
12. Tschiersch, B.; Hofmann, A.; Krauss, V.; Dorn, R.; Korge, G.; Reuter, G. The protein encoded by the Drosophila posi-tion-effect variegation suppressor gene Su(var)3-9 combines domains of antagonistic regulators of homeotic gene com-plexes. *EMBO J.* **1994**, *13*, 3822–3831. [CrossRef]
13. Rea, S.; Eisenhaber, F.; O'Carroll, D.; Strahl, B.D.; Sun, Z.-W.; Schmid, M.; Opravil, S.; Mechtler, K.; Ponting, C.P.; Allis, C.D.; et al. Regulation of chromatin structure by site-specific histone H3 methyltransferases. *Nature* **2000**, *406*, 593–599. [CrossRef]
14. O'Carroll, D.; Scherthan, H.; Peters, A.H.F.M.; Opravil, S.; Haynes, A.R.; Laible, G.; Rea, S.; Schmid, M.; Lebersorger, A.; Jerratsch, M.; et al. Isolation and Characterization of Suv39h2, a Second Histone H3 Methyltransferase Gene That Displays Testis-Specific Expression. *Mol. Cell. Biol.* **2000**, *20*, 9423–9433. [CrossRef]
15. Nakayama, J.-I.; Rice, J.C.; Strahl, B.D.; Allis, C.D.; Grewal, S.I.S. Role of Histone H3 Lysine 9 Methylation in Epigenetic Control of Heterochromatin Assembly. *Science* **2001**, *292*, 110–113. [CrossRef] [PubMed]
16. Schotta, G.; Ebert, A.; Reuter, G. SU(VAR)3-9 is a Conserved Key Function in Heterochromatic Gene Silencing. *Genetica* **2003**, *117*, 149–158. [CrossRef] [PubMed]
17. Cheng, X.; Collins, R.E.; Zhang, X. Structural and Sequence Motifs of Protein (Histone) Methylation Enzymes. *Annu. Rev. Biophys. Biomol. Struct.* **2005**, *34*, 267–294. [CrossRef] [PubMed]
18. Dillon, S.C.; Zhang, X.; Trievel, R.C.; Cheng, X. The SET-domain protein superfamily: Protein lysine methyltransferases. *Genome Biol.* **2005**, *6*, 227. [CrossRef] [PubMed]
19. Wu, H.; Min, J.; Lunin, V.V.; Antoshenko, T.; Dombrovski, L.; Zeng, H.; Allali-Hassani, A.; Campagna-Slater, V.; Vedadi, M.; Arrowsmith, C.; et al. Structural Biology of Human H3K9 Methyltransferases. *PLoS ONE* **2010**, *5*, e8570. [CrossRef]
20. Taverna, S.D.; Li, H.; Ruthenburg, A.J.; Allis, C.D.; Patel, D.J. How chromatin-binding modules interpret histone modifications: Lessons from professional pocket pickers. *Nat. Struct. Mol. Biol.* **2007**, *14*, 1025–1040. [CrossRef] [PubMed]
21. Patel, D.J.; Wang, Z. Readout of Epigenetic Modifications. *Annu. Rev. Biochem.* **2013**, *82*, 81–118. [CrossRef]
22. Wang, T.; Xu, C.; Liu, Y.; Fan, K.; Li, Z.; Sun, X.; Ouyang, H.; Zhang, X.; Zhang, J.; Li, Y.; et al. Crystal Structure of the Human SUV39H1 Chromodomain and Its Recognition of Histone H3K9me2/3. *PLoS ONE* **2012**, *7*, e52977. [CrossRef]
23. Luo, M. Chemical and Biochemical Perspectives of Protein Lysine Methylation. *Chem. Rev.* **2018**, *118*, 6656–6705. [CrossRef]
24. Peters, A.H.; Kubicek, S.; Mechtler, K.; O'Sullivan, R.J.; Derijck, A.A.; Perez-Burgos, L.; Kohlmaier, A.; Opravil, S.; Tachibana, M.; Shinkai, Y.; et al. Partitioning and Plasticity of Repressive Histone Methylation States in Mammalian Chromatin. *Mol. Cell* **2003**, *12*, 1577–1589. [CrossRef]
25. Chin, H.G.; Patnaik, D.; Estève, P.-O.; Jacobsen, S.E.; Pradhan, S. Catalytic Properties and Kinetic Mechanism of Human Recombinant Lys-9 Histone H3 Methyltransferase SUV39H1: Participation of the Chromodomain in Enzymatic Catalysis. *Biochemistry* **2006**, *45*, 3272–3284. [CrossRef]
26. Kudithipudi, S.; Schuhmacher, M.K.; Kebede, A.F.; Jeltsch, A. The SUV39H1 Protein Lysine Methyltransferase Methylates Chromatin Proteins Involved in Heterochromatin Formation and VDJ Recombination. *ACS Chem. Biol.* **2017**, *12*, 958–968. [CrossRef]
27. Rathert, P.; Dhayalan, A.; Murakami, M.; Zhang, X.; Tamas, R.; Jurkowska, R.; Komatsu, Y.; Shinkai, Y.; Cheng, X.; Jeltsch, A. Protein lysine methyltransferase G9a acts on non-histone targets. *Nat. Chem. Biol.* **2008**, *4*, 344–346. [CrossRef]
28. Schuhmacher, M.K.; Kudithipudi, S.; Kusevic, D.; Weirich, S.; Jeltsch, A. Activity and specificity of the human SUV39H2 protein lysine methyltransferase. *Biochim. Biophys. Acta (BBA) Bioenerg.* **2015**, *1849*, 55–63. [CrossRef]
29. Nishioka, K.; Chuikov, S.; Sarma, K.; Erdjument-Bromage, H.; Allis, C.D.; Tempst, P.; Reinberg, D. Set9, a novel histone H3 methyltransferase that facilitates transcription by precluding histone tail modifications required for heterochromatin formation. *Genes Dev.* **2002**, *16*, 479–489. [CrossRef]
30. Binda, O.; Leroy, G.; Bua, D.J.; Garcia, B.A.; Gozani, O.; Richard, S. Trimethylation of histone H3 lysine 4 impairs meth-ylation of histone H3 lysine 9: Regulation of lysine methyltransferases by physical interaction with their substrates. *Epigenetics* **2010**, *5*, 767–775. [CrossRef]
31. Crooks, G.E.; Hon, G.; Chandonia, J.-M.; Brenner, S.E. WebLogo: A Sequence Logo Generator. *Genome Res.* **2004**, *14*, 1188–1190. [CrossRef] [PubMed]
32. Rice, J.C.; Briggs, S.D.; Ueberheide, B.; Barber, C.M.; Shabanowitz, J.; Hunt, D.F.; Shinkai, Y.; Allis, C. Histone Methyltransferases Direct Different Degrees of Methylation to Define Distinct Chromatin Domains. *Mol. Cell* **2003**, *12*, 1591–1598. [CrossRef]
33. Allali-Hassani, A.; Wasney, G.A.; Siarheyeva, A.; Hajian, T.; Arrowsmith, C.; Vedadi, M. Fluorescence-Based Methods for Screening Writers and Readers of Histone Methyl Marks. *J. Biomol. Screen.* **2011**, *17*, 71–84. [CrossRef] [PubMed]
34. Rathert, P.; Zhang, X.; Freund, C.; Cheng, X.; Jeltsch, A. Analysis of the Substrate Specificity of the Dim-5 Histone Lysine Methyltransferase Using Peptide Arrays. *Chem. Biol.* **2008**, *15*, 5–11. [CrossRef]
35. Weirich, S.; Kudithipudi, S.; Jeltsch, A. Specificity of the SUV4–20H1 and SUV4–20H2 protein lysine methyltransferases and methylation of novel substrates. *J. Mol. Biol.* **2016**, *428*, 2344–2358. [CrossRef]
36. Iglesias, N.; Currie, M.A.; Jih, G.; Paulo, J.A.; Siuti, N.; Kalocsay, M.; Gygi, S.P.; Moazed, D. Automethylation-induced conforma-tional switch in Clr4 (Suv39h) maintains epigenetic stability. *Nat. Cell Biol.* **2018**, *560*, 504–508. [CrossRef]
37. Khella, M.S.; Bröhm, A.; Weirich, S.; Jeltsch, A. Mechanistic Insights into the Allosteric Regulation of the Clr4 Protein Lysine Methyltransferase by Autoinhibition and Automethylation. *Int. J. Mol. Sci.* **2020**, *21*, 8832. [CrossRef]

38. Piao, L.; Nakakido, M.; Suzuki, T.; Dohmae, N.; Nakamura, Y.; Hamamoto, R. Automethylation of SUV39H2, an oncogenic histone lysine methyltransferase, regulates its binding affinity to substrate proteins. *Oncotarget* **2016**, *7*, 22846–22856. [CrossRef]
39. Müller, M.M.; Fierz, B.; Bittova, L.; Liszczak, G.; Muir, T.W. A two-state activation mechanism controls the histone methyltransferase Suv39h1. *Nat. Chem. Biol.* **2016**, *12*, 188–193. [CrossRef]
40. Johnson, W.L.; Yewdell, W.T.; Bell, J.C.; McNulty, S.; Duda, Z.; O'Neill, R.J.; Sullivan, B.A.; Straight, A.F. RNA-dependent stabilization of SUV39H1 at constitutive heterochromatin. *eLife* **2017**, *6*. [CrossRef] [PubMed]
41. Shirai, A.; Kawaguchi, T.; Shimojo, H.; Muramatsu, D.; Ishida-Yonetani, M.; Nishimura, Y.; Kimura, H.; Nakayama, J.-I.; Shinkai, Y. Impact of nucleic acid and methylated H3K9 binding activities of Suv39h1 on its heterochromatin assembly. *eLife* **2017**, *6*, 6. [CrossRef]
42. Porro, A.; Feuerhahn, S.; Delafontaine, J.; Riethman, H.; Rougemont, J.; Lingner, J. Functional characterization of the TERRA transcriptome at damaged telomeres. *Nat. Commun.* **2014**, *5*, 5379. [CrossRef]
43. Bannister, A.; Zegerman, P.; Partridge, J.; Miska, E.; Thomas, J.O.; Allshire, R.; Kouzarides, T. Selective recognition of methylated lysine 9 on histone H3 by the HP1 chromo domain. *Nat. Cell Biol.* **2001**, *410*, 120–124. [CrossRef] [PubMed]
44. Lachner, M.; O'Carroll, D.; Rea, S.; Mechtler, K.; Jenuwein, T. Methylation of histone H3 lysine 9 creates a binding site for HP1 proteins. *Nat. Cell Biol.* **2001**, *410*, 116–120. [CrossRef] [PubMed]
45. Yamamoto, K.; Sonoda, M. Self-interaction of heterochromatin protein 1 is required for direct binding to histone methyltransferase, SUV39H1. *Biochem. Biophys. Res. Commun.* **2003**, *301*, 287–292. [CrossRef]
46. Muramatsu, D.; Kimura, H.; Kotoshiba, K.; Tachibana, M.; Shinkai, Y. Pericentric H3K9me3 formation by HP1 interaction-defective histone methyltransferase Suv39h1. *Cell Struct. Funct.* **2016**, *41*, 145–152. [CrossRef] [PubMed]
47. Wang, D.; Zhou, J.; Liu, X.; Lu, D.; Shen, C.; Du, Y.; Wei, F.-Z.; Song, B.; Lu, X.; Yu, Y.; et al. Methylation of SUV39H1 by SET7/9 results in heterochromatin relaxation and genome instability. *Proc. Natl. Acad. Sci. USA* **2013**, *110*, 5516–5521. [CrossRef]
48. Djeghloul, D.; Kuranda, K.; Kuzniak, I.; Barbieri, D.; Naguibneva, I.; Choisy, C.; Bories, J.-C.; Dosquet, C.; Pla, M.; Vanneaux, V.; et al. Age-Associated Decrease of the Histone Methyltransferase SUV39H1 in HSC Perturbs Heterochromatin and B Lymphoid Differentiation. *Stem Cell Rep.* **2016**, *6*, 970–984. [CrossRef]
49. He, Y.; Korboukh, I.; Jin, J.; Huang, J. Targeting protein lysine methylation and demethylation in cancers. *Acta Biochim. Biophys. Sin.* **2011**, *44*, 70–79. [CrossRef]
50. Dong, C.; Wu, Y.; Wang, Y.; Wang, C.; Kang, T.; Rychahou, P.; Chi, Y.-I.; Evers, B.M.; Zhou, B.P. Interaction with Suv39H1 is critical for Snail-mediated E-cadherin repression in breast cancer. *Oncogene* **2013**, *32*, 1351–1362. [CrossRef]
51. Li, B.; Zheng, Y.; Yang, L. The Oncogenic Potential of SUV39H2: A Comprehensive and Perspective View. *J. Cancer* **2019**, *10*, 721–729. [CrossRef] [PubMed]
52. Kim, A.-R.; Sung, J.Y.; Rho, S.B.; Kim, Y.-N.; Yoon, K. Suppressor of Variegation 3-9 Homolog 2, a Novel Binding Protein of Translationally Controlled Tumor Protein, Regulates Cancer Cell Proliferation. *Biomol. Ther.* **2019**, *27*, 231–239. [CrossRef] [PubMed]
53. Uhlén, M.; Fagerberg, L.; Hallström, B.M.; Lindskog, C.; Oksvold, P.; Mardinoglu, A.; Sivertsson, Å.; Kampf, C.; Sjöstedt, E.; Asplund, A.; et al. Tissue-based map of the human proteome. *Science* **2015**, *347*, 1846–1884. [CrossRef]
54. Peters, A.H.; O'Carroll, D.; Scherthan, H.; Mechtler, K.; Sauer, S.; Schöfer, C.; Weipoltshammer, K.; Pagani, M.; Lachner, M.; Kohlmaier, A.; et al. Loss of the Suv39h Histone Methyltransferases Impairs Mammalian Heterochromatin and Genome Stability. *Cell* **2001**, *107*, 323–337. [CrossRef]
55. Peters, A.H.; Mermoud, J.E.; O'Carroll, D.; Pagani, M.; Schweizer, D.; Brockdorff, N.; Jenuwein, T. Histone H3 lysine 9 methylation is an epigenetic imprint of facultative heterochromatin. *Nat. Genet.* **2001**, *30*, 77–80. [CrossRef]
56. Bulut-Karslioglu, A.; De La Rosa-Velázquez, I.A.; Ramirez, F.; Barenboim, M.; Onishi-Seebacher, M.; Arand, J.; Galán, C.; Winter, G.; Engist, B.; Gerle, B.; et al. Suv39h-Dependent H3K9me3 Marks Intact Retrotransposons and Silences LINE Elements in Mouse Embryonic Stem Cells. *Mol. Cell* **2014**, *55*, 277–290. [CrossRef]
57. Jørgensen, S.; Schotta, G.; Sørensen, C.S. Histone H4 Lysine 20 methylation: Key player in epigenetic regulation of genomic integrity. *Nucleic Acids Res.* **2013**, *41*, 2797–2806. [CrossRef]
58. Schotta, G.; Lachner, M.; Sarma, K.; Ebert, A.; Sengupta, R.; Reuter, G.; Reinberg, D.; Jenuwein, T. A silencing pathway to induce H3-K9 and H4-K20 trimethylation at constitutive heterochromatin. *Genes Dev.* **2004**, *18*, 1251–1262. [CrossRef]
59. Schotta, G.; Sengupta, R.; Kubicek, S.; Malin, S.; Kauer, M.; Callén, E.; Celeste, A.; Pagani, M.; Opravil, S.; De La Rosa-Velazquez, I.A.; et al. A chromatin-wide transition to H4K20 monomethylation impairs genome integrity and programmed DNA rearrangements in the mouse. *Genes Dev.* **2008**, *22*, 2048–2061. [CrossRef]
60. Fritsch, L.; Robin, P.; Mathieu, J.R.; Souidi, M.; Hinaux, H.; Rougeulle, C.; Harel-Bellan, A.; Ameyar-Zazoua, M.; Ait-Si-Ali, S. A Subset of the Histone H3 Lysine 9 Methyltransferases Suv39h1, G9a, GLP, and SETDB1 Participate in a Multimeric Complex. *Mol. Cell* **2010**, *37*, 46–56. [CrossRef]
61. Nishioka, K.; Rice, J.C.; Sarma, K.; Erdjument-Bromage, H.; Werner, J.; Wang, Y.; Chuikov, S.; Valenzuela, P.; Tempst, P.; Steward, R.; et al. PR-Set7 Is a Nucleosome-Specific Methyltransferase that Modifies Lysine 20 of Histone H4 and Is Associated with Silent Chromatin. *Mol. Cell* **2002**, *9*, 1201–1213. [CrossRef]
62. Fang, J.; Feng, Q.; Ketel, C.S.; Wang, H.; Cao, R.; Xia, L.; Erdjument-Bromage, H.; Tempst, P.; Simon, J.A.; Zhang, Y. Purification and Functional Characterization of SET8, a Nucleosomal Histone H4-Lysine 20-Specific Methyltransferase. *Curr. Biol.* **2002**, *12*, 1086–1099. [CrossRef]

63. Kudithipudi, S.; Dhayalan, A.; Kebede, A.F.; Jeltsch, A. The SET8 H4K20 protein lysine methyltransferase has a long recognition sequence covering seven amino acid residues. *Biochimie* **2012**, *94*, 2212–2218. [CrossRef] [PubMed]
64. Oda, H.; Okamoto, I.; Murphy, N.; Chu, J.; Price, S.M.; Shen, M.; Torres-Padilla, M.E.; Heard, E.; Reinberg, D. Monomethylation of Histone H4-Lysine 20 Is Involved in Chromosome Structure and Stability and Is Essential for Mouse Development. *Mol. Cell. Biol.* **2009**, *29*, 2278–2295. [CrossRef]
65. Bradley, S.P.; Kaminski, D.A.; Peters, A.H.; Jenuwein, T.; Stavnezer, J. The histone methyltransferase Suv39h1 increases class switch recombination specifically to IgA. *J. Immunol.* **2006**, *177*, 1179–1188. [CrossRef]
66. Kuang, F.L.; Luo, Z.; Scharff, M.D. H3 trimethyl K9 and H3 acetyl K9 chromatin modifications are associated with class switch recombination. *Proc. Natl. Acad. Sci. USA* **2009**, *106*, 5288–5293. [CrossRef] [PubMed]
67. McLean, C.; Karemaker, I.; van Leeuwen, F. The emerging roles of DOT1L in leukemia and normal development. *Leukemia* **2014**, *28*, 2131–2138. [CrossRef]
68. Jones, B.; Su, H.; Bhat, A.; Lei, H.; Bajko, J.; Hevi, S.; Baltus, G.A.; Kadam, S.; Zhai, H.; Valdez, R.; et al. The Histone H3K79 Methyltransferase Dot1L Is Essential for Mammalian Development and Heterochromatin Structure. *PLoS Genet.* **2008**, *4*, e1000190. [CrossRef]
69. Yaseen, I.; Choudhury, M.; Sritharan, M.; Khosla, S. Histone methyltransferase SUV 39H1 participates in host defense by methylating mycobacterial histone-like protein HupB. *EMBO J.* **2018**, *37*, 183–200. [CrossRef]
70. Piao, L.; Suzuki, T.; Dohmae, N.; Nakamura, Y.; Hamamoto, R. SUV39H2 methylates and stabilizes LSD1 by inhibiting polyubiquitination in human cancer cells. *Oncotarget* **2015**, *6*, 16939–16950. [CrossRef]
71. Sone, K.; Piao, L.; Nakakido, M.; Ueda, K.; Jenuwein, T.; Nakamura, Y.; Hamamoto, R. Critical role of lysine 134 methylation on histone H2AX for γ-H2AX production and DNA repair. *Nat. Commun.* **2014**, *5*, 5691. [CrossRef]
72. Schuhmacher, M.K.; Kudithipudi, S.; Jeltsch, A. Investigation of H2A methylation by the SUV39H2 protein lysine methyltransferase. *FEBS Lett.* **2016**, *590*, 1713–1719. [CrossRef]
73. Ting, D.T.; Lipson, D.; Paul, S.; Brannigan, B.W.; Akhavanfard, S.; Coffman, E.J.; Contino, G.; Deshpande, V.; Iafrate, A.J.; Letovsky, S.; et al. Aberrant Overexpression of Satellite Repeats in Pancreatic and Other Epithelial Cancers. *Science* **2011**, *331*, 593–596. [CrossRef] [PubMed]
74. Zhu, Q.; Pao, G.M.; Huynh, A.M.; Suh, H.; Tonnu, N.; Nederlof, P.M.; Gage, F.H.; Verma, I.M. BRCA1 tumour suppression occurs via heterochromatin-mediated silencing. *Nat. Cell Biol.* **2011**, *477*, 179–184. [CrossRef] [PubMed]
75. Alagoz, M.; Katsuki, Y.; Ogiwara, H.; Ogi, T.; Shibata, A.; Kakarougkas, A.; Jeggo, P. SETDB1, HP1 and SUV39 promote repositioning of 53BP1 to extend resection during homologous recombination in G2 cells. *Nucleic Acids Res.* **2015**, *43*, 7931–7944. [CrossRef] [PubMed]
76. Ayrapetov, M.K.; Gursoy-Yuzugullu, O.; Xu, C.; Xu, Y.; Price, B.D. DNA double-strand breaks promote methylation of histone H3 on lysine 9 and transient formation of repressive chromatin. *Proc. Natl. Acad. Sci. USA* **2014**, *111*, 9169–9174. [CrossRef] [PubMed]
77. Zheng, H.; Chen, L.; Pledger, W.J.; Fang, J.; Chen, J. p53 promotes repair of heterochromatin DNA by regulating JMJD2b and SUV39H1 expression. *Oncogene* **2013**, *33*, 734–744. [CrossRef] [PubMed]
78. García-Cao, M.; O'Sullivan, R.; Peters, A.H.F.M.; Jenuwein, T.; Blasco, M.A. Epigenetic regulation of telomere length in mammalian cells by the Suv39h1 and Suv39h2 histone methyltransferases. *Nat. Genet.* **2003**, *36*, 94–99. [CrossRef]
79. Allan, R.S.; Zueva, E.; Cammas, F.; Schreiber, H.A.; Masson, V.; Belz, G.; Roche, D.; Maison, C.; Quivy, J.-P.; Almouzni, G.; et al. An epigenetic silencing pathway controlling T helper 2 cell lineage commitment. *Nat. Cell Biol.* **2012**, *487*, 249–253. [CrossRef]
80. Scarola, M.; Comisso, E.; Pascolo, R.; Chiaradia, R.; Marion, R.M.; Schneider, C.; Blasco, M.A.; Schoeftner, S.; Benetti, R. Epigenetic silencing of Oct4 by a complex containing SUV39H1 and Oct4 pseudogene lncRNA. *Nat. Commun.* **2015**, *6*, 7631. [CrossRef]
81. Zhang, W.; Li, J.; Suzuki, K.; Qu, J.; Wang, P.; Zhou, J.; Liu, X.; Ren, R.; Xu, X.; Ocampo, A.; et al. A Werner syndrome stem cell model unveils heterochromatin alterations as a driver of human aging. *Science* **2015**, *348*, 1160–1163. [CrossRef] [PubMed]
82. Chu, Y.; Chen, Y.; Guo, H.; Li, M.; Wang, B.; Shi, D.; Cheng, X.; Guan, J.; Wang, X.; Xue, C.; et al. SUV39H1 regulates the progression of MLL-AF9-induced acute myeloid leukemia. *Oncogene* **2020**, *39*, 7239–7252. [CrossRef]
83. Braig, M.; Lee, S.; Loddenkemper, C.; Rudolph, C.; Peters, A.H.; Schlegelberger, B.; Stein, H.; Dörken, B.; Jenuwein, T.; Schmitt, C.A. Oncogene-induced senescence as an initial barrier in lymphoma development. *Nat. Cell Biol.* **2005**, *436*, 660–665. [CrossRef] [PubMed]
84. Reimann, M.; Lee, S.; Loddenkemper, C.; Dörr, J.R.; Tabor, V.; Aichele, P.; Stein, H.; Dörken, B.; Jenuwein, T.; Schmitt, C.A. Tumor Stroma-Derived TGF-β Limits Myc-Driven Lymphomagenesis via Suv39h1-Dependent Senescence. *Cancer Cell* **2010**, *17*, 262–272. [CrossRef] [PubMed]
85. Yang, Y.-J.; Han, J.-W.; Youn, H.-D.; Cho, E.-J. The tumor suppressor, parafibromin, mediates histone H3 K9 methylation for cyclin D1 repression. *Nucleic Acids Res.* **2009**, *38*, 382–390. [CrossRef]
86. Zheng, Y.; Li, B.; Wang, J.; Xiong, Y.; Wang, K.; Qi, Y.; Sun, H.; Wu, L.; Yang, L. Identification of SUV39H2 as a potential oncogene in lung adenocarcinoma. *Clin. Epigenetics* **2018**, *10*, 129. [CrossRef] [PubMed]
87. Mutonga, M.; Tamura, K.; Malnassy, G.; Fulton, N.; de Albuquerque, A.; Hamamoto, R.; Stock, W.; Nakamura, Y.; Alachkar, H. Targeting Suppressor of Variegation 3-9 Homologue 2 (SUV39H2) in Acute Lymphoblastic Leukemia (ALL). *Transl. Oncol.* **2015**, *8*, 368–375. [CrossRef]
88. Piao, L.; Yuan, X.; Zhuang, M.; Qiu, X.; Xu, X.; Kong, R.; Liu, Z. Histone methyltransferase SUV39H2 serves oncogenic roles in osteosarcoma. *Oncol. Rep.* **2018**, *41*, 325–332. [CrossRef]

89. Wang, R.; Cheng, L.; Yang, X.; Chen, X.; Miao, Y.; Qiu, Y.; Zhou, Z. Histone methyltransferase SUV39H2 regulates cell growth and chemosensitivity in glioma via regulation of hedgehog signaling. *Cancer Cell Int.* **2019**, *19*, 269. [CrossRef]
90. Weirich, S.; Kudithipudi, S.; Jeltsch, A. Somatic cancer mutations in the MLL1 histone methyltransferase modulate its enzymatic activity and dependence on the WDR5/RBBP5/ASH2L complex. *Mol. Oncol.* **2017**, *11*, 373–387. [CrossRef]
91. Shuai, W.; Wu, J.; Chen, S.; Liu, R.; Ye, Z.; Kuang, C.; Fu, X.; Wang, G.; Li, Y.; Peng, Q.; et al. SUV39H2 promotes colorectal cancer proliferation and metastasis via tri-methylation of the SLIT1 promoter. *Cancer Lett.* **2018**, *422*, 56–69. [CrossRef] [PubMed]
92. Hunter, R.G.; Murakami, G.; Dewell, S.; Seligsohn, M.; Baker, M.E.R.; Datson, N.; McEwen, B.S.; Pfaff, D.W. Acute stress and hippocampal histone H3 lysine 9 trimethylation, a retrotransposon silencing response. *Proc. Natl. Acad. Sci. USA* **2012**, *109*, 17657–17662. [CrossRef] [PubMed]
93. Balan, S.; Iwayama, Y.; Ohnishi, T.; Fukuda, M.; Shirai, A.; Yamada, A.; Weirich, S.; Schuhmacher, M.K.; Vijayan, D.K.; Endo, T.; et al. A loss of function variant in SUV39H2 identified in autism spectrum disorder causes altered H3K9-trimethylation and dysregulation of protocadherin β cluster genes in the developing brain. *Mol. Psychiatry* **2021**. [CrossRef] [PubMed]
94. Jagannathan, V.; Bannoehr, J.; Plattet, P.; Hauswirth, R.; Drögemüller, C.; Drögemüller, M.; Wiener, D.J.; Doherr, M.; Owczarek-Lipska, M.; Galichet, A.; et al. A Mutation in the SUV39H2 Gene in Labrador Retrievers with Hereditary Nasal Parakeratosis (HNPK) Provides Insights into the Epigenetics of Keratinocyte Differentiation. *PLoS Genet.* **2013**, *9*, e1003848. [CrossRef]
95. Lyons, D.B.; Magklara, A.; Goh, T.; Sampath, S.C.; Schaefer, A.; Schotta, G.; Lomvardas, S. Heterochromatin-Mediated Gene Silencing Facilitates the Diversification of Olfactory Neurons. *Cell Rep.* **2014**, *9*, 884–892. [CrossRef]

Review

Structure, Activity, and Function of the Protein Lysine Methyltransferase G9a

Coralie Poulard [1,2,3,*], Lara M. Noureddine [1,2,3,4], Ludivine Pruvost [1,2,3] and Muriel Le Romancer [1,2,3]

1 Cancer Research Cancer of Lyon, Université de Lyon, F-69000 Lyon, France; noureddinelara@gmail.com (L.M.N.); ludivine.pruvost@etu.univ-lyon1.fr (L.P.); Muriel.LEROMANCER-CHERIFI@lyon.unicancer.fr (M.L.R.)
2 Inserm U1052, Centre de Recherche en Cancérologie de Lyon, F-69000 Lyon, France
3 CNRS UMR5286, Centre de Recherche en Cancérologie de Lyon, F-69000 Lyon, France
4 Laboratory of Cancer Biology and Molecular Immunology, Faculty of Sciences, Lebanese University, Hadat-Beirut 90565, Lebanon
* Correspondence: coralie.poulard@lyon.unicancer.fr

Abstract: G9a is a lysine methyltransferase catalyzing the majority of histone H3 mono- and dimethylation at Lys-9 (H3K9), responsible for transcriptional repression events in euchromatin. G9a has been shown to methylate various lysine residues of non-histone proteins and acts as a coactivator for several transcription factors. This review will provide an overview of the structural features of G9a and its paralog called G9a-like protein (GLP), explore the biochemical features of G9a, and describe its post-translational modifications and the specific inhibitors available to target its catalytic activity. Aside from its role on histone substrates, the review will highlight some non-histone targets of G9a, in order gain insight into their role in specific cellular mechanisms. Indeed, G9a was largely described to be involved in embryonic development, hypoxia, and DNA repair. Finally, the involvement of G9a in cancer biology will be presented.

Keywords: G9a; GLP; H3K9 methylation; protein lysine methylation; EHMT2; EHMT1; protein post-translational modification; cancer

1. Introduction

Protein lysine methylation is a dynamic post-translational modification (PTM) regulating protein stability and function. Lysine methylation of histone proteins can modulate transcriptional activity without affecting the DNA sequence itself, enabling dynamic gene transcription patterns in response to environmental stimuli [1]. Lysine methylation is deposited by writer enzymes called protein lysine methyltransferases (PKMTs), removed by eraser enzymes called lysine demethylases (PKDMs) and interpreted by reader proteins that bind to lysine methylation marks. PKMTs catalyze the transfer of the methyl group from the S-adenosyl-l-methionine (AdoMet) donor to the ε-nitrogen of a lysine residue on protein substrates [1]. The lysine ε-amino group of proteins can accept up to three methyl groups, resulting in either mono-, di-, or trimethyl lysines. To date, more than 50 PKMTs have been reported, with sequence and product specificity. Two PKMT families have been identified: the SET lysine methyltransferases containing the majority of PKMTs [2] and the Seven β-strand methyltransferase (7βS) or class I family [3]. Histones are methylated on several lysine residues. A growing number of reports also describe the methylation of non-histone proteins on lysine residues [1].

G9a was identified and sequenced in the 1990s [4]. It belongs to the SET PKMT family. G9a was extensively studied as a key enzyme in the mono- and dimethylation of lysine 9 of histone H3 (H3K9me1 and H3K9me2, respectively) in euchromatin [5]. Since the H3K9me2 mark is associated with transcriptional repression, G9a was primarily considered to be an epigenetic repressor [5–7]. Its role as a coactivator of several transcription factors emerged more recently [8–12]. Though G9a is the most commonly used term for this lysine methyltransferase, it is also known as lysine methyltransferase-1C (KMT1C), euchromatic histone N-methyltransferase 2 (EHMT2), or BAT8 (HLA-B associated transcript 8).

The current review will provide an overview of the structural features of the protein with a particular focus on its paralog GLP (G9a-like protein). The biochemical features of G9a will also be detailed with a special emphasis on the key PTMs affecting G9a and regulating its activity and function. Finally, among the large number of G9a substrates described, including histone and non-histone substrates, the present report will focus on their involvement in specific physiological pathways and their connection to cancer.

2. Structural Features

2.1. Structure and Domain Architecture

In human cells, G9a exists as two isoforms: a full-length isoform of 1210 amino acids (called isoform A) derived from 24 exons of the G9a gene and a splice variant of 1176 amino acids (isoform B) that arises from the excision of exon 10 (Figure 1a). The alternative splicing of G9a is conserved in different species, tissues, and cell lines [13]. Even if the two isoforms are ubiquitously found in different tissues, the ratio between them varies. For example, isoform A is preponderant in the kidney, thymus, and testis, and, interestingly, is more abundant in epithelial cell lines compared to mesenchymal cell lines and more transformed cell lines [13]. Mauger et al. reported that the two isoforms display similar methyltransferase activities and subcellular localizations. Likewise, Fiszbein et al. showed that isoform B expression increased during neuronal differentiation [14]. They did not report any change in G9a catalytic activity following exon 10 inclusion, but demonstrated that exon 10 inclusion increases G9a nuclear localization in a neuronal cell line [14]. Mouse G9a is also subjected to alternative splicing. Full-length mouse G9a protein contains 1263 amino acids and shares more than 90% homology with human G9a [15].

G9a belongs to the Su(var)3-9 family of methyltransferases, which was first identified in Drosophila melanogaster [16]. The main characteristic of this family of proteins is the presence of a highly conserved SET domain [17]. SET, an acronym for Su(var)3-9, Enhancer-of-zeste and Trithorax, is a long sequence of 130 to 140 amino acids, characterized in 1998, that has a unique structural fold [17]. The SET domain is composed of a series of β strands that fold into three sheets and surround a knot-like structure [18]. The conserved core of the SET domain is flanked by a pre-SET (nSET) domain providing structural stability by interacting with different surfaces of the core SET domain, and a post-SET (cSET) domain responsible of the formation of a hydrophobic channel via an aromatic residue [19]. Neither pre-SET nor post-SET domains are conserved across KTM SET domains, as they vary in size and tertiary structure [20]. In the core SET domain, G9a contains an inserted i-SET domain (Figure 1a). The i-SET domain forms a rigid docking platform and a substrate binding groove with the post-SET domain in three-dimensional structures [21]. The G9a SET domain contains four structural zinc fingers for proper folding and enzymatic activity. A cluster of three Zn^{2+} ions is chelated by nine cysteines, whereas the fourth Zn^{2+} ion, adjacent to the S-adenosylmethionine (SAM)-binding site, is chelated by four cysteines [22]. The binding of AdoMet and the protein substrate occurs on opposite sides of the SET domain. AdoMet binds and positions its methyl group at the base of the channel, while the side chain of the target lysine protrudes into the channel [20]. Within the SET domain, the tyrosine residue Y1154 was demonstrated to be essential for the catalytic activity of G9a [23]. The tyrosine may allow deprotonation of the positively charged ammonium group in order to favor methylation.

G9a also contains a cysteine-rich region, a polyglutamate region and seven ankyrin repeats of 33 amino acids (Figure 1a). The ankyrin repeat domain was reported to be a mono- and dimethyllysine binding module, a reader domain important for protein-protein interactions [24]. The specificity of the G9a ankyrin repeat domain is comparable to the specificity of other groups of reader proteins recognizing methyl binding protein modules, such as the chromodomain, the tudor domain, or the PHD finger domain [24]. G9a was the first protein described to harbor within a single polypeptide, the signal to catalyze and read the same epigenetic marks, H3K9me1, and H3K9me2 [24].

A nuclear localization signal was identified in the N-terminal region of human G9a [25], and amino acids 1-280 of human G9a were shown to act as a coactivator domain in transient reporter gene assays [10] (Figure 1a).

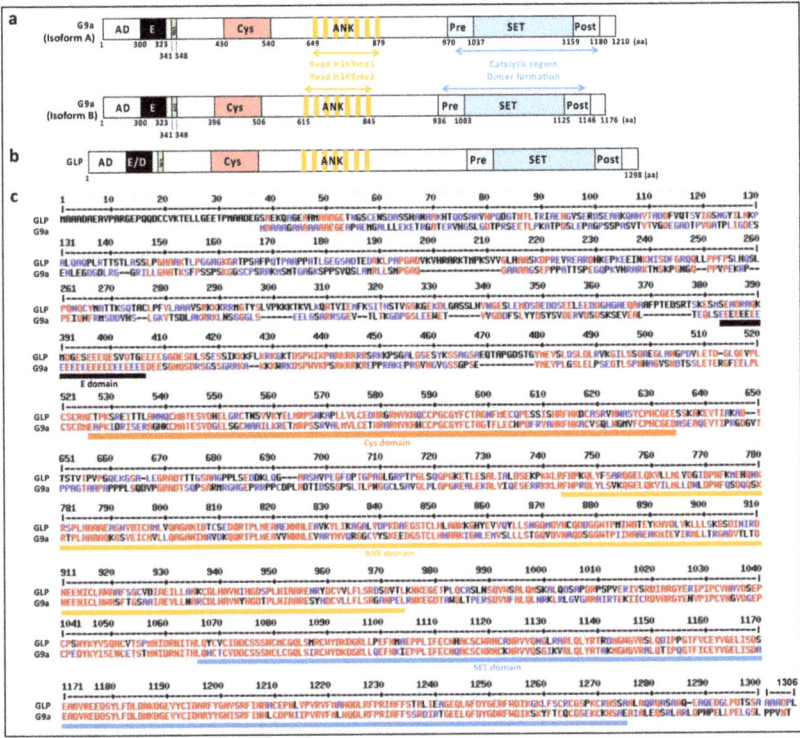

Figure 1. Schematic representation of the structure and domains of human G9a (**a**) and GLP (**b**). G9a and GLP contain different domains: an activation domain (AD), a Cys-rich region (Cys), an ankyrin repeat domain (ANK), and a SET domain composed of a core SET domain associated with pre- and a post-SET domains. G9a and GLP contain a nuclear localization signal (NLS). G9a also contains a Glu-rich region (E) and GLP a Glu/Asp-rich region (E/D). (**c**) Sequence alignment of G9a (NP_006700.3) and GLP (NP_079033.4). The alignment was performed using the MultAlin program [26] (http://multalin.toulouse.inra.fr/multalin) (accessed on 7 October 2021). Amino acids with 100% and >60% conservation are shown in red and blue, respectively.

2.2. GLP, a G9a Paralog

A paralog of G9a was identified and called G9a-like protein (GLP), though it is also termed lysine methyltransferase-1D (KMT1D) or euchromatic histone N-methyltransferase 1 (EHMT1) (Figure 1b). G9a and GLP share 45% sequence identity and around 70% sequence similarity (Figure 1c) [2]. They differ primarily in the N-terminus, and present a high level of conservation in the SET domain with over 80% shared sequence identity (Figure 1c) [27]. The main difference in structure between the two proteins concerns the E-rich domain of G9a, which is composed of a sequence of repeated glutamic and aspartic acid residues in the case of GLP (Figure 1b,c). In addition, binding affinities of the ankyrin domains of G9a and GLP for H3K9 differ, as GLP and G9a preferentially bind to mono- and dimethylated H3K9, respectively [24,28].

G9a and GLP form homo- and heterodimers via their SET domains in complex with ZNF644 and WIZ [6,29–31]. In the endogenous complex, they act mainly as heterodimers in a large variety of human cells [6]. However, in vitro, independently of each other, G9a and GLP are able to catalyze lysine methylation by forming homodimers. Extensive research has focused on G9a, albeit GLP seems to be equally important for most biological phenomena ascribed to G9a. Indeed, GLP generally possesses similar catalytic activities as G9a [29]. However, the individual effects of G9a and GLP are hard to study, as G9a depletion destabilizes GLP [6,32].

3. Biochemical Features

3.1. Sequence Specificity

The majority of studies conducted on G9a sequence specificity focused on Histone H3. In vitro, the minimum substrate recognition site of seven amino acids of H3 is composed of residues 6 to 11 (TARKSTG), with a consensus methylation site encompassing RK/ARK [33]. The arginine residue adjacent to the lysine residue is essential for G9a activity [33]. G9a preferentially acts when a hydrophobic amino acid is positioned before the arginine residue, such as alanine. After the lysine residue, G9a favors a hydrophilic residue followed by a hydrophobic one. This G9a recognition site is present in several non-histone proteins, as well as on its N-terminal domain [34–36].

Several biochemical studies have shown that specific PTMs affect the catalytic activity of G9a. For instance, phosphorylation of S10 or T11 of H3 impairs G9a catalytic efficacy [33,36]. In addition, R8 of H3 can be methylated by the arginine methyltransferase PRMT5 in vivo, and this event impairs methylation of H3K9 by G9a [36]. Indeed, a decrease in methylation of over 80% was reported for peptides carrying an asymmetric dimethylation of R8, a methylation mark catalyzed by PRMT5 [36].

3.2. Product Specificity

G9a mainly catalyzes mono- and dimethylation events, as illustrated with H3K9 [6,24]. However, several reports demonstrated that G9a also generates, after a long incubation time, trimethylation of H3K9 (H3K9me3) [25,37]. Investigations on G9a-deficient cells demonstrated that G9a is the major H3K9me1 and H3K9me2 methyltransferase of euchromatin [5].

Biochemically, the specificity of G9a methylation for a particular state is largely due to a tyrosine residue in its active site. Indeed, Y1067 controls whether G9a catalyzes mono-, di- or trimethylation of lysines; Y1067 mutation to F1067 allowing G9a trimethylation of H3K9 [21]. Mechanistically, Y1067 forms hydrogen bonds with the nitrogen atom of the ε-amino group of the target lysine residue [21].

3.3. Regulation

3.3.1. PTMs

As for most proteins, G9a is subjected by many PTMs that regulate its ability to bind new partners and impact its cellular functions (Figure 2). Further details about their cellular features will be given in the corresponding sections below.

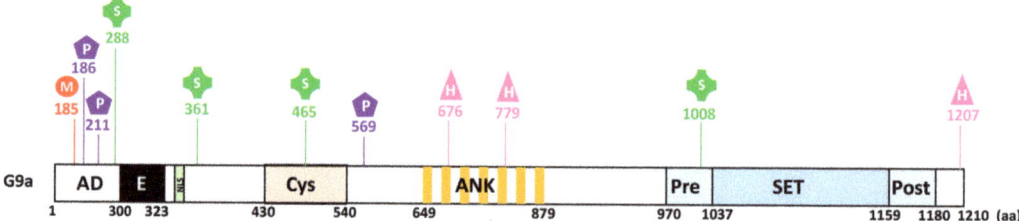

Figure 2. G9a undergoes several post-translational modifications including methylation (M), phosphorylation (P), sumoylation (S), and hydroxylation (H). The numbers indicate amino acid (aa) residues.

G9a was shown to be auto-methylated on lysine 185 (K185) and phosphorylated by the Aurora kinase B (AurKB) on the adjacent threonine 186 (T186) in the N-terminal domain of the protein [35] (Figure 2). Heterochromatin protein 1 proteins (HP1α, HP1β, HP1γ) and CDYL (chromodomain Y-like) were identified as specific partners that bind methylated G9a [34,35]. These proteins contain chromodomains functioning as methyl-lysine binding modules. Of note, a similar methylation and phosphorylation switch on adjacent residues was previously demonstrated for the histone H3 [38,39]. H3K9me2 methylated by G9a recruits HP1 proteins, whereas H3 phosphorylated on S10 by AurKB has an opposite effect [38,39]. Like G9a, GLP is also auto-methylated on lysine 205 (K205) and phosphorylated by AurKB on threonine 206 (T206) [32]. Both G9a and GLP auto-methylation sites can be demethylated by the KDM4 lysine demethylase family [40]. Sampath et al. found no evidence of a role for G9a auto-methylation in the regulation of G9a enzymatic activity [35].

Additionally, G9a was shown to be phosphorylated on two serine residues involved in DNA damage repair, namely Serine 211 (S211) phosphorylated by casein kinase 2 (CK2) and serine 569 (S569) phosphorylated by ATM kinase (Figure 2) [41,42]. Interestingly, phosphorylation of G9a on S211 does not change its methyltransferase activity and G9a catalytic inhibitor does not affect G9a phosphorylation on S569 [41,42].

G9a is sumoylated in skeletal myoblasts in order to regulate its transcriptional activity [43]. This event acts as a signal for the recruitment of the histone acetyltransferase PCAF (p300/CBP-associated factor) to E2F1 target genes, implicated in cell cycle progression by increasing the level of histone H3 lysine 9 acetylation [43].

Casciello et al. demonstrated that G9a stability is regulated by proline hydroxylation catalyzed by oxygen sensors, as inhibition of the latter increased protein stability [44]. Authors showed that G9a hydroxylation is detected in normoxic conditions, whereas it is not detected under hypoxia. Proline hydroxylation occurs on proline residues 676 (P676) and 1207 (P1207) in consensus hydroxylation motifs LXXLAP and leads to efficient degradation by the proteasome (Figure 2) [44]. G9a is also hydroxylated in the ankyrin repeat domain of G9a on asparagine 779 (N779) by the asparaginyl hydroxylase factor inhibiting HIF (FIH) (Figure 2) [45]. This event impedes G9a binding to methylated H3K9 products and to di- and trimethylated H3K9. Hydroxylation of N779 destabilizes the interaction of H3K9me2 with the ankyrin repeat domain of G9a by disrupting the structural pocket that facilitates methyl binding [24,45]. Likewise, GLP is hydroxylated on N867 [45].

3.3.2. Stability

G9a protein stability relies on the presence of GLP, as GLP depletion also decreases G9a expression [6,32]. Using $G9a^{-/-}$ and $GLP^{-/-}$ embryonic stem cells, Tachibana et al. reported that G9a is more stable in the G9a/GLP heteromeric complex. This observation did not apply to GLP [6]. The protein WIZ was reported to be a key partner of both G9a and GLP to stabilize the G9a/GLP heteromeric complex [30]. Both WIZ and GLP depletion decreases G9a protein levels, suggesting that the WIZ/G9a/GLP complex protects G9a from degradation [30]. Later, Bian et al. mapped the specific sequence of WIZ interacting with G9a/GLP. They showed that WIZ only interacts directly with the NTD of GLP [31]. Its interaction with G9a might be indirect and mediated by the fact that G9a and GLP form heterodimers. WIZ contains multiple zinc finger motifs, targeting the G9a/GLP complex to chromatin in order to mediate H3K9 methylation [31].

3.4. Substrates

3.4.1. Histone Substrates

In 2001, Tachibana et al. identified the first substrates of G9a as histone proteins [46] (Table 1). They demonstrated that G9a was able to add methyl groups to H3 on lysine 9 and lysine 27 [46]. Since then, G9a has largely been described as the major PKMT catalyzing the mono- and dimethylation of H3K9 [5], and, to a lesser extent, H3K9 trimethylation [25,37]. Though H3K9 methylation is well known for its role in transcriptional silencing [6,47], the impact of H3K27 methylation by G9a emerged more recently. Wu et al. demonstrated in

2011 that even though H3K27me2/3 is not affected in G9a$^{-/-}$ ES cells, H3K27me1 levels were clearly lower in these cells [48]. G9a also methylates H3 on lysine 56 (H3K56me1) in order to maintain proper DNA replication [49], a methylation event that was shown to be induce by DNA damage [41].

Table 1. List of histone substrates of G9a and their biological outcome. Nd: not determined.

Histone Types	Sites	Biological Outcome	References
Histone H3	H3K9me1 H3K9me2 H3K9me3	Transcriptional repression Heterochromatin formation	[5,25,37]
Histone H3	H3K27me1	Transcriptional repression Heterochromatin formation	[46,48]
Histone H3	H3K56me1	DNA replication	[49]
Histone H1.2	H1.2K187me	nd	[51]
Histone H1.4	H1.4K26me1 H1.4K26me2	Transcriptional repression Chromatin structure	[50]

G9a methylates histone H1 in a variant-specific manner. Human cells have 11 H1 variants, two of which were shown to be methylated by G9a, namely isotype 2 (H1.2) and isotype 4 (H1.4) [50,51]. H1.4 was reported to be mono- and dimethylated on H1.4K26. This event provides a recognition site for HP1 binding, establishing a proper chromatin surface and suggesting a role for H1.4K26me1/2 in transcriptional repression [50]. G9a methylates H1.2 on K187 in vitro and in vivo. However, H1.2K187me2 is not recognized by HP1 proteins, demonstrating selective recognition by these proteins [51]. Weiss et al. demonstrated that G9a does not directly bind to methylated histone variants, suggesting a different mechanism from that observed in H3K9me1/2 to achieve methylation [51].

3.4.2. Non-Histone Substrates

G9a also methylates a large number of non-histone proteins involved in several biological functions listed in Table 2. Most of these are linked with transcriptional regulation, as G9a methylates numerous transcription factors, chromatin remodeling factors, and coregulators.

3.5. Inhibitors

Among the numerous G9a inhibitors, there are three different types: (i) substrate competitive inhibitors, (ii) SAM cofactor competitive inhibitors and (iii) inhibitors by ejection of Zn^{2+} ions. Substrate competitive inhibitors act by binding to G9a substrate binding sites, while SAM inhibitors prevent G9a-mediated methylation by interacting with SAM binding sites on G9a [52]. Most of these inhibitors also impact GLP [53].

3.5.1. Substrate Competitive Inhibitors

Substrate competitive inhibitors specifically bind to the substrate binding site of G9a. The first substrate competitive inhibitor discovered was BIX01294, a quinazolin derivative able to inhibit H3K9me2 [70]. Many studies then sought to optimize this inhibitor by enhancing its G9a specificity, efficacy and by reducing cell toxicity. Based on Structure-Activity Relationship studies (SAR), modifications of BIX01294 provided more specific and powerful G9a inhibitors including UNC0224, UNC0321, UNC0638, UNC0646 [52]. The majority of G9a substrate competitive inhibitors impede G9a activity by interacting with two G9a aspartate residues in the SET domain (D1074 and D1083) [71,72]. Recently, by adding and expanding the 1,4 benzodiazepine cycle, Milite et al. improved UNC0638 potency and named it EML741 [73].

Table 2. List of substrates of G9a categorized by their biological functions. Nd: not determined.

Functions	Substrates	Site	Biological Outcome	References
Transcription Factors	C/EBPb	K39	Inhibits transcriptional activity by repressing C/EBPb transactivation	[54]
	MyoD	K104me1/2	Inhibits MyoD transcriptional activity	[55]
	MEF2D	K267me1/2	Inhibits MEF2D transcriptional activity by preventing its recruitment on chromatin	[56]
	p53	K373me2	Inhibits transcriptional activity and p53-dependent apoptosis	[57]
	ERα	K235me2	Induces transcriptional activity by recruiting the PHF20/MOF HAT complex	[58]
	Foxo1	K273me1/2	Induces Foxo1 degradation	[59]
	KLF12	K313	nd	[36]
Chromatin remodeling factors and coregulators	G9a	K185me2/3	Induces specific glucocorticoid receptor transcriptional activity by recruiting HP1γ	[32,34,35]
	GLP	K205me2	Induces specific glucocorticoid receptor transcriptional activity by recruiting HP1γ	[32]
	Sirt1	K662	nd	[60]
	Pontin	K265, K267, K268, K274, K281, K285	Induces HIF-1 transcriptional activity by enhancing p300 recruitment	[61]
	Reptin	K67me1	Inhibits HIF transcriptional activity by recruiting corepressors	[62]
	HDAC1	K432	nd	[36]
	HIFα	K674me1/2	Inhibits HIF-1 transcriptional activity	[63]
	CSB	K170, K297, K448, K1054	nd	[36]
	MTA1	K532me1	Inhibits transcription by recruiting the assembly of the NuRD repressive complex	[64]
	ATF7IP (hAM)	K16me3	Induces transgene silencing by recruiting MPP8	[65]

Table 2. Cont.

Functions	Substrates	Site	Biological Outcome	References
Chromatin binding protein	CDYL1	K135me3	Decreases its interaction with H3K9me3	[36]
	WIZ	K305me3	nd	[36]
DNA methyl-transferases	DNMT1	K70me2	nd	[36]
	DNMT3	K47me2	Inhibits transcription by recruiting MPP8/DNMT3/G9a/GLP repressive complex	[66]
Others	Acinus	K654me2	nd	[36]
	MDC1	K45me2	Induces ATM accumulation on damage sites	[67]
	Plk1	K209me1	Antagonizes T210 phosphorylation to inhibit Plk1 activity on DNA replication	[68]
	Lig1	K126me2/3	Maintenance in DNA methylation by promoting UHRF1 recruitment to replication foci	[69]

3.5.2. SAM Competitive Inhibitors

The cofactor SAM is the methyl donor essential for G9a-mediated methylation. SAM competitive inhibitors compete with SAM to bind to the SAM binding site of G9a. The first inhibitor of this class to be identified by Kubicek et al. was BIX01338, discovered around the same time as BIX01294 [70]. Analogous inhibitors were then synthetized with similar structures, such as BRD9536 and BRD4770 [74]. However, this type of inhibitor remains less specific than substrate competitive inhibitors, as it also downregulates the enzymatic activity of several other PKMTs [52].

3.5.3. Inhibition by Ejection of Structural Zn^{2+}

Lastly, Lenstra et al. reported that structural zinc ions are essential to maintain the enzymatic activity of the methyltransferases G9a/GLP [22]. By using selenium- or sulfur-containing proteins able to eject the fourth structural zinc ions, they demonstrated that G9a methyltransferase activity could be inhibited. Molecules used clinically such as ebselen, disulfiram, and cisplatin work specifically as inhibitors of G9a and GLP. These findings may offer new perspectives to develop further G9a-specific inhibitors [22].

4. Cellular Features

4.1. Connection with Chromatin Regulation

4.1.1. G9a Corepressor Functions

As mentioned above, G9a is a coregulator with an essential role in repression of gene transcription. Functionally, G9a is involved in several mechanisms, primarily the methylation of the histone H3 N-terminal tail in order to close chromatin (Table 1).

- G9a in Euchromatin

Numerous studies have shown that G9a is recruited to specific target genes as a corepressor by transcription factors, such as CCAAT displacement protein/cut (CDP/cut) [75], growth factor independent 1 (Gfi1) [76], positive regulatory domain I-binding factor 1 (PRDI-BF1) [77], neuron restrictive silencing factor (NRSF) (also known as REST) [78], multi-domain protein UHRF1 [79], and the noncoding RNA Air [80], in order to remodel chromatin structure. G9a also represses active gene transcription by recruiting other corepressors. For example, in euchromatin, G9a interacts with Polycomb Repressive Complex 2-proteins, including the PKMT EZH2, in order to transcriptionally silence specific regions within the genome (Figure 3a) [81].

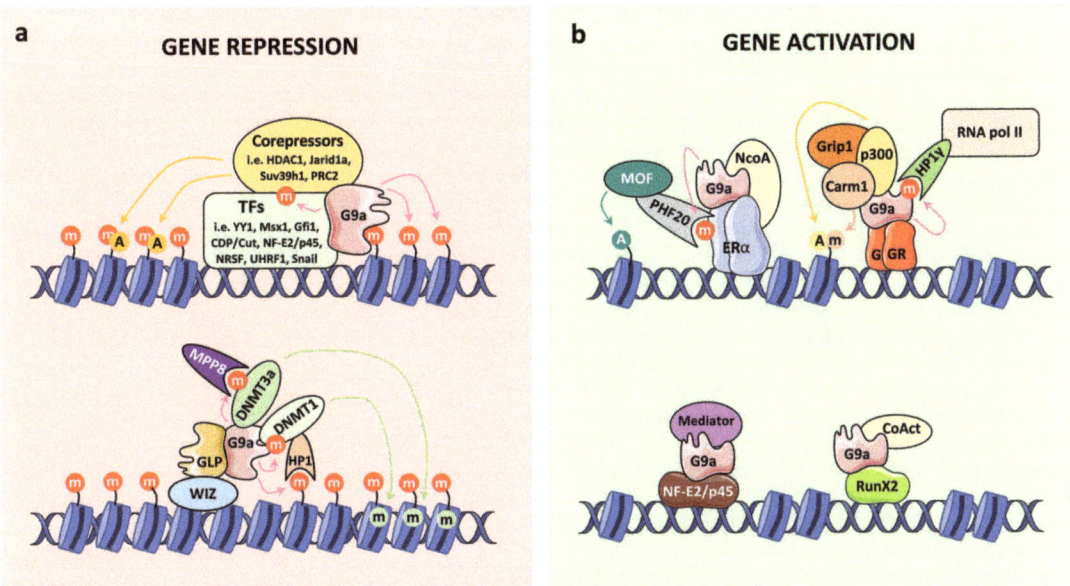

Figure 3. G9a acts as a transcriptional coregulator, either as a corepressor (**a**) or coactivator (**b**). (**a**) After G9a recruitment by some transcription factors (TFs), G9a methylates histones (red circles) leading to chromatin remodeling and gene repression. G9a also recruits corepressor proteins (i.e., other PKMTs and chromatin remodelers) and DNA methyltransferases (i.e., DNMT3a and DNMT1) in order to fully repress transcription via histone modifications (i.e., acetylation (orange circles) and DNA methylation (green circles)). Of note, G9a also methylates some TFs and DNA methyltransferases modulating their functions. (**b**) Conversely, G9a recruitment by the glucocorticoid receptor (GR), estrogen receptor (ERα), RunX2 and NF-E2/p45 leads to gene activation through the recruitment of specific coactivators (CoAct) (i.e., histone acetyltransferases and methyltransferases) and the transcription machinery (i.e., Mediator complex or RNA polymerase II).

- G9a in heterochromatin

In heterochromatin, G9a drives silencing mechanisms by serving as a platform for the formation of repressive complexes. Methylation of H3K9 leads to the recruitment of proteins such as HP1, which can bind to methylated H3K9 via their chromodomains [38,39]. This recruitment is crucial for heterochromatin formation and gene silencing [82]. In addition, G9a also recognizes H3K9 methylation via its ankyrin repeat in order to work as a scaffold for the recruitment of other corepressors [24]. It was shown for instance that G9a interacts with the PKMT Suv39h and SETDB1 in specific regions of heterochromatin to maintain chromosomal stability (Figure 3a) [83].

- G9a and DNA methylation

Other mechanisms underlying G9a repressive function have been identified. For example, the ankyrin repeat domain of G9a was reported to contribute to DNA methylation-mediated repression of transcription by recruiting DNA methyltransferases (DNMT3a and DNMT3b), and by recognizing the H3K9me2 histone mark [24,84]. A specific residue of the ankyrin repeat domain (Asp905) has also been associated with this co-repressive function by maintaining H3K9me2 levels and establishing DNA methylation [85]. In addition, Chang et al. demonstrated that G9a dimethylates DNMT3a on K47, allowing its recognition by the MPP8 chromodomain [66]. This event results in a silencing complex containing DNMT3a/MPP8/G9a on chromatin that could in part explain the co-occurrence of DNA methylation and H3K9 methylation in chromatin (Figure 3a). Additionally, Smallwood et al reported that HP1 proteins, the readers of H3K9 methylation, target DNMT1 enzyme to euchromatic sites, providing a basis for the generation of CpG methylation [86]. Finally, DNMT1 is methylated by G9a reinforcing the whole model [36] (Figure 3a).

4.1.2. G9a Coactivator Functions

In addition to the well-studied and established co-repressive function of G9a, reports have emerged on its function as a coactivator, by contributing to the activation of gene expression [9–12,32,87,88].

It was suggested that different binding partners may play critical roles in the switch between the coactivator and corepressor functions of G9a. Indeed, G9a stabilizes the occupancy of the Mediator complex on the promoter of the adult β globin gene in a NF-E2/p45-dependent manner to exert its coactivator function, while it recruits the H3K4 demethylase Jarid1a to the promoter of the embryonic β globin gene and results in transcription repression [12,89] (Figure 3b). It has also been shown that G9a is recruited to the promoter or enhancer regions of its positively regulated target genes, indicating that G9a may act directly on their expression [8–12,32,87–89]. In addition, G9a was reported to bind to RNA polymerase II, indicating that G9a may be involved in the establishment of a preinitiation or initiation complex during transcription [12].

The G9a activation domain (AD) (amino acid 1–280 in human G9a) was first identified by Dr. Stallcup's group using transient reporter gene assay [10] (Figure 1). G9a AD is sufficient and required for its coactivator function [10] and contains an autonomous activation domain [9]. Recently, we demonstrated the importance of G9a auto-methylation in the G9a AD for its coactivator function. Indeed, auto-methylation of G9a (K185) is required for its coactivator function with the glucocorticoid receptor (GR), by facilitating the binding of HP1γ and the subsequent recruitment of RNA pol II [32]. Inversely, G9a phosphorylation (T186) by AurKB antagonizes these effects (Figure 3b). Thus, these adjacent modifications regulate coactivator functions and contribute to determining whether G9a act as a coactivator or corepressor [32]. At the physiological level, we demonstrated that the coactivator activity of G9a regulates migration of the lung cancer cell line, A549 [32], and GC-induced cell death in leukemia [32,88]. In addition, G9a was reported to function as a scaffold protein to recruit the coactivators p300 and CARM1 on a subset of GR target genes, leading to transcriptional activation [8,9].

G9a also acts as a coactivator by specifically methylating the estrogen receptor alpha (ERα) on K235 [58]. This event is recognized by the Tudor domain of PHF20, which recruits the MOF histone acetyltransferase complex in order to acetylate H4K16 and promote active transcription (Figure 3b). Through this mechanism, G9a regulates a specific subset of ERα target genes [58].

4.2. Cellular Roles and Functions

4.2.1. Embryonic Development

Most PKMTs are essential for the formation of healthy embryo, as they remodel histones and control chromatin packaging and transcriptional accessibility along the genome [1]. Hence, it came as no surprise that G9a knockout impacted embryonic development [5]. Embryo of mice genetically engineered to be G9a-deficient displayed delayed development, growth arrest by the earliest stages monitored, and were no longer viable by embryonic day 9.5 [5]. Histones extracted from G9a-deficient embryos showed a strong decrease in H3K9me2 [5,6] Later studies, then reported the importance of G9a in specific developing tissues and organs based on different analyses.

- Germ Cell Development

Germ line-specific G9a knockout mice were shown to be sterile due to a drastic loss of mature gametes [90]. In addition, completion of meiosis was not observed in either gender. In G9a-deficient germ cells, H3K9me1/2 decreased during meiosis, suggesting that gene silencing induced by G9a is crucial for proper meiotic prophase progression [90].

- Cardiac Development

Engineered mice in which GLP was knocked out and G9a knocked down in cardiomyocytes showed neonatal lethality and atrioventricular septal defects, strongly implicating G9a and GLP in cardiomyocyte function for atrioventricular septum formation [91]. However, cardiomyocyte-specific G9a knockout mice were normal and the loss of G9a induced only a slight decrease in H3K9me2 levels in cardiomyocytes, indicating that adequate H3K9me2 can be performed by enzymes other than G9a in cardiomyocytes [91].

- Neuronal Development

Neuron-specific deficiency of G9a did not reveal obvious neuronal developmental or architectural defects [92]. However, these mice displayed various abnormal phenotypes, including defects in cognition and adaptive behaviors, such as difficulties in learning, motivation and environmental adaptation [92]. Authors demonstrated that multiple non-adult neuronal and non-neuronal progenitor genes were derepressed in the forebrain of these mice deficient for G9a [92]. Using pharmacological inhibition of G9a/GLP activity, it was demonstrated that G9a/GLP are required in the dorsal hippocampus for the transcriptional switch from short-term to long-term spatial memory formation [93]. Repression of G9a and H3K9 methylation has been described in postmortem nucleus accumbens of human cocaine addicts, indicating a clinical relevance of G9a in human addiction [94]. Through extended analyses, Maze et al. demonstrated a role for G9a in neuronal subtype identity in the adult central nervous system, and a critical function for G9a and H3K9 methylation in the regulation of behavioral responses to environmental stimuli [95].

- Bone Formation

G9a protein levels and H3K9me2 were reported to increase during developmental progression in tooth and growth plate cartilage [96]. G9a methyltransferase activity regulates cell proliferation and differentiation in dental mesenchyme in order to promote proper tooth development [96].

Using two different models of conditional G9a knockout mice, G9a was shown to be involved in cranial bone formation, since mutant mice had severe defects in cranial vault bones with opened fontanelles [97,98]. Mechanistically, the effect of G9a on cranial bone formation relies on its function as repressor of Twist expression during osteoblastic differentiation and as coactivator of RunX2 [97,98]. Stallcup's group demonstrated that G9a is able to enhance RunX2-mediated transcription in transient reporter gene assays by acting as a coactivator of RunX2 [11]. RunX2 is a key transcription factor of bone-forming cells by regulating osteoblastic differentiation [99]. Later, Ideno et al. showed that G9a enhances RunX2 transcriptional activity in mesodermal cells through binding and activation of RunX2 [97].

- Other Mechanisms

G9a knockdown or inhibition through pharmacological inhibitors in adult erythroid cells induces re-emergence of a fetal gene program, illustrated by the switch in expression from adult to fetal β-globin isoforms [12,89] (Figure 3).

Conditional knockout of G9a in the skeletal muscle lineage highlighted that G9a has little effect on skeletal myogenesis [100].

Targeted depletion of G9a in the developing mouse retina generated disorganized tissues [101]. According to the authors this was due to the fact that retinal progenitor cells depleted for G9a were highly proliferative and were not able to mature into the specialized components of the retina [101]. Similar results were obtained in zebrafish embryos knocked down for G9a using morpholino antisense oligos [102].

4.2.2. Hypoxia

In mammalian cell lines, G9a activity was reported to increase under hypoxic conditions, concomitant to an increase in total H3K9me2 levels, resulting in gene silencing [103]. In G9a$^{-/-}$ mouse embryonic stem cells under hypoxic conditions, the level of H3K9me2 was significantly lower, demonstrating that G9a was involved in hypoxia-induced H3K9me2 [103]. The hypoxic upregulation of G9a was attributed to specific PTMs (Figure 4). As described previously, G9a is hydroxylated at residues P676 and P1207 by PHD1 in order to target G9a toward proteasome degradation via ubiquitinylation [44]. Hypoxia induces PHD1 inhibition and a subsequent upregulation of G9a, leading to an increase in H3K9me2 and the silencing of a specific subset of target genes. Casciello et al demonstrated that G9a inhibition decreases proliferation, migration, and in vivo tumor growth [44]. Likewise, in ovarian cancer, FIH reaction was limited under hypoxia, leading to a reduced expression of metastasis-suppressor genes via H3K9 methylation [45]. Mechanistically, FIH induces hydroxylation of G9a on N779, impairing its ability to bind mono- and dimethylated H3K9, and thus methylate H3K9 [45] (Figure 4).

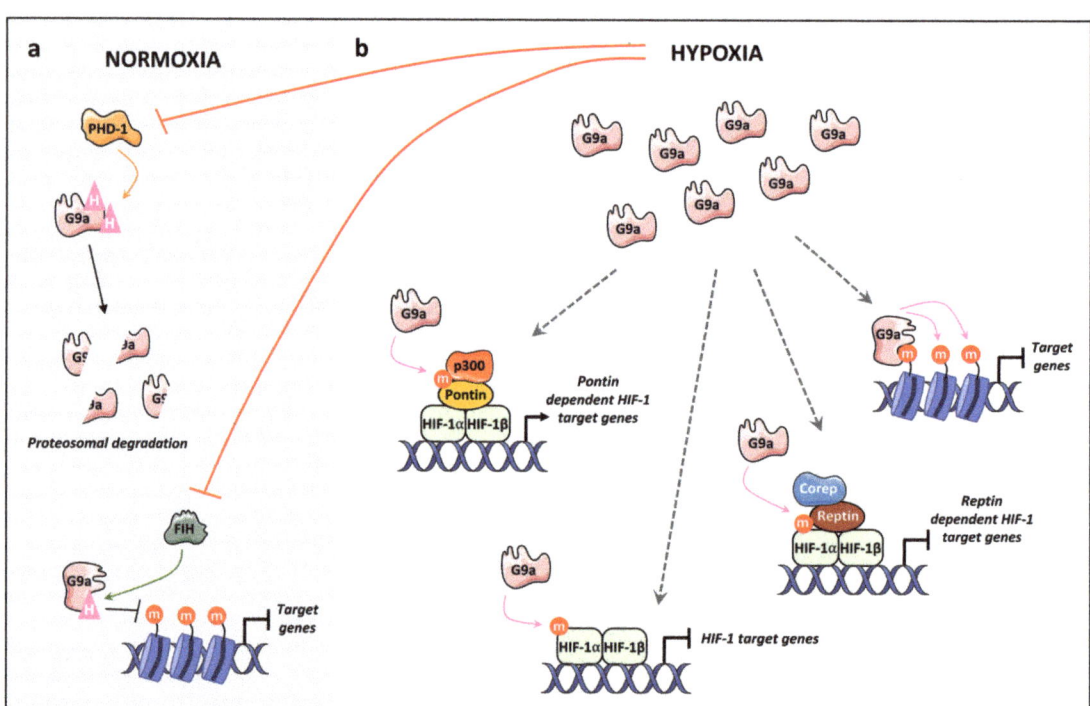

Figure 4. The role of G9a in hypoxia. (**a**) In normoxia, PHD-1 hydroxylates G9a on P676 and P1207 leading to proteosomal degradation. Likewise, FIH hydroxylates G9a on N779 impairing its ability to bind to H3K9me1/2 products. These hydroxylation processes are inhibited under hypoxic conditions resulting in an increase in the global level of G9a protein and H3K9me2. (**b**) In addition, under hypoxia, G9a methylates histones and non-histone targets. Hypoxia increases G9a-dependent H3K9me2 at the promoter regions of several genes leading to their repression. In addition, G9a methylates HIF-1α coregulators Pontin and Reptin during hypoxic stress, leading to the activation and repression of HIF-1α target genes. Finally, HIF-1 methylation by G9a suppresses HIF-1α transcriptional activity under hypoxia.

However, the role of G9a under hypoxia is likely more extensive, as G9a methylates many protein substrates involved in hypoxia, namely Pontin, Reptin, and HIF-1α [61–63] (Figure 4). Bao et al. demonstrated that HIF-1α, a master regulator of the hypoxic response, is mono- and dimethylated by G9a on K674 [63]. They demonstrated that G9a is able to methylate HIF-1α in an oxygen-independent manner. However, endogenous HIF-1α is unstable and degraded under normoxic conditions, indicating that HIF-1α is unlikely to be methylated in normoxia [63]. HIF-1αK674me1/2 suppresses HIF-1α transcriptional activity under hypoxia and expression of its downstream target genes (Figure 4). These authors also demonstrated that HIFα methylation by G9a decreases HIF-1-dependent migration of glioblastoma cells [63]. In addition, G9a methylates Reptin and Pontin, two chromatin remodelers involved in hypoxia, known to bind to HIF-1 proteins [61,62]. Under hypoxia, G9a monomethylates Reptin on K67 (K67me1), this methylation negatively regulates a subset of hypoxia target genes via the recruitment of Reptin K67me1 to their promoters and an enhanced binding to HIF-1α [62]. In addition, Reptin K67me1 leads to the recruitment of corepressors such as HDAC1 to hypoxia-responsive gene promoters in order to decrease HIF-1α transcriptional activity [62] (Figure 4). Conversely, under hypoxia, G9a methylates Pontin on six lysine residues (K265, K267, K268, K274, K281, K285), enhancing p300 coactivator recruitment on the promoters of HIF-1α target genes, resulting in an increase in HIF-1 transcriptional activity [62] (Figure 4). Although Reptin and Pontin share similarities in their structures, they act as coactivator or corepressor of HIF-1 depending on their subset of target genes in order to modulate cellular responses to hypoxia [61,62].

The ability of G9a to repress genes under hypoxic conditions suggests a key role for G9a in cell survival processes in this condition, especially in solid tumors where hypoxia is a common microenvironmental state.

4.2.3. DNA Damage and DNA Repair

Two reports demonstrated that G9a was recruited to DNA-damage sites, mainly through G9a phosphorylation [41,42]. G9a is phosphorylated by casein kinase 2 (CK2) at S211 in response to DNA double-strand breaks (DSBs), promoting G9a recruitment to sites of DNA damage by increasing its interaction with chromatin, where it can directly interact with replication protein A (RPA) [42]. In turn, binding of G9a to RPA modulates RPA and Rad51 foci formation, allowing efficient homologous recombination of DSBs and cell survival [42]. In parallel, Ginjala et al. demonstrated that G9a is phosphorylated by ATM kinase on S569 [41]. This event also leads to its recruitment to sites of DNA breaks. Authors demonstrated that the catalytic activity of G9a is critical for early recruitment of 53BP1 and BRCA1 to DNA lesions, but dispensable for their late recruitment. Induction of DSBs leads to an increase in H3K9me2 and H3K56me1 in their neighboring chromatin, two histone targets of G9a [41]. Inhibition of the catalytic activity of G9a decreases these modifications, suggesting that G9a could be recruited to DNA breaks in order to induce local histone methylation and subsequent local transcriptional silencing. Finally, using GFP-based reporters of homologous repair (HR) or non-homologous end-joining repair (NHEJ), they demonstrated that the catalytic activity of G9a impairs both mechanisms, HR and NHEJ [41]. Moreover, phosphorylation of S211 and S569 appears to be essential for proper DNA repair [41,42].

G9a may also methylate specific non-histone proteins involved in DNA repair mechanisms, such as Polo-loke kinase 1 (Plk1) and p53 [57,68]. Plk1 phosphorylation on T210 is required during DNA damage repair and checkpoint recovery [104]. Recently Li et al. demonstrated that the activity of Plk1 is controlled by a switch between methylation and phosphorylation, as for G9a and GLP [68]. Authors showed that under DNA damage stress conditions, the interaction between G9a and Plk1 is enhanced and G9a monomethylation on K209 of Plk1 is increased [68]. Interestingly, Plk1 methylation by G9a is not necessary for its recruitment to DNA lesions or for the assembly of the DNA repair machinery via RPA and Rad51 recruitment. However, this methylation is crucial for the timely removal of this DNA repair machinery from DNA lesions, which is essential for the proper completion of DNA damage repair [68]. The tumor suppressor p53 was also demonstrated to be a substrate for G9a on K373 [57]. However, p53 methylation seems to be link with inactive p53, as the level of methylated p53 during DNA damage does not change even though the total level of p53 increases dramatically [57]. This data is consistent with the fact that catalytic inhibition of G9a using inhibitors under low DNA damage conditions impairs DNA DSB repair in a p53-independent manner [105]. However, it is interesting to note that G9a dimethylation of p53 at K373 increases Plk1 expression and promotes colorectal cancer [106].

These reports clearly demonstrate the relevance of G9a in the maintenance of genome integrity, implicating G9a in cancer biology.

5. G9a in Cancer
5.1. G9a Oncogenic Role

Recently, dysregulations in the PTMs of both DNA and histones were shown to contribute to cancer initiation and progression [107]. These epigenetic modifications, which result in altered chromatin structure and gene expression were reported in different types of cancers [108] (Figure 3). G9a was overexpressed in breast, gastric, ovarian, cervical, endometrial, prostate, lung, colorectal, liver, urinary bladder, and brain cancers, as well as in hematological malignancies, melanoma, and cholangiocarcinoma, leading to aberrant H3K9 methylation [109–122]. One of the main reasons for this increase in G9a expression and H3K9 methylation is hypoxia [103]. The molecular mechanisms associated with this phenomenon are described in a previous section (Figure 4). Furthermore, high levels of G9a expression were associated with poor prognosis and shorter survival in cancer patients [57,123–127]. G9a involvement in cancer biology is likely due to its pivotal role in tumor cell proliferation, survival, and metastasis primarily by controlling several transcription programs (Table 3).

5.1.1. Breast Cancer

High G9a-mediated H3K9 methylation triggers the proliferation and progression of breast cancer (Table 3) [109,128,129]. For instance, G9a overexpression was shown to downregulate the expression of some tumor suppressor genes, such as ARNTL, CEACAM7, GATA2, HHEX, KLRG1, and OGN. Blocking G9a methyltransferase activity was sufficient to re-express these genes, and consequently inhibit breast cancer cell proliferation and migration in vitro and tumor growth in vivo [44]. G9a was also demonstrated to interact with MYC and suppress its target genes by favoring H3K9me2, in order to stimulate MYC-dependent breast tumor growth [129]. G9a may also contribute to enhancing breast tumor metastasis by silencing several genes implicated in epithelial-mesenchymal transition (EMT), namely the two anti-metastatic tumor suppressor genes, desmocollin 3 (DSC3), belonging to the cadherin superfamily, and the protease inhibitor MASPIN, which were transcriptionally reactivated in a dose-dependent manner upon inhibition of G9a activity, concomitantly to a significant decrease in global H3K9 dimethylation [130]. In addition, in EMT, G9a was shown to repress the expression of E-cadherin, a cell adhesion factor, upon association with the SNAIL transcription factor and to induce H3K9me2 of its promoter [131]. Depletion of G9a restored E-cadherin expression and inhibited breast cancer cell migration and invasion in vitro and in vivo [131]. G9a also silenced the expression of the type-II cadherin CDH10 through histone methylation, stimulating hypoxia-mediated cellular motility; and its inhibition prevented cellular movement and breast cancer cell colonization in the lungs [123]. G9a methyltransferase activity was further reported to (i) collaborate with the transcription factor YY1 and HDAC1 to disrupt cellular iron homeostasis by repressing ferroxidase hephaestin, resulting in iron accumulation and breast cancer progression [109], (ii) induce breast cancer cell autophagy by modulating the AMPK-mTOR pathways [132], and (iii) promote breast cancer recurrence through the suppression of pro-inflammatory genes [133].

5.1.2. Gastric Cancer

In gastric cancer, G9a activation reduces apoptosis and promotes tumor cell growth (Table 3) [134]. For instance, blocking the catalytic activity of G9a reduces cell growth and autophagy by downregulating the mechanistic target of rapamycin (mTOR) pathways. Authors showed that G9a activates mTOR through H3K9 monomethylation at the mTOR promoter [125]. G9a inhibition by (i) kaempferol, a flavonoid present in fruits and vegetables [135], (ii) SH003, an herbal formulation [136], or (iii) cinnamaldehyde (CA), the bioactive ingredient in Cinnamomum [137], stimulated autophagic gastric cancer cell death. Increase in H3K9 methylation under hypoxia also mediated the silencing of the tumor suppressor gene, runt-domain transcription factor 3 (RUNX3) [138]. Finally, G9a overexpression was shown to upregulate the expression of ITGB3, an integrin family member, in an enzyme-independent manner inducing gastric cancer metastasis [139].

5.1.3. Human Reproductive Cancers

Alterations in G9a expression were also associated with human reproductive cancers (Table 3). In ovarian cancer (OCa), high G9a expression levels were correlated with late stage, high grade, and a decreased overall survival in OCa patients [111,140]. An elevation in the level of G9a was observed in vitro in invasive cell lines ES-2, SKOV-3, TOV-21G, OV-90, and OVCAR-3, and in vivo in metastatic lesions in comparison with less aggressive tumor cells and primary tumors [111]. Depletion of G9a inhibited cellular adhesion, migration, invasion, and anoikis-resistance of OCa cell lines in vitro and suppressed OCa metastasis in vivo [111]. Further investigations revealed that several tumor suppressor genes were repressed in OCa by G9a, such as DUSP5, SPRY4, CDH1, and PPP1R15A. PARP inhibitor-resistant high-grade serous ovarian carcinoma (HGSOC) displayed an increase in H3K9me2 associated with an increase in the overall expression of G9a [140]. Similar observations were made in vivo on patient-derived xenografts, indicating that a high G9a expression maintains resistance to PARP inhibitors [140]. Interestingly, inhibition of G9a displayed synergistic anti-tumor effects in combination with DNA methylation inhibitors in OCa cell lines, where authors induced cell death by upregulating endogenous retroviruses (ERVs), consequently activating the viral immune response [141].

In cervical cancer, G9a induces the expression of angiogenic factors including angiogenin, interleukin-8, and C-X-C motif chemokine ligand-16, prompting angiogenesis and cancer cell invasion, and decreasing patient survival [142]. Interestingly, depletion of G9a decreased the expression of oncogenic proteins such as Bcl-2, Mcl-1, and Survivin, and increased the expression of E-cadherin inhibiting cell adhesion and invasion [112].

Likewise, in endometrial cancer, G9a-mediated H3K9 methylation induced tumor invasion in vitro and in vivo via the silencing of the E-cadherin [113]. Indeed, G9a depletion reduces H3K9me2 levels, restores E-cadherin expression and decreases E-cadherin promoter DNA methyltransferase recruitment. G9a expression is higher in endometrial cancer tissues and its expression is correlated with deep myometrial invasion [113].

Finally, in prostate cancer, high G9a expression was associated with high pathological grade and poor overall survival. In this model, G9a promoted cancer proliferation by inhibiting PI3K/AKT/mTOR pathway [114].

5.1.4. Lung Cancer

In lung cancer, G9a possesses proliferative and metastatic properties (Table 3) [114]. Highly invasive lung cancer cell lines were reported to display higher G9a protein levels, in comparison with weakly invasive cells. Overexpressing G9a increased cell motility and invasiveness [143]. Different reports demonstrated that G9a induced tumor growth, invasion, and migration by (i) silencing specific EMT-regulating genes, including caspase-1 and the epithelial cell adhesion molecule Ep-CAM [124,144], (ii) mediating the Snail2-induced E-cadherin suppression [145], and/or (iii) activating the focal adhesion kinase signaling pathway [146]. Depletion of G9a abolished lung cancer cell migration and invasion in vitro and metastasis in vivo [124,144,146]. G9a also induced cell proliferation through the activation of the WNT signaling pathway by suppressing WNT signaling inhibitors like DKK1, APC2, and WIFI [121]. Moreover, G9a was shown to play an important role in maintaining lung cancer cell stemness by maintaining DNA methylation of multiple lung cancer stem cell genes and their subsequent expression [147].

5.1.5. Colorectal Cancer

In colorectal cancer (CRC), high levels of G9a are associated with tumor initiation, maintenance, and proliferation (Table 3) [59,106,148]. In primary CRC patient samples, transcriptome profiling revealed the co-enrichment of G9a and H3K9me2 of multiple genes involved in the negative regulation of the WNT signaling pathway, in repression of EMT and extracellular matrix organization, leading to their repression in CRC [148]. G9a also methylates two non-histone substrates involved in CRC cell proliferation, FOXO1 (Forkhead family transcription factor) and p53 [59,106]. FOXO1 is methylated by G9a on K273, increasing the interaction between FOXO1 and the E3 ligase SKP2. This event decreases FOXO1 protein stability and promotes cellular proliferation in colon cancer [59]. These authors also demonstrated that G9a protein expression is increased in human colon cancer patient tissue samples associated with a decrease in FOXO1 protein level [59]. Likewise, G9a-mediated p53 dimethylation at lysine 373 was shown to increase Plk1 expression and consequently CRC cell growth [106].

5.1.6. Hepatocellular Carcinoma

In hepatocellular carcinoma (HCC), targeting G9a is suggested as a novel therapy for HCC treatment as it drives tumorigenesis and aggressiveness (Table 3) [149,150]. Indeed, G9a is upregulated in HCC, which leads to the epigenetic silencing of the retinoic acid receptor responder protein 3 (RARRES3) tumor suppressor gene, thus triggering HCC proliferation and metastasis in vitro and in vivo [116]. Moreover, G9a was shown to enhance metastasis formation through an epigenetic regulation of EMT, as it interacts with SNAIL2 and HDACs at the E-cadherin promoter in order to inhibit E-cadherin transcription [151]. A recent study showed that G9a contributes to HCC initiation by escaping p53-induced apoptosis in DNA-damaged hepatocytes via the repression of Bcl-G expression, a pro-apoptotic Bcl-2 family member [152].

5.1.7. Urinary Bladder Cancer

G9a was reported to be upregulated or amplified in urinary bladder cancer (UBC) [153]. G9a represents a promising therapeutic target for UBC as various G9a inhibitors decrease cell proliferation and increase cell death through the endoplasmic reticulum stress pathway [153]. Likewise, targeting G9a and DNMT methyltransferase activity with a novel dual inhibitor called CM-272 induces cell apoptosis and immunogenic cell death [153].

5.1.8. Hematological Cancers

G9a is upregulated in hematological malignancies, for which G9a inhibitors have been identified as promising targets for patient management (Table 3) [154–158]. In T-lymphoblastic leukemia cells (T-ALL), inhibiting G9a activity suppresses cellular proliferation and induces apoptosis by downregulating the expression of Bcl-2 and upregulating the expression of Bax and caspase-3 [155]. Likewise, in chronic lymphocytic leukemia, targeting G9a and GLP was shown to stimulate cancer cell death [154]. In multiple myeloma, G9a fosters ReIB-dependent cancer growth and survival, whereas its depletion reduces the expression of ReIB and increases the expression of pro-apoptotic genes, such as Bim and BMF [118]. In acute myeloid leukemia (AML), G9a inhibition attenuates the transcriptional activity of the leukemogenic transcription factor HoxA9 and thus promotes AML proliferation, progression, and self-renewal [157]. In childhood acute lymphoblastic leukemia, G9a is reported to enhance the ability of cancer cells to migrate [159].

Table 3. Role of G9a in Cancer Biology.

G9a Roles	Cancer Types	G9a Biological Roles	References
Oncogenic	Breast Cancer	Suppresses tumor suppressor genes Enhances EMT Disrupts iron homeostasis Inhibits autophagy	[44,130] [123,131] [109] [132]
	Gastric Cancer	Suppresses tumor suppressor genes Inhibits apoptosis and autophagy Promotes metastasis	[138] [125,135–137] [139]
	Ovarian Cancer	Promotes metastasis Suppresses tumor suppressor genes Maintains PARP-inhibitor resistance	[111] [45,111] [140]
	Cervical Cancer	Induces angiogenesis Enhances tissue invasion	[142] [112]
	Endometrial Cancer	Enhances tissue invasion	[113]
	Prostate Cancer	Stimulates proliferation	[114]
	Lung Cancer	Enhances EMT Activates WNT signaling pathway Maintains lung cancer stemness Supports resistance to radiotherapy	[124,144,145,160] [121] [147] [161]
	Colorectal Cancer	Stimulates proliferation Enhances self-renewal and stemness Promotes resistance to chemotherapy	[59,106] [148] [162]
	Liver Cancer	Suppresses tumor suppressor genes Enhances EMT Inhibits cell apoptosis	[116] [151] [152]
	Bladder Cancer	Inhibits cell apoptosis and autophagy	[122,153]
	Brain Cancer	Stimulates proliferation Inhibits autophagy	[117,163] [164]
	Hematological malignancies	Enhances self-renewal and stemness Promotes migration Inhibits apoptosis and stimulates proliferation	[157] [159] [118,155]
	Skin Cancer	Promotes progression	[119,165]
	Head and Neck Cancer	Enhances EMT	[166]
	Bile duct Cancer	Suppresses tumor suppressor genes	[120]
Anti-oncogenic	Lung Cancer	Inhibits cancer progression	[167]
	Brain Cancer	Inhibits HIF-induced migration Inhibits cancer stemness	[63]

5.1.9. Other Cancers

G9a represents an intriguing target in various other types of cancers (Table 3). In medulloblastoma, G9a drives H3K9me1/2/3 at the promoter of ubiquitin-specific protease 37 (USP37) to repress its gene expression [163]. USP37 controls cell proliferation by regulating the stability of the cyclin-dependent kinase inhibitor 1B (CDKN1B/p27Kip1) in cell cycle. Thus, blocking G9a inhibits cellular proliferation and tumorigenic potential of medulloblastoma cells [163]. Pre- or post-treatment of glioma cells with a G9a inhibitor sensitizes these cells to Temozolomide (TMZ), the first line therapy for glioblastoma patients, and increases its cytotoxicity [168]. Interestingly, authors demonstrated that the G9a inhibitor reprograms glioma cells and glioma stem-like cells to increase sensitivity to TMZ [164,168]. As previously described in breast cancer, HCC, and lung cancer, G9a interacts with SNAIL in order to mediate repression of E-cadherin and EMT in head and neck squamous cell carcinoma (HNSCC) [166]. Additionally, G9a was associated with cholangiocarcinoma, a highly malignant epithelial tumor of the biliary tree, where G9a-mediated H3K9 methylation suppressed the expression of the tumor suppressor gene LATS2, leading to the subsequent activation of the oncogenic YAP signaling pathway [120]. Recently in melanoma, elevated G9a levels promoted cancer progression through the activation of the WNT/β-catenin signaling by epigenetic silencing of the WNT antagonist DKK1 gene [165], or through the upregulation of the Notch1 signaling pathway, that further stimulates PI3K/AKT pathway [119].

5.2. G9a Tumor Suppressive Role

In stark contrast to its oncogenic roles, several studies demonstrated that G9a also promotes tumor suppressive functions. For example, G9a depletion increased the aggressiveness of lung tumor propagating cells (TPC) and accelerated disease progression and metastasis [167]. Inhibition of G9a derepresses genes that regulate the extracellular matrix. Patients with high levels of G9a displayed a better survival in early-stage lung cancer [167]. Interestingly, in glioblastoma, G9a inhibited HIF-1α-mediated migration via the methylation of the alpha subunit at lysine 674 [63].

6. Outlook

Over the last three decades since G9a was discovered, extensive studies were conducted to gain further insight into its physiological and pathophysiological roles. Aside from its key role in epigenetic repression through H3K9 methylation, G9a displays many biological functions, notably in gene expression, associated with its methylation of histone and non-histone substrates. Furthermore, a growing body of evidence indicates that G9a acts as a coregulator of transcription factors and steroid receptors, and could hence endorse other functions through these properties. Owing to its broad implication in biological activities, dysregulation of G9a expression is common to many types of cancers, and, as such, G9a represents a promising target for anti-cancer agents. Indeed, many inhibitors of G9a inhibitors have been synthetized and characterized, and could represent interesting therapeutic agents.

Author Contributions: C.P., L.M.N. and L.P. wrote the manuscript. M.L.R. revised it. All authors have read and agreed to the published version of the manuscript.

Funding: M.L.R., L.M.N., L.P. and C.P.'s laboratory is funded with grants from "La Ligue contre le Cancer", the "Fondation ARC Cancer" and the association "Cancer du sein, parlons en". L.M.N. was supported by a fellowship from AZM & Saadeh Association and Lebanese University.

Acknowledgments: We thank B. Manship for proofreading the manuscript. The illustrations were created by using Servier Medical Art.

Conflicts of Interest: The authors declare no conflict of interest.

References

1. Wu, Z.; Connolly, J.; Biggar, K.K. Beyond Histones—The Expanding Roles of Protein Lysine Methylation. *FEBS J.* **2017**, *284*, 2732–2744. [CrossRef] [PubMed]
2. Dillon, S.C.; Zhang, X.; Trievel, R.C.; Cheng, X. The SET-Domain Protein Superfamily: Protein Lysine Methyltransferases. *Genome Biol.* **2005**, *6*, 1–10. [CrossRef] [PubMed]
3. Falnes, P.; Jakobsson, M.E.; Davydova, E.; Ho, A.; Malecki, J. Protein Lysine Methylation by Seven-β-Strand Methyltransferases. *Biochem. J.* **2016**, *473*, 1995–2009. [CrossRef] [PubMed]
4. Milner, C.M.; Campbell, R.D. The G9a Gene in the Human Major Histocompatibility Complex Encodes a Novel Protein Containing Ankyrin-like Repeats the Class III Region of the Human Major Histocompatibility Complex Spans Approx. *Biochem. J.* **1993**, *290*, 811–818. [CrossRef]
5. Tachibana, M.; Sugimoto, K.; Nozaki, M.; Ueda, J.; Ohta, T.; Ohki, M.; Fukuda, M.; Takeda, N.; Niida, H.; Kato, H.; et al. G9a Histone Methyltransferase Plays a Dominant Role in Euchromatic Histone H3 Lysine 9 Methylation and Is Essential for Early Embryogenesis. *Genes Dev.* **2002**, *16*, 1779–1791. [CrossRef]
6. Tachibana, M.; Ueda, J.; Fukuda, M.; Takeda, N.; Ohta, T.; Iwanari, H.; Sakihama, T.; Kodama, T.; Hamakubo, T.; Shinkai, Y. Histone Methyltransferases G9a and GLP Form Heteromeric Complexes and Are Both Crucial for Methylation of Euchromatin at H3-K9. *Genes Dev.* **2005**, *19*, 815–826. [CrossRef]
7. Tachibana, M.; Matsumura, Y.; Fukuda, M.; Kimura, H.; Shinkai, Y. G9a/GLP Complexes Independently Mediate H3K9 and DNA Methylation to Silence Transcription. *EMBO J.* **2008**, *27*, 2681–2690. [CrossRef]
8. Bittencourt, D.; Wu, D.Y.; Jeong, K.W.; Gerke, D.S.; Herviou, L.; Ianculescu, I.; Chodankar, R.; Siegmund, K.D.; Stallcup, M.R. G9a Functions as a Molecular Scaffold for Assembly of Transcriptional Coactivators on a Subset of Glucocorticoid Receptor Target Genes. *Proc. Natl. Acad. Sci. USA* **2012**, *109*, 19673–19678. [CrossRef]
9. Lee, D.Y.; Northrop, J.P.; Kuo, M.H.; Stallcup, M.R. Histone H3 Lysine 9 Methyltransferase G9a Is a Transcriptional Coactivator for Nuclear Receptors. *J. Biol. Chem.* **2006**, *281*, 8476–8485. [CrossRef]
10. Purcell, D.J.; Jeong, K.W.; Bittencourt, D.; Gerke, D.S.; Stallcup, M.R. A Distinct Mechanism for Coactivator versus Corepressor Function by Histone Methyltransferase G9a in Transcriptional Regulation. *J. Biol. Chem.* **2011**, *286*, 41963–41971. [CrossRef]
11. Purcell, D.J.; Khalid, O.; Ou, C.Y.; Little, G.H.; Frenkel, B.; Baniwal, S.K.; Stallcup, M.R. Recruitment of Coregulator G9a by Runx2 for Selective Enhancement or Suppression of Transcription. *J. Cell. Biochem.* **2012**, *113*, 2406–2414. [CrossRef] [PubMed]
12. Chaturvedi, C.P.; Hosey, A.M.; Palii, C.; Perez-Iratxeta, C.; Nakatani, Y.; Ranish, J.A.; Dilworth, F.J.; Brand, M. Dual Role for the Methyltransferase G9a in the Maintenance of β-Globin Gene Transcription in Adult Erythroid Cells. *Proc. Natl. Acad. Sci. USA* **2009**, *106*, 18303–18308. [CrossRef] [PubMed]
13. Mauger, O.; Klinck, R.; Chabot, B.; Muchardt, C.; Allemand, E.; Batschébatsch´batsché, E. Alternative Splicing Regulates the Expression of G9A and SUV39H2 Methyltransferases, and Dramatically Changes SUV39H2 Functions. *Nucleic Acids Res.* **2015**, *43*, 1869–1882. [CrossRef]
14. Fiszbein, A.; Giono, L.E.; Quaglino, A.; Berardino, B.G.; Sigaut, L.; von Bilderling, C.; Schor, I.E.; Steinberg, J.H.E.; Rossi, M.; Pietrasanta, L.I.; et al. Alternative Splicing of G9a Regulates Neuronal Differentiation. *Cell Rep.* **2016**, *14*, 2797–2808. [CrossRef]
15. Shankar, S.R.; Bahirvani, A.G.; Rao, V.K.; Bharathy, N.; Ow, J.R.; Taneja, R. G9a, a Multipotent Regulator of Gene Expression. *Epigenetics* **2013**, *8*, 16–22. [CrossRef] [PubMed]
16. Schotta, G.; Ebert, A.; Krauss, V.; Fischer, A.; Hoffmann, J.; Rea, S.; Jenuwein, T.; Dorn, R.; Reuter, G. Central Role of Drosophila SU(VAR)3–9 in Histone H3-K9 Methylation and Heterochromatic Gene Silencing. *EMBO J.* **2002**, *21*, 1121–1131. [CrossRef]
17. Jenuwein, T.; Laible, G.; Dorn, R.; Reuter, G. SET Domain Proteins Modulate Chromatin Domains in Eu-and Heterochromatin. *Cell. Mol. Life Sci* **1998**, *54*, 80–93. [CrossRef]
18. Taylor, W.R.; Xiao, B.; Gamblin, S.J.; Lin, K. A Knot or Not a Knot? SETting the Record 'Straight' on Proteins. *Comput. Biol. Chem.* **2003**, *27*, 11–15. [CrossRef]
19. Trievel, R.C.; Beach, B.M.; Dirk, L.M.A.; Houtz, R.L.; Hurley, J.H. Structure and Catalytic Mechanism of a SET Domain Protein Methyltransferase. *Cell* **2002**, *111*, 91–103. [CrossRef]
20. Qian, C.; Zhou, M.M. SET Domain Protein Lysine Methyltransferases: Structure, Specificity and Catalysis. *Cell. Mol. Life Sci.* **2006**, *63*, 2755–2763. [CrossRef] [PubMed]
21. Wu, H.; Min, J.; Lunin, V.V.; Antoshenko, T.; Dombrovski, L.; Zeng, H.; Allali-Hassani, A.; Campagna-Slater, V.; Vedadi, M.; Arrowsmith, C.H.; et al. Structural Biology of Human H3K9 Methyltransferases. *PLoS ONE* **2010**, *5*, e8570. [CrossRef]
22. Lenstra, D.C.; al Temimi, A.H.K.; Mecinović, J. Inhibition of Histone Lysine Methyltransferases G9a and GLP by Ejection of Structural Zn(II). *Bioorganic Med. Chem. Lett.* **2018**, *28*, 1234–1238. [CrossRef]
23. Schapira, M. Structural Chemistry of Human SET Domain Protein Methyltransferases. *Curr. Chem. Genom.* **2011**, *5*, 85–94. [CrossRef]
24. Collins, R.E.; Northrop, J.P.; Horton, J.R.; Lee, D.Y.; Zhang, X.; Stallcup, M.R.; Cheng, X. The Ankyrin Repeats of G9a and GLP Histone Methyltransferases Are Mono- and Dimethyllysine Binding Modules. *Nat. Struct. Mol. Biol.* **2008**, *15*, 245–250. [CrossRef]
25. Estève, P.O.; Patnaik, D.; Chin, H.G.; Benner, J.; Teitell, M.A.; Pradhan, S. Functional Analysis of the N- and C-Terminus of Mammalian G9a Histone H3 Methyltransferase. *Nucleic Acids Res.* **2005**, *33*, 3211–3223. [CrossRef] [PubMed]
26. Corpet, F. Multiple Sequence Alignment with Hierarchical Clustering. *Nucleic Acids Res.* **1988**, *16*, 10881. [CrossRef] [PubMed]

27. Chang, Y.; Zhang, X.; Horton, J.R.; Upadhyay, A.K.; Spannhoff, A.; Liu, J.; Snyder, J.P.; Bedford, M.T.; Cheng, X. Structural Basis for G9a-like Protein Lysine Methyltransferase Inhibition by BIX-01294. *Nat. Struct. Mol. Biol.* **2009**, *16*, 312–317. [CrossRef]
28. Liu, N.; Zhang, Z.; Wu, H.; Jiang, Y.; Meng, L.; Xiong, J.; Zhao, Z.; Zhou, X.; Li, J.; Li, H.; et al. Recognition of H3K9 Methylation by GLP Is Required for Efficient Establishment of H3K9 Methylation, Rapid Target Gene Repression, and Mouse Viability. *Genes Dev.* **2015**, *29*, 379–393. [CrossRef]
29. Shinkai, Y.; Tachibana, M. H3K9 Methyltransferase G9a and the Related Molecule GLP. *Genes Dev.* **2011**, *25*, 781–788. [CrossRef] [PubMed]
30. Ueda, J.; Tachibana, M.; Ikura, T.; Shinkai, Y. Zinc Finger Protein Wiz Links G9a/GLP Histone Methyltransferases to the Co-Repressor Molecule CtBP. *J. Biol. Chem.* **2006**, *281*, 20120–20128. [CrossRef]
31. Bian, C.; Chen, Q.; Yu, X. The Zinc Finger Proteins ZNF644 and WIZ Regulate the G9A/GLP Complex for Gene Repression. *eLife* **2015**, *4*, e05606. [CrossRef] [PubMed]
32. Poulard, C.; Bittencourt, D.; Wu, D.; Hu, Y.; Gerke, D.S.; Stallcup, M.R. A Post-translational Modification Switch Controls Coactivator Function of Histone Methyltransferases G9a and GLP. *EMBO Rep.* **2017**, *18*, 1442–1459. [CrossRef] [PubMed]
33. Chin, H.G.; Pradhan, M.; Estève, P.O.; Patnaik, D.; Evans, T.C.; Pradhan, S. Sequence Specificity and Role of Proximal Amino Acids of the Histone H3 Tail on Catalysis of Murine G9a Lysine 9 Histone H3 Methyltransferase. *Biochemistry* **2005**, *44*, 12998–13006. [CrossRef] [PubMed]
34. Chin, H.G.; Estève, P.O.; Pradhan, M.; Benner, J.; Patnaik, D.; Carey, M.F.; Pradhan, S. Automethylation of G9a and Its Implication in Wider Substrate Specificity and HP1 Binding. *Nucleic Acids Res.* **2007**, *35*, 7313–7323. [CrossRef] [PubMed]
35. Sampath, S.C.; Marazzi, I.; Yap, K.L.; Sampath, S.C.; Krutchinsky, A.N.; Mecklenbräuker, I.; Viale, A.; Rudensky, E.; Zhou, M.M.; Chait, B.T.; et al. Methylation of a Histone Mimic within the Histone Methyltransferase G9a Regulates Protein Complex Assembly. *Mol. Cell* **2007**, *27*, 596–608. [CrossRef] [PubMed]
36. Rathert, P.; Dhayalan, A.; Murakami, M.; Zhang, X.; Tamas, R.; Jurkowska, R.; Komatsu, Y.; Shinkai, Y.; Cheng, X.; Jeltsch, A. Protein Lysine Methyltransferase G9a Acts on Non-Histone Targets. *Nat. Chem. Biol.* **2008**, *4*, 344–346. [CrossRef]
37. Patnaik, D.; Hang, G.C.; Estève, P.O.; Benner, J.; Jacobsen, S.E.; Pradhan, S. Substrate Specificity and Kinetic Mechanism of Mammalian G9a Histone H3 Methyltransferase. *J. Biol. Chem.* **2004**, *279*, 53248–53258. [CrossRef]
38. Fischle, W.; Tseng, B.S.; Dormann, H.L.; Ueberheide, B.M.; Garcia, B.A.; Shabanowitz, J.; Hunt, D.F.; Funabiki, H.; Allis, C.D. Regulation of HP1–Chromatin Binding by Histone H3 Methylation and Phosphorylation. *Nature* **2005**, *438*, 1116–1122. [CrossRef] [PubMed]
39. Hirota, T.; Lipp, J.J.; Toh, B.-H.; Peters, J.-M. Histone H3 Serine 10 Phosphorylation by Aurora B Causes HP1 Dissociation from Heterochromatin. *Nature* **2005**, *438*, 1176–1180. [CrossRef]
40. Poulard, C.; Baulu, E.; Lee, B.H.; Pufall, M.A.; Stallcup, M.R. Increasing G9a Automethylation Sensitizes B Acute Lymphoblastic Leukemia Cells to Glucocorticoid-Induced Death. *Cell Death Dis.* **2018**, *9*, 1–13. [CrossRef]
41. Ginjala, V.; Rodriguez-Colon, L.; Ganguly, B.; Gangidi, P.; Gallina, P.; Al-Hraishawi, H.; Kulkarni, A.; Tang, J.; Gheeya, J.; Simhadri, S.; et al. Protein-Lysine Methyltransferases G9a and GLP1 Promote Responses to DNA Damage. *Sci. Rep.* **2017**, *7*, 1–12. [CrossRef]
42. Yang, Q.; Zhu, Q.; Lu, X.; Du, Y.; Cao, L.; Shen, C.; Hou, T.; Li, M.; Li, Z.; Liu, C.; et al. G9a Coordinates with the RPA Complex to Promote DNA Damage Repair and Cell Survival. *Proc. Natl. Acad. Sci. USA* **2017**, *114*, E6054–E6063. [CrossRef]
43. Srinivasan, S.; Shankar, S.R.; Wang, Y.; Taneja, R. SUMOylation of G9a Regulates Its Function as an Activator of Myoblast Proliferation. *Cell Death Dis.* **2019**, *10*, 1–15. [CrossRef]
44. Casciello, F.; Al-Ejeh, F.; Kelly, G.; Brennan, D.J.; Ngiow, S.F.; Young, A.; Stoll, T.; Windloch, K.; Hill, M.M.; Smyth, M.J.; et al. G9a Drives Hypoxia-Mediated Gene Repression for Breast Cancer Cell Survival and Tumorigenesis. *Proc. Natl. Acad. Sci. USA* **2017**, *114*, 7077–7082. [CrossRef]
45. Kang, J.; Shin, S.H.; Yoon, H.; Huh, J.; Shin, H.W.; Chun, Y.S.; Park, J.W. FIH Is an Oxygen Sensor in Ovarian Cancer for G9a/GLP-Driven Epigenetic Regulation of Metastasis-Related Genes. *Cancer Res.* **2018**, *78*, 1184–1199. [CrossRef] [PubMed]
46. Tachibana, M.; Sugimoto, K.; Fukushima, T.; Shinkai, Y. SET Domain-Containing Protein, G9a, Is a Novel Lysine-Preferring Mammalian Histone Methyltransferase with Hyperactivity and Specific Selectivity to Lysines 9 and 27 of Histone H3. *J. Biol. Chem.* **2001**, *276*, 25309–25317. [CrossRef]
47. Richards, E.J.; Elgin, S.C.R. Epigenetic Codes for Heterochromatin Formation and Silencing: Rounding up the Usual Suspects. *Cell* **2002**, *108*, 489–500. [CrossRef]
48. Wu, H.; Chen, X.; Xiong, J.; Li, Y.; Li, H.; Ding, X.; Liu, S.; Chen, S.; Gao, S.; Zhu, B. Histone Methyltransferase G9a Contributes to H3K27 Methylation in vivo. *Cell Res.* **2011**, *21*, 365–367. [CrossRef] [PubMed]
49. Yu, Y.; Song, C.; Zhang, Q.; DiMaggio, P.A.; Garcia, B.A.; York, A.; Carey, M.F.; Grunstein, M. Histone H3 Lysine 56 Methylation Regulates DNA Replication through Its Interaction with PCNA. *Mol. Cell* **2012**, *46*, 7–17. [CrossRef]
50. Trojer, P.; Zhang, J.; Yonezawa, M.; Schmidt, A.; Zheng, H.; Jenuwein, T.; Reinberg, D. Dynamic Histone H1 Isotype 4 Methylation and Demethylation by Histone Lysine Methyltransferase G9a/KMT1C and the Jumonji Domain-Containing JMJD2/KDM4 Proteins. *J. Biol. Chem.* **2009**, *284*, 8395–8405. [CrossRef]
51. Weiss, T.; Hergeth, S.; Zeissler, U.; Izzo, A.; Tropberger, P.; Zee, B.M.; Dundr, M.; Garcia, B.A.; Daujat, S.; Schneider, R. Histone H1 Variant-Specific Lysine Methylation by G9a/KMT1C and Glp1/KMT1D. *Epigenetics Chromatin* **2010**, *3*, 1–13. [CrossRef]
52. Cao, H.; Li, L.; Deying, Y.; Liming, Z.; Yewei, X.; Yu, B.; Liao, G.; Chen, J. Recent Progress in Histone Methyltransferase (G9a) Inhibitors as Anticancer Agents. *Eur. J. Med. Chem.* **2019**, *179*, 537–546. [CrossRef]

53. Xiong, Y.; Li, F.; Babault, N.; Wu, H.; Dong, A.; Zeng, H.; Chen, X.; Arrowsmith, C.H.; Brown, P.J.; Liu, J.; et al. Structure-Activity Relationship Studies of G9a-like Protein (GLP) Inhibitors. *Bioorg. Med. Chem.* **2017**, *25*, 4414–4423. [CrossRef]
54. Pless, O.; Kowenz-Leutz, E.; Knoblich, M.; Lausen, J.; Beyermann, M.; Walsh, M.J.; Leutz, A. G9a-Mediated Lysine Methylation Alters the Function of CCAAT/Enhancer-Binding Protein-β. *J. Biol. Chem.* **2008**, *283*, 26357–26363. [CrossRef]
55. Ling, B.M.T.; Bharathy, N.; Chung, T.-K.; Kok, W.K.; Li, S.; Tan, Y.H.; Rao, V.K.; Gopinadhan, S.; Sartorelli, V.; Walsh, M.J.; et al. Lysine Methyltransferase G9a Methylates the Transcription Factor MyoD and Regulates Skeletal Muscle Differentiation. *Proc. Natl. Acad. Sci. USA* **2012**, *109*, 841–846. [CrossRef] [PubMed]
56. Choi, J.; Jang, H.; Kim, H.; Lee, J.H.; Kim, S.T.; Cho, E.J.; Youn, H.D. Modulation of Lysine Methylation in Myocyte Enhancer Factor 2 during Skeletal Muscle Cell Differentiation. *Nucleic Acids Res.* **2014**, *42*, 224–234. [CrossRef] [PubMed]
57. Huang, J.; Dorsey, J.; Chuikov, S.; Zhang, X.; Jenuwein, T.; Reinberg, D.; Berger, S.L. G9a and Glp Methylate Lysine 373 in the Tumor Suppressor P53. *J. Biol. Chem.* **2010**, *285*, 9636–9641. [CrossRef] [PubMed]
58. Zhang, X.; Peng, D.; Xi, Y.; Yuan, C.; Sagum, C.A.; Klein, B.J.; Tanaka, K.; Wen, H.; Kutateladze, T.G.; Li, W.; et al. G9a-Mediated Methylation of ERα Links the PHF20/MOF Histone Acetyltransferase Complex to Hormonal Gene Expression. *Nat. Commun.* **2016**, *7*, 1–12. [CrossRef] [PubMed]
59. Chae, Y.C.; Kim, J.Y.; Park, J.W.; Kim, K.B.; Oh, H.; Lee, K.H.; Seo, S.B. FOXO1 Degradation via G9a-Mediated Methylation Promotes Cell Proliferation in Colon Cancer. *Nucleic Acids Res.* **2019**, *47*, 1692–1705. [CrossRef] [PubMed]
60. Moore, K.E.; Carlson, S.M.; Camp, N.D.; Cheung, P.; James, R.G.; Chua, K.F.; Wolf-Yadlin, A.; Gozani, O. A General Molecular Affinity Strategy for Global Detection and Proteomic Analysis of Lysine Methylation. *Mol. Cell* **2013**, *50*, 444–456. [CrossRef]
61. Lee, J.S.; Kim, Y.; Bhin, J.; Shin, H.-J.R.; Nam, H.J.; Lee, H.; Yoon, J.-B.; Binda, O.; Hwang, D.; Baek, S.H. Hypoxia-Induced Methylation of a Pontin Chromatin Remodeling Factor. *Proc. Natl. Acad. Sci. USA* **2011**, *108*, 13510–13515. [CrossRef] [PubMed]
62. Lee, J.S.; Kim, Y.; Kim, I.S.; Kim, B.; Choi, H.J.; Lee, J.M.; Shin, H.J.R.; Kim, J.H.; Kim, J.Y.; Seo, S.B.; et al. Negative Regulation of Hypoxic Responses via Induced Reptin Methylation. *Mol. Cell* **2010**, *39*, 71–85. [CrossRef] [PubMed]
63. Bao, L.; Chen, Y.; Lai, H.T.; Wu, S.Y.; Wang, J.E.; Hatanpaa, K.J.; Raisanen, J.M.; Fontenot, M.; Lega, B.; Chiang, C.M.; et al. Methylation of Hypoxia-Inducible Factor (HIF)-1α by G9a/GLP Inhibits HIF-1 Transcriptional Activity and Cell Migration. *Nucleic Acids Res.* **2018**, *46*, 6576–6591. [CrossRef] [PubMed]
64. Nair, S.S.; Li, D.Q.; Kumar, R. A Core Chromatin Remodeling Factor Instructs Global Chromatin Signaling through Multivalent Reading of Nucleosome Codes. *Mol. Cell* **2013**, *49*, 704–718. [CrossRef]
65. Tsusaka, T.; Kikuchi, M.; Shimazu, T.; Suzuki, T.; Sohtome, Y.; Akakabe, M.; Sodeoka, M.; Dohmae, N.; Umehara, T.; Shinkai, Y. Tri-Methylation of ATF7IP by G9a/GLP Recruits the Chromodomain Protein MPP8. *Epigenetics Chromatin* **2018**, *11*, 1–16. [CrossRef]
66. Chang, Y.; Sun, L.; Kokura, K.; Horton, J.R.; Fukuda, M.; Espejo, A.; Izumi, V.; Koomen, J.M.; Bedford, M.T.; Zhang, X.; et al. MPP8 Mediates the Interactions between DNA Methyltransferase Dnmt3a and H3K9 Methyltransferase GLP/G9a. *Nat. Commun.* **2011**, *2*, 1–10. [CrossRef]
67. Watanabe, S.; Iimori, M.; Chan, D.V.; Hara, E.; Kitao, H.; Maehara, Y. MDC1 Methylation Mediated by Lysine Methyltransferases EHMT1 and EHMT2 Regulates Active ATM Accumulation Flanking DNA Damage Sites. *Sci. Rep.* **2018**, *8*, 1–10. [CrossRef]
68. Li, W.; Wang, H.-Y.; Zhao, X.; Duan, H.; Cheng, B.; Liu, Y.; Zhao, M.; Shu, W.; Mei, Y.; Wen, Z.; et al. A Methylation-Phosphorylation Switch Determines Plk1 Kinase Activity and Function in DNA Damage Repair. *Sci. Adv.* **2019**, *5*, eaau7566. [CrossRef]
69. Ferry, L.; Fournier, L.; Tsusaka, T.; Adelmant, G.; Shimazu, T.; Matano, S.; Kirsh, O.; Amouroux, R.; Dohmae, N.; Suzuki, T.; et al. Methylation of DNA Ligase 1 by G9a/GLP Recruits UHRF1 to Replicating DNA and Regulates DNA Methylation. *Mol. Cell* **2017**, *67*, 550–565. [CrossRef]
70. Kubicek, S.; O'Sullivan, R.J.; August, E.M.; Hickey, E.R.; Zhang, Q.; Teodoro, M.L.L.; Rea, S.; Mechtler, K.; Kowalski, J.A.; Homon, C.A.; et al. Reversal of H3K9me2 by a Small-Molecule Inhibitor for the G9a Histone Methyltransferase. *Mol. Cell* **2007**, *25*, 473–481. [CrossRef]
71. Liu, F.; Chen, X.; Allali-Hassani, A.; Quinn, A.M.; Wasney, G.A.; Dong, A.; Barsyte, D.; Kozieradzki, I.; Senisterra, G.; Chau, I.; et al. Discovery of a 2,4-Diamino-7-Aminoalkoxyquinazoline as a Potent and Selective Inhibitor of Histone Lysine Methyltransferase G9a. *J. Med. Chem.* **2009**, *52*, 7950–7953. [CrossRef] [PubMed]
72. Vedadi, M.; Barsyte-Lovejoy, D.; Liu, F.; Rival-Gervier, S.; Allali-Hassani, A.; Labrie, V.; Wigle, T.J.; DiMaggio, P.A.; Wasney, G.A.; Siarheyeva, A.; et al. A Chemical Probe Selectively Inhibits G9a and GLP Methyltransferase Activity in Cells. *Nat. Chem. Biol.* **2011**, *7*, 566–574. [CrossRef] [PubMed]
73. Milite, C.; Feoli, A.; Horton, J.R.; Rescigno, D.; Cipriano, A.; Pisapia, V.; Viviano, M.; Pepe, G.; Amendola, G.; Novellino, E.; et al. Discovery of a Novel Chemotype of Histone Lysine Methyltransferase EHMT1/2 (GLP/G9a) Inhibitors: Rational Design, Synthesis, Biological Evaluation, and Co-Crystal Structure. *J. Med. Chem.* **2019**, *62*, 2666–2689. [CrossRef] [PubMed]
74. Yuan, Y.; Wang, Q.; Paulk, J.; Kubicek, S.; Kemp, M.M.; Adams, D.J.; Shamji, A.F.; Wagner, B.K.; Schreiber, S.L. A Small-Molecule Probe of the Histone Methyltransferase G9a Induces Cellular Senescence in Pancreatic Adenocarcinoma. *ACS Chem. Biol.* **2012**, *7*, 1152–1157. [CrossRef] [PubMed]
75. Nishio, H.; Walsh, M.J. CCAAT Displacement Protein/Cut Homolog Recruits G9a Histone Lysine Methyltransferase to Repress Transcription. *Proc. Natl. Acad. Sci. USA* **2004**, *101*, 11257–11262. [CrossRef]

76. Duan, Z.; Zarebski, A.; Montoya-Durango, D.; Grimes, H.L.; Horwitz, M. Gfi1 Coordinates Epigenetic Repression of $P21^{Cip/WAF1}$ by Recruitment of Histone Lysine Methyltransferase G9a and Histone Deacetylase 1. *Mol. Cell. Biol.* **2005**, *25*, 10338–10351. [CrossRef]
77. Győry, I.; Wu, J.; Fejér, G.; Seto, E.; Wright, K.L. PRDI-BF1 Recruits the Histone H3 Methyltransferase G9a in Transcriptional Silencing. *Nat. Immunol.* **2004**, *5*, 299–308. [CrossRef]
78. Roopra, A.; Qazi, R.; Schoenike, B.; Daley, T.J.; Morrison, J.F. Localized Domains of G9a-Mediated Histone Methylation Are Required for Silencing of Neuronal Genes. *Mol. Cell* **2004**, *14*, 727–738. [CrossRef]
79. Kim, J.K.; Estève, P.O.; Jacobsen, S.E.; Pradhan, S. UHRF1 Binds G9a and Participates in P21 Transcriptional Regulation in Mammalian Cells. *Nucleic Acids Res.* **2009**, *37*, 493–505. [CrossRef]
80. Nagano, T.; Mitchell, J.A.; Sanz, L.A.; Pauler, F.M.; Ferguson-Smith, A.C.; Feil, R.; Fraser, P. The Air Noncoding RNA Epigenetically Silences Transcription by Targeting G9a to Chromatin. *Science* **2008**, *322*, 1717–1720. [CrossRef]
81. Mozzetta, C.; Pontis, J.; Fritsch, L.; Robin, P.; Portoso, M.; Proux, C.; Margueron, R.; Ait-Si-Ali, S. The Histone H3 Lysine 9 Methyltransferases G9a and GLP Regulate Polycomb Repressive Complex 2-Mediated Gene Silencing. *Mol. Cell* **2014**, *53*, 277–289. [CrossRef] [PubMed]
82. Lomberk, G.; Wallrath, L.; Urrutia, R. The Heterochromatin Protein 1 Family. *Genome Biol.* **2006**, *7*, 1–8. [CrossRef] [PubMed]
83. Fritsch, L.; Robin, P.; Mathieu, J.R.R.; Souidi, M.; Hinaux, H.; Rougeulle, C.; Harel-Bellan, A.; Ameyar-Zazoua, M.; Ait-Si-Ali, S. A Subset of the Histone H3 Lysine 9 Methyltransferases Suv39h1, G9a, GLP, and SETDB1 Participate in a Multimeric Complex. *Mol. Cell* **2010**, *37*, 46–56. [CrossRef] [PubMed]
84. Epsztejn-Litman, S.; Feldman, N.; Abu-Remaileh, M.; Shufaro, Y.; Gerson, A.; Ueda, J.; Deplus, R.; Fuks, F.; Shinkai, Y.; Cedar, H.; et al. De Novo DNA Methylation Promoted by G9a Prevents Reprogramming of Embryonically Silenced Genes. *Nat. Struct. Mol. Biol.* **2008**, *15*, 1176–1183. [CrossRef] [PubMed]
85. Bittencourt, D.; Lee, B.H.; Gao, L.; Gerke, D.S.; Stallcup, M.R. Role of Distinct Surfaces of the G9a Ankyrin Repeat Domain in Histone and DNA Methylation during Embryonic Stem Cell Self-Renewal and Differentiation. *Epigenetics Chromatin* **2014**, *7*, 1–12. [CrossRef]
86. Smallwood, A.; Estève, P.O.; Pradhan, S.; Carey, M. Functional Cooperation between HP1 and DNMT1 Mediates Gene Silencing. *Genes Dev.* **2007**, *21*, 1169–1178. [CrossRef] [PubMed]
87. Oh, S.T.; Kim, K.B.; Chae, Y.C.; Kang, J.Y.; Hahn, Y.; Seo, S.B. H3K9 Histone Methyltransferase G9a-Mediated Transcriptional Activation of P21. *FEBS Lett.* **2014**, *588*, 685–691. [CrossRef]
88. Poulard, C.; Kim, H.N.; Fang, M.; Kruth, K.; Gagnieux, C.; Gerke, D.S.; Bhojwani, D.; Kim, Y.M.; Kampmann, M.; Stallcup, M.R.; et al. Relapse-Associated AURKB Blunts the Glucocorticoid Sensitivity of B Cell Acute Lymphoblastic Leukemia. *Proc. Natl. Acad. Sci. USA* **2019**, *116*, 3052–3061. [CrossRef]
89. Chaturvedi, C.P.; Somasundaram, B.; Singh, K.; Carpenedo, R.L.; Stanford, W.L.; Dilworth, F.J.; Brand, M. Maintenance of Gene Silencing by the Coordinate Action of the H3K9 Methyltransferase G9a/KMT1C and the H3K4 Demethylase Jarid1a/KDM5A. *Proc. Natl. Acad. Sci. USA* **2012**, *109*, 18845–18850. [CrossRef]
90. Tachibana, M.; Nozaki, M.; Takeda, N.; Shinkai, Y. Functional Dynamics of H3K9 Methylation during Meiotic Prophase Progression. *EMBO J.* **2007**, *26*, 3346–3359. [CrossRef]
91. Inagawa, M.; Nakajima, K.; Makino, T.; Ogawa, S.; Kojima, M.; Ito, S.; Ikenishi, A.; Hayashi, T.; Schwartz, R.J.; Nakamura, K.; et al. Histone H3 Lysine 9 Methyltransferases, G9a and GLP Are Essential for Cardiac Morphogenesis. *Mech. Dev.* **2013**, *130*, 519–531. [CrossRef] [PubMed]
92. Schaefer, A.; Sampath, S.C.; Intrator, A.; Min, A.; Gertler, T.S.; Surmeier, D.J.; Tarakhovsky, A.; Greengard, P. Control of Cognition and Adaptive Behavior by the GLP/G9a Epigenetic Suppressor Complex. *Neuron* **2009**, *64*, 678–691. [CrossRef] [PubMed]
93. Nicolay-Kritter, K.; Lassalle, J.; Guillou, J.L.; Mons, N. The Histone H3 Lysine 9 Methyltransferase G9a/GLP Complex Activity Is Required for Long-Term Consolidation of Spatial Memory in Mice. *Neurobiol. Learn. Mem.* **2021**, *179*, 107406. [CrossRef]
94. Maze, I.; Covington, H.E.; Dietz, D.M.; LaPlant, Q.; Renthal, W.; Russo, S.J.; Mechanic, M.; Mouzon, E.; Neve, R.L.; Haggarty, S.J.; et al. Essential Role of the Histone Methyltransferase G9a in Cocaine-Induced Plasticity. *Science* **2010**, *327*, 213–216. [CrossRef]
95. Maze, I.; Chaudhury, D.; Dietz, D.M.; von Schimmelmann, M.; Kennedy, P.J.; Lobo, M.K.; Sillivan, S.E.; Miller, M.L.; Bagot, R.C.; Sun, H.; et al. G9a Influences Neuronal Subtype Specification in Striatum. *Nat. Neurosci.* **2014**, *17*, 533–539. [CrossRef] [PubMed]
96. Kamiunten, T.; Ideno, H.; Shimada, A.; Arai, Y.; Terashima, T.; Tomooka, Y.; Nakamura, Y.; Nakashima, K.; Kimura, H.; Shinkai, Y.; et al. Essential Roles of G9a in Cell Proliferation and Differentiation during Tooth Development. *Exp. Cell Res.* **2017**, *357*, 202–210. [CrossRef]
97. Ideno, H.; Nakashima, K.; Komatsu, K.; Araki, R.; Abe, M.; Arai, Y.; Kimura, H.; Shinkai, Y.; Tachibana, M.; Nifuji, A. G9a Is Involved in the Regulation of Cranial Bone Formation through Activation of Runx2 Function during Development. *Bone* **2020**, *137*, 115332. [CrossRef] [PubMed]
98. Higashihori, N.; Lehnertz, B.; Sampaio, A.; Underhill, T.M.; Rossi, F.; Richman, J.M. Methyltransferase G9A Regulates Osteogenesis via Twist Gene Repression. *J. Dent. Res.* **2017**, *96*, 1136–1144. [CrossRef]
99. Komori, T. Regulation of Proliferation, Differentiation and Functions of Osteoblasts by Runx2. *Int. J. Mol. Sci.* **2019**, *20*, 1694. [CrossRef]
100. Zhang, R.H.; Judson, R.N.; Liu, D.Y.; Kast, J.; Rossi, F.M.V. The Lysine Methyltransferase Ehmt2/G9a Is Dispensable for Skeletal Muscle Development and Regeneration. *Skelet. Muscle* **2016**, *6*, 1–10. [CrossRef]

101. Katoh, K.; Yamazaki, R.; Onishi, A.; Sanuki, R.; Furukawa, T. G9a Histone Methyltransferase Activity in Retinal Progenitors Is Essential for Proper Differentiation and Survival of Mouse Retinal Cells. *J. Neurosci.* **2012**, *32*, 17658–17670. [CrossRef]
102. Olsen, J.B.; Wong, L.; Deimling, S.; Miles, A.; Guo, H.; Li, Y.; Zhang, Z.; Greenblatt, J.F.; Emili, A.; Tropepe, V. G9a and ZNF644 Physically Associate to Suppress Progenitor Gene Expression during Neurogenesis. *Stem Cell Rep.* **2016**, *7*, 454–470. [CrossRef]
103. Chen, H.; Yan, Y.; Davidson, T.L.; Shinkai, Y.; Costa, M. Hypoxic Stress Induces Dimethylated Histone H3 Lysine 9 through Histone Methyltransferase G9a in Mammalian Cells. *Cancer Res.* **2006**, *66*, 9009–9016. [CrossRef]
104. Macůrek, L.; Lindqvist, A.; Lim, D.; Lampson, M.A.; Klompmaker, R.; Freire, R.; Clouin, C.; Taylor, S.S.; Yaffe, M.B.; Medema, R.H. Polo-like Kinase-1 Is Activated by Aurora A to Promote Checkpoint Recovery. *Nature* **2008**, *455*, 119–123. [CrossRef]
105. Agarwal, P.; Jackson, S.P. G9a Inhibition Potentiates the Anti-Tumour Activity of DNA Double-Strand Break Inducing Agents by Impairing DNA Repair Independent of P53 Status. *Cancer Lett.* **2016**, *380*, 467–475. [CrossRef]
106. Zhang, J.; Wang, Y.; Shen, Y.; He, P.; Ding, J.; Chen, Y. G9a Stimulates CRC Growth by Inducing P53 Lys373 Dimethylation-Dependent Activation of Plk1. *Theranostics* **2018**, *8*, 2884–2895. [CrossRef]
107. Sharma, S.; Kelly, T.K.; Jones, P.A. Epigenetics in Cancer. *Carcinogenesis* **2010**, *31*, 27–36. [CrossRef] [PubMed]
108. Campbell, M.J.; Turner, B.M. Altered Histone Modifications in Cancer. *Adv. Exp. Med. Biol.* **2013**, *754*, 81–107. [CrossRef] [PubMed]
109. Wang, Y.; Zhang, J.; Su, Y.; Shen, Y.; Jiang, D.; Hou, Y.; Geng, M.; Ding, J.; Chen, Y. G9a Regulates Breast Cancer Growth by Modulating Iron Homeostasis through the Repression of Ferroxidase Hephaestin. *Nat. Commun.* **2017**, *8*, 1–14. [CrossRef] [PubMed]
110. Chen, P.; Qian, Q.; Zhu, Z.; Shen, X.; Yu, S.; Yu, Z.; Sun, R.; Li, Y.; Guo, D.; Fan, H. Increased Expression of EHMT2 Associated with H3K9me2 Level Contributes to the Poor Prognosis of Gastric Cancer. *Oncol. Lett.* **2020**, *20*, 1734–1742. [CrossRef]
111. Hua, K.-T.; Wang, M.-Y.; Chen, M.-W.; Wei, L.-H.; Chen, C.-K.; Ko, C.-H.; Jeng, Y.-M.; Sung, P.-L.; Jan, Y.-H.; Hsiao, M.; et al. The H3K9 Methyltransferase G9a Is a Marker of Aggressive Ovarian Cancer That Promotes Peritoneal Metastasis. *Mol. Cancer* **2014**, *13*, 1–13. [CrossRef]
112. Chen, G.; Yu, X.; Zhang, M.; Zheng, A.; Wang, Z.; Zuo, Y.; Liang, Q.; Jiang, D.; Chen, Y.; Zhao, L.; et al. Inhibition of Euchromatic Histone Lysine Methyltransferase 2 (EHMT2) Suppresses the Proliferation and Invasion of Cervical Cancer Cells. *Cytogenet. Genome Res.* **2019**, *158*, 205–212. [CrossRef]
113. Hsiao, S.M.; Chen, M.W.; Chen, C.A.; Chien, M.H.; Hua, K.T.; Hsiao, M.; Kuo, M.L.; Wei, L.H. The H3K9 Methyltransferase G9a Represses E-Cadherin and Is Associated with Myometrial Invasion in Endometrial Cancer. *Ann. Surg. Oncol.* **2015**, *22*, 1556–1565. [CrossRef]
114. Fan, H.T.; Shi, Y.Y.; Lin, Y.; Yang, X.P. EHMT2 Promotes the Development of Prostate Cancer by Inhibiting PI3K/AKT/MTOR Pathway. *Eur. Rev. Med. Pharmacol. Sci.* **2019**, *23*, 7808–7815. [CrossRef]
115. Qin, J.; Zeng, Z.; Luo, T.; Li, Q.; Hao, Y.; Chen, L. Clinicopathological Significance of G9A Expression in Colorectal Carcinoma. *Oncol. Lett.* **2018**, *15*, 8611–8619. [CrossRef] [PubMed]
116. Wei, L.; Chiu, D.K.C.; Tsang, F.H.C.; Law, C.T.; Cheng, C.L.H.; Au, S.L.K.; Lee, J.M.F.; Wong, C.C.L.; Ng, I.O.L.; Wong, C.M. Histone Methyltransferase G9a Promotes Liver Cancer Development by Epigenetic Silencing of Tumor Suppressor Gene RARRES3. *J. Hepatol.* **2017**, *67*, 758–769. [CrossRef]
117. Guo, A.S.; Huang, Y.Q.; Ma, X.D.; Lin, R.S. Mechanism of G9a Inhibitor BIX-01294 Acting on U251 Glioma Cells. *Mol. Med. Rep.* **2016**, *14*, 4613–4621. [CrossRef] [PubMed]
118. Zhang, X.Y.; Rajagopalan, D.; Chung, T.H.; Hooi, L.; Toh, T.B.; Tian, J.S.; Rashid, M.B.M.A.; Sahib, N.R.B.M.; Gu, M.; Lim, J.J.; et al. Frequent Upregulation of G9a Promotes RelB-Dependent Proliferation and Survival in Multiple Myeloma. *Exp. Hematol. Oncol.* **2020**, *9*, 8. [CrossRef] [PubMed]
119. Dang, N.N.; Jiao, J.; Meng, X.; An, Y.; Han, C.; Huang, S. Abnormal Overexpression of G9a in Melanoma Cells Promotes Cancer Progression via Upregulation of the Notch1 Signaling Pathway. *Aging* **2020**, *12*, 2393–2407. [CrossRef]
120. Ma, W.; Han, C.; Zhang, J.; Song, K.; Chen, W.; Kwon, H.; Wu, T. The Histone Methyltransferase G9a Promotes Cholangiocarcinogenesis through Regulation of the Hippo Pathway Kinase LATS2 and YAP Signaling Pathway. *FASEB J.* **2020**, *34*, 1283–1297. [CrossRef]
121. Zhang, K.; Wang, J.; Yang, L.; Yuan, Y.-C.; Tong, T.R.; Wu, J.; Yun, X.; Bonner, M.; Pangeni, R.; Liu, Z.; et al. Targeting Histone Methyltransferase G9a Inhibits Growth and Wnt Signaling Pathway by Epigenetically Regulating HP1α and APC2 Gene Expression in Non-Small Cell Lung Cancer. *Mol. Cancer* **2018**, *17*, 1–15. [CrossRef]
122. Cui, J.; Sun, W.; Hao, X.; Wei, M.; Su, X.; Zhang, Y.; Su, L.; Liu, X. EHMT2 Inhibitor BIX-01294 Induces Apoptosis through PMAIP1-USP9X-MCL1 Axis in Human Bladder Cancer Cells. *Cancer Cell Int.* **2015**, *15*, 1–9. [CrossRef]
123. Casciello, F.; Al-Ejeh, F.; Miranda, M.; Kelly, G.; Baxter, E.; Windloch, K.; Gannon, F.; Lee, J.S. G9a-Mediated Repression of CDH10 in Hypoxia Enhances Breast Tumour Cell Motility and Associates with Poor Survival Outcome. *Theranostics* **2020**, *10*, 4515–4529. [CrossRef]
124. Chen, M.-W.; Hua, K.-T.; Kao, H.-J.; Chi, C.-C.; Wei, L.-H.; Johansson, G.; Shiah, S.-G.; Chen, P.-S.; Jeng, Y.-M.; Cheng, T.-Y.; et al. H3K9 Histone Methyltransferase G9a Promotes Lung Cancer Invasion and Metastasis by Silencing the Cell Adhesion Molecule Ep-CAM. *Cancer Res.* **2010**, *70*, 7830–7840. [CrossRef]
125. Yin, C.; Ke, X.; Zhang, R.; Hou, J.; Dong, Z.; Wang, F.; Zhang, K.; Zhong, X.; Yang, L.; Cui, H. G9a Promotes Cell Proliferation and Suppresses Autophagy in Gastric Cancer by Directly Activating MTOR. *FASEB J.* **2019**, *33*, 14036–14050. [CrossRef]

126. Zhong, X.; Chen, X.; Guan, X.; Zhang, H.; Ma, Y.; Zhang, S.; Wang, E.; Zhang, L.; Han, Y. Overexpression of G9a and MCM7 in Oesophageal Squamous Cell Carcinoma Is Associated with Poor Prognosis. *Histopathology* **2015**, *66*, 192–200. [CrossRef]
127. Ho, J.C.; Abdullah, L.N.; Pang, Q.Y.; Jha, S.; Chow, E.K.H.; Yang, H.; Kato, H.; Poellinger, L.; Ueda, J.; Lee, K.L. Inhibition of the H3K9 Methyltransferase G9A Attenuates Oncogenicity and Activates the Hypoxia Signaling Pathway. *PLoS ONE* **2017**, *12*, e0188051. [CrossRef]
128. Liu, X.R.; Zhou, L.H.; Hu, J.X.; Liu, L.M.; Wan, H.P.; Zhang, X.Q. UNC0638, a G9a Inhibitor, Suppresses Epithelial-Mesenchymal Transition-Mediated Cellular Migration and Invasion in Triple Negative Breast Cancer. *Mol. Med. Rep.* **2018**, *17*, 2239–2244. [CrossRef] [PubMed]
129. Tu, W.B.; Shiah, Y.J.; Lourenco, C.; Mullen, P.J.; Dingar, D.; Redel, C.; Tamachi, A.; Ba-Alawi, W.; Aman, A.; Al-awar, R.; et al. MYC Interacts with the G9a Histone Methyltransferase to Drive Transcriptional Repression and Tumorigenesis. *Cancer Cell* **2018**, *34*, 579–595. [CrossRef] [PubMed]
130. Wozniak, R.J.; Klimecki, W.T.; Lau, S.S.; Feinstein, Y.; Futscher, B.W. 5-Aza-2′-Deoxycytidine-Mediated Reductions in G9A Histone Methyltransferase and Histone H3 K9 Di-Methylation Levels Are Linked to Tumor Suppressor Gene Reactivation. *Oncogene* **2007**, *26*, 77–90. [CrossRef] [PubMed]
131. Dong, C.; Wu, Y.; Yao, J.; Wang, Y.; Yu, Y.; Rychahou, P.G.; Evers, B.M.; Zhou, B.P. G9a Interacts with Snail and Is Critical for Snail-Mediated E-Cadherin Repression in Human Breast Cancer. *J. Clin. Investig.* **2012**, *122*, 1469–1486. [CrossRef] [PubMed]
132. Zhang, J.; Yao, D.; Jiang, Y.; Huang, J.; Yang, S.; Wang, J. Synthesis and Biological Evaluation of Benzimidazole Derivatives as the G9a Histone Methyltransferase Inhibitors That Induce Autophagy and Apoptosis of Breast Cancer Cells. *Bioorg. Chem.* **2017**, *72*, 168–181. [CrossRef]
133. Mabe, N.W.; Garcia, N.M.G.; Wolery, S.E.; Newcomb, R.; Meingasner, R.C.; Vilona, B.A.; Lupo, R.; Lin, C.C.; Chi, J.T.; Alvarez, J.V. G9a Promotes Breast Cancer Recurrence through Repression of a Pro-Inflammatory Program. *Cell Rep.* **2020**, *33*, 108341. [CrossRef]
134. Lin, X.; Huang, Y.; Zou, Y.; Chen, X.; Ma, X. Depletion of G9a Gene Induces Cell Apoptosis in Human Gastric Carcinoma. *Oncol. Rep.* **2016**, *35*, 3041–3049. [CrossRef] [PubMed]
135. Kim, T.W.; Lee, S.Y.; Kim, M.; Cheon, C.; Ko, S.-G. Kaempferol Induces Autophagic Cell Death via IRE1-JNK-CHOP Pathway and Inhibition of G9a in Gastric Cancer Cells. *Cell Death Dis.* **2018**, *9*, 1–14. [CrossRef]
136. Kim, T.W.; Cheon, C.; Ko, S.-G. SH003 Activates Autophagic Cell Death by Activating ATF4 and Inhibiting G9a under Hypoxia in Gastric Cancer Cells. *Cell Death Dis.* **2020**, *11*, 1–14. [CrossRef] [PubMed]
137. Kim, T.W. Cinnamaldehyde Induces Autophagy-Mediated Cell Death through ER Stress and Epigenetic Modification in Gastric Cancer Cells. *Acta Pharmacol. Sin.* **2021**, *2021*, 1–12. [CrossRef]
138. Lee, S.H.; Kim, J.; Kim, W.-H.; Lee, Y.M. Hypoxic Silencing of Tumor Suppressor RUNX3 by Histone Modification in Gastric Cancer Cells. *Oncogene* **2009**, *28*, 184–194. [CrossRef]
139. Hu, L.; Zang, M.; Wang, H.; Zhang, B.; Wang, Z.; Fan, Z.; Wu, H.; Li, J.; Su, L.; Yan, M.; et al. G9A Promotes Gastric Cancer Metastasis by Upregulating ITGB3 in a SET Domain-Independent Manner. *Cell Death Dis.* **2018**, *9*, 1–14. [CrossRef]
140. Watson, Z.L.; Yamamoto, T.M.; McMellen, A.; Kim, H.; Hughes, C.J.; Wheeler, L.J.; Post, M.D.; Behbakht, K.; Bitler, B.G. Histone Methyltransferases EHMT1 and EHMT2 (GLP/G9A) Maintain PARP Inhibitor Resistance in High-Grade Serous Ovarian Carcinoma. *Clin. Epigenetics* **2019**, *11*, 1–16. [CrossRef]
141. Liu, M.; Thomas, S.L.; DeWitt, A.K.; Zhou, W.; Madaj, Z.B.; Ohtani, H.; Baylin, S.B.; Liang, G.; Jones, P.A. Dual Inhibition of DNA and Histone Methyltransferases Increases Viral Mimicry in Ovarian Cancer Cells. *Cancer Res.* **2018**, *78*, 5754–5766. [CrossRef] [PubMed]
142. Chen, R.-J.; Shun, C.-T.; Yen, M.-L.; Chou, C.-H.; Lin, M.-C.; Chen, R.-J.; Shun, C.-T.; Yen, M.-L.; Chou, C.-H.; Lin, M.-C. Methyltransferase G9a Promotes Cervical Cancer Angiogenesis and Decreases Patient Survival. *Oncotarget* **2017**, *8*, 62081–62098. [CrossRef] [PubMed]
143. Watanabe, H.; Soejima, K.; Yasuda, H.; Kawada, I.; Nakachi, I.; Yoda, S.; Naoki, K.; Ishizaka, A. Deregulation of Histone Lysine Methyltransferases Contributes to Oncogenic Transformation of Human Bronchoepithelial Cells. *Cancer Cell Int.* **2008**, *8*, 1–12. [CrossRef]
144. Huang, T.; Zhang, P.; Li, W.; Zhao, T.; Zhang, Z.; Chen, S.; Yang, Y.; Feng, Y.; Li, F.; Shirley Liu, X.; et al. G9A Promotes Tumor Cell Growth and Invasion by Silencing CASP1 in Non-Small-Cell Lung Cancer Cells. *Cell Death Dis.* **2017**, *8*, e2726. [CrossRef]
145. Hu, Y.; Zheng, Y.; Dai, M.; Wu, J.; Yu, B.; Zhang, H.; Kong, W.; Wu, H.; Yu, X. Snail2 Induced E-Cadherin Suppression and Metastasis in Lung Carcinoma Facilitated by G9a and HDACs. *Cell Adhes. Migr.* **2019**, *13*, 285–292. [CrossRef]
146. Sun, T.; Zhang, K.; Pangeni, R.P.; Wu, J.; Li, W.; Du, Y.; Guo, Y.; Chaurasiya, S.; Arvanitis, L.; Raz, D.J. G9a Promotes Invasion and Metastasis of Non–Small Cell Lung Cancer through Enhancing Focal Adhesion Kinase Activation via NF-κB Signaling Pathway. *Mol. Cancer Res.* **2021**, *19*, 429–440. [CrossRef] [PubMed]
147. Pangeni, R.P.; Yang, L.; Zhang, K.; Wang, J.; Li, W.; Guo, C.; Yun, X.; Sun, T.; Wang, J.; Raz, D.J. G9a Regulates Tumorigenicity and Stemness through Genome-Wide DNA Methylation Reprogramming in Non-Small Cell Lung Cancer. *Clin. Epigenetics* **2020**, *12*, 1–17. [CrossRef] [PubMed]
148. Bergin, C.J.; Zouggar, A.; Haebe, J.R.; Masibag, A.N.; Desrochers, F.M.; Reilley, S.Y.; Agrawal, G.; Benoit, Y.D. G9a Controls Pluripotent-like Identity and Tumor-Initiating Function in Human Colorectal Cancer. *Oncogene* **2020**, *40*, 1191–1202. [CrossRef]

149. Bárcena-Varela, M.; Caruso, S.; Llerena, S.; Álvarez-Sola, G.; Uriarte, I.; Latasa, M.U.; Urtasun, R.; Rebouissou, S.; Alvarez, L.; Jimenez, M.; et al. Dual Targeting of Histone Methyltransferase G9a and DNA-Methyltransferase 1 for the Treatment of Experimental Hepatocellular Carcinoma. *Hepatology* **2019**, *69*, 587–603. [CrossRef]
150. Yokoyama, M.; Chiba, T.; Zen, Y.; Oshima, M.; Kusakabe, Y.; Noguchi, Y.; Yuki, K.; Koide, S.; Tara, S.; Saraya, A.; et al. Histone Lysine Methyltransferase G9a Is a Novel Epigenetic Target for the Treatment of Hepatocellular Carcinoma. *Oncotarget* **2017**, *8*, 21315–21326. [CrossRef]
151. Hu, Y.; Zheng, Y.; Dai, M.; Wang, X.; Wu, J.; Yu, B.; Zhang, H.; Cui, Y.; Kong, W.; Wu, H.; et al. G9a and Histone Deacetylases Are Crucial for Snail2-Mediated E-Cadherin Repression and Metastasis in Hepatocellular Carcinoma. *Cancer Sci.* **2019**, *110*, 3442–3452. [CrossRef]
152. Nakatsuka, T.; Tateishi, K.; Kato, H.; Fujiwara, K.; Yamamoto, K.; Kudo, Y.; Nakagawa, H.; Tanaka, Y.; Ijichi, H.; Ikenoue, T.; et al. Inhibition of Histone Methyltransferase G9a Attenuates Liver Cancer Initiation by Sensitizing DNA-Damaged Hepatocytes to P53-Induced Apoptosis. *Cell Death Dis.* **2021**, *12*, 1–13. [CrossRef]
153. Segovia, C.; San José-Enériz, E.; Munera-Maravilla, G.; Martínez-Fernández, M.; Garate, L.; Miranda, E.; Vilas-Zornoza, A.; Lodewijk, I.; Rubio, C.; Segrelles, C.; et al. Inhibition of a G9a/DNMT Network Triggers Immune-Mediated Bladder Cancer Regression. *Nat. Med.* **2019**, *25*, 1073–1081. [CrossRef] [PubMed]
154. Alves-Silva, J.C.; de Carvalho, J.L.; Rabello, D.A.; Serejo, T.R.T.; Rego, E.M.; Neves, F.A.R.; Lucena-Araujo, A.R.; Pittella-Silva, F.; Saldanha-Araujo, F. GLP Overexpression Is Associated with Poor Prognosis in Chronic Lymphocytic Leukemia and Its Inhibition Induces Leukemic Cell Death. *Investig. New Drugs* **2018**, *36*, 955–960. [CrossRef]
155. Huang, Y.; Zou, Y.; Lin, L.; Ma, X.; Huang, X. Effect of BIX-01294 on Proliferation, Apoptosis and Histone Methylation of Acute T Lymphoblastic Leukemia Cells. *Leuk. Res.* **2017**, *62*, 34–39. [CrossRef] [PubMed]
156. Kondengaden, S.M.; Luo, L.F.; Huang, K.; Zhu, M.; Zang, L.; Bataba, E.; Wang, R.; Luo, C.; Wang, B.; Li, K.K.; et al. Discovery of Novel Small Molecule Inhibitors of Lysine Methyltransferase G9a and Their Mechanism in Leukemia Cell Lines. *Eur. J. Med. Chem.* **2016**, *122*, 382–393. [CrossRef]
157. Lehnertz, B.; Pabst, C.; Su, L.; Miller, M.; Liu, F.; Yi, L.; Zhang, R.; Krosl, J.; Yung, E.; Kirschner, J.; et al. The Methyltransferase G9a Regulates HoxA9-Dependent Transcription in AML. *Genes Dev.* **2014**, *28*, 317–327. [CrossRef]
158. San José-Enériz, E.; Agirre, X.; Rabal, O.; Vilas-Zornoza, A.; Sanchez-Arias, J.A.; Miranda, E.; Ugarte, A.; Roa, S.; Paiva, B.; Estella-Hermoso de Mendoza, A.; et al. Discovery of First-in-Class Reversible Dual Small Molecule Inhibitors against G9a and DNMTs in Hematological Malignancies. *Nat. Commun.* **2017**, *8*, 1–10. [CrossRef]
159. Madrazo, E.; Ruano, D.; Abad, L.; Alonso-Gómez, E.; Sánchez-Valdepeñas, C.; González-Murillo, Á.; Ramírez, M.; Redondo-Muñoz, J. G9a Correlates with VLA-4 Integrin and Influences the Migration of Childhood Acute Lymphoblastic Leukemia Cells. *Cancers* **2018**, *10*, 325. [CrossRef]
160. Nagaraja, S.G.S.; Subramanian, U.; Nagarajan, D. Radiation-Induced H3K9 Methylation on E-Cadherin Promoter Mediated by ROS/Snail Axis: Role of G9a Signaling during Lung Epithelial-Mesenchymal Transition. *Toxicol. Vitr.* **2021**, *70*, 105037. [CrossRef] [PubMed]
161. Li, Y.; Chen, Z.; Cao, K.; Zhang, L.; Ma, Y.; Yu, S.; Jin, H.; Liu, X.; Li, W. G9a Regulates Cell Sensitivity to Radiotherapy via Histone H3 Lysine 9 Trimethylation and Ccdc8 in Lung Cancer. *OncoTargets Ther.* **2021**, *14*, 3721–3728. [CrossRef]
162. Luo, C.W.; Wang, J.Y.; Hung, W.C.; Peng, G.; Tsai, Y.L.; Chang, T.M.; Chai, C.Y.; Lin, C.H.; Pan, M.R. G9a Governs Colon Cancer Stem Cell Phenotype and Chemoradioresistance through PP2A-RPA Axis-Mediated DNA Damage Response. *Radiother. Oncol.* **2017**, *124*, 395–402. [CrossRef]
163. Dobson, T.H.W.; Hatcher, R.J.; Swaminathan, J.; Das, C.M.; Shaik, S.; Tao, R.-H.; Milite, C.; Castellano, S.; Taylor, P.H.; Sbardella, G.; et al. Regulation of USP37 Expression by REST-Associated G9a-Dependent Histone Methylation. *Mol. Cancer Res.* **2017**, *15*, 1073–1084. [CrossRef] [PubMed]
164. Ciechomska, I.A.; Przanowski, P.; Jackl, J.; Wojtas, B.; Kaminska, B. BIX01294, an Inhibitor of Histone Methyltransferase, Induces Autophagy-Dependent Differentiation of Glioma Stem-like Cells. *Sci. Rep.* **2016**, *6*, 1–15. [CrossRef]
165. Kato, S.; Weng, Q.Y.; Insco, M.L.; Chen, K.Y.; Muralidhar, S.; Pozniak, J.; Diaz, J.M.S.; Drier, Y.; Nguyen, N.; Lo, J.A.; et al. Gain-of-Function Genetic Alterations of G9a Drive Oncogenesis. *Cancer Discov.* **2020**, *10*, 980–997. [CrossRef]
166. Liu, S.; Ye, D.; Guo, W.; Yu, W.; He, Y.; Hu, J.; Wang, Y.; Zhang, L.; Liao, Y.; Song, H.; et al. G9a Is Essential for EMT-Mediated Metastasis and Maintenance of Cancer Stem Cell-like Characters in Head and Neck Squamous Cell Carcinoma. *Oncotarget* **2015**, *6*, 6887–6901. [CrossRef] [PubMed]
167. Rowbotham, S.P.; Li, F.; Dost, A.F.M.; Louie, S.M.; Marsh, B.P.; Pessina, P.; Anbarasu, C.R.; Brainson, C.F.; Tuminello, S.J.; Lieberman, A.; et al. H3K9 Methyltransferases and Demethylases Control Lung Tumor-Propagating Cells and Lung Cancer Progression. *Nat. Commun.* **2018**, *9*, 1–13. [CrossRef] [PubMed]
168. Ciechomska, I.A.; Marciniak, M.P.; Jackl, J.; Kaminska, B. Pre-Treatment or Post-Treatment of Human Glioma Cells with BIX01294, the Inhibitor of Histone Methyltransferase G9a, Sensitizes Cells to Temozolomide. *Front. Pharmacol.* **2018**, *9*, 1271. [CrossRef]

Review

Structure, Activity and Function of the SETDB1 Protein Methyltransferase

Mariam Markouli [†], Dimitrios Strepkos [†] and Christina Piperi *

Department of Biological Chemistry, Medical School, National and Kapodistrian University of Athens, 11527 Athens, Greece; myriam.markouli@gmail.com (M.M.); smd1700150@uoa.gr (D.S.)
* Correspondence: cpiperi@med.uoa.gr; Tel.: +30-210-7462610
† Equal contribution.

Abstract: The SET Domain Bifurcated Histone Lysine Methyltransferase 1 (SETDB1) is a prominent member of the Suppressor of Variegation 3–9 (SUV39)-related protein lysine methyltransferases (PKMTs), comprising three isoforms that differ in length and domain composition. SETDB1 is widely expressed in human tissues, methylating Histone 3 lysine 9 (H3K9) residues, promoting chromatin compaction and exerting negative regulation on gene expression. SETDB1 has a central role in normal physiology and nervous system development, having been implicated in the regulation of cell cycle progression, inactivation of the X chromosome, immune cells function, expression of retroelements and formation of promyelocytic leukemia (PML) nuclear bodies (NB). SETDB1 has been frequently deregulated in carcinogenesis, being implicated in the pathogenesis of gliomas, melanomas, as well as in lung, breast, gastrointestinal and ovarian tumors, where it mainly exerts an oncogenic role. Aberrant activity of SETDB1 has also been implicated in several neuropsychiatric, cardiovascular and gastrointestinal diseases, including schizophrenia, Huntington's disease, congenital heart defects and inflammatory bowel disease. Herein, we provide an update on the unique structural and biochemical features of SETDB1 that contribute to its regulation, as well as its molecular and cellular impact in normal physiology and disease with potential therapeutic options.

Keywords: SETDB1; methyltransferase; epigenetics; cancer; schizophrenia; Huntington's disease; Rett syndrome; Prader–Willi syndrome; congenital heart diseases; inflammatory bowel disease

Citation: Markouli, M.; Strepkos, D.; Piperi, C. Structure, Activity and Function of the SETDB1 Protein Methyltransferase. *Life* **2021**, *11*, 817. https://doi.org/10.3390/life11080817

Academic Editors: Albert Jeltsch and Arunkumar Dhayalan

Received: 27 July 2021
Accepted: 9 August 2021
Published: 11 August 2021

Publisher's Note: MDPI stays neutral with regard to jurisdictional claims in published maps and institutional affiliations.

Copyright: © 2021 by the authors. Licensee MDPI, Basel, Switzerland. This article is an open access article distributed under the terms and conditions of the Creative Commons Attribution (CC BY) license (https://creativecommons.org/licenses/by/4.0/).

1. Introduction

The SET Domain Bifurcated Histone Lysine Methyltransferase 1 (SETDB1) belongs to the family of the Suppressor of Variegation 3–9 (SUV39) proteins [1], representing a member of the group of SET domain-containing protein lysine methyltransferases (PKMTs) which are implicated in epigenetic regulation. SETDB1 is characterized by a highly conserved bifurcated SET domain which contains an intercepting sequence of approximately 150 amino acids [2]. The SET domain was first discovered in the *Suppressor of Variegation 3–9 (SUV3–9)*, *Enhancer of Zeste* (*EZ*) and *Trithorax* genes of *Drosophila* sp. [3]. Subsequent research detected SET domains in more than 40 species, including *Saccharomyces cerevisiae* (*SET1* gene), *S. pombe* (*Clr4*+ gene) and humans (*SETDB1* as well as other SET domain-containing HKMT genes) [3,4]. Moreover, in *Caenorhabditis elegans*, the *YNCA* gene product exerts high similarity to SETDB1, also containing a bifurcated SET domain. Furthermore, SETDB1 orthologs have been studied in several other species, a few of which are *Mus musculus*, *Rattus norvegicus*, *Danio rerio*, *Bos taurus* and *Macaca mullata*, among others [5].

SETDB1 is involved in a variety of physiologic as well as pathologic processes. Its ability to interact with many different genes at the same time is a product of epigenetic regulation. The term epigenetics is used to describe changes in gene expression without involving alterations in the DNA sequence itself. It includes modifications in proteins called histones, around which the DNA is wrapped, which can influence gene expression by affecting the level of chromatin compaction and, thus, its transcriptional accessibility.

Histones form an octamer that is included in the structure of chromatin and is always comprised of histone H2A, H2B, H3 and H4 duplicates. In this way, SETDB1 is able to influence the expression of a multitude of genes by di-/or trimethylating the lysine 9 (K9) residue of the H3 protein located throughout different chromatin regions. It is therefore evident that SETDB1 functions as a chromatin regulator, mediating H3K9 di-/trimethylation. This histone mark functions as a repressive histone post-translational modification by indirectly increasing chromatin's compaction after recruiting Heterochromatin protein 1 (HP1), thus decreasing its accessibility and affecting gene expression. In more detail, the compacted form of chromatin prevents transcription factors from binding, leading to repression of gene expression or transcription [6].

SETDB1 is highly involved in a variety of physiologic functions, interacting with several proteins and transcription factors such as ETS-related gene (ERG) partners to control cell growth and differentiation. In this way, SETDB1 participates in the regulation of important cellular functions.

A complex interplay of direct and indirect interactions with other enzymes and signaling pathways has been detected to regulate SETDB1 activity in normal cells. However, SETDB1 has also been strongly implicated in the pathogenesis of multiple diseases, including neurological disorders, cardiovascular, gastrointestinal diseases and most notably, tumor development and progression. Tumorigenesis is a great example of this enzyme's functional variability since SETDB1 may acquire both a tumor-suppressive function in some tissues, as well as an oncogenic function in others, and thus can repress either tumor-suppressing or tumor-promoting genes, respectively.

In the following sections, we provide a detailed description of the structural and biochemical characteristics of SETDB1, addressing its physiologic cell functions and connection with diseases.

2. Structural Features of SETDB1

The human *SETDB1* gene (OMIM 604396), alternatively known as ESET, KG1T, KIAA0067, KMT1E or TDRD21, is located on the chromosome 1q21.3 and encodes the SETDB1 protein, composed of 1291 amino acids [7]. The *SETDB1* gene consists of 23 exons and is expressed in several human tissues such as the testes, ovaries, appendix, brain, spleen, lymph nodes and thyroid gland [8].

The domain composition of SETDB1 includes an N-terminal part which contains three Tudor domains and a methyl-CpG-binding domain (MBD) as well as a C-terminus with a pre-SET, a SET and a post-SET domain [9]. The three Tudor domains are crucial for the formation of complexes with proteins that regulate transcriptional activity via chromatin modifications, such as Histone Deacetylase 1/2 (HDAC1/2) and Kruppel-associated box-Zinc Finger Proteins-KRAB-Associated Protein-1 (KRAB-ZFP-KAP-1) [10] (Figure 1A). This is achieved by the triple Tudor domain binding to H3 tails, which contain the combination of H3K14 acetylation and H3K9 methylation [11]. Moreover, the Tudor domains regulate snRNP processing in Cajal bodies [12]. The MBD domain contains two arginine residues that contribute to DNA binding and is responsible for coupling DNA methylation with H3K9 trimethylation by interacting with DNA Methyltransferase 3 (DNMT3) and inducing gene silencing [13,14]. Additionally, the N-terminal part contains two nuclear export signals (NES) and two nuclear localization signals (NLS), which regulate the localization of SETDB1 [15].

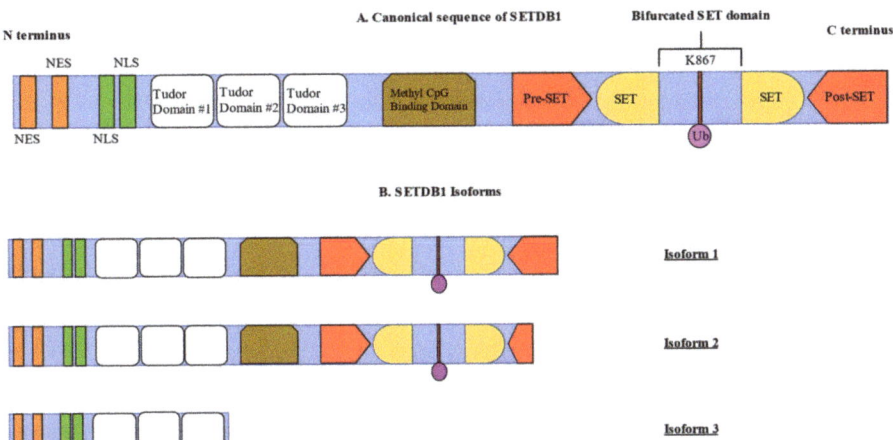

Figure 1. SETDB domain composition and isoform architecture. (**A**) The "canonical" sequence of SETDB1 is made up of an N-terminal part which contains two Nuclear Export Signal (NES) domains, two Nuclear Localization Signal (NLS) domains, the three Tudor domains and a Methyl CpG Binding (MBD) domain. The C-terminus of the SETDB1 protein contains the pre-SET, bifurcated SET and post-SET domains. The intercepting sequence of amino acids, which splits the SET domain into two parts, also becomes ubiquitinated at the K867 residue, a post-translational modification that is crucial for the protein's full functionality. (**B**) SETDB1 exists in three isoforms, only two of which exhibit enzymatic activity. Therefore, isoform 1 is the complete SETDB1 protein, while isoform 2 contains the same domains but is shorter than the first isoform due to alternative splicing. The third isoform lacks all the domains of the C-terminus, exhibiting no enzymatic activity.

The presence of pre-, post-SET and SET domains at the C- terminus is of paramount importance for the activity of protein methyltransferases. The SET domain is arranged in a helix formation which is linked to an anti-parallel two-stranded β-sheet by loops of different lengths which consist of amino acids that intercept the bifurcated SET domain [16]. This intercepting chain of amino acids, preserved through evolution, was shown to significantly regulate the activity of the SETDB1 protein. The ubiquitination at the lysine residue 867 of SETDB1 mediated by Ubiquitin-Conjugating Enzyme E2 (UBE2E) was demonstrated as a prerequisite for its full methyltransferase activity [17,18].

SETDB1 exists in three isoforms, with isoform 1 being considered as the "canonical" sequence, including all the necessary domains for full enzyme activity. Isoform 2 contains the same domains as isoform 1. However, it is produced by alternate splicing of an in-frame splice site which is present in the 3′ coding region, resulting in C-terminal truncation in the post-SET domain, thus producing a shorter protein form. Lastly, isoform 3 lacks all the domains of the C- terminus that are required for enzyme activity (Figure 1B) [19].

3. Biochemical Features of SETDB1

The main role of SETDB1 is gene silencing by di- and trimethylating H3K9 residues. In this reaction, SETDB1 utilizes the cofactor S-adenosylmethionine (SAM) as a methyl group donor, which binds to the substrate-binding site of SET [20]. Furthermore, the human homolog of murine ATFa-associated modulator (hAM) is able to induce the conversion of H3 lysine dimethylation to trimethylation and promote the gene-repressive activity of SETDB1 via a SAM-dependent mechanism. This is achieved by binding to SETDB1 thus, forming a SETDB1/hAM complex [21]. The interaction of SETDB1 with hAM, although not a prerequisite for the enzyme's function, increases its activity [21]. There is evidence that SETDB1 is guided to histone H3 by a factor that is recruited by KRAB zinc-finger proteins, namely, TRIM28/TIF1B. Furthermore, the KRAB–ZFP–KAP-1 complex is also responsible for guiding SETDB1 to H3 in repetitive elements and retrotransposons, as well as other target genes [22–24]. This is achieved by KRAB–ZFP interacting with SETDB1 in a sequence-specific manner after SETDB1 interaction with KAP1. KAP1 recruits SETDB1, as well as HP1 and the NuRD histone deacetylase complex [25]. The binding of SETDB1 to H3K9 is achieved via the three Tudor domains, which detect regions of H3 histones containing both K14 acetylation and K9 methylation [11]. Following the methylation of H3K9 by SETDB1, HP1 is recruited to chromatin and alters its structure from a euchromatic to a heterochromatic state [26]. This process is critical for the formation of Heterochromatin since HP1, in turn, works to recruit other proteins, which further establish Heterochromatin formation [27].

Concerning the regulation of SETDB1 expression and activity, several mechanisms have been proposed, which are based on the presence of NES and NLS motifs on SETDB1 protein. These two motifs can control the nuclear levels of SETDB1. Another interaction that can regulate the nuclear levels of SETDB1 is its degradation by the proteasome; however, the importance of this mechanism remains to be explored [28]. Moreover, SETDB1 gains its full enzyme activity after it has been monoubiquitinated at the K867 site, revealing an additional regulatory mechanism [17].

Furthermore, SETDB1 appears to interact with several complexes that can regulate its activity or mediate its functions. In more detail, its activity has been shown to be tightly bound to that of MBD1. SETDB1 can form a complex with MBD1 and ATF7IP, which represses the transcription and couples DNA methylation with H3K9 trimethylation [26,29]. In fact, loss of ATF7IP had similar effects to SETDB1 loss in regards to H3K9 trimethylation and gene transcription levels [30]. MBD1 loss also results in the loss of H3K9me3 in many genomic loci [31]. This indicates that both MBD1 and ATF7IP are crucial for the activity of SETDB1. Although the complete regulatory pathway by which ATF7IP regulates SETDB1 is not fully understood, it has been suggested that ATF7IP contributes to the conversion of H3K9 dimethylation to H3K9 trimethylation by SETDB1 [21]. Further studies have demonstrated that ATF7IP increases the stability of nuclear SETDB1 [30]. The nuclear localization of SETDB1 is regulated by ATF7IP, which binds to the NES motifs and antagonizes their action. Nuclear retention of SETDB1 then upregulates the level of its monoubiquitination, enhancing its activity. Thus, ATF7IP directly acts to ensure SETDB1's nuclear retention, also promoting its activation while it remains in the nucleus [32]. These results agree with previous studies demonstrating that SETDB1 is mainly located in the cytoplasm of normal and cancer cells while also being able to shuttle between the nucleus and the cytoplasm. Moreover, SETDB1's transport from the nucleus to the cytoplasm was attributed to the action of Chromosome Region Maintenance 1 (CRM1) nuclear export protein [28], suggesting that the main regulatory mechanisms of SETDB1 activity involve post-transcriptional modifications. In agreement, miRNA targeting of the SETDB1 mRNA has been suggested as an alternative way of post-transcriptional modulation of SETDB1 expression. The miRNA-621, -29 and -381-3p have been shown to downregulate the activity of SETDB1. Interestingly, SETDB1 regulation can also occur on the transcriptional level, and the proteins c-Myc, Specificity Protein 1 (SP1), Specificity Protein 3 (SP3) and TCF4 have been shown to bind to the SETDB1 promoter, inducing its expression [20].

Apart from the abovementioned SETDB1/KAP1/KRAB–Zfp, SETDB1/hAM and SETDB1/DNMT3A/B complexes, SETDB1 can also form large complexes with HDAC1/2, and the transcriptional corepressor mSin3A/B, in order to achieve transcriptional gene repression [33]. In addition, MBD1 attracts SETDB1 to Chromatin assembly factor (CAF-1), thus forming an MBD1/SETDB1/CAF-1 complex, which is specific to the S-phase and facilitates H3K9 methylation and stable Heterochromatin formation [31]. Finally, SETDB1 interacts with the human silencing hub (HUSH) complex, which is required to mediate Heterochromatin formation and gene silencing [21,30].

4. Physiologic Functions of SETDB1 and Cellular Features

4.1. SETDB1 Directly and Indirectly Affects Major Cellular Functions

SETDB1's cellular functions are mostly related to the trimethylation of H3K9, a repressive mark. Thus, SETDB1 generates a "closed", more compact and inaccessible chromatin to transcription factors from an "open" and "relaxed", easily accessible chromatin [10]. SETDB1 can also cause the deposition of other repressive marks on histone tails in an indirect way by cross-talking with other repressive enzymes. One such indirect pathway of gene silencing is the interaction of SETDB1 with the Polycomb Repressive Complex 2 (PRC2), which possesses histone methyltransferase activity and trimethylates Histone 3 Lysine 27 (H3K27), establishing another repressive mark [34]. Therefore, by increasing the enzymatic activity of PRC2, SETDB1 has the potential to interact indirectly with more pathways and repress a wider variety of genes. The change in chromatin architecture caused by SETDB1, especially in gene promoters, leads to the silencing of a vast array of genes. In this way, SETDB1 participates in several cellular functions, including regulation of the cell cycle and cell proliferation [35–40], suppression of retroelements [41], regulation of immune cell function [42], maintenance of X chromosome inactivation [28], control of nervous system development [43] and formation of PML-NBs [44]. Additionally, SETDB1 has been implicated in the restriction of pre-adipocyte differentiation [45].

4.2. SETDB1's Role in Cell Division

Moreover, it plays a major role in the regulation of cell division and proliferation. Many studies have shown that SETDB1 can interfere with the stability of p53, a central regulator of cell cycle progression and apoptosis, by inhibiting its effects through methylation, thus promoting cell proliferation [46]. Furthermore, SETDB1 has also been linked to increased Protein Kinase B (Akt) activity which, as part of the Inositol trisphosphate (IP3)/Akt pathway, is a crucial regulator of cell proliferation and survival [35]. It has been demonstrated that SETDB1 activates Akt by trimethylating it on the K64 position, leading to TRAF6-mediated Akt ubiquitination in xenograft models [36]. A study by Guo et al. showed that SETDB1 methylates Akt at the K140 and K142 positions, leading to Akt activation, acting in synergy with the PI3K pathway. In this context, the absence of Akt methylation was shown to attenuate its activity [36]. Lastly, the effects of SETDB1 on the cell cycle are attributed to its interaction with central cell cycle regulators, such as Cyclin D1 and c-myc [37].

4.3. SETDB1 in ERV Regulation and Its Implications

Another major cellular effect of SETDB1 is the regulation of retroelement expression. Endogenous Retroviruses (ERVs) represent a subcategory of viral retroelements, containing long-terminal repeats which are dispersed among the euchromatic regions of the mammalian DNA [38]. Their transcription has been associated with retrotransposition, a mechanism that causes increased genome instability, leading to many spontaneous mutations [39]. SETDB1 can inhibit ERVs expression, thus minimizing their potential to alter DNA. In accordance, Tan et al. demonstrated that reconstitution of ERV expression in SETDB1-knockout mice upregulates the expression of other neighboring genes. A high percentage of these genes create chimeric transcripts with ERVs or possess ERVs proximally to their initiation sites within a 10 kb range [40]. This role of SETDB1 was observed not only in early embryonic cells but also in further differentiated somatic cells [47]. These results demonstrate that SETDB1 has a continuous role in the repression of ERV expression even after the early developmental stages. Another interesting finding by Fukuda et al. was the connection of SETDB1 with Retroelement Silencing Factor 1 (RESF1), a provirus silencing factor. Their study showed that Resf1 knockout mouse embryonic stem cells had decreased SETDB1 enrichment on provirus and ERV sites. This interaction implicates that RESF1 may also play a part in the SETDB1-mediated repression of ERVs by regulating the action of SETDB1 [48].

A more recent study was able to link the silencing of ERVs by SETDB1 to the regulation of CD4+ T cell differentiation [43]. In more detail, SETDB1 was demonstrated to be essential for both the acquisition and the maintenance of T helper 2 (Th2) response by CD4+ T cells. This study showed that Th2-differentiated, SETDB1-knockdown CD4+ T cells were unable to maintain their Th2 differentiation when exposed to Th1-inducing signals. Takikita et al.

demonstrated that SETDB1 is crucial for the selection of single-positive T cells. They observed that SETDB1 deletion resulted in decreased Extracellular Signal-Regulated Kinase (ERK) activity, a protein that is of paramount importance for the development of T cells and is activated by the T Cell Receptor (TCR). This effect was a result of FcγRIIB derepression, which in turn inhibited ERK activation [41]. Finally, SETDB1 has been implicated in the maintenance of primordial germ cells, with decreased SETDB1 activity being associated with depletion of the primordial germ cell pool in males [49].

Oogenesis is also influenced by SETDB1, with its depletion in maternal gametes leading to embryonic growth arrest during pre-implantation [50]. The absence of SETDB1 in oocytes induced DNA damage through the reactivation of ERVs, leading to meiosis defects [51]. A similar result was observed in spermatogenic cells, where the loss of SETDB1 resulted in the reactivation of ERVs and caused early meiotic arrest [52].

Lastly, the interaction of SETDB1 with ATF7IP explained above is in part responsible for the maintenance of X chromosome inactivation [26]. It has been shown that depletion of SETDB1 and MBD1 leads to re-expression of Xi genes due to Xi chromosome decompaction, making heterochromatic regions switch into a euchromatic state [26]. Sun et al. found that the decompaction of Xi chromatin upon the loss of SETDB1 was partly due to reactivation of an Endogenous Retrovirus-Related Mammalian-apparent LTR-Retrotransposons (ERVL-MaLR) element and the gene Interleukin 1 Receptor Accessory Protein-Like 1 (IL1RAPL1) [53]. Concerning reproductive system physiology, SETDB1 also seems to be necessary for female identity maintenance in *Drosophila* sp. germ cells since H3K9me3 allows for the suppression of genes normally expressed in testes. SETDB1 loss thus results in the ectopic expression of testicular genes [54].

4.4. SETDB1 in the Regulation of the Inflammatory Response

In another study, SETDB1 was found to suppress the expression of Toll-Like Receptor 4 (TLR4)-induced proinflammatory mediators, such as IL-6 and IL-12β in macrophages. This effect was attributed to the regulatory effect that SETDB1 had on Nuclear Factor Kappa Beta (NF-κB). Wild-type SETDB1 mice exhibited decreased NF-κB recruitment on the promoter of IL-6, possibly due to the chromatin remodeling effect of H3K9me3. Upon lipopolysaccharide (LPS) stimulation, the H3K9 demethylase LSD2/KDM1B/AOF1 was recruited to promoters of these proinflammatory mediators, decreasing the levels of H3K9me3 and allowing recruitment of NF-κB. This interaction suggests that SETDB1 may act as a gatekeeper for the organism's inflammatory response, contributing to the balance between repression and activation of proinflammatory cytokines [55].

4.5. SETDB1 Regulates the Formation of PML-NBs

Moreover, SETDB1 has been shown to be involved in the formation of PML-NBs, which play a major role in apoptosis, the maintenance of embryonic stem cell pluripotency, DNA damage response and cellular stress as well as tumor growth inhibition. PML-NBs are also involved in the recruitment of proteins such as Death Domain Associated Protein (Daxx), Small Ubiquitin-Like Modifier 1 (SUMO-1), Speckled 100 KDa (Sp100) and CREB-Binding Protein (CBP) [15,56]. A SUMO-interaction motif on the sequence of SETDB1, which binds sumoylated KAP1 and SP3, increases its methylating activity, finally promoting the stabilization of PML-NBs [57].

4.6. SETDB1 Coordinates the Development of the Nervous System

On top of the abovementioned physiologic functions, SETDB1 has been demonstrated to participate in the early development of the nervous system. During early embryogenesis, SETDB1 is responsible for maintaining the expression of pluripotency-associated transcription factors while also inhibiting the transcription of trophectoderm differentiation markers [42]. Furthermore, during brain development, SETDB1 takes part in the delicate balance of the amount of neural and astrocytic cells that will be generated. Thus, in the early stages, SETDB1 participates in the inhibition of astrocyte-related genes such as Glial

Fibrillary Acidic Protein (GFAP) and SRY box transcription factor 9 (Sox9). In the later stages of brain development, SETDB1 levels decrease, resulting in de-repression of these genes and increased production of astrocytes [40] (Figure 2).

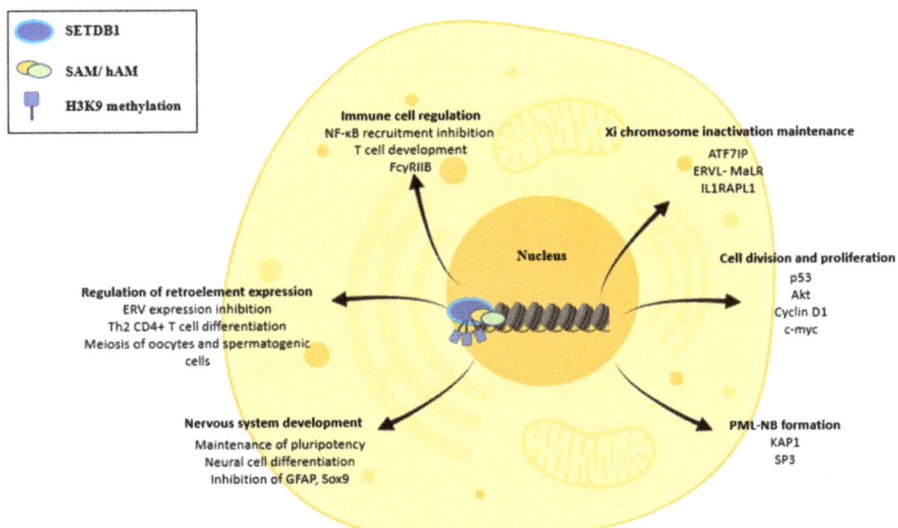

Figure 2. Cellular effects of SETDB1. SETDB1 can trimethylate Histone 3, Lysine 9 (H3K9), using S-Adenosyl methionine (SAM) as a methyl-group donor and human homolog of murine ATFa-associated modulator (hAM) as an inducer, in order to change chromatin composition, inducing compaction of chromatin and gene expression inhibition. In this way, SETDB1 is involved in many cellular functions. It can increase cell division by interfering with p53 and Akt activity, as well as interact with Cyclin D1 and c-myc to induce cell proliferation. SETDB1 can also participate in ERV element suppression, promoting genome stability, differentiation of CD4+ T cells to T helper 2 (Th2) cells and regulation of meiosis in oocytes and spermatogenic cells. In addition, SETDB1 is implicated in immune response regulations by preventing NF-κB recruitment and influencing T-cell development. SETDB1 is also crucial for the formation of PML-NBs. SETDB1 is also critical for the maintenance of X chromosome inactivation (Xi). Lastly, SETDB1 regulates the development of the nervous system by promoting pluripotency and suppressing differentiation markers in early embryogenesis.

5. Connection of SETDB1 with Tumorigenesis

Aberrant expression and activity of SETDB1 have been implicated in the pathophysiology of various diseases. Most importantly, it has been extensively associated with tumorigenesis. Moreover, it plays an important role in neuropsychiatric and genetic disorders, as well as in some cardiovascular and gastrointestinal diseases (Figure 3).

When it comes to tumorigenesis, SETDB1 downregulates important tumor suppressor genes through histone methylation, acting primarily as an oncogene but also rarely as a tumor suppressor (Figure 3). Another mechanism of SETDB1 involvement in cancer is the suppression of tumor-intrinsic immunogenicity and evasion of the immune response through inhibition of genome regions enriched with transposable elements that would trigger the host's immune responses if activated [58].

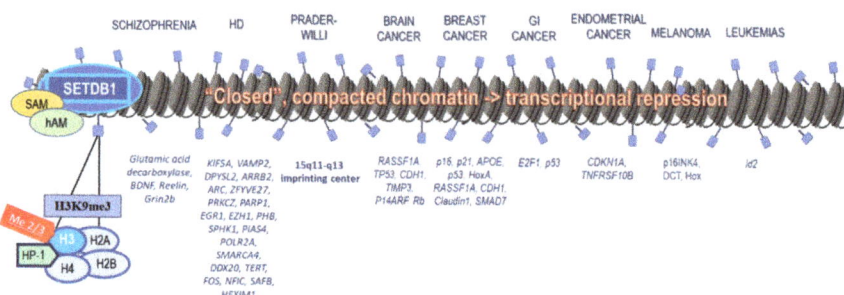

Figure 3. SETDB1 and gene silencing in disease pathogenesis. SETDB1 is capable of epigenetically modifying the Lysine 9 (K9) of histone H3 of the histone octamer through di- or trimethylation. The methyl-group donor during this process is SAM, and hAM works to induce this process. Upon methylation of H3K9 by SETDB1, Heterochromatin Protein (HP1) is recruited, ultimately changing the structure of the chromatin from eu- to Heterochromatin. These changes create a more compact and "closed" chromatin state, which is transcriptionally silenced. Transcriptional inhibition through gene promoter methylation is therefore the main mechanism of SETDB1 implication in a vast array of diseases, and most notably cancer, where SETDB1 is frequently upregulated and mostly serves as an oncogene, silencing tumor suppressor genes.

5.1. Brain and Head–Neck Cancer

SETDB1 is considered a key mediator of H3K9 trimethylation in CNS tumors and frequently associates with *IDH1* and *BRAF* mutations [39,40]. Increased nuclear SETDB1 expression has been detected in glioma tissues and correlates with high histological grades [44,59,60], as well as enhanced resistance to chemotherapeutic drugs [61]. Elevated SETDB1 expression has also been observed in metastatic head and neck cancers and nasopharyngeal carcinomas, being associated with decreased survival time [62]. SETDB1 upregulation enhances cellular proliferation, invasion and migration, possibly by favoring the transition from the G1 to the S cell cycle phase [63]. It may contribute to brain tumorigenesis by methylating tumor suppressor genes [58], such as *Ras association domain family 1 isoform A (RASSF1A)* [64], *TP53* [65], *E-Cadherin (CDH1)* [64], *P14 alternate reading frame (P14ARF)* [64], *Metalloproteinase inhibitor 3 (TIMP3)* [64] and *Retinoblastoma protein (Rb)* [64].

Small interfering RNAs (siRNAs) or histone methyltransferase inhibitors that inhibit SETDB1 have been applied in glioma cell lines and significantly decrease cell proliferation and migration while promoting apoptosis [44,60]. Unfortunately, only non-specific inhibitors against SETDB1 are currently available, including chaetocin, mithramycin A, 3'-deazaneplanocin A (DZNep), paclitaxel (PTX) and miR-381-3p inhibitors [66]. Of note, the chemotherapeutic drug PTX has been shown to downregulate SETDB1 activity through p53 expression and effectively reduce glioma cell growth and brain metastases [67]. Nanoparticle delivery systems are expected to overcome some serious adverse effects, as well as the problem of its low penetration through the Blood–Brain Barrier (BBB), allowing PTX to resurface as a potential therapeutic option against brain cancer.

5.2. Lung Cancer and Malignant Pleural Mesothelioma

In Non-Small Cell Lung Cancers (NSCLC), upregulated SETDB1 expression favors tumor progression through interaction and methylation of p53 and Akt (AKT Serine/Threonine Kinase 1) [68–70], resulting in poorer prognosis and tumor recurrence in patients with stage I NSCLC [71,72]. According to a recent study, SETDB1-derived circular RNA (circSETDB1) was significantly increased in lung adenocarcinoma hypoxia-induced exosomes and was associated with disease stage, whereas *circSETDB1* knockdown notably inhibited in vitro malignant growth [73]. SETDB1 also activates the Wingless-related integration site (WNT) pathway, causing the accumulation of nuclear β-catenin and induction of a cancerous phenotype [74]. On the contrary, highly metastatic lung adenocarcinomas exhibit decreased SETDB1 activity [75], suggesting that SETDB1 may act as a key oncogene only in the initial stages of NSCLC. SETDB1 targeting in lung cancer has been attempted through the use of the methyltransferase inhibitor DZNep, which downregulates SETDB1 expression and H3K9me3 levels, decreasing lung cancer cell growth and increasing apoptosis [76]. Piperlongumine was also demonstrated to reduce SETDB1 expression, ultimately resulting in the death of lung cancer cell lines [77], along with other chemotherapeutic agents, such as PTX, doxorubicin and cisplatin [66].

The Malignant Pleural Mesotheliomas (MPM) are characterized by a high frequency of SETDB1 mutations, resulting in a non-functional SETDB1 protein [78]. Young-age MPMs often exhibit p53 mutations that lead to chromosomal loss and a near-haploid state, with subsequent genome reduplication and inactivation of SETDB1 [79].

5.3. Breast Cancer

Aberrant expression of SETDB1 has been observed in breast cancer (BC) [80], contributing to tumor progression [68]. SETDB1 promotes the Internal Ribosome Entry Segment (IRES)-guided translation of *c-MYC* and *Cyclin D1* (CCND1) oncogenes. Silencing of *SETDB1* drastically decreases the transcription of cell cycle-progression genes, such as phosphorylated *RB, Cyclin A2* and *Cyclin E1*, but also downregulates BMI1, one of the downstream targets of MYC, and increases p21 and p16 expression, contributing to cell cycle arrest and senescence [37]. BC cells lacking SETDB1 exhibited decreased BC type 1 susceptibility protein (BRCA1), a telomere protective and an alternative lengthening of telomeres (ALT)-promoting oncogene involved in the majority of familial BCs [81]. Finally, SETDB1 silencing in BC cells inhibited tumor metastasis through regulation of Mothers against decapentaplegic homolog 7 (SMAD7) expression, which antagonized the transforming growth factor beta (TGF-β) pathway. On the contrary, cells overexpressing SETDB1 were characterized by decreased epithelial markers, such as β-catenin and E-cadherin, but elevated mesenchymal markers, such as vimentin, promoting migration and invasion [82,83]. SETDB1 has been involved in epithelial–mesenchymal transition (EMT) after binding to *Snail* promoter in triple-negative BCs [82]. Hox antisense intergenic RNA (HOTAIR), a functional Long Non-Coding RNA (lncRNA), aids in BC progression when overexpressed by indirectly inhibiting miR-7. This causes the upregulation of SETDB1, c-Myc and suppression of E-cadherin, enhancing the EMT process [84–86]. Lastly, activated SETDB1 interaction with ΔNp63 in BCs, a p63 isoform without an N-terminal transactivation domain, redirects SETDB1 to specific tumor suppressor genes, such as *p53, Apolipoprotein E (APOE)* and *Homeobox A (HoxA)*, causing chromatin changes and gene silencing [87].

Regarding BC treatment, various TGF-β pathway inhibitors target SETDB1, such as SMAD7, which appears to prevent metastasis [83]. Moreover, Cardamonin suppresses SETDB1 and inhibits BC cell growth while also downregulating BC inflammatory mediators that are tied to increased aggressiveness, chemotherapy resistance, poor patient survival and stem cell phenotypes [88]. Lastly, HOTAIR inhibitors, including calycosin, delphinidin-3-glucoside, genistein and BML-284 have been proposed as potential targets in BC treatment [86]. As mentioned above, HOTAIR has been shown to enhance SETDB1, c-myc and STAT3 [84] but suppress E-cadherin [82,89] in favor of EMT [86], thus justifying its potential inhibition in the treatment against BC.

5.4. Gastrointestinal Cancers

In CRC, SETDB1 overexpression positively correlates with increased histological grade and stage and associates with poor prognosis [90–92]. It contributes to the H3K9 histone methylation of the *p53* promoter, inducing p53 dysregulation in CRC cells [13]. It also binds directly to the Signal Transducer and Activator of Transcription 1 (STAT1) promoter, resulting in the enhanced function of CCND1/Cyclin-dependent kinase 6 (CCND1/CDK6) complex and a shift from the G0/G1 to the S phase [93]. SETDB1 depletion restores the transcriptional status of affected genes back to normal, allowing for cell re-differentiation and cancer cell transformation to normal-like post-miotic cells [94], especially when combined with cytotoxic drugs such as 5-fluorouracil, oxaliplatin and irinotecan.

In Hepatocellular Carcinoma (HCC), SETDB1 upregulation is linked to metastasis and poor prognosis after interaction with p53 and dimethylation of K370 [95–98]. In pancreatic disease, SETDB1 is needed for exocrine regeneration of the pancreas after cerulein-mediated acute pancreatitis, and its absence leads to significant pancreatic atrophy or apoptosis [99]. In mouse Pancreatic Ductal Adenocarcinoma (PDA), SETDB1 directly binds p53 and regulates its expression. On the other hand, in the early, non-aggressive stages of PDA, SETDB1 may acquire a tumor-suppressive role since it protects cells from KRAS-induced PDA, even after double p53 allele loss. Additionally, SETDB1 deletion induces the accelerated development of pancreatic intraepithelial neoplasia and acinar-to-ductal metaplasia, again suggesting its anti-oncogenic properties. Of importance, overexpression of miR-621, which downregulates SETDB1 and p53 activity, enhanced HCC cell radiosensitivity [100,101], while the H3K9 methylation inhibitor Mithramycin A was shown to significantly reduce HCC tumor growth [36,102,103].

5.5. Reproductive System Cancers

Ovarian serous cancer (SOC) is characterized by increased circSETDB1 levels, which are associated with lymph node metastasis, advanced clinical stage, and chemoresistance as well as a shorter patient survival [104,105]. In advanced ovarian cancer, TGF-β- induced epigenetic silencing of epithelial genes, including *CDH1*, is mediated through SETDB1, leading to EMT and metastasis. Additionally, TGF-β activation recruits SMAD2 and 3 to the IL-2 promoter. SETDB1 binding to SMAD3 methylates and suppresses T cell receptor-induced IL-2 transcription, presenting an additional mechanism of SETDB1 involvement in ovarian cancer tumorigenesis [75,106].

SETDB1 has also been involved in endometrial carcinoma by inhibiting *p53* and favoring tumorigenesis [107]. SETDB1 is also overly expressed in prostate cancer (PC) tissues, especially when they are androgen-independent. *SETDB1* knockdown was shown to promote G0/G1 phase arrest, decrease colony formation and suppress cancer growth and migration [108]. It affects genomic stability by interacting with URI (Unconventional Prefoldin RPB5 Interactor protein) that represses retrotransposons [109]. LINE-1 retroelements are derepressed in PC so that URI dysfunction impairs the SETDB1-controlled repressive function of KAP1 on retroelements, favoring genomic rearrangements [110].

5.6. Melanoma

In melanomas, SETDB1 is positively correlated with several prognostic factors, including high mitotic counts, advanced invasion depth (Clark levels), involvement of the epidermis and p16INK4 methylation [111,112]. SETDB1 expression was further associated with the *BRAFV600E* mutation in favor of melanoma development [113,114]. The tumorigenic effects of SETDB1 are attributed to the regulation of Thrombospondin 1 (THBS1), which promotes melanoma invasiveness and metastasis as well as downregulation of the expression of DOPAchrome tautomerase (DCT), an enzyme that participates in melanin synthesis [102]. Mithramycin A was shown to reduce SETDB1 expression and tumor growth in melanomas with upregulated SETDB1 levels [36,102,103]. CAS 935693-62-2, a small molecule SETDB1 inhibitor, decreased the number of viable cells overexpressing SETDB1 [111].

5.7. Hematologic Cancers

SETDB1 functions as a tumor suppressor in Acute Myeloid Leukemia (AML) by promoter histone methylation and repression of tumorigenic genes [115], such as Sineoculis homeobox homolog 1 (*Six1*), HoxA9 and Dedicator of Cytokinesis 1 (*Dock1*) [116,117]. AML patients exhibit reduced SETDB1 activity, whereas increased SETDB1 levels were positively correlated with more favorable patient survival. However, SETDB1 may be needed for the initiation of AML pathogenesis and early progression since it methylates and represses retrotransposons, thus rescuing AML cells from the patient's innate immune response that is initiated when sensing retrotransposons as "non-self" [118].

Regarding Acute Promyelocytic Leukemia (APL), an aggressive subtype of AML, SETDB1 is a stable member and responsible for the integrity of PML-NBs which are found interspersed in chromatin, regulating transcription, apoptosis and DNA damage responses [119]. SETDB1 regulates PML-NB-associated genes, including Inhibitor of DNA binding 2 *(Id2)*, which is decreased in APL [120,121]. Of note, SETDB1 function may be mediated by the polymerase associated factor *(PAF1)* complex, which regulates *HoxA9* and *Meis1* and other key genes responsible for leukemogenesis [117].

The therapeutic potential of SETDB1 inhibition in AML was evidenced by the use of UNC0638, an H3K9me2/3 inhibitor that caused myeloid leukemia cell cytotoxicity, but also cKit+ hematopoietic stem cell line expansion in healthy bone marrow cells [116]. In APL, arsenic trioxide (As_2O_3) use in mice resulted in PML degradation, as well as the significant reduction of SETDB1 levels with PML-NBs disassembly and increased Id2 expression [15].

5.8. Osteosarcoma

Deletions of the 6q16.3 region of their tumor-suppressor Glutamate Ionotropic Receptor Kainate Type Subunit 2 *(GRIK2)* gene in osteosarcomas have been shown to interfere with SETDB1 binding since this deleted region contains a SETDB1 binding site [122]. SETDB1 normally causes H3K9 methylation and downregulation of *GRIK2* expression. This deletion, therefore, results in overexpression of the anti-oncogenic GRIK2, apoptosis and decreased proliferation and migration while also revealing a possible tumorigenic role of SETDB1 in osteosarcomas [123].

6. SETDB1 Connection to Other Diseases

SETDB1 has also been reported to be involved in several other diseases, mostly neuropsychiatric disorders. It is also implicated in a series of genetic diseases, as well as congenital cardiovascular diseases and Inflammatory Bowel Disease (IBD).

6.1. SETDB1 Association with Neuropsychiatric Diseases

SETDB1 plays a major role in the pathogenesis of several neuropsychiatric disorders, such as schizophrenia and autism spectrum disorder, as well as in neurodevelopmental diseases. Increased levels of H3K9me2 and SETDB1 have been found postmortem in patients with a history of schizophrenia, and enhanced methyltransferase activity has been associated with a particular clinical phenotype, consisting of positive family history, longer duration, negative symptoms difficult to treat and thus a poorer disease prognosis [124]. Upregulation of H3K9me2 marks has been observed in schizophrenia biomarker genes, such as *Glutamic acid decarboxylase, Brain-derived neurotrophic factor (BDNF)* and *Reelin* [125,126]. Notably, decreased levels of these biomarkers have also been associated with chronic disease, which bears a worse prognosis and a positive family history. Moreover, SETDB1 was shown to methylate H3K9 in the ventral striatum and hippocampus, thus regulating genes, such as the *NMDA receptor subunit N-methyl D-aspartate receptor subtype 2B (NR2B/Grin2b)*. In more detail, *Grin2B* repression is involved in the pathogenesis of schizophrenia and bipolar disorder [127,128]. Lastly, neurodegeneration and memory deficits in Frontotemporal Dementia (FTD) and Amyotrophic Lateral Sclerosis (ALS) are linked to a global downregulation of the H3K9me3 mark, further demonstrating the implication of SETDB1 in the pathogenesis of neuropsychiatric diseases [129]. When it comes to treatment options affecting SETDB1 function in neurocognitive disorders, SETDB1 activity enhancement has anti-depressive effects, and H3K9me3 inhibition with an elevation of BDNF expression was shown to prevent perioperative neurocognitive disorders [130], thus demonstrating the need for further research on the therapeutic modulation of SETDB1 activity.

In neurodevelopment, the impact of SETDB1 is crucial, as evidenced by a de novo 1q21.3 deletion present in patients with intellectual disability (ID) or typical autism spectrum disorder (ASD) that affects the *SETDB1* gene, among others [131]. Additional mutations that impair epigenetic modifications have been detected in patients with ASD, including an in-frame 3 bp deletion in the *SETDB1* gene [132,133]. The influence of SETDB1 on neurodevelopment may further be affected by substances, such as alcohol and nicotine. Alcohol consumption during pregnancy has been shown to lead to dose-dependent fetal epigenetic abnormalities, which result in fetal neurobehavioral deficits in the context of Fetal Alcohol Spectrum Disorder (FASD). SETDB1 is responsible for alcohol-induced epigenetic changes in the fetal DNA. Acute exposure of the fetus to alcohol leads to SETDB1 downregulation, whereas prolonged exposure for more than 7 days increases the enzyme's levels [134]. In accordance, other studies report significantly upregulated SETDB1 and H3K9me2 levels in the hypothalamus of offspring exposed to alcohol in utero [135]. Administration of choline, which can mitigate the behavioral effects of alcohol exposure [136], can normalize SETDB1 mRNA levels [137], further confirming the association of SETDB1 with the pathogenesis of FASD. On the other hand, human cells exposed to nicotine in vitro demonstrated decreased SETDB1 levels, as well as decreased GLP and G9 methyltransferase levels [138]. The H3K9me2 levels were also downregulated, implying that nicotine may overall be able to antagonize the chromatin-condensing effects of SETDB1.

6.2. SETDB1 Association with Genetic Diseases

SETDB1 has also been implicated in the pathophysiology of a series of genetic diseases, such as Huntington's disease (HD) and Rett, Prader–Willi and Cockayne syndromes. In HD patients, SETDB1 expression is significantly elevated, pointing scientific interest towards approaches that downregulate SETDB1-promoter activity as potential beneficial therapeutic schemes [139]. Experimental studies have shown that Huntingtin (HTT) binds to ATF7IP, a SETDB1 regulator, resulting in low H3K9me3 levels, whereas loss of HTT upregulates H3K9me3 marks mainly on genes affecting neuronal differentiation. Interestingly, genetic variations of *ATF7IP* seem to correlate with HD's age of onset [140], adding ATF7IP to the list of potential targets for reducing H3K9me3 levels upregulated by the mutant HTT [141].

Furthermore, a wide variety of genes largely occupied by H3K9me3 seem to be involved in the pathogenesis of HD, such as synapse-associated genes: *Kinesin heavy chain isoform 5A (KIF5A), Vesicle-associated membrane protein 2 (VAMP2), Dihydropyrimidinase-related protein 2 (DPYSL2)* and *arrestin beta-2 (ARRB2)*; cytoskeleton regulation genes: *Activity-regulated cytoskeleton-associated protein (ARC), Zinc finger, FYVE domain containing 27 (ZFYVE27), Protein kinase C, zeta (PRKCZ)*; protein metabolism genes: *Poly (ADP-ribose) polymerase 1 (PARP1), Early growth response protein 1 (EGR1), Enhancer of Zeste Homolog 1 (EZH1)* and *Polyhydroxybutyrate (PHB)*; immune response genes: *Sphingosine kinase 1 (SPHK1), protein inhibitor of activated STAT protein gamma (PIAS4)*; DNA replication and repair genes: *E2F6,RNA polymerase II subunit A (POLR2A), SWI/SNF Related, Matrix Associated, Actin-Dependent Regulator of Chromatin, Subfamily A, Member 4 (SMARCA4), DEAD-box helicase 20 (DDX20)* and *DNA topoisomerase II alpha (TOP2A), Telomerase reverse transcriptase (TERT)* and transcriptional regulation genes: *FOS, Nuclear factor 1 C-type (NFIC) Scaffold attachment factor B (SAFB)* and *Hexamethylene Bis-Acetamide-Inducible Protein 1 (HEXIM1)*. At last, SETDB1 involvement in the HD phenotype was further confirmed with the ocular expression of mHTT in a *Drosophila melanogaster* model with HD, which led to progressive eye degeneration and ommatidium disruption that was exacerbated by SETDB1 overexpression. On the contrary, SETDB1 deletion saved the affected eye [142].

Collectively, increased SETDB1 activity along with the establishment of H3K9me3 marks are suggested to contribute to the suppression of several genes implicated in the pathophysiology of HD. HDAC inhibitors (HDACi) have been shown to improve neuronal survival in HD [143,144] and have been suggested as potential treatment options since the SETDB1 Tudor domain interacts with HDAC 1/2 to achieve transcriptional repression, while mHTT itself reduces the activity of histone acetyltransferases, causing histone deacetylation and thus gene repression [145]. The use of HDACi helps to restore normal transcription and prevents histone deacetylation in the presence of mHTT [145,146]. Examples of HDACi include butyrates nogalamycin, which restores the normal histone H3K9 trimethylation and acetylation balance; cystamine, which decreases *Htt* aggregates and mithramycin, which can induce SETDB1 suppression [147]. The 5-allyloxy-2-(pyrrolidine-1-yl) quinoline (APQ) is another newly discovered SETDB1 inhibitor that reduces H3K9me3 levels and improves motor and neuropathological symptoms in an HD model [148]. Overall, HDAC and SETDB1 inhibitors have both increased HD patient survival, but further research is needed to validate their effects in human clinical trials [142].

SETDB1 is also associated with Rett syndrome, which is caused by mutations in *Methyl-CpG-binding Protein 2 (MECP2)* gene, leading to histone modification dysregulation that causes Heterochromatin formation. *MECP2* knockdown in mice rendered them incapable of tolerating increased H3K9 levels and deteriorated their Rett phenotype, whereas normal mice were able to deal with increased H3K9 methylation, suggesting a possible correlation between Rett syndrome and the H3K9 methylation mark, which is related to SETDB1 activity [149].

Prader–Willi syndrome is a chromosomal deletion of 15q11-q13 of the paternal chromosome, causing a characteristic phenotype that includes intellectual disability, obesity, hyperphagia and hypogonadism [150]. Cruvinel et al. demonstrated that *SETDB1* knockdown in Prader–Willi-specific-induced pluripotent cells (iPSCs) decreased the H3K9me3 levels on the Small Nucleolar RNA 116 (SNORD116) cluster and increased the cluster's transcriptional activity [151], which is normally silenced in Prader–Willi patients. *SETDB1* knockdown also resulted in decreased methylation of the 15q11-q13 imprinting center, which regulates imprinting [152], but was not able to upregulate the Small Nuclear Ribonucleoprotein Polypeptide N (SNRPN) cluster transcriptional activity, which is implicated in disease pathogenesis. However, the knockdown of the ZNF274 transcription factor, which interacts with SETDB1, decreased the H3K9me3 pattern in the Prader–Willi imprinting center and reactivated both SNORD116 as well as the SNRPN clusters [153]. It is evident that SETDB1 is increased in Prader–Willi and silences the SNORD116 cluster as well as the unaffected maternal chromosome.

Cockayne syndrome has been associated with *CSA* or *CSB* gene mutations and results in accelerated aging. In cells lacking CSB, unrepaired DNA damage leads to persistent activation of the poly-ADP ribose polymerase (PARP) so that the cell's nicotinamide adenine dinucleotide is used and subsequently depleted, resulting in mitochondrial dysfunction. Induction of SETDB1 expression in CSB-deficient cells decreased PAR and restored mitochondrial function. This suggests that CSB defects in Cockayne syndrome are strongly related to SETDB1 downregulation and Heterochromatin loss, allowing for PAR buildup from freely-transcribed regions and thus, mitochondrial dysfunction from freely-transcribed regions and, finally, mitochondrial dysfunction [154].

6.3. SETDB1 Association with Cardiovascular Diseases

Studies have shown that disruption of the interaction between SETDB1 and JARID2 could explain how the latter participates in the occurrence of congenital heart defects, such as ventricular septal defect (VSD), double outlet right ventricle (DORV) and hypertrabeculation causing ventricular noncompaction. More specifically, Jarid2 regulates normal cardiac development by silencing *Notch1* through SETDB1 recruitment at its enhancer region, which induces H3K9 trimethylation. Therefore, *Jarid2* deletions in mice resulted in the development of cardiac defects similar to VSD [155], DORV and hypertrabeculation with impaired ventricular compaction, explaining that possible dysregulation of the SETDB1/JARID2 association could form the basis for the development of the abovementioned anomalies [156,157].

6.4. SETDB1 Association with Gastrointestinal Diseases

Rare missense variants of SETDB1 have been identified in Inflammatory Bowel Disease (IBD) patients and are associated with its pathogenesis. Physiologically, SETDB1 participates in intestinal homeostasis, and deletion of *Setdb1* in the intestinal epithelial cells has been shown to impair their differentiation. This results in further loss of transporters responsible for nutrient absorption with subsequent barrier breakdown as well as osmotic fluid shifts, promoting mortality due to metabolic dysfunctions, such as severe dehydration and hypoglycemia. Moreover, *Setdb1* deletion results in de-silencing of ERVs and activation of the innate immune response. Progressive inflammation with DNA damage results in p53 accumulation and intestinal epithelium cell death, further destroying the intestinal barrier but also diminishing the stem cell compartment. Not only deletions but also slight or transient SETDB1 dysregulation due to environmental factors may promote intestinal inflammation. All this justifies the potential implication of the rare *SETDB1* variants observed in IBD patients with disease pathogenesis [158].

7. Outlook and Directions for Future Research

Taken altogether, *SETDB1* presents a central regulator of many cellular functions, beginning from early development. Its unique structure with the bifurcated SET domain, as well as the entire structural composition, enable its repressive function, alternating its location between the cytoplasm and nucleus and methylating the H3K9 residues on histone tails to induce chromatin compaction. The collection of cellular effects of SETDB1 make apparent that its role in cell homeostasis is unprecedented, playing a pivotal role in the regulation of the cell cycle along with cell proliferation, the suppression of retroelements which are associated with T cell function, the regulation of immune cell function, the formation of PML-NB bodies, the maintenance of X chromosome inactivation and the development of the nervous system. The array of interacting signaling pathways regulated by SETDB1 has not yet been fully elucidated; however, it has been suggested as a master regulator in many crucial cellular functions.

Aberrant SETDB1 activity has been ultimately linked to disease onset, including nervous, cardiovascular and gastrointestinal system disorders, as well as numerous inherited genetic syndromes. SETDB1 is, however, most significantly involved in tumorigenesis by repressing tumor suppressor genes after establishing the H3K9me3 mark. Altogether, SETDB1 activity results in higher aggressiveness and worse cancer prognosis and has therefore been regarded as an oncogene. Further research on the use of SETDB1 inhibitors to combat aggressive cancer subtypes could help maximize the effects of current therapeutic regimens. First, a deeper understanding of the enzyme's intracellular effects and affected genes is needed since there is evidence that SETDB1 may also act as a tumor suppressor in some stages of cancer development. The complex interplay of SETDB1 with other epigenetic enzymes also needs more in-depth investigation in order to minimize off-target side effects from its therapeutic targeting. This will allow for the successful implementation of SETDB1 activity manipulations in patient- and disease phenotype-specific treatments to improve patient prognosis and survival rates.

Author Contributions: Conceptualization, C.P.; methodology, M.M. and D.S.; software, M.M. and D.S.; validation, M.M., D.S. and C.P.; formal analysis, M.M. and D.S.; investigation, M.M., D.S. and C.P.; resources, C.P.; data curation, M.M., D.S. and C.P.; writing—original draft preparation, M.M., D.S. and C.P.; writing—review and editing, M.M., D.S. and C.P.; visualization, M.M. and D.S.; supervision, C.P.; project administration, C.P.; funding acquisition, C.P. All authors have equally contributed to this manuscript. All authors have read and agreed to the published version of the manuscript.

Funding: This research received no external funding.

Informed Consent Statement: Not applicable.

Data Availability Statement: Not applicable.

Conflicts of Interest: The authors declare no conflict of interest.

Abbreviations

SET Domain Bifurcated Histone Lysine Methyltransferase 1; SETDB1, Suppressor of Variegation 3-9; SUV39, Protein lysine methyltransferases; PKMTs, Enhancer of Zeste; EZ, H3 lysine 9; H3K9, ETS-related gene; ERG, Methyl-CpG-binding domain; MBD, Histone Deacetylase $\frac{1}{2}$; HDAC1/2, Kruppel-associated box-Zinc Finger Proteins-KRAB-Associated Protein-1; KRAB–ZFP–KAP-1, DNA Methyltransferase 3; DNMT3, Nuclear export signals; NES, Nuclear localization signals; NLS, Ubiquitin-Conjugating Enzyme E2; UBE2E, S-adenosylmethionine; SAM, homolog of murine ATFa-associated modulator; hAM, Heterochromatin protein 1; HP1, Chromosome Region Maintenance 1; CRM1, Specificity Protein 1; SP1, Specificity Protein 3; SP3, Chromatin assembly factor-1; CAF-1, human silencing hub; HUSH, Polycomb Repressive Complex 2; PRC2, Histone 3 Lysine 27; H3K27, Protein Kinase B; Akt, Inositol trisphosphate; IP3, Endogenous Retroviruses; ERVs, Retroelement Silencing Factor 1; RESF1, T helper 2; Th2, Extracellular Signal-Regulated Kinase; ERK, T-Cell Receptor; TCR, Endogenous Retrovirus-Related Mammalian-Apparent LTR-Retrotransposons; ERVL-MaLR, Interleukin 1 Receptor Accessory Protein-Like 1; IL1RAPL1, Toll-Like Receptor 4; TLR4, Nuclear Factor Kappa Beta; NF-κB, lipo-polysaccharide; LPS, Death Domain-Associated Protein; Daxx, Small Ubiquitin-Like Modifier 1; SUMO-1, Speckled 100 KDa; Sp100, CREB Binding Protein; CBP, Glial Fibrillary Acidic Protein; GFAP, SRY Box transcription factor 9; Sox9, Ras association domain family 1 isoform A; RASSF1A, E-Cadherin; CDH1, P14 alternate reading frame; P14ARF, Metalloproteinase inhibitor 3; TIMP3, Retinoblastoma protein; Rb, Small interfering RNAs; siRNAs, 3′-deazaneplanocin A; DZNep, paclitaxel; PTX, Blood–Brain Barrier; BBB, Non-Small Cell Lung Cancers; NSCLC, SETDB1-derived circular RNA; circSETDB1, Wingless-related integration site; WNT, Malignant Pleural Mesotheliomas; MPM, breast cancer; BC, Internal Ribosome Entry Segment; IRES, Cyclin D1; CCND1, BC type 1 susceptibility protein; BRCA1, alternative lengthening of telomeres; ALT, Mothers against decapentaplegic homolog 7; SMAD7, transforming growth factor beta; TGF-β, epithelial mesenchymal transition; EMT, Hox antisense intergenic RNA; HOTAIR, Long Non-Coding RNA; lncRNA, Apolipoprotein E; APOE, Homeobox A; HoxA, Signal Transducer and Activator of Transcription 1; STAT1, CCND1/Cyclin-dependent kinase 6; CCND1/CDK6, Hepatocellular Carcinoma; HCC, Pancreatic Ductal Adenocarcinoma; PDA, Ovarian serous cancer; SOC, Prostate cancer; PC, URI; Unconventional Pre-folding RPB5 Interactor protein, Thrombospondin 1; THBS1, DOPAchrome tautomerase; DCT, Sineo-culis homeobox homolog 1; Six1, Dedicator of Cytokinesis 1; Dock1, Acute Promyelocytic Leukemia; APL, Polymerase associated factor; PAF1, arsenic trioxide; As2O3, Glutamate Ionotropic Receptor Kainate Type Subunit 2; GRIK2, inflammatory bowel disease; IBD, brain-derived neurotrophic factor; BDNF, N-methyl D-aspartate receptor subtype 2B; NR2B/Grin2b, frontotemporal dementia; FTD, Amyotrophic Lateral Sclerosis; ALS, Brain-derived neurotrophic factor; BDNF, intellectual disability; ID, autism spectrum disorder; ASD, fetal alcohol spectrum disorder; FASD, Huntington's disease; HD, Huntingtin; HTT, Kinesin heavy chain isoform 5A; KIF5A, Vesicle-associated membrane protein 2; VAMP2, Dihydropyrimidinase-related protein 2; DPYSL2, arrestin beta-2; ARRB2, activity-regulated cy-toskeleton-associated protein; ARC, Zinc finger FYVE domain containing 27; ZFYVE27, Protein kinase C zeta; PRKCZ), Poly (ADP-ribose) polymerase 1; PARP1, early growth response protein 1; EGR1, Enhancer Of Zeste Homolog 1; EZH1 and Polyhy-droxybutyrate; PHB, Sphingosine kinase 1; SPHK1, protein inhibitor of activated STAT protein gamma; PIAS4, RNA polymerase II subunit A; POLR2A, Subfamily A, Member 4; SMARCA4, DEAD-box helicase 20; DDX20, DNA topoisomerase II alpha; TOP2A, Telomerase reverse transcriptase; TERT, Nuclear factor 1 C-type; NFIC, Scaffold attachment factor B; SAFB, Hexamethylene Bis-Acetamide-Inducible Protein 1; HEXIM1, HDAC inhibitors; HDACi, Methyl-CpG-binding Protein 2; MECP2, Small Nucleolar RNA 116; SNORD116, Small Nuclear Ribonucleoprotein Polypeptide N; SNRP, poly-ADP ribose polymerase; PARP, ventricular septal defect; VSD, double outlet right ventricle; DORV.

References

1. Albini, S.; Zakharova, V.; Ait-Si-Ali, S. Chapter 3—Histone Modifications. In *Translational Epigenetics*; Palacios, D., Ed.; Academic Press: Cambridge, MA, USA, 2019; Volume 11, pp. 47–72. ISBN 25425358.
2. Markouli, M.; Strepkos, D.; Chlamydas, S.; Piperi, C. Histone lysine methyltransferase SETDB1 as a novel target for central nervous system diseases. *Prog. Neurobiol.* **2020**, *200*, 101968. [CrossRef]

3. Jenuwein, T.; Laible, G.; Dorn, R.; Reuter, G. SET domain proteins modulate chromatin domains in eu- and heterochromatin. *Cell. Mol. Life Sci.* **1998**, *54*, 80–93. [CrossRef]
4. Stassen, M.J.; Bailey, D.; Nelson, S.; Chinwalla, V.; Harte, P.J. The Drosophila trithorax proteins contain a novel variant of the nuclear receptor type DNA binding domain and an ancient conserved motif found in other chromosomal proteins. *Mech. Dev.* **1995**, *52*, 209–223. [CrossRef]
5. SETDB1 Orthologs—NCBI. Available online: https://www.ncbi.nlm.nih.gov/gene/9869/ortholog/?scope=117570 (accessed on 21 July 2021).
6. Greer, E.L.; Shi, Y. Histone methylation: A dynamic mark in health, disease and inheritance. *Nat. Rev. Genet.* **2012**, *13*, 343–357. [CrossRef]
7. SET DOMAIN PROTEIN, BIFURCATED, 1; SETDB1—OMIM. Available online: https://www.omim.org/entry/604396#references (accessed on 21 July 2021).
8. SET Domain Bifurcated Histone Lysine Methyltransferase 1 [Homo Sapiens (Human)]—NCBI. Available online: https://www.ncbi.nlm.nih.gov/gene/9869 (accessed on 21 July 2021).
9. Torrano, J.; Al Emran, A.; Hammerlindl, H.; Schaider, H. Emerging roles of H3K9me3, SETDB1 and SETDB2 in therapy-induced cellular reprogramming. *Clin. Epigenet.* **2019**, *11*, 43. [CrossRef]
10. Schultz, D.C.; Ayyanathan, K.; Negorev, D.; Maul, G.G.; Rauscher, F.J., III. SETDB1: A novel KAP-1-associated histone H3, lysine 9-specific methyltransferase that contributes to HP1-mediated silencing of euchromatic genes by KRAB zinc-finger proteins. *Genes Dev.* **2002**, *16*, 919–932. [CrossRef]
11. Jurkowska, R.Z.; Qin, S.; Kungulovski, G.; Tempel, W.; Liu, Y.; Bashtrykov, P.; Stiefelmaier, J.; Jurkowski, T.P.; Kudithipudi, S.; Weirich, S.; et al. H3K14ac is linked to methylation of H3K9 by the triple Tudor domain of SETDB1. *Nat. Commun.* **2017**, *8*, 2057. [CrossRef] [PubMed]
12. Terns, M.P.; Terns, R.M. Macromolecular complexes: SMN—The master assembler. *Curr. Biol.* **2001**, *11*, R862–R864. [CrossRef]
13. Chen, K.; Zhang, F.; Ding, J.; Liang, Y.; Zhan, Z.; Zhan, Y.; Chen, L.-H.; Ding, Y. Histone Methyltransferase SETDB1 Promotes the Progression of Colorectal Cancer by Inhibiting the Expression of TP53. *J. Cancer* **2017**, *8*, 3318–3330. [CrossRef] [PubMed]
14. Li, H.; Rauch, T.; Chen, Z.-X.; Szabó, P.E.; Riggs, A.D.; Pfeifer, G.P. The Histone Methyltransferase SETDB1 and the DNA Methyltransferase DNMT3A Interact Directly and Localize to Promoters Silenced in Cancer Cells. *J. Biol. Chem.* **2006**, *281*, 19489–19500. [CrossRef] [PubMed]
15. Cho, S.; Park, J.S.; Kang, Y.-K. Regulated nuclear entry of over-expressed Setdb1. *Genes Cells* **2013**, *18*, 694–703. [CrossRef]
16. Wu, H.; Min, J.; Lunin, V.V.; Antoshenko, T.; Dombrovski, L.; Zeng, H.; Allali-Hassani, A.; Campagna-Slater, V.; Vedadi, M.; Arrowsmith, C.H.; et al. Structural biology of human H3K9 methyltransferases. *PLoS ONE* **2010**, *5*, e8570. [CrossRef] [PubMed]
17. Ishimoto, K.; Kawamata, N.; Uchihara, Y.; Okubo, M.; Fujimoto, R.; Gotoh, E.; Kakinouchi, K.; Mizohata, E.; Hino, N.; Okada, Y.; et al. Ubiquitination of Lysine 867 of the Human SETDB1 Protein Upregulates Its Histone H3 Lysine 9 (H3K9) Methyltransferase Activity. *PLoS ONE* **2016**, *11*, e0165766. [CrossRef] [PubMed]
18. Yang, L.; Xia, L.; Wu, D.Y.; Wang, H.; Chansky, H.A.; Schubach, W.H.; Hickstein, D.D.; Zhang, Y. Molecular cloning of ESET, a novel histone H3-specific methyltransferase that interacts with ERG transcription factor. *Oncogene* **2002**, *21*, 148–152. [CrossRef] [PubMed]
19. Blackburn, M.L.; Chansky, H.A.; Zielinska-Kwiatkowska, A.; Matsui, Y.; Yang, L. Genomic structure and expression of the mouse ESET gene encoding an ERG-associated histone methyltransferase with a SET domain. *Biochim. Biophys. Acta Gene Struct. Expr.* **2003**, *1629*, 8–14. [CrossRef]
20. Lazaro-Camp, V.J.; Salari, K.; Meng, X.; Yang, S. SETDB1 in cancer: Overexpression and its therapeutic implications. *Am. J. Cancer Res.* **2021**, *11*, 1803–1827.
21. Wang, H.; An, W.; Cao, R.; Xia, L.; Erdjument-Bromage, H.; Chatton, B.; Tempst, P.; Roeder, R.G.; Zhang, Y. mAM Facilitates Conversion by ESET of Dimethyl to Trimethyl Lysine 9 of Histone H3 to Cause Transcriptional Repression. *Mol. Cell* **2003**, *12*, 475–487. [CrossRef]
22. Ecco, G.; Cassano, M.; Kauzlaric, A.; Duc, J.; Coluccio, A.; Offner, S.; Imbeault, M.; Rowe, H.M.; Turelli, P.; Trono, D. Transposable Elements and Their KRAB-ZFP Controllers Regulate Gene Expression in Adult Tissues. *Dev. Cell* **2016**, *36*, 611–623. [CrossRef]
23. Tie, C.H.; Fernandes, L.; Conde, L.; Robbez-Masson, L.; Sumner, R.P.; Peacock, T.; Rodriguez-Plata, M.T.; Mickute, G.; Gifford, R.; Towers, G.J.; et al. KAP1 regulates endogenous retroviruses in adult human cells and contributes to innate immune control. *EMBO Rep.* **2018**, *19*, e45000. [CrossRef]
24. Wolf, G.; Yang, P.; Füchtbauer, A.C.; Füchtbauer, E.-M.; Silva, A.M.; Park, C.; Wu, W.; Nielsen, A.L.; Pedersen, F.S.; Macfarlan, T.S. The KRAB zinc finger protein ZFP809 is required to initiate epigenetic silencing of endogenous retroviruses. *Genes Dev.* **2015**, *29*, 538–554. [CrossRef]
25. Zhu, Y.; Sun, D.; Jakovcevski, M.; Jiang, Y. Epigenetic mechanism of SETDB1 in brain: Implications for neuropsychiatric disorders. *Transl. Psychiatry* **2020**, *10*, 115. [CrossRef]
26. Minkovsky, A.; Sahakyan, A.; Rankin-Gee, E.; Bonora, G.; Patel, S.; Plath, K. The Mbd1-Atf7ip-Setdb1 pathway contributes to the maintenance of X chromosome inactivation. *Epigenet. Chromatin* **2014**, *7*, 12. [CrossRef] [PubMed]
27. Nozawa, R.-S.; Nagao, K.; Masuda, H.-T.; Iwasaki, O.; Hirota, T.; Nozaki, N.; Kimura, H.; Obuse, C. Human POGZ modulates dissociation of HP1alpha from mitotic chromosome arms through Aurora B activation. *Nat. Cell Biol.* **2010**, *12*, 719–727. [CrossRef] [PubMed]

28. Tachibana, K.; Gotoh, E.; Kawamata, N.; Ishimoto, K.; Uchihara, Y.; Iwanari, H.; Sugiyama, A.; Kawamura, T.; Mochizuki, Y.; Tanaka, T.; et al. Analysis of the subcellular localization of the human histone methyltransferase SETDB1. *Biochem. Biophys. Res. Commun.* **2015**, *465*, 725–731. [CrossRef] [PubMed]
29. Ichimura, T.; Watanabe, S.; Sakamoto, Y.; Aoto, T.; Pujita, N.; Nakao, M. Transcriptional repression and heterochromatin formation by MBD1 and MCAF/AM family proteins. *J. Biol. Chem.* **2005**, *280*, 13928–13935. [CrossRef] [PubMed]
30. Timms, R.T.; Tchasovnikarova, I.A.; Antrobus, R.; Dougan, G.; Lehner, P.J. ATF7IP-Mediated Stabilization of the Histone Methyltransferase SETDB1 Is Essential for Heterochromatin Formation by the HUSH Complex. *Cell Rep.* **2016**, *17*, 653–659. [CrossRef]
31. Sarraf, S.A.; Stancheva, I. Methyl-CpG binding protein MBD1 couples histone H3 methylation at lysine 9 by SETDB1 to DNA replication and chromatin assembly. *Mol. Cell* **2004**, *15*, 595–605. [CrossRef]
32. Tsusaka, T.; Shimura, C.; Shinkai, Y. ATF7IP regulates SETDB1 nuclear localization and increases its ubiquitination. *EMBO Rep.* **2019**, *20*, e48297. [CrossRef]
33. Yang, L.; Mei, Q.; Zielinska-Kwiatkowska, A.; Matsui, Y.; Blackburn, M.L.; Benedetti, D.; Krumm, A.A.; Taborsky, G.J., Jr.; Chansky, H.A. An ERG (ets-related gene)-associated histone methyltransferase interacts with histone deacetylases 1/2 and transcription co-repressors mSin3A/B. *Biochem. J.* **2003**, *369*, 651–657. [CrossRef]
34. Fei, Q.; Yang, X.; Jiang, H.; Wang, Q.; Yu, Y.; Yu, Y.; Yi, W.; Zhou, S.; Chen, T.; Lu, C.; et al. SETDB1 modulates PRC2 activity at developmental genes independently of H3K9 trimethylation in mouse ES cells. *Genome Res.* **2015**, *25*, 1325–1335. [CrossRef]
35. Manning, B.D.; Toker, A. AKT/PKB Signaling: Navigating the Network. *Cell* **2017**, *169*, 381–405. [CrossRef]
36. Guo, J.; Dai, X.; Laurent, B.; Zheng, N.; Gan, W.; Zhang, J.; Guo, A.; Yuan, M.; Liu, P.; Asara, J.M.; et al. AKT methylation by SETDB1 promotes AKT kinase activity and oncogenic functions. *Nat. Cell Biol.* **2019**, *21*, 226–237. [CrossRef]
37. Xiao, J.-F.; Sun, Q.-Y.; Ding, L.-W.; Chien, W.; Liu, X.-Y.; Mayakonda, A.; Jiang, Y.-Y.; Loh, X.-Y.; Ran, X.-B.; Doan, N.B.; et al. The c-MYC–BMI1 axis is essential for SETDB1-mediated breast tumourigenesis. *J. Pathol.* **2018**, *246*, 89–102. [CrossRef]
38. Maksakova, I.A.; Romanish, M.T.; Gagnier, L.; Dunn, C.A.; van de Lagemaat, L.N.; Mager, D.L. Retroviral elements and their hosts: Insertional mutagenesis in the mouse germ line. *PLoS Genet.* **2006**, *2*, e2. [CrossRef] [PubMed]
39. Matsui, T.; Leung, D.; Miyashita, H.; Maksakova, I.A.; Miyachi, H.; Kimura, H.; Tachibana, M.; Lorincz, M.C.; Shinkai, Y. Proviral silencing in embryonic stem cells requires the histone methyltransferase ESET. *Nature* **2010**, *464*, 927–931. [CrossRef]
40. Tan, S.-L.; Nishi, M.; Ohtsuka, T.; Matsui, T.; Takemoto, K.; Kamio-Miura, A.; Aburatani, H.; Shinkai, Y.; Kageyama, R. Essential roles of the histone methyltransferase ESET in the epigenetic control of neural progenitor cells during development. *Development* **2012**, *139*, 3806–3816. [CrossRef] [PubMed]
41. Takikita, S.; Muro, R.; Takai, T.; Otsubo, T.; Kawamura, Y.I.; Dohi, T.; Oda, H.; Kitajima, M.; Oshima, K.; Hattori, M.; et al. A Histone Methyltransferase ESET Is Critical for T Cell Development. *J. Immunol.* **2016**, *197*, 2269–2279. [CrossRef] [PubMed]
42. Lohmann, F.; Loureiro, J.; Su, H.; Fang, Q.; Lei, H.; Lewis, T.; Yang, Y.; Labow, M.; Li, E.; Chen, T.; et al. KMT1E Mediated H3K9 Methylation Is Required for the Maintenance of Embryonic Stem Cells by Repressing Trophectoderm Differentiation. *Stem Cells* **2010**, *28*, 201–212. [CrossRef] [PubMed]
43. Adoue, V.; Binet, B.; Malbec, A.; Fourquet, J.; Romagnoli, P.; van Meerwijk, J.P.M.; Amigorena, S.; Joffre, O.P. The Histone Methyltransferase SETDB1 Controls T Helper Cell Lineage Integrity by Repressing Endogenous Retroviruses. *Immunity* **2019**, *50*, 629–644.e8. [CrossRef] [PubMed]
44. Sepsa, A.; Levidou, G.; Gargalionis, A.; Adamopoulos, C.; Spyropoulou, A.; Dalagiorgou, G.; Thymara, I.; Boviatsis, E.; Themistocleous, M.S.; Petraki, K.; et al. Emerging role of linker histone variant H1x as a biomarker with prognostic value in astrocytic gliomas. A multivariate analysis including trimethylation of H3K9 and H4K20. *PLoS ONE* **2015**, *10*, e0115101. [CrossRef]
45. Zhang, J.; Matsumura, Y.; Kano, Y.; Yoshida, H.; Kawamura, T.; Hirakawa, H.; Inagaki, T.; Tanaka, T.; Kimura, H.; Yanagi, S.; et al. Ubiquitination-dependent and -independent repression of target genes by SETDB1 reveal a context-dependent role for its methyltransferase activity during adipogenesis. *Genes Cells* **2021**, *26*, 513–529. [CrossRef] [PubMed]
46. Strepkos, D.; Markouli, M.; Klonou, A.; Papavassiliou, A.G.; Piperi, C. Histone methyltransferase SETDB1: A common denominator of tumorigenesis with therapeutic potential. *Cancer Res.* **2020**, *81*, 525–534. [CrossRef] [PubMed]
47. Kato, M.; Takemoto, K.; Shinkai, Y. A somatic role for the histone methyltransferase Setdb1 in endogenous retrovirus silencing. *Nat. Commun.* **2018**, *9*, 1683. [CrossRef] [PubMed]
48. Fukuda, K.; Okuda, A.; Yusa, K.; Shinkai, Y. A CRISPR knockout screen identifies SETDB1-target retroelement silencing factors in embryonic stem cells. *Genome Res.* **2018**, *28*, 846–858. [CrossRef] [PubMed]
49. Liu, S.; Brind'Amour, J.; Karimi, M.M.; Shirane, K.; Bogutz, A.; Lefebvre, L.; Sasaki, H.; Shinkai, Y.; Lorincz, M.C. Setdb1 is required for germline development and silencing of H3K9me3-marked endogenous retroviruses in primordial germ cells. *Genes Dev.* **2014**, *28*, 2041–2055. [CrossRef]
50. Eymery, A.; Liu, Z.; Ozonov, E.A.; Stadler, M.B.; Peters, A.H.F.M. The methyltransferase Setdb1 is essential for meiosis and mitosis in mouse oocytes and early embryos. *Development* **2016**, *143*, 2767–2779. [CrossRef]
51. Kim, J.; Zhao, H.; Dan, J.; Kim, S.; Hardikar, S.; Hollowell, D.; Lin, K.; Lu, Y.; Takata, Y.; Shen, J.; et al. Maternal Setdb1 Is Required for Meiotic Progression and Preimplantation Development in Mouse. *PLoS Genet.* **2016**, *12*, e1005970. [CrossRef]
52. Cheng, E.-C.; Hsieh, C.-L.; Liu, N.; Wang, J.; Zhong, M.; Chen, T.; Li, E.; Lin, H. The Essential Function of SETDB1 in Homologous Chromosome Pairing and Synapsis during Meiosis. *Cell Rep.* **2021**, *34*. [CrossRef]

53. Sun, Z.; Chadwick, B.P. Loss of SETDB1 decompacts the inactive X chromosome in part through reactivation of an enhancer in the IL1RAPL1 gene. *Epigenet. Chromatin* **2018**, *11*, 45. [CrossRef]
54. Smolko, A.E.; Shapiro-Kulnane, L.; Salz, H.K. The H3K9 methyltransferase SETDB1 maintains female identity in Drosophila germ cells. *Nat. Commun.* **2018**, *9*, 4155. [CrossRef]
55. Hachiya, R.; Shiihashi, T.; Shirakawa, I.; Iwasaki, Y.; Matsumura, Y.; Oishi, Y.; Nakayama, Y.; Miyamoto, Y.; Manabe, I.; Ochi, K.; et al. The H3K9 methyltransferase Setdb1 regulates TLR4-mediated inflammatory responses in macrophages. *Sci. Rep.* **2016**, *6*, 28845. [CrossRef] [PubMed]
56. Dellaire, G.; Bazett-Jones, D.P. PML nuclear bodies: Dynamic sensors of DNA damage and cellular stress. *Bioessays* **2004**, *26*, 963–977. [CrossRef]
57. Cho, S.; Park, J.S.; Kang, Y.-K. Dual functions of histone-lysine N-methyltransferase Setdb1 protein at promyelocytic leukemia-nuclear body (PML-NB): Maintaining PML-NB structure and regulating the expression of its associated genes. *J. Biol. Chem.* **2011**, *286*, 41115–41124. [CrossRef]
58. Griffin, G.K.; Wu, J.; Iracheta-Vellve, A.; Patti, J.C.; Hsu, J.; Davis, T.; Dele-Oni, D.; Du, P.P.; Halawi, A.G.; Ishizuka, J.J.; et al. Epigenetic silencing by SETDB1 suppresses tumour intrinsic immunogenicity. *Nature* **2021**, *595*, 309–314. [CrossRef] [PubMed]
59. Venneti, S.; Felicella, M.M.; Coyne, T.; Phillips, J.J.; Gorovets, D.; Huse, J.T.; Kofler, J.; Lu, C.; Tihan, T.; Sullivan, L.M.; et al. Histone 3 lysine 9 trimethylation is differentially associated with isocitrate dehydrogenase mutations in oligodendrogliomas and high-grade astrocytomas. *J. Neuropathol. Exp. Neurol.* **2013**, *72*, 298–306. [CrossRef]
60. Spyropoulou, A.; Gargalionis, A.; Dalagiorgou, G.; Adamopoulos, C.; Papavassiliou, K.A.; Lea, R.W.; Piperi, C.; Papavassiliou, A.G. Role of Histone Lysine Methyltransferases SUV39H1 and SETDB1 in Gliomagenesis: Modulation of Cell Proliferation, Migration, and Colony Formation. *Neuromol. Med.* **2014**, *16*, 70–82. [CrossRef]
61. Ramachandran, A.; Gong, E.M.; Pelton, K.; Ranpura, S.A.; Mulone, M.; Seth, A.; Gomez, P., III; Adam, R.M. FosB regulates stretch-induced expression of extracellular matrix proteins in smooth muscle. *Am. J. Pathol.* **2011**, *179*, 2977–2989. [CrossRef]
62. Özdaş, S. Knockdown of SET Domain, Bifurcated 1 suppresses head and neck cancer cell viability and wound-healing ability in vitro. *Turk. J. Biol.* **2019**, *43*, 281–292. [CrossRef] [PubMed]
63. Huang, J.; Huang, W.; Liu, M.; Zhu, J.; Jiang, D.; Xiong, Y.; Zhen, Y.; Yang, D.; Chen, Z.; Peng, L.; et al. Enhanced expression of SETDB1 possesses prognostic value and promotes cell proliferation, migration and invasion in nasopharyngeal carcinoma. *Oncol. Rep.* **2018**, *40*, 1017–1025. [CrossRef] [PubMed]
64. Xu, J.; Liu, C.; Yu, X.; Jin, C.; Fu, D.; Ni, Q. Activation of multiple tumor suppressor genes by MBD1 siRNA in pancreatic cancer cell line BxPC-3. *Zhonghua Yi Xue Za Zhi* **2008**, *88*, 1948–1951.
65. Jiang, Y.; Jakovcevski, M.; Bharadwaj, R.; Connor, C.; Schroeder, F.A.; Lin, C.L.; Straubhaar, J.; Martin, G.; Akbarian, S. Setdb1 histone methyltransferase regulates mood-related behaviors and expression of the NMDA receptor subunit NR2B. *J. Neurosci.* **2010**, *30*, 7152–7167. [CrossRef] [PubMed]
66. Batham, J.; Lim, P.S.; Rao, S. SETDB-1: A Potential Epigenetic Regulator in Breast Cancer Metastasis. *Cancers* **2019**, *11*, 1143. [CrossRef]
67. Noh, H.-J.; Kim, K.-A.; Kim, K.-C. p53 Down-regulates SETDB1 gene expression during paclitaxel induced-cell death. *Biochem. Biophys. Res. Commun.* **2014**, *446*, 43–48. [CrossRef] [PubMed]
68. Rodriguez-Paredes, M.; Martinez de Paz, A.; Simó-Riudalbas, L.; Sayols, S.; Moutinho, C.; Moran, S.; Villanueva, A.; Vázquez-Cedeira, M.; Lazo, P.A.; Carneiro, F.; et al. Gene amplification of the histone methyltransferase SETDB1 contributes to human lung tumorigenesis. *Oncogene* **2014**, *33*, 2807–2813. [CrossRef]
69. Wang, G.; Long, J.; Gao, Y.; Zhang, W.; Han, F.; Xu, C.; Sun, L.; Yang, S.-C.; Lan, J.; Hou, Z.; et al. SETDB1-mediated methylation of Akt promotes its K63-linked ubiquitination and activation leading to tumorigenesis. *Nat. Cell Biol.* **2019**, *21*, 214–225. [CrossRef] [PubMed]
70. Fei, Q.; Shang, K.; Zhang, J.; Chuai, S.; Kong, D.; Zhou, T.; Fu, S.; Liang, Y.; Li, C.; Chen, Z.; et al. Histone methyltransferase SETDB1 regulates liver cancer cell growth through methylation of p53. *Nat. Commun.* **2015**, *6*, 8651. [CrossRef]
71. Inoue, Y.; Matsuura, N.; Kurabe, N.; Kahyo, T.; Mori, H.; Kawase, A.; Karayama, M.; Inui, N.; Funai, K.; Shinmura, K.; et al. Clinicopathological and Survival Analysis of Japanese Patients with Resected Non-Small-Cell Lung Cancer Harboring *NKX2-1*, *SETDB1*, *MET*, *HER2*, *SOX2*, *FGFR1*, or *PIK3CA* Gene Amplification. *J. Thorac. Oncol.* **2015**, *10*, 1590–1600. [CrossRef]
72. Lafuente-Sanchis, A.; Zúñiga, Á.; Galbis, J.M.; Cremades, A.; Estors, M.; Martínez-Hernández, N.J.; Carretero, J. Prognostic value of ERCC1, RRM1, BRCA1 and SETDB1 in early stage of non-small cell lung cancer. *Clin. Transl. Oncol.* **2016**, *18*, 798–804. [CrossRef]
73. Xu, L.; Liao, W.-L.; Lu, Q.-J.; Zhang, P.; Zhu, J.; Jiang, G.-N. Hypoxic tumor-derived exosomal circular RNA SETDB1 promotes invasive growth and EMT via the miR-7/Sp1 axis in lung adenocarcinoma. *Mol. Ther. Nucleic Acids* **2021**, *23*, 1078–1092. [CrossRef]
74. Na, H.-H.; Noh, H.-J.; Cheong, H.-M.; Kang, Y.; Kim, K.-C. SETDB1 mediated FosB expression increases the cell proliferation rate during anticancer drug therapy. *BMB Rep.* **2016**, *49*, 238–243. [CrossRef]
75. Wu, P.-C.; Lu, J.-W.; Yang, J.-Y.; Lin, I.-H.; Ou, D.-L.; Lin, Y.-H.; Chou, K.-H.; Huang, W.-F.; Wang, W.-P.; Huang, Y.-L.; et al. H3K9 Histone Methyltransferase, KMT1E/SETDB1, Cooperates with the SMAD2/3 Pathway to Suppress Lung Cancer Metastasis. *Cancer Res.* **2014**, *74*, 7333. [CrossRef]
76. Lee, J.-K.; Kim, K.-C. DZNep, inhibitor of S-adenosylhomocysteine hydrolase, down-regulates expression of SETDB1 H3K9me3 HMTase in human lung cancer cells. *Biochem. Biophys. Res. Commun.* **2013**, *438*, 647–652. [CrossRef]

77. Park, J.-A.; Na, H.-H.; Jin, H.-O.; Kim, K.-C. Increased Expression of FosB through Reactive Oxygen Species Accumulation Functions as Pro-Apoptotic Protein in Piperlongumine Treated MCF7 Breast Cancer Cells. *Mol. Cells* **2019**, *42*, 884–892. [CrossRef]
78. Kang, H.C.; Kim, H.K.; Lee, S.; Mendez, P.; Kim, J.W.; Woodard, G.; Yoon, J.-H.; Jen, K.-Y.; Fang, L.T.; Jones, K.; et al. Whole exome and targeted deep sequencing identify genome-wide allelic loss and frequent SETDB1 mutations in malignant pleural mesotheliomas. *Oncotarget* **2016**, *7*, 8321–8331. [CrossRef]
79. Hmeljak, J.; Sanchez-Vega, F.; Hoadley, K.A.; Shih, J.; Stewart, C.; Heiman, D.; Tarpey, P.; Danilova, L.; Drill, E.; Gibb, E.A.; et al. Integrative Molecular Characterization of Malignant Pleural Mesothelioma. *Cancer Discov.* **2018**, *8*, 1548–1565. [CrossRef]
80. Liu, L.; Kimball, S.; Liu, H.; Holowatyj, A.; Yang, Z.-Q. Genetic alterations of histone lysine methyltransferases and their significance in breast cancer. *Oncotarget* **2015**, *6*, 2466–2482. [CrossRef]
81. Gauchier, M.; Kan, S.; Barral, A.; Sauzet, S.; Agirre, E.; Bonnell, E.; Saksouk, N.; Barth, T.K.; Ide, S.; Urbach, S.; et al. SETDB1-dependent heterochromatin stimulates alternative lengthening of telomeres. *Sci. Adv.* **2019**, *5*, eaav3673. [CrossRef] [PubMed]
82. Yang, W.; Ying, S.U.; Chenjian, H.O.U.; Chen, L.; Zhou, D.; Kehan, R.E.N.; Zhou, Z.; Zhang, R.; Xiuping, L.I.U. SETDB1 induces epithelial-mesenchymal transition in breast carcinoma by directly binding with Snail promoter. *Oncol. Rep.* **2019**, *41*, 1284–1292. [CrossRef] [PubMed]
83. Ryu, T.Y.; Kim, K.; Kim, S.-K.; Oh, J.-H.; Min, J.-K.; Jung, C.-R.; Son, M.-Y.; Kim, D.-S.; Cho, H.-S. SETDB1 regulates SMAD7 expression for breast cancer metastasis. *BMB Rep.* **2019**, *52*, 139–144. [CrossRef] [PubMed]
84. Zhang, H.; Cai, K.; Wang, J.; Wang, X.; Cheng, K.; Shi, F.; Jiang, L.; Zhang, Y.; Dou, J. MiR-7, Inhibited Indirectly by LincRNA HOTAIR, Directly Inhibits SETDB1 and Reverses the EMT of Breast Cancer Stem Cells by Downregulating the STAT3 Pathway. *Stem Cells* **2014**, *32*, 2858–2868. [CrossRef]
85. Ma, L.; Young, J.; Prabhala, H.; Pan, E.; Mestdagh, P.; Muth, D.; Teruya-Feldstein, J.; Reinhardt, F.; Onder, T.T.; Valastyan, S.; et al. miR-9, a MYC/MYCN-activated microRNA, regulates E-cadherin and cancer metastasis. *Nat. Cell Biol.* **2010**, *12*, 247–256. [CrossRef]
86. Mozdarani, H.; Ezzatizadeh, V.; Rahbar Parvaneh, R. The emerging role of the long non-coding RNA HOTAIR in breast cancer development and treatment. *J. Transl. Med.* **2020**, *18*, 152. [CrossRef] [PubMed]
87. Regina, C.; Compagnone, M.; Peschiaroli, A.; Lena, A.; Annicchiarico-Petruzzelli, M.; Piro, M.C.; Melino, G.; Candi, E. Setdb1, a novel interactor of ∆Np63, is involved in breast tumorigenesis. *Oncotarget* **2016**, *7*, 28836–28848. [CrossRef] [PubMed]
88. Jia, D.; Tan, Y.; Liu, H.; Ooi, S.; Li, L.; Wright, K.; Bennett, S.; Addison, C.L.; Wang, L. Cardamonin reduces chemotherapy-enriched breast cancer stem-like cells in vitro and in vivo. *Oncotarget* **2016**, *7*, 771–785. [CrossRef]
89. Creighton, C.J.; Chang, J.C.; Rosen, J.M. Epithelial-Mesenchymal Transition (EMT) in Tumor-Initiating Cells and Its Clinical Implications in Breast Cancer. *J. Mammary Gland Biol. Neoplasia* **2010**, *15*, 253–260. [CrossRef] [PubMed]
90. Wang, C.; Rauscher, F.J.; Cress, W.D.; Chen, J. Regulation of E2F1 function by the nuclear corepressor KAP1. *J. Biol. Chem.* **2007**, *282*, 29902–29909. [CrossRef]
91. Yokoe, T.; Toiyama, Y.; Okugawa, Y.; Tanaka, K.; Ohi, M.; Inoue, Y.; Mohri, Y.; Miki, C.; Kusunoki, M. KAP1 Is Associated With Peritoneal Carcinomatosis in Gastric Cancer. *Ann. Surg. Oncol.* **2010**, *17*, 821–828. [CrossRef]
92. Ho, Y.-J.; Lin, Y.-M.; Huang, Y.-C.; Chang, J.; Yeh, K.-T.; Lin, L.-I.; Gong, Z.; Tzeng, T.-Y.; Lu, J.-W. Significance of histone methyltransferase SETDB1 expression in colon adenocarcinoma. *APMIS* **2017**, *125*, 985–995. [CrossRef]
93. Yu, L.; Ye, F.; Li, Y.-Y.; Zhan, Y.-Z.; Liu, Y.; Yan, H.-M.; Fang, Y.; Xie, Y.-W.; Zhang, F.-J.; Chen, L.-H.; et al. Histone methyltransferase SETDB1 promotes colorectal cancer proliferation through the STAT1-CCND1/CDK6 axis. *Carcinogenesis* **2019**. [CrossRef] [PubMed]
94. Lee, S.; Lee, C.; Hwang, C.Y.; Kim, D.; Han, Y.; Hong, S.N.; Kim, S.-H.; Cho, K.-H. Network Inference Analysis Identifies SETDB1 as a Key Regulator for Reverting Colorectal Cancer Cells into Differentiated Normal-Like Cells. *Mol. Cancer Res.* **2020**, *18*, 118. [CrossRef]
95. Longerich, T. Dysregulation of the epigenetic regulator SETDB1 in liver carcinogenesis—More than one way to skin a cat. *Chin. Clin. Oncol.* **2016**, *5*, 72. [CrossRef]
96. Wong, C.-M.; Wei, L.; Law, C.-T.; Ho, D.W.-H.; Tsang, F.H.-C.; Au, S.L.-K.; Sze, K.M.-F.; Lee, J.M.-F.; Wong, C.C.-L.; Ng, I.O.-L. Up-regulation of histone methyltransferase SETDB1 by multiple mechanisms in hepatocellular carcinoma promotes cancer metastasis. *Hepatology* **2016**, *63*, 474–487. [CrossRef]
97. Brosh, R.; Rotter, V. When mutants gain new powers: News from the mutant p53 field. *Nat. Rev. Cancer* **2009**, *9*, 701–713. [CrossRef]
98. Muller, P.A.J.; Vousden, K.H. Mutant p53 in cancer: New functions and therapeutic opportunities. *Cancer Cell* **2014**, *25*, 304–317. [CrossRef] [PubMed]
99. Ogawa, S.; Fukuda, A.; Matsumoto, Y.; Hanyu, Y.; Sono, M.; Fukunaga, Y.; Masuda, T.; Araki, O.; Nagao, M.; Yoshikawa, T.; et al. et al SETDB1 Inhibits p53-Mediated Apoptosis and is Required for Formation of Pancreatic Ductal Adenocarcinomas in Mice. *Gastroenterology* **2020**, *159*, 682–696.e13. [CrossRef] [PubMed]
100. Parpart, S.; Roessler, S.; Dong, F.; Rao, V.; Takai, A.; Ji, J.; Qin, L.-X.; Ye, Q.-H.; Jia, H.-L.; Tang, Z.-Y.; et al. Modulation of miR-29 expression by α-fetoprotein is linked to the hepatocellular carcinoma epigenome. *Hepatology* **2014**, *60*, 872–883. [CrossRef] [PubMed]

101. Ding, W.; Dang, H.; You, H.; Steinway, S.; Takahashi, Y.; Wang, H.-G.; Liao, J.; Stiles, B.; Albert, R.; Rountree, C.B. miR-200b restoration and DNA methyltransferase inhibitor block lung metastasis of mesenchymal-phenotype hepatocellular carcinoma. *Oncogenesis* **2012**, *1*, e15. [CrossRef] [PubMed]
102. Federico, A. Molecular and functional characterization of the role of the histone methyltransferase SETDB1 in malignant melanoma. *Fac. Nat. Sci. Fac. Math.* **2019**. [CrossRef]
103. Wu, M.; Fan, B.; Guo, Q.; Li, Y.; Chen, R.; Lv, N.; Diao, Y.; Luo, Y. Knockdown of SETDB1 inhibits breast cancer progression by miR-381-3p-related regulation. *Biol. Res.* **2018**, *51*, 39. [CrossRef]
104. Vo, J.N.; Cieslik, M.; Zhang, Y.; Shukla, S.; Xiao, L.; Zhang, Y.; Wu, Y.-M.; Dhanasekaran, S.M.; Engelke, C.G.; Cao, X.; et al. The Landscape of Circular RNA in Cancer. *Cell* **2019**, *176*, 869–881.e13. [CrossRef]
105. Wang, W.; Wang, J.; Zhang, X.; Liu, G. Serum circSETDB1 is a promising biomarker for predicting response to platinum-taxane-combined chemotherapy and relapse in high-grade serous ovarian cancer. *Onco Targets Ther.* **2019**, *12*, 7451–7457. [CrossRef]
106. Wakabayashi, Y.; Tamiya, T.; Takada, I.; Fukaya, T.; Sugiyama, Y.; Inoue, N.; Kimura, A.; Morita, R.; Kashiwagi, I.; Takimoto, T.; et al. Histone 3 lysine 9 (H3K9) methyltransferase recruitment to the interleukin-2 (IL-2) promoter is a mechanism of suppression of IL-2 transcription by the transforming growth factor-β-smad pathway. *J. Biol. Chem.* **2011**, *286*, 35456–35465. [CrossRef] [PubMed]
107. Dou, Y.; Kawaler, E.A.; Cui Zhou, D.; Gritsenko, M.A.; Huang, C.; Blumenberg, L.; Karpova, A.; Petyuk, V.A.; Savage, S.R.; Satpathy, S.; et al. Proteogenomic Characterization of Endometrial Carcinoma. *Cell* **2020**, *180*, 729–748.e26. [CrossRef]
108. Sun, Y.; Wei, M.; Ren, S.-C.; Chen, R.; Xu, W.-D.; Wang, F.-B.; Lu, J.; Shen, J.; Yu, Y.-W.; Hou, J.-G.; et al. Histone methyltransferase SETDB1 is required for prostate cancer cell proliferation, migration and invasion. *Asian J. Androl.* **2014**, *16*, 319–324. [CrossRef] [PubMed]
109. Robbez-Masson, L.; Tie, C.H.C.; Rowe, H.M. Cancer cells, on your histone marks, get SETDB1, silence retrotransposons, and go! *J. Cell Biol.* **2017**, *216*, 3429–3431. [CrossRef] [PubMed]
110. Mita, P.; Savas, J.N.; Briggs, E.M.; Ha, S.; Gnanakkan, V.; Yates, J.R., III; Robins, D.M.; David, G.; Boeke, J.D.; Garabedian, M.J.; et al. URI Regulates KAP1 Phosphorylation and Transcriptional Repression via PP2A Phosphatase in Prostate Cancer Cells. *J. Biol. Chem.* **2016**, *291*, 25516–25528. [CrossRef]
111. Orouji, E.; Federico, A.; Larribère, L.; Novak, D.; Lipka, D.B.; Assenov, Y.; Sachindra, S.; Hüser, L.; Granados, K.; Gebhardt, C.; et al. Histone methyltransferase SETDB1 contributes to melanoma tumorigenesis and serves as a new potential therapeutic target. *Int. J. Cancer* **2019**, *145*, 3462–3477. [CrossRef] [PubMed]
112. Kostaki, M.; Manona, A.D.; Stavraka, I.; Korkolopoulou, P.; Levidou, G.; Trigka, E.-A.; Christofidou, E.; Champsas, G.; Stratigos, A.J.; Katsambas, A.; et al. High-frequency p16INK4A promoter methylation is associated with histone methyltransferase SETDB1 expression in sporadic cutaneous melanoma. *Exp. Dermatol.* **2014**, *23*, 332–338. [CrossRef]
113. Ceol, C.J.; Houvras, Y.; Jane-Valbuena, J.; Bilodeau, S.; Orlando, D.A.; Battisti, V.; Fritsch, L.; Lin, W.M.; Hollmann, T.J.; Ferré, F.; et al. The histone methyltransferase SETDB1 is recurrently amplified in melanoma and accelerates its onset. *Nature* **2011**, *471*, 513–517. [CrossRef]
114. Idilli, A.I.; Precazzini, F.; Mione, M.C.; Anelli, V. Zebrafish in Translational Cancer Research: Insight into Leukemia, Melanoma, Glioma and Endocrine Tumor Biology. *Genes* **2017**, *8*, 236. [CrossRef]
115. Koide, S.; Takubo, K.; Oshima, M.; Miyagi, S.; Saraya, A.; Wang, C.; Matsui, H.; Kimura, H.; Shinkai, Y.; Suda, T.; et al. Histone methyltransferase Setdb1 regulates energy metabolism in hematopoietic stem and progenitor cells. *Exp. Hematol.* **2015**, *43*, S73. [CrossRef]
116. Ropa, J.; Saha, N.; Hu, H.; Peterson, L.F.; Talpaz, M.; Muntean, A.G. SETDB1 mediated histone H3 lysine 9 methylation suppresses MLL-fusion target expression and leukemic transformation. *Haematologica* **2019**, *105*. [CrossRef] [PubMed]
117. Ropa, J.; Saha, N.; Chen, Z.; Serio, J.; Chen, W.; Mellacheruvu, D.; Zhao, L.; Basrur, V.; Nesvizhskii, A.I.; Muntean, A.G. PAF1 complex interactions with SETDB1 mediate promoter H3K9 methylation and transcriptional repression of Hoxa9 and Meis1 in acute myeloid leukemia. *Oncotarget* **2018**, *9*, 22123–22136. [CrossRef]
118. Cuellar, T.L.; Herzner, A.-M.; Zhang, X.; Goyal, Y.; Watanabe, C.; Friedman, B.A.; Janakiraman, V.; Durinck, S.; Stinson, J.; Arnott, D.; et al. Silencing of retrotransposons by SETDB1 inhibits the interferon response in acute myeloid leukemia. *J. Cell Biol.* **2017**, *216*, 3535–3549. [CrossRef]
119. Kang, Y.-K. SETDB1 in early embryos and embryonic stem cells. *Curr. Issues Mol. Biol.* **2015**, *17*, 1–10. [CrossRef] [PubMed]
120. Ruzinova, M.B.; Benezra, R. Id proteins in development, cell cycle and cancer. *Trends Cell Biol.* **2003**, *13*, 410–418. [CrossRef]
121. Zhou, J.-D.; Ma, J.-C.; Zhang, T.-J.; Li, X.-X.; Zhang, W.; Wu, D.-H.; Wen, X.-M.; Xu, Z.-J.; Lin, J.; Qian, J. High bone marrow ID2 expression predicts poor chemotherapy response and prognosis in acute myeloid leukemia. *Oncotarget* **2017**, *8*, 91979–91989. [CrossRef] [PubMed]
122. Mayers, J.G.; Gokgoz, N.; Wunder, J.S.; Andrulis, I.L. Abstract 5507: Investigation of the effects of alterations in the glutamate receptor, GRIK2 on osteosarcoma tumorigenesis. *Cancer Res.* **2017**, *77*, 5507. [CrossRef]
123. Mayers, J.G.; Gokgoz, N.; Wunder, J.S.; Andrulis, I.L. Abstract B39: Investigation of the effects of alterations in the glutamate receptor, GRIK2, on osteosarcoma tumorigenesis. *Clin. Cancer Res.* **2018**, *24*, B39. [CrossRef]
124. Chase, K.A.; Gavin, D.P.; Guidotti, A.; Sharma, R.P. Histone methylation at H3K9: Evidence for a restrictive epigenome in schizophrenia. *Schizophr. Res.* **2013**, *149*, 15–20. [CrossRef]

125. Hossein Fatemi, S.; Stary, J.M.; Earle, J.A.; Araghi-Niknam, M.; Eagan, E. GABAergic dysfunction in schizophrenia and mood disorders as reflected by decreased levels of glutamic acid decarboxylase 65 and 67 kDa and Reelin proteins in cerebellum. *Schizophr. Res.* **2005**, *72*, 109–122. [CrossRef]
126. Jindal, R.D.; Pillai, A.K.; Mahadik, S.P.; Eklund, K.; Montrose, D.M.; Keshavan, M.S. Decreased BDNF in patients with antipsychotic naïve first episode schizophrenia. *Schizophr. Res.* **2010**, *119*, 47–51. [CrossRef]
127. Avramopoulos, D.; Lasseter, V.K.; Fallin, M.D.; Wolyniec, P.S.; McGrath, J.A.; Nestadt, G.; Valle, D.; Pulver, A.E. Stage II follow-up on a linkage scan for bipolar disorder in the Ashkenazim provides suggestive evidence for chromosome 12p and the GRIN2B gene. *Genet. Med.* **2007**, *9*, 745–751. [CrossRef] [PubMed]
128. Bharadwaj, R.; Peter, C.J.; Jiang, Y.; Roussos, P.; Vogel-Ciernia, A.; Shen, E.Y.; Mitchell, A.C.; Mao, W.; Whittle, C.; Dincer, A.; et al. Conserved higher-order chromatin regulates NMDA receptor gene expression and cognition. *Neuron* **2014**, *84*, 997–1008. [CrossRef] [PubMed]
129. Jury, N.; Abarzua, S.; Diaz, I.; Guerra, M.V.; Ampuero, E.; Cubillos, P.; Martinez, P.; Herrera-Soto, A.; Arredondo, C.; Rojas, F.; et al. Widespread loss of the silencing epigenetic mark H3K9me3 in astrocytes and neurons along with hippocampal-dependent cognitive impairment in C9orf72 BAC transgenic mice. *Clin. Epigenet.* **2020**, *12*, 32. [CrossRef] [PubMed]
130. Wu, T.; Sun, X.-Y.; Yang, X.; Liu, L.; Tong, K.; Gao, Y.; Hao, J.-R.; Cao, J.; Gao, C. Histone H3K9 Trimethylation Downregulates the Expression of Brain-Derived Neurotrophic Factor in the Dorsal Hippocampus and Impairs Memory Formation During Anaesthesia and Surgery. *Front. Mol. Neurosci.* **2019**, *12*, 246. [CrossRef] [PubMed]
131. Xu, Q.; Goldstein, J.; Wang, P.; Gadi, I.K.; Labreche, H.; Rehder, C.; Wang, W.-P.; McConkie, A.; Xu, X.; Jiang, Y.-H. Chromosomal microarray analysis in clinical evaluation of neurodevelopmental disorders-reporting a novel deletion of SETDB1 and illustration of counseling challenge. *Pediatr. Res.* **2016**, *80*, 371–381. [CrossRef]
132. Iossifov, I.; O'Roak, B.J.; Sanders, S.J.; Ronemus, M.; Krumm, N.; Levy, D.; Stessman, H.A.; Witherspoon, K.T.; Vives, L.; Patterson, K.E.; et al. The contribution of de novo coding mutations to autism spectrum disorder. *Nature* **2014**, *515*, 216–221. [CrossRef]
133. Cukier, H.N.; Lee, J.M.; Ma, D.; Young, J.I.; Mayo, V.; Butler, B.L.; Ramsook, S.S.; Rantus, J.A.; Abrams, A.J.; Whitehead, P.L.; et al. The expanding role of MBD genes in autism: Identification of a MECP2 duplication and novel alterations in MBD5, MBD6, and SETDB1. *Autism Res.* **2012**, *5*, 385–397. [CrossRef]
134. Veazey, K.J.; Parnell, S.E.; Miranda, R.C.; Golding, M.C. Dose-dependent alcohol-induced alterations in chromatin structure persist beyond the window of exposure and correlate with fetal alcohol syndrome birth defects. *Epigenet. Chromatin* **2015**, *8*, 39. [CrossRef]
135. Basavarajappa, B.S.; Subbanna, S. Epigenetic Mechanisms in Developmental Alcohol-Induced Neurobehavioral Deficits. *Brain Sci.* **2016**, *6*, 12. [CrossRef]
136. Thomas, J.D.; Idrus, N.M.; Monk, B.R.; Dominguez, H.D. Prenatal choline supplementation mitigates behavioral alterations associated with prenatal alcohol exposure in rats. *Birth Defects Res. A Clin. Mol. Teratol.* **2010**, *88*, 827–837. [CrossRef]
137. Bekdash, R.A.; Zhang, C.; Sarkar, D.K. Gestational choline supplementation normalized fetal alcohol-induced alterations in histone modifications, DNA methylation, and proopiomelanocortin (POMC) gene expression in β-endorphin-producing POMC neurons of the hypothalamus. *Alcohol. Clin. Exp. Res.* **2013**, *37*, 1133–1142. [CrossRef] [PubMed]
138. Chase, K.A.; Sharma, R.P. Nicotine induces chromatin remodelling through decreases in the methyltransferases GLP, G9a, Setdb1 and levels of H3K9me2. *Int. J. Neuropsychopharmacol.* **2013**, *16*, 1129–1138. [CrossRef]
139. Ryu, H.; Lee, J.; Hagerty, S.W.; Soh, B.Y.; McAlpin, S.E.; Cormier, K.A.; Smith, K.M.; Ferrante, R.J. ESET/SETDB1 gene expression and histone H3 (K9) trimethylation in Huntington's disease. *Proc. Natl. Acad. Sci. USA* **2006**, *103*, 19176–19181. [CrossRef] [PubMed]
140. Valcárcel-Ocete, L.; Alkorta-Aranburu, G.; Iriondo, M.; Fullaondo, A.; García-Barcina, M.; Fernández-García, J.M.; Lezcano-García, E.; Losada-Domingo, J.M.; Ruiz-Ojeda, J.; Álvarez de Arcaya, A.; et al. Exploring Genetic Factors Involved in Huntington Disease Age of Onset: E2F2 as a New Potential Modifier Gene. *PLoS ONE* **2015**, *10*, e0131573. [CrossRef]
141. Irmak, D.; Fatima, A.; Gutiérrez-Garcia, R.; Rinschen, M.M.; Wagle, P.; Altmüller, J.; Arrigoni, L.; Hummel, B.; Klein, C.; Frese, C.K.; et al. Mechanism suppressing H3K9 trimethylation in pluripotent stem cells and its demise by polyQ-expanded huntingtin mutations. *Hum. Mol. Genet.* **2018**, *27*, 4117–4134. [CrossRef] [PubMed]
142. Lee, J.; Hwang, Y.J.; Kim, Y.; Lee, M.Y.; Hyeon, S.J.; Lee, S.; Kim, D.H.; Jang, S.J.; Im, H.; Min, S.-J.; et al. Remodeling of heterochromatin structure slows neuropathological progression and prolongs survival in an animal model of Huntington's disease. *Acta Neuropathol.* **2017**, *134*, 729–748. [CrossRef]
143. Ferrante, R.J.; Kubilus, J.K.; Lee, J.; Ryu, H.; Beesen, A.; Zucker, B.; Smith, K.; Kowall, N.W.; Ratan, R.R.; Luthi-Carter, R.; et al. Histone deacetylase inhibition by sodium butyrate chemotherapy ameliorates the neurodegenerative phenotype in Huntington's disease mice. *J. Neurosci.* **2003**, *23*, 9418–9427. [CrossRef] [PubMed]
144. Ryu, H.; Lee, J.; Olofsson, B.A.; Mwidau, A.; Dedeoglu, A.; Escudero, M.; Flemington, E.; Azizkhan-Clifford, J.; Ferrante, R.J.; Ratan, R.R. Histone deacetylase inhibitors prevent oxidative neuronal death independent of expanded polyglutamine repeats via an Sp1-dependent pathway. *Proc. Natl. Acad. Sci. USA* **2003**, *100*, 4281–4286. [CrossRef]

145. Steffan, J.S.; Bodai, L.; Pallos, J.; Poelman, M.; McCampbell, A.; Apostol, B.L.; Kazantsev, A.; Schmidt, E.; Zhu, Y.-Z.; Greenwald, M.; et al. Histone deacetylase inhibitors arrest polyglutamine-dependent neurodegeneration in Drosophila. *Nature* **2001**, *413*, 739–743. [CrossRef] [PubMed]
146. McCampbell, A.; Taye, A.A.; Whitty, L.; Penney, E.; Steffan, J.S.; Fischbeck, K.H. Histone deacetylase inhibitors reduce polyglutamine toxicity. *Proc. Natl. Acad. Sci. USA* **2001**, *98*, 15179–15184. [CrossRef]
147. Simó-Riudalbas, L.; Esteller, M. Targeting the histone orthography of cancer: Drugs for writers, erasers and readers. *Br. J. Pharmacol.* **2015**, *172*, 2716–2732. [CrossRef] [PubMed]
148. Hwang, Y.J.; Hyeon, S.J.; Kim, Y.; Lim, S.; Lee, M.Y.; Kim, J.; Londhe, A.M.; Gotina, L.; Kim, Y.; Pae, A.N.; et al. Modulation of SETDB1 activity by APQ ameliorates heterochromatin condensation, motor function, and neuropathology in a Huntington's disease mouse model. *J. Enzym. Inhib. Med. Chem.* **2021**, *36*, 856–868. [CrossRef]
149. Jiang, Y.; Matevossian, A.; Guo, Y.; Akbarian, S. Setdb1-mediated histone H3K9 hypermethylation in neurons worsens the neurological phenotype of Mecp2-deficient mice. *Neuropharmacology* **2011**, *60*, 1088–1097. [CrossRef]
150. Kim, Y.; Wang, S.E.; Jiang, Y.-H. Epigenetic therapy of Prader-Willi syndrome. *Transl. Res.* **2019**, *208*, 105–118. [CrossRef]
151. Cruvinel, E.; Budinetz, T.; Germain, N.; Chamberlain, S.; Lalande, M.; Martins-Taylor, K. Reactivation of maternal SNORD116 cluster via SETDB1 knockdown in Prader-Willi syndrome iPSCs. *Hum. Mol. Genet.* **2014**, *23*, 4674–4685. [CrossRef] [PubMed]
152. Buiting, K.; Saitoh, S.; Gross, S.; Dittrich, B.; Schwartz, S.; Nicholls, R.D.; Horsthemke, B. Inherited microdeletions in the Angelman and Prader–Willi syndromes define an imprinting centre on human chromosome 15. *Nat. Genet.* **1995**, *9*, 395–400. [CrossRef]
153. Langouët, M.; Glatt-Deeley, H.R.; Chung, M.S.; Dupont-Thibert, C.M.; Mathieux, E.; Banda, E.C.; Stoddard, C.E.; Crandall, L.; Lalande, M. Zinc finger protein 274 regulates imprinted expression of transcripts in Prader-Willi syndrome neurons. *Hum. Mol. Genet.* **2017**, *27*, 505–515. [CrossRef] [PubMed]
154. Lee, J.-H.; Demarest, T.G.; Babbar, M.; Kim, E.W.; Okur, M.N.; De, S.; Croteau, D.L.; Bohr, V.A. Cockayne syndrome group B deficiency reduces H3K9me3 chromatin remodeler SETDB1 and exacerbates cellular aging. *Nucleic Acids Res.* **2019**, *47*, 8548–8562. [CrossRef]
155. Olson, E.N. Gene regulatory networks in the evolution and development of the heart. *Science* **2006**, *313*, 1922–1927. [CrossRef] [PubMed]
156. Mysliwiec, M.R.; Bresnick, E.H.; Lee, Y. Endothelial Jarid2/Jumonji is required for normal cardiac development and proper Notch1 expression. *J. Biol. Chem.* **2011**, *286*, 17193–17204. [CrossRef] [PubMed]
157. Mysliwiec, M.R.; Carlson, C.D.; Tietjen, J.; Hung, H.; Ansari, A.Z.; Lee, Y. Jarid2 (Jumonji, AT rich interactive domain 2) regulates NOTCH1 expression via histone modification in the developing heart. *J. Biol. Chem.* **2012**, *287*, 1235–1241. [CrossRef] [PubMed]
158. Južnić, L.; Peuker, K.; Strigli, A.; Brosch, M.; Herrmann, A.; Häsler, R.; Koch, M.; Matthiesen, L.; Zeissig, Y.; Löscher, B.-S.; et al. SETDB1 is required for intestinal epithelial differentiation and the prevention of intestinal inflammation. *Gut* **2021**, *70*, 485–498. [CrossRef] [PubMed]

Review

NSD1: A Lysine Methyltransferase between Developmental Disorders and Cancer

Samantha Tauchmann and Juerg Schwaller *

University Children's Hospital, Department of Biomedicine, University of Basel, 4031 Basel, Switzerland; Samantha.Tauchmann@unibas.ch
* Correspondence: J.Schwaller@unibas.ch; Tel.: +41-61-265-3517 or +41-61-265-3504

Abstract: Recurrent epigenomic alterations associated with multiple human pathologies have increased the interest in the nuclear receptor binding SET domain protein 1 (NSD1) lysine methyltransferase. Here, we review the current knowledge about the biochemistry, cellular function and role of NSD1 in human diseases. Several studies have shown that NSD1 controls gene expression by methylation of lysine 36 of histone 3 (H3K36me1/2) in a complex crosstalk with de novo DNA methylation. Inactivation in flies and mice revealed that NSD1 is essential for normal development and that it regulates multiple cell type-specific functions by interfering with transcriptional master regulators. In humans, putative loss of function NSD1 mutations characterize developmental syndromes, such as SOTOS, as well as cancer from different organs. In pediatric hematological malignancies, a recurrent chromosomal translocation forms a NUP98-NSD1 fusion with SET-dependent leukemogenic activity, which seems targetable by small molecule inhibitors. To treat or prevent diseases driven by aberrant NSD1 activity, future research will need to pinpoint the mechanistic correlation between the NSD1 gene dosage and/or mutational status with development, homeostasis, and malignant transformation.

Keywords: NSD1; H3K36; SOTOS; cancer; NUP98-NSD1; AML

Citation: Tauchmann, S.; Schwaller, J. NSD1: A Lysine Methyltransferase between Developmental Disorders and Cancer. *Life* **2021**, *11*, 877. https://doi.org/10.3390/life11090877

Academic Editors: Albert Jeltsch and Arunkumar Dhayalan

Received: 14 July 2021
Accepted: 23 August 2021
Published: 25 August 2021

Publisher's Note: MDPI stays neutral with regard to jurisdictional claims in published maps and institutional affiliations.

Copyright: © 2021 by the authors. Licensee MDPI, Basel, Switzerland. This article is an open access article distributed under the terms and conditions of the Creative Commons Attribution (CC BY) license (https://creativecommons.org/licenses/by/4.0/).

1. Introduction

Gene expression is controlled by temporarily and spatially coordinated modification of chromatin. Hereby, the N-terminal tails of the histone octamers formed by H2A, H2B, H3, and H4 undergo post-translational modifications including methylation, phosphorylation, acetylation, ubiquitylation and sumoylation executed by proteins acting as "writers" of an epigenetic code [1]. Histone lysine methyltransferases (KMTs) have been characterized as critical regulators of multiple cellular processes including DNA replication, DNA damage response, cell cycle progression or cytokinesis. Genetic lesions (mutations, translocations) as well as altered gene expression functionally affecting KMTs are recurrently found in various human malignancies but also in developmental disorders [2]. An increasing number of compounds that selectively target aberrantly activated KMTs have been developed and underwent clinical trials as novel cancer therapeutics [3]. In this review, we summarize the current knowledge on the nuclear receptor binding SET domain protein 1 (NSD1, aka KMT3B), a H3 lysine 36 (H3K36) methyltransferase that has recently gained attention because of its critical role in several human pathologies, such as germline developmental syndromes and cancers.

2. Identification and Structure of NSD1

NSD1 was discovered in a yeast two hybrid screen for proteins associated with the ligand-binding domain (LBD) of the retinoic acid receptor alpha (RARa). NSD1 was shown to interact directly with the LBD of several nuclear receptors, including the retinoic acid (RAR), thyroid (TR), retinoid X (RXR), and estrogen (ER) receptors. These interactions are mediated by two distinct nuclear receptor interaction domains (NID) in NSD1, NID^{-L} and

NID^{+L}. NID^{-L} interacts with RAR and TR when a ligand is absent, whereas NID^{+L} binds RAR, TR, RXR and ER when a ligand is present, indicating that NSD1 controls repression or activation of target genes by distinct binding to nuclear receptors [4]. Similarly, a yeast-two-hybrid screen, using the LBD of the androgen receptor (AR) and the orphan receptor TR4 as baits, allowed for the detection of a human androgen receptor-associated protein of 267 Kd (ARA267) that showed the highest homology to mouse NSD1. ARA-267 (which turned out to be NSD1) was shown to be widely expressed in different tissues, with highest levels in lymph nodes. Functional studies have suggested its primary role as co-activator of AR controlled transcription [5].

The NSD1 gene maps to human chromosome 5q35.3, close to the telomere, with an 8088 bp open reading frame (ORF) [6]. Interrogation of ensembl.org indicates the existence of three NSD1 isoforms produced by alternative splicing, one long isoform and two shorter ones, with additional potential smaller isoforms that have been computationally mapped [7]. NSD1 isoform 1 (NSD1(204), Q96L73-1; ARA267-beta) with an ORF starting at exon 2 and ending at exon 23 has been chosen as the canonical sequence and is 2696 amino acids (aa) long, resulting in 296 kDa [6,7]. NSD1 isoform 2 (NSD1(202), Q96L73-2; ARA267-alpha) is 2427aa and 267 kDa and differs at the 5'UTR, compared to isoform 1 where 1-269aa are missing [8]. Furthermore, through an mRNA splicing event, a 740 bp long intron within exon 2 is removed, leading to an additional exon with 90 bp (exon 3), resulting in a total length of 24 exons. NSD1 isoform 3 (NSD1(201), Q96L73-3) is similar to isoform 2 with a 740 bp spliced intron; however, it differs by lacking 310-412aa, thereby resulting in a smaller intron between exon 1 and 2, with 841 bp. [7] Furthermore, exon 24 has a length of 1931 bp, which is smaller compared to isoform 2 that has a 6379 bp long exon 24. However, the ORF for both, isoform 2 and 3, starts at exon 2 and ends at exon 24, resulting in the same length and size of the protein (Figure 1A). Notably, the three isoforms (204, 202, 201) encode for proteins that contain all of the functionally characterized NSD1 domains, suggesting that variations close to the 5' end of the ORF may be linked to regulation of gene expression.

Interrogating public databases suggests ubiquitous NSD1 expression (or its related homologs) in most tissues from various organisms. Somehow higher NSD1 mRNA levels seem to be expressed in normal brain, pancreas, male reproductive tract, and hematopoietic organs such as the bone marrow and lymphoid tissues [9]. Significant NSD1 mRNA expression in bone marrow polymorphonuclear cells, CD4, CD8 and NK cells is also supported by genevisible.com [10]. Integrated expression analysis in normal tissues and cell lines indicates abundant NSD1 protein expression in B-lymphocytes, CD8 T cells, platelets, fetal brain, retina, fetal gut, rectum, liver, adipocytes, pancreas, placenta and ovaries [11]. However, there seems to be an overall low tissue specificity for NSD1 protein expression.

The NSD1 protein contains two NIDs, two proline-tryptophan-tryptophan-proline (PWWP) domains, five plant homeodomains (PHD), an atypical (C5HCH) plant homeo-domain (PHD) finger and a catalytic domain (CD) composed of a pre-SET (AWS), Su(var)3–9, Enhancer-of-zeste, Trithorax (SET) and post-SET domain [6]. The aa sequences from both the PWWP-I and PHD-II domains are 100% identical between mouse and human NSD1, while the SET domain is 99% identical. A 97% homology between human and mouse was found for PHD-I and PHD-III. PWWP-II was 95% conserved whereas the NID^{-L} and NID^{+L} showed the least identity, with 88 and 83%, respectively [6] (Figure 1B).

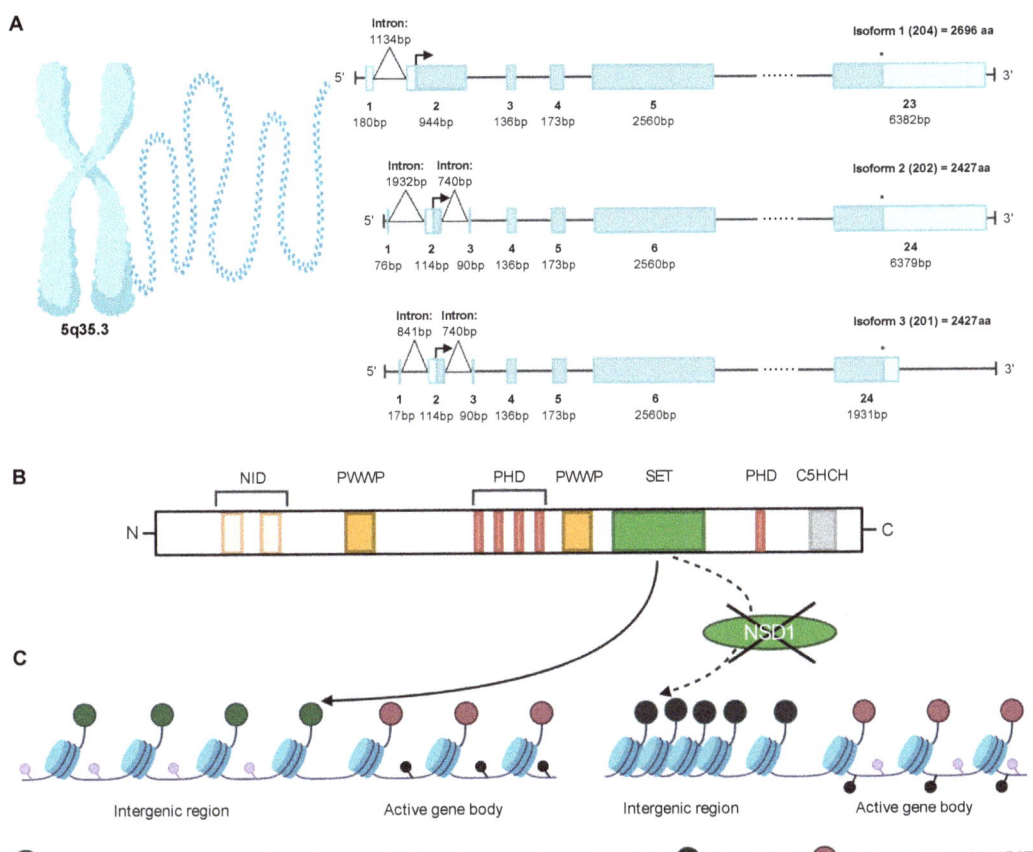

Figure 1. NSD1 gene and protein architecture and function. (**A**) Exon structure of the three different *NSD1* isoforms. Isoform 1 (204) contains 23 exons, whereas isoform 2 (202) and 3 (201) contain 24 exons. Open reading frame is shown by an arrow as start and asterisk at the end. (**B**) All three major NSD1 isoforms contain two nuclear receptor interacting domains (NID), two proline-tryptophan-tryptophan-proline (PWWP), five plant homodomain zinc fingers (PHD), the catalytic Su(var)3-9, enhancer-of-zeste, Trithorax (SET) and the C-terminal C5HCH (Cys-His) domain. (**C**) The NSD1 SET domain methylates H3K36me1/2 predominantly at intergenic regions allowing recruitment of DNMT3A and facilitating H3K36me3 by other KMTs allowing recruitment of DNMT3B to active gene bodies. Reduced NSD1 catalytic activity results in loss of H3K36me2 marks, which allows spreading of PRC2-mediated H3K27me3 marks at intergenic regions and redistribution of DNTM3A-mediated DNA methylation to active gene bodies.

NSD1 is a member of a SET-containing methyltransferase protein family, which contains two additional members, NSD2 and NSD3. Both are significantly smaller than NSD1 due to the absence of the NID^{-L} and NID^{+L} in the N-terminus. NSD2, also called Wolf-Hirschhorn Syndrome Candidate 1 (WHSC1) or Multiple Myeloma SET domain protein (MMSET) is located on the short arm of chromosome 4 (4p16.3), a locus targeted by a recurrent t(4;14)(p16;q32) translocation found in up to 20% of patients with multiple myeloma. NSD2 contains a PWWP domain, a SET domain, PHD zinc fingers and a high mobility group (HMG) box with 75% homology to NSD1 [6]. Similarly, NSD3, also called WHSC1L1, contains a PWWP, SET and PHD zinc finger domains but lacks the HMG box and is therefore only 68% identical to NSD1. NSD3 was mapped to the short arm of chromosome 8 (8p11.2), a locus involved in cancer-associated amplifications and translocations, such as t(8;11)(p11;p15) associated with myelodysplastic syndromes (MDS) and acute myeloid leukemia (AML) [12].

3. NSD1 Is an Epigenetic Regulator Writing and Reading Chromatin Marks

3.1. The SET Domain Mediates the Catalytic Activity

NSD1-3 have been functionally characterized as histone methyltransferases (HMT) due to its conserved catalytic SET domain involved in methylation of histone 3-K4, -K9, -K27, -K36, and -K79, and methylation of histone 4-K20 [13]. Members of the NSD family seem to differ from other protein lysine methyltransferases (PKMTs) as in the absence of a ligand, the SET histone binding site is closed, preventing any access to the catalytic groove [14]. In general, SET domains are approximately 130 aa long and contain binding sites for the lysine ligand and the co-factor S-adenosylmethionine (SAM), which donates methyl groups. The C-terminal post-SET domain can form a loop, thereby regulating substrate binding by forming one side of the SAM binding pocket [15]. The PWWP domains of NSD1 are critical for binding to H3K36me marks but also to DNA, whereas the PHD zinc fingers are needed for interactions with other methylated histones, such as H3K4 and H3K9 [16].

Several, mostly in vitro studies, reported other histones (H4K20) and non-histone proteins as potential NSD1 substrates. Berdasco et al. found that loss of NSD1 by 5′-CpG island DNA hypermethylation interferes with histone lysine methylation not only by decreasing the levels of H3K36me3 but also of H4K20me3 [17]. Lu et al. suggested that NSD1 acts (in tandem with the F-box and leucine-rich repeat protein 11 (FBXL11) demethylase) as a regulator of the NFκB signaling pathway indicated by reversible methylation of K218 and K221 of NFκB-p65. However, these observations were based on associations upon NSD1 overexpression or knockdown, and not validated in biochemical assays [18]. Using a biochemical approach, others were unable to validate NSD1-SET mediated methylation of H4K20 and NFκB-p65. However, they found in addition to H3K36, H1 linker histones, in particular H1.5 (K168) but also H1.2 (K168) and H1.3 (K169) as well as H4 (K44), as potential NSD1 substrates. Furthermore, they identified peptides of 50 non-histone proteins recognized by NSD1-SET. NSD1 methylation on two of those non-histone proteins, the chromatin remodeler ATRX (K1033) and the small nuclear RNA-binding protein U3 (K189), could be validated in vitro [19].

3.2. NSD1 Chromatin Modification and Regulation

Methylation of histone H3K36 occurs in three states mono-, di- and trimethylation and is primarily described as a hallmark of active transcription. Several KMTs were shown to be recruited by RNA polymerase II and deposit H3K36me3 over gene bodies essential for transcriptional elongation, whereas H3K36me2 is enriched at intergenic regions or promoters [20,21].

Functional studies have shown that NSD1 catalyzes mono- and dimethylation of H3K36 specifically. NSD2 leads to mono- and dimethylation of H3K36, whereas it prefers to catalyze dimethylation compared to monomethylation. Interestingly, H3K36me2 marks are not only set by NSD1-3 but also by ASH1L (ASH1 Like Histone Lysine Methyltransferase), whereas SETD2 (SET Domain Containing 2, Histone Lysine Methyltransferase) is the only enzyme able to introduce K36 methylation up to the trimethylation stage (H3K36me3) [22]. Previous studies have shown that NSD1 exhibits an autoinhibitory state that is relieved by binding to nucleosomes enabling dimethylation of histone H3 at Lys36 (H3K36) [23]. To better understand H3K36 recognition by NSD proteins, Li et al. recently solved the cryo-electron microscopy structures of mononucleosome-bound NSD2 and NSD3 [24]. They observed that binding of NSD2 and NSD3 causes DNA near the linker region to unwrap, facilitating insertion of the catalytic core between the histone octamer and the unwrapped DNA segment. Multiple DNA- and histone-specific contacts between NSD and the nucleosome precisely defined the position of the enzyme on the nucleosome.

Yuan et al. suggested that H2A mono ubiquitination (ubH2A) impairs the enzymatic activity of HMTs including NSD1, indicating another layer of complexity in NSD1 regulation [25]. Notably, ubH2A can recruit the Polycomb Repressive Complex 2 (PRC2). PRC2 regulates gene expression by methylation of lysine 27 of histone 3 (H3K27) marks through its enzymatic component EZH2 (Enhancer Of Zeste 2 Polycomb Repressive Complex 2 Subunit). The different degrees of H3K27 methylation (H3K27me1/me2/me3) have distinct genomic distributions: H3K27me1 is enriched within gene bodies of actively transcribed genes; H3K27me2 is abundant, marking 50–70% of total histone H3 and covering inter- and intragenic regions. H3K27me3 (present on 5–10% of histone H3) is strongly enriched at sites overlapping with PRC2 binding and is considered the hallmark of PRC2-mediated gene repression [26]. Streubel et al. found that genetic inactivation of *Nsd1* leads to genome-wide expansion of H3K27me3 not only at PRC2 target genes but also as de novo accumulation within broad H3K27me2 marked domains. Thus, NSD1-mediated H3K36me2 seems crucial to restrict PRC2 activity by preventing uncontrolled deposition of H3K27me3 [27].

3.3. Functional Interaction with DNA Methyltransferases

In addition to PRC2, epigenomic regulation by NSD1 also involves DNA methyltransferases (DNMTs), which methylate CpG dinucleotides. In total, there are five different DNMTs, of which three play a role in DNA methylation. DNMT1 is important to maintain methylation during DNA replication and acts in response to DNA damage, while DNMT3A and DNMT3B are responsible for de novo methylation [28]. DNMT3 enzymes are recruited through their PWWP domain to methylated H3K36 [29]. DNMT3B colocalizes selectively with H3K36me3 and methylates active gene bodies to enhance gene expression. DNMT3A binds more strongly to H3K36me2 than to H3K36me3 and preferentially methylates intergenic chromatin, which often co-occurs with PRC2-mediated H3K27me2 as well as NSD1-mediated-H3K36me2 [30,31]. Functional studies in ES cells revealed that ablation of NSD1 results in redistribution of DNMT3A to H3K36me gene bodies and reduced methylation of intergenic DNA [32]. Likewise, expression of a H3K36M mutant (not recognized by NSD1), resulted in an increase in H3K27me3 at intergenic regions and redistribution of PRC2 resulting in aberrant gene expression [31] (Figure 1C).

3.4. Regulation of Gene Expression

Depletion of NSD1 leads to both up- and down-regulation of gene expression, indicating NSD1 functions as transcriptional co-activator and co-repressor. In earlier studies, distinct stretches of the NSD1 ORF sequence were tested for their transcriptional activity by fusing them to a GAL4 DNA binding domain, which identified a region (1084–1400 aa) with a significant repressive activity in vitro. This suggested that NSD1 has a silencing domain that functions autonomously, which might act as corepressor for unliganded TR and RAR [4]. Although the mechanisms of gene repression by NSD1 are not fully understood, experimental work suggested that transcription is impaired through binding of the NSD1 C5HCH domain (adjacent to the C-terminus of PHD-V) to the C2HR zinc finger motif of ZNF496 (aka NSD1 interacting zinc finger protein 1, NIZP1) tethered on RNA polymerase II promoters [33,34]. In contrast, only expression of an N-terminal stretch of NSD1 (1–731) fused to the estrogen-receptor alpha DNA binding domain showed strong transcriptional activation in yeast but not mammalian cells [4]. More recent work demonstrated that loss of NSD1 increases H3K27ac associated with active enhancers in mESCs. NSD1 was shown to recruit the histone deacetylase 1 (HDAC1), which can deacetylate H3K27ac. Hence, inactivation of HDAC1 recapitulated increased H3K27ac similar to loss of NSD1 [35]. Overall, although these studies provided some insights into the role of NSD1 as a transcriptional co-repressor, its function as a co-activator, particularly in the context of specific nuclear receptors remains poorly understood.

4. Cellular Functions of NSD1

Earlier in vitro studies showed that NSD1 overexpression allowed NIH-3T3 fibroblasts to grow in reduced serum levels, whereas vector-transfected control cells did not. Overexpression of *Schizosaccharomyces pombe* SET2, which contains a SET domain but no PHD or PWWP domains, conferred reduced serum dependence, indicating that the catalytic NSD1 activity is able to modulate serum dependence [36].

4.1. Modeling NSD1 Activity in the Fly

To better understand the function of NSD1 in vivo, gain- and loss-of-function studies in various organisms have been performed. Ubiquitous NSD (the fly NSD1 homolog) overexpression in *Drosophila melanogaster* caused developmental delay and reduced body size at the larval stage, resulting in pupal lethality. Targeted overexpression in various tissues led to significant alterations that rescued RNAi-based NSD knockdown. NSD overexpression enhanced the transcription of pro-apoptotic genes and led to caspase activation. Notably, NSD-overexpression associated wing atrophy was reduced by a loss-of-function mutation in Jun N-terminal (JNK) kinase [37]. NSD1 overexpression in *Drosophila* imaginal discs induced organ atrophy. Interestingly, ectopic expression of the DNA replication-related element-binding factor (DREF) resulted in increased NSD expression [38]. DREF proteins are central regulators of cell proliferation; however, whether the human homolog ZBED1 (zinc finger BED-type-containing 1) regulates NSD1 expression remains unknown. Pan-glial, but not pan-neuronal NSD overexpression induced apoptosis in *Drosophila* larval brain cells. However, pan-glial NSD overexpression also induced caspase-3 cleavage in neuronal cells. Among the various glial cell types, NSD overexpression in only astrocytic glia induced apoptosis and abnormal learning defects in the larval stage. These observations in *Drosophila* suggested that aberrant NSD expression may result in neurodevelopmental disorders through functional interference with astrocytes [39]. In contrast, NSD deletion by CRISPR/Cas9-mediated knock-out resulted in an increase in the body size of *Drosophila* larvae. Although the NSD mutant flies survived to adulthood, their fecundity was dramatically decreased. NSD lacking flies also showed neurological dysfunctions, such as lower memory performance and motor defects, and a diminished extracellular signal-regulated kinase activity [40]. Collectively, these functional studies in the fly suggested that NSD is a central regulator of proliferation and, cell and/or body size.

4.2. Modeling NSD1 Activity in the Mouse

To gain insight into the biological functions of NSD1 in mammals, Losson and colleagues have generated mice carrying a floxed *Nsd1* exon 5 containing the nuclear factor interaction domain. Ubiquitous inactivation (*Actin-iCre;Nsd1$^{f/f}$*) embryos displayed a high incidence of apoptosis and failed to complete gastrulation, indicating that NSD1 is essential for early post-implantation development [41]. More recent work, using the same *Nsd1$^{f/f}$* allele, showed that conditional targeted ablation in primordial germ cells (*Tnap-iCre;Nsd1$^{f/f}$*) resulted in male sterility associated with absence of mature spermatozoa and loss of testicular germ cells in adult testis and epididymis. A similar effect was seen when DNMT3A was conditionally ablated in germ cells. Male mutant mice presented with impaired spermatogenesis due to loss of methylation at two out of three paternally imprinted loci in spermatogonia [42]. Molecular studies confirmed previous findings that NSD1 safeguards a subset of genes against H3K27me3-associated transcriptional silencing. In contrast, H3K36me2 in oocytes is predominantly dependent on the SETD2 HMT coinciding with H3K36me3. Hence, in contrast to males, *Nsd1$^{-/-}$* females are fertile. These studies showed that NSD1 plays a critical role in the maturation of mouse gametes by regulating distinct profiles of H3K36 methylation [43]. A third study using the floxed *Nsd1* mouse allele generated by Losson et al. inactivated the gene in the hematopoietic system. Unexpectedly homozygous ablation during late fetal liver hematopoiesis (*Vav-iCre;Nsd1$^{f/f}$*) resulted in a fully penetrant hematological malignancy phenocopying many aspects of human acute erythroleukemia. Functional studies revealed that lack of *Nsd1* impairs terminal differentiation of erythroblasts, which could be rescued by expression of wildtype, but not a catalytically inactive SET-domain NSD1^{N1918Q} mutant. Interestingly, NSD1, but not the inactive mutant, significantly increased the occupancy of the erythroid transcriptional master regulator GATA1 at target genes and their expression. These studies identified NSD1 as a novel regulator of GATA1-controlled erythroid differentiation [44]. Very recently, Zou and coworkers used the same floxed murine *Nsd1* allele for targeted activation of the gene in mesenchymal progenitor cells (*Prx1-iCre;Nsd1*$^{fl/fl}$). Ablation of *Nsd1* in mesenchymal progenitors resulted in impaired cartilage development, skeletal growth defects, and impaired fracture healing. Chondrogenic differentiation was impaired, which was associated with reduced H3K36me2 marks and lower expression of critical mediators including the SRY-box transcription factor 9 (SOX9). Interestingly, in chondrocytes NSD1 seems to bind the promoter and to control expression of the hypoxia-inducible factor 1alpha (HIF1alpha), a well-known regulator of SOX9 [45]. Importantly, *Sox9* overexpression rescued the chondrogenic differentiation effects of *Nsd1*$^{-/-}$ cells. Collectively, these data suggest that NSD1 controls chondrogenic differentiation by direct (H3K36me2) and indirect (HIF1A) regulation of SOX9 [46]

Piper and colleagues used a CRISPR/Cas9 strategy to inactivate exon 3 of *Nsd1*. Although they did not find any major morphologic defects in *Nsd1*$^{+/-}$ brains, the animals exhibited deficits in social behavior without significant learning or memory deficits. *Nsd1*$^{-/-}$ E9.5 embryos had a smaller prosencephalon compared to heterozygous and wild-type animals, with abnormal morphology and aberrant formation of the luminal cavity of the brain [47]. Taken together, NSD1 inactivation studies in *Drosophila* and mice showed that NSD1 is essential for normal development and that it regulates a wide variety of cellular functions, of which many seem to be cell type-specific, most likely by controlling the activity of distinct transcriptional master regulators.

5. Role of NSD1 in Human Diseases

5.1. Aberrant NSD1 Activity Is a Hallmark of Developmental Syndromes

Germline lesions (including missense, truncating and splice-site mutations and submicroscopic deletions) potentially resulting in loss-of-function of the NSD1 protein have been linked to a developmental syndrome called SOTOS [48]. SOTOS is a childhood overgrowth syndrome characterized by a distinctive facial appearance, physical overgrowth with height and head circumference >97th percentile, advanced bone age and learning disabilities [49]. Interestingly, microduplications of 5q35.2–q35.3 encompassing the *NSD1* gene locus have been reported in rare patients with a clinically reversed SOTOS syndrome. These individuals are characterized by short stature, microcephaly, learning disability or mild to moderate intellectual disability, and distinctive facial features. These observations suggest that the NSD1 gene dosage determines the phenotype of these developmental syndromes [50].

Analysis of a cohort of >700 individuals with overgrowth and intellectual disability revealed a putative causal mutation in less than 15 genes in almost half of the individuals [51]. Notably, epigenetic regulation was a prominent biological process not only represented by NSD1 but also by five additional genes including PRC2 complex proteins (EZH2, EED), H1.5 linker histone (HIST1H1E), the de novo DNMT3A methyltransferase, and the chromatin remodeler CHD8. Other patients had mutations in genes controlling cellular growth (PTEN, AKT3, PIK3CA, MTOR, PPP2R5D). The PI3K/AKT pathway is a central regulator of growth by increased cell metabolism, survival, and turnover, as well as protein synthesis. As deregulated cellular growth is a hallmark of cancer, and certain human overgrowth syndromes are associated with increased cancer risk, it is not unexpected that the majority of the mutated genes in overgrowth syndromes including NSD1, EZH2, DNMT3A, PTEN, CHD8, HIST1H1E, MTOR, PIK3CA are also frequently altered in human cancers [51]. Interestingly, overgrowth-related PIK3CA mutations were shown to exhibit a striking allele dose-dependent stemness phenotype in human pluripotent stem cells (PSC) [52,53]. Whether NSD1 mutations affect PSC stemness remains unknown.

Analysis of genome-wide DNA methylation of SOTOS syndrome patients revealed a highly specific signature able to differentiate patients with pathogenic NSD1 mutations from controls, benign NSD1 variants and clinically overlapping syndromes. This $NSD1^{+/-}$ DNA methylation signature encompasses genes that function in cellular morphogenesis and neuronal differentiation reflecting cardinal features of SOTOS syndrome [54]. SOTOS-related DNA methylation signatures were used to model epigenetic clocks that predict biological age. The so-called Horvath epigenetic clock model revealed that NSD1 loss-of-function mutations substantially accelerate epigenetic aging [55].

5.2. Aberrant NSD1 in Human Cancers

The first evidence linking NSD1 genetic aberrations to cancer came from cloning of a cytogenetically silent t(5;11)(q35;115) chromosomal translocation associated with pediatric de novo MDS or aggressive AML that leads to fusion of the N-terminal domains of the nucleopore 98 (NUP98) protein to the C-terminal part (including the SET) of NSD1 [56]. Importantly, in most patients, additional genetic lesions are found in NUP98-NSD1$^+$ AML cells of which activating FLT3-ITD mutations are by far the most prevalent, present in about 80% of the cases [57]. Reconstitution of lethally irradiated mice with bone marrow retrovirally overexpressing the NUP98-NSD1 fusion (in presence or absence of a functionally cooperating FLT3-ITD mutation) was reported to induce an AML-like disease in mice [58,59]. Functional studies suggested that the NUP98-NSD1 fusion binds genomic elements adjacent to the *HoxA7* and *HoxA9* loci and maintains histone H3K36 methylation and histone acetylation, preventing transcriptional repression of the *HoxA* gene cluster during differentiation. Structure functional analysis indicated that the phenylalanine-glycine (FG) repeats of the NUP98 moiety as well as the NSD1-SET domain are necessary for its transforming activity [58]. Targeted sequencing of a large number of genes associated with hematologic malignancies revealed rare and potentially deleterious *NSD1* mutations in AML patients suggesting that not only gain but also loss of NSD1 can contribute to transformation of hematopoietic cells [60].

Analysis of cancer-associated aberrant CpG promoter methylation revealed epigenetic silencing of NSD1 in human brain tumor cell lines associated with reduced H3K36 methylation [17]. While NSD1 overexpression impaired colony growth in semi-solid medium and proliferation of cancer cells, RNAi-mediated knock-down increased proliferation, suggesting a role of a tumor suppressor [17]. Frequent NSD1 epigenetic silencing was also found in human clear cell renal cell carcinoma (ccRCC). Notably, tumors harboring NSD1 promoter methylation were of higher grade and stage, and NSD1 promoter methylation correlated with somatic mutations in the SETD2 H3K36me3 HMT. Interestingly, ccRCC with epigenetic NSD1 silencing displayed a specific genome-wide methylome signature consistent with the NSD1 mutation methylome signature observed in SOTOS syndrome [61]. Comprehensive genomic characterization of human head and neck squamous cell carcinomas (HNSCC) identified inactivating NSD1 mutations and focal homozygous deletions in up to 10% of the patients [62]. Further studies revealed recurrent mutations including a K36M oncomutation in multiple H3 histone genes. Interestingly, direct in vitro inhibition of NSD2 and SETD2 by H3K36M has been described, whereas inhibition of NSD1 was only found in steady-state kinetic analysis using inhibitory H3 (27-43) peptide containing K36M [63,64]. Notably, along with previously described NSD1 mutations, they corresponded to a specific DNA methylation cluster. In addition, the K36M substitution and NSD1 defects converged on altering methylation of H3K36, subsequently blocking cellular differentiation and promoting oncogenesis [62]. Extensive genetic analysis of HNSCCs revealed that, similar to what has been experimentally observed in ES cells, loss of function NSD1 mutations are responsible for reduced intergenic H3K36me2 marks, followed by loss of DNA methylation and gain of H3K27me3 in the affected genomic regions. Those regions seem enriched in cis-regulatory elements, and subsequent loss of H3K27ac correlated with reduced expression of putative target genes [65]. In addition to HNSCC, H3.3 K36M mutations are recurrently found in several rare human cancers including chondroblastomas and poorly differentiated sarcomas. Comparison of the epigenomic and transcriptomic landscape of mesenchymal cells experimentally depleted of H3K36me2 indicated recapitulation of H3K36M's effect on H3K27me3 redistribution and gene expression [66]. Notably, transgenic mice overexpressing H3.3K36M in the hematopoietic system developed a lethal phenotype characterized by blocked erythroid differentiation that was very similar to that reported upon conditional *Nsd1* inactivation again supporting the converting consequences on epigenomic regulation [44,67].

A similar hypomethylated tumor subtype enriched for inactivating NSD1 mutations and deletions was also found in lung squamous cell carcinoma (LUSC). NSD1-altered HNSCC and LUSC correlated at the DNA methylation and gene expression levels, featuring ectopic expression of developmental transcription factors and genes that are also hypomethylated in SOTOS syndrome. Reduced expression of NSD1 was also reported to be part of an epigenetic gene signature able to distinguish non-malignant tumor from tissue of prostate cancer. Surprisingly, metastatic lesions appeared to express significantly higher NSD1 levels than primary tumors [68]. Highly prevalent NSD1 mutations were also found in testicular germ cell tumors, and low NSD1 expression was associated with resistance to cisplatin [69]. However, the functional significance of NSD1 alterations in human urogenital cancers remains to be investigated.

Comprehensive genomic analysis of 21 tumor types originating from >6000 samples revealed that the degrees of overall methylation in CpG island and demethylation in intergenic regions, defined as the 'backbone', are highly variable between different tumors [70]. Interestingly, NSD1 mutations showed the most significant association with backbone DNA demethylation not only in HNSCC but also in other cancers. In fact, bi-allelic NSD1 aberrations by mutation or gene copy loss showed the highest backbone demethylation [70]. A computational search for cancer predisposition genes based on the Knudson's two-hit hypothesis using genome data of ~10,000 tumors identified genes including *NSD1* that may contribute to cancer through a combination of rare germline variants and somatic loss-of-heterozygosity (LOH). Interestingly, rare germline variants in such genes may contribute substantially to cancer risk, particularly of ovarian carcinomas, but also other cancers [71].

Researchers also explored the correlation between allele frequency of somatic variants and total gene expression of the affected gene using matched tumor and normal RNA and DNA sequencing data from almost 400 individuals across 10 cancer types. They defined higher allele frequency of somatic variants in cancer-implicated genes. This study revealed that somatic alleles bearing premature terminating variants (PTVs) in cancer implicated genes seemed to be less degraded via nonsense-mediated mRNA decay, possibly favoring truncated proteins. Notably, NSD1 appeared as a gene with more than five somatic variants and PTVs with high allele frequency [72].

Collectively, increased NSD1-SET activity drives a particular hematological cancer, whereas loss-of function mutation or impaired expression characterize a wide variety of mostly solid human cancers (Figure 2).

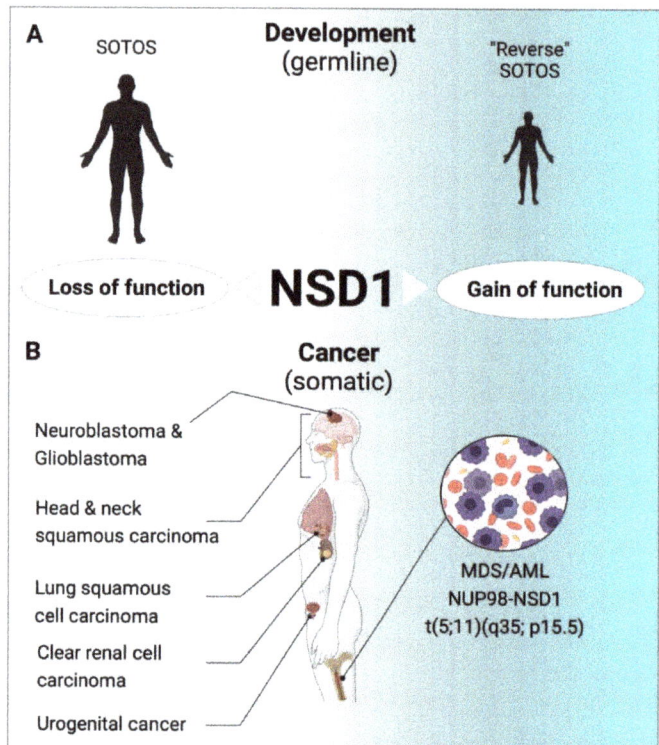

Figure 2. Role of NSD1 in human diseases. (**A**) Inactivating mutations in NSD1 are the molecular hallmark of SOTOS syndrome, a developmental disorder characterized by a distinctive facial appearance, physical overgrowth advanced bone age and learning disabilities [48,49]. "Reverse" SOTOS Syndrome is characterized by a short stature, microcephaly, and learning disability, and is associated with microdeletions of 5q35 carrying NSD1 [50]. (**B**) Putative loss of function mutations of NSD1 are among the most prevalent lesions in human head and neck and lung squamous cell carcinomas, neuroblastomas and glioblastomas [17,61,62,65,66,68]. NSD1 gene silencing was found in human clear cell renal cell carcinoma, and urogenital cancers [69,71]. In pediatric myeloid malignancies (de novo MDS and AML) the chromosomal translocation t(5;11)(q35;115) results in expression of a NUP98-NSD1 fusion gene with SET-dependent leukemogenic activity [56–59].

6. Therapeutic Interference with NSD1

Several strategies have been explored to selectively interfere with NSD1 activity. Earlier work characterized a small molecule methyltransferase inhibitor (BIX-01294) able to modulate H3K9 methylation. BIX.01294 was characterized as a G9a inhibitor by binding to the histone-tail groove in the SET domain [73]. Notably, BIX-01294 was also found to differentially inhibit NSD1, NDS2 and NSD3 in vitro based on the structural conserved catalytic SET domain but the molecule clearly lacks any NSD1 specificity [74].

NSD1 contains several PHD zinc fingers, whereby the PHD-V C5HCH domain serves as a binding site for protein–protein interactions. This region is particularly interesting as it has been shown to be involved in dysregulated Hox gene activation in AML and occurrence of point mutations in SOTOS Syndrome [16]. PHD-V C5HCH recruits a transcriptional repressor, resulting in a direct finger–finger interaction with the C2HR domain of Nizp1 [34]. The consequences of this binding are not clear; therefore, interfering with this interaction would be interesting to elucidate the biological and pathological relevance. Targeting PHD fingers has been considered to pharmacologically interfere with protein function; however, the affinity of compounds to specifically target a particular region is not advanced enough to be implemented in vivo. Berardi et al. designed a computational and experimental pipeline to investigate the druggability by using a 3D model of the PHD-V C5HCH domain of NSD1 with the C2HR domain of Nizp1 [34]. Applying a structure-base in silico screening following NMR validation, they found three structurally related molecules that were able to bind to the PHD-V C5HCH domain of NSD1: type II topoisomerase inhibitor mitoxantrone, chloroquine and quinacrine. Even if these compounds are interesting to target the NSD1/Nizp1 interaction, the consequences of derepressing transcription and selective inhibition are not clear and more functional studies have to be performed before this can be translated into the clinic.

Using a luminescence screening platform that quantifies S-adenosyl homocysteine (which is produced during methyl transfer from S-adenosylmethionine used by NSD1 and other HMTs), researchers identified suramin and other scaffolds as potential inhibitors of the enzymatic NSD1 HMT activity [75]. A computational strategy incorporating ligand contact information into classical alignment-based comparisons applied to SET containing proteins revealed additional scaffolds that inhibited NSD1 activity [76].

More recently, Grembecka, Cierpicki and colleagues employed a fragment-based screening strategy to identify and optimize first-in-class irreversible small-molecule inhibitors of the NSD1 SET domain [77]. Structural analysis revealed that NSD1 in complex with covalently bound ligands results in a conformational change in the autoinhibitory loop of the SET domain and formation of a channel-like pocket suitable for targeting with small molecules. Importantly, their lead-compound ("BT5") demonstrated on-target activity in NUP98-NSD1 immortalized cells associated with reduction of H3K36me2 and downregulation of critical target genes, such as the *HOXA* gene cluster and *MEIS1*. Notably, BT5 also impaired the clonogenic growth of primary NUP98-NSD1$^+$ AML cells but not leukemic cells carrying an KMT2A-MLLT1 fusion or normal human CD34$^+$ hematopoietic stem and progenitor cells. The discovery of this compound provides a platform for the development of potent and selective NSD1-SET inhibitors [77].

7. Outlook

It is well established that NSD1 regulates gene expression programs through H3K36 methylation in a complex crosstalk between activating and repressive histone marks, as well as DNA methylation. In addition, NSD1 is a target of recurrent germline or somatically acquired loss and gain-of-function alterations associated with developmental syndromes (e.g., SOTOS) and various human cancers. However, many open questions remain; in particular, it is currently poorly understood how a putative loss of function mutation or reduced expression results in the observed developmental and cancer phenotype.

Molecular characterization of SOTOS patient-derived DNA confirmed the connection between NSD1 loss-of-function mutations and aberrant DNA CpG methylation [54]. However, it seems unclear whether SOTOS is based on simple NSD1 haploinsufficiency, or whether particular mutants eventually have dominant-negative activity, functionally impairing the protein expressed from the unmutated allele. Some studies have suggested an increased risk for SOTOS patients to develop cancer, raising the question about the role of NSD1 mutations in this context. As the cancer risk in SOTOS patients is small, one wonders whether further reduction of the NSD1 gene dosage (by, e.g., epigenetically silencing of the wildtype allele) could be involved [78]. A better characterization of the gene dosage and protein activity relationship in developmental syndromes is necessary to explore whether the presence of NSD1 mutations can serve more than as a diagnostic marker but eventually also provide some translational opportunities [79].

As outlined before, predicted loss-of-function mutations or epigenetic silencing of NSD1 have been described in a variety of human cancers. Investigating HNSSC or lung squamous cell carcinomas revealed that mutations did not abrogate NSD1 expression in most samples. Notably, interrogation of the cancer cell line encyclopedia (CCLE) indicates that only a very small number (5/1457) of human cancer cell lines completely lost NSD1 expression at the mRNA level [80,81]. Currently, it remains unclear how a single NSD1 point mutation will contribute to malignant transformation. Is further reduction of the NSD1 gene dosage, e.g., by loss of heterozygosity (LOH), necessary to significantly enhance malignant transformation? Notably, heterozygous $Nsd1^{+/-}$ mice do not develop any pathologies and express normal $Nsd1$ mRNA and protein levels [44]. In addition, we observed that shRNA-mediated knockdown experiments only significantly affected growth of various human and mouse cells after reduction of NSD1 mRNA levels over 50% (unpublished data). Hence, a systematic analysis of the functional NSD1 gene dose in malignant transformation is necessary. It also remains unresolved whether genetic alterations are early or late events in cancer development.

In addition, the critical downstream effectors of the tumor suppressive activity of NSD1 remain unknown. Although recent molecular analysis of human HNSCC cancer cell lines with and without NSD1 mutations (generated by CRISPR/Cas9 genome editing) revealed aberrant regulation of genes related to oxidative phosphorylation, MYC, mTORC1 or RAS signaling and other pathways, the impact on the cell biology has not been addressed and no particular transformation effector genes have been validated [65]. In addition, further studies are necessary to show whether the disease phenotypes with aberrant functional NSD1 dose are purely the consequence of its chromatin regulatory role or whether yet to be identified non-chromatin NSD1 substrate proteins are critically involved [3].

Notably, one of the most significantly down-regulated pathways in HNSSC cells, carrying engineered NSD1 mutations, was interferon alpha/gamma signaling [65]. Earlier studies identified human HNSCC and lung squamous cell carcinoma enriched for NSD1 inactivating mutations and deletions that displayed an immune-cold phenotype characterized by low degree of infiltration by tumor-associated leukocytes (macrophages, $CD8^+$ T cells) as well as low expression of immune checkpoint ligands and receptors (PD1, PDL1, PDCD1LG2) [82]. Interestingly, tumors formed by lung cancer cell lines with shRNA-mediated reduced NSD1 expression in immunodeficient mice contained also less tumor-infiltrating T cells and were associated with reduced expression of various cytokines and chemokines [82]. Another study proposed that a chemokine expression signature allows classification of HNSCC into high and low $CD8^+$ T cell-infiltrated tumor phenotypes (TCIP-H vs. TCIP-L) associated with different clinical outcome. Notably about 20% of TCIP-L tumors carried loss of function NSD1 mutations [83]. These observations suggest that human cancers may escape the immune system through acquisition of NSD1 mutations. Further work is necessary to dissect the cellular and molecular circuits of cell-autonomous from non-autonomous consequences of aberrant NSD1 activity in human diseases. Interestingly, in vitro functional studies performed with human brain and breast cancer cells lines found a potential link of reduced expression or mutations of NSD1 to drug resistance; however, its general significance for cancer therapy remains to be validated [84,85].

When fused to the N-terminus of NUP98, the NSD1-SET gains transforming activities in hematopoietic cells, resulting in myelodysplasia and AML, and the presence of a NUP98-NSD1 (and other NUP98-fusions) is often associated with primary resistance to chemotherapy [86,87]. Functional studies suggested that transformation by these fusions involves the NUP98-GFLG repeats recruiting a large WDR82-SET1A/B-COMPASS protein complex to promote H3K4 trimethylation and favor active transcription [88]. The fusions may also directly interact with KMT2A (aka MLL1) to reach critical target gene loci such as the HOX-A gene cluster regulated by the fusion partner like NSD1 that favors transcription by H3K36 methylation [58,89]. These findings strongly suggest that targeted inactivation of the NSD1-SET domain shows anti-leukemic activity in NUP98-NSD1$^+$ hematological malignancies.

Although selective NSD1-SET inhibitors are highly relevant for aggressive NUP98-NSD1$^+$ pediatric AML, one has to take into consideration that loss-of-function mutations of NSD1 are much more prevalent in human cancers. Will such NSD1-SET inhibitors also block NSD1's role as a tumor suppressor? Significantly reduced NSD1 activity may functionally affect transcription factors controlling maturation of hematopoietic cells (and eventually also cells from other tissues). In the best-case scenario, some reduction of the NSD1-SET might be sufficient to induce differentiation of NUP98-NSD1-transformed myeloid cells, whereas significant side effects (as observed in gene targeted mice) may only develop upon complete inactivation over a longer time period that will most likely never be reached by such compounds.

The future NSD1 research agenda should aim to (i) mechanistically determine the gene dosage–phenotype correlation in germline syndromes with aberrant NSD1 activity, (ii) identify the cellular and molecular mechanisms of malignant transformation by altered NSD1 activity (mutations, epigenetic silencing), and (iii) optimize and validate small molecule NSD1-SET inhibitors for therapy of pediatric AML, driven by the NUP98-NSD1 fusion gene, and research for strategies to selectively interfere in situations when reduced NSD1 activity is the driving force.

Author Contributions: S.T. and J.S. both conceptualized and wrote the manuscript. Both authors have read and agreed to the published version of the manuscript.

Funding: Research on NSD1 was supported by Cancer Research Switzerland (KFS-4258-08-2017).

Institutional Review Board Statement: Not applicable.

Informed Consent Statement: Not applicable.

Acknowledgments: The authors thank Albert Jeltsch and Jonathan Séguin for critical reading the manuscript. All figures were created with BioRender.com (accessed on 25 August 2021).

Conflicts of Interest: The authors declare no conflict of interest.

References

1. Jenuwein, T.; Allis, C.D. Translating the histone code. *Science* **2001**, *293*, 1074–1080. [CrossRef]
2. Husmann, D.; Gozani, O. Histone lysine methyltransferases in biology and disease. *Nat. Struct. Mol. Biol.* **2019**, *26*, 880–889. [CrossRef]
3. Bhat, K.P.; Kaniskan, H.Ü.; Jin, J.; Gozani, O. Epigenetics and beyond: Targeting writers of protein lysine methylation to treat disease. *Nat. Rev. Drug Discov.* **2021**, *20*, 265–286. [CrossRef]
4. Huang, N.; Baur, E.V.; Garnier, J.; Lerouge, T.; Vonesch, J.; Lutz, Y.; Chambon, P.; Losson, R. Two distinct nuclear receptor interaction domains in NSD1, a novel SET protein that exhibits characteristics of both corepressors and coactivators. *EMBO J.* **1998**, *17*, 3398–3412. [CrossRef] [PubMed]
5. Sampson, E.R.; Yeh, S.Y.; Chang, H.-C.; Tsai, M.Y.; Wang, X.; Ting, H.J.; Chang, C. Identification and characterization of androgen receptor associated coregulators in prostate cancer cells. *J. Boil. Regul. Homeost. Agents* **2001**, *15*, 123–129.
6. Kurotaki, N.; Harada, N.; Yoshiura, K.-I.; Sugano, S.; Niikawa, N.; Matsumoto, N. Molecular characterization of NSD1, a human homologue of the mouse Nsd1 gene. *Gene* **2001**, *279*, 197–204. [CrossRef]
7. Ensembl Gene: NSD1 ENSG00000165671. Available online: http://www.ensembl.org/Homo_sapiens/Gene/Summary?db=core;g=ENSG00000165671;r=5:177133025-177300213;t=ENST00000347982 (accessed on 17 June 2021).
8. Wang, X.; Yeh, S.; Wu, G.; Hsu, C.-L.; Wang, L.; Chiang, T.; Yang, Y.; Guo, Y.; Chang, C. Identification and Characterization of a Novel Androgen Receptor Coregulator ARA267-α in Prostate Cancer Cells. *J. Biol. Chem.* **2001**, *276*, 40417–40423. [CrossRef]
9. The Human Protein Atlas. Available online: https://www.proteinatlas.org/ENSG00000165671-NSD1/tissue (accessed on 13 July 2021).
10. GeneCards: The Human Gene Database: NSD1. Available online: https://www.genecards.org/cgi-bin/carddisp.pl?gene=NSD1#protein_expression (accessed on 13 July 2021).
11. GeneCards the Human Gene Database. Available online: https://www.genecards.org/cgi-bin/carddisp.pl?gene=NSD1 (accessed on 17 June 2021).
12. Rosati, R.; La Starza, R.; Veronese, A.; Aventin, A.; Schwienbacher, C.; Vallespi, T.; Negrini, M.; Martelli, M.F.; Mecucci, C. NUP98 is fused to the NSD3 gene in acute myeloid leukemia associated with t(8;11)(p11.2;p15). *Blood* **2002**, *99*, 3857–3860. [CrossRef]
13. Murn, J.; Shi, Y. The winding path of protein methylation research: Milestones and new frontiers. *Nat. Rev. Mol. Cell Biol.* **2017**, *18*, 517–527. [CrossRef]
14. Morishita, M.; di Luccio, E. Structural insights into the regulation and the recognition of histone marks by the SET domain of NSD1. *Biochem. Biophys. Res. Commun.* **2011**, *412*, 214–219. [CrossRef]
15. Amin, N.; Nietlispach, D.; Qamar, S.; Coyle, J.; Chiarparin, E.; Williams, G. NMR backbone resonance assignment and solution secondary structure determination of human NSD1 and NSD2. *Biomol. NMR Assign.* **2016**, *10*, 315–320. [CrossRef]

16. Pasillas, M.P.; Shah, M.; Kamps, M.P. NSD1 PHD domains bind methylated H3K4 and H3K9 using interactions disrupted by point mutations in human sotos syndrome. *Hum. Mutat.* **2011**, *32*, 292–298. [CrossRef] [PubMed]
17. Berdasco, M.; Ropero, S.; Setien, F.; Fraga, M.; Lapunzina, P.; Losson, R.; Alaminos, M.; Cheung, N.-K.; Rahman, N.; Esteller, M. Epigenetic inactivation of the Sotos overgrowth syndrome gene histone methyltransferase NSD1 in human neuroblastoma and glioma. *Proc. Natl. Acad. Sci. USA* **2009**, *106*, 21830–21835. [CrossRef]
18. Lu, T.; Jackson, M.W.; Wang, B.; Yang, M.; Chance, M.R.; Miyagi, M.; Gudkov, A.V.; Stark, G.R. Regulation of NF κB by NSD1/FBXL11-dependent reversible lysine methylation of p65. *Cytokine* **2009**, *48*, 19–20. [CrossRef]
19. Kudithipudi, S.; Lungu, C.; Rathert, P.; Happel, N.; Jeltsch, A. Substrate Specificity Analysis and Novel Substrates of the Protein Lysine Methyltransferase NSD1. *Chem. Biol.* **2014**, *21*, 226–237. [CrossRef]
20. Li, J.; Ahn, J.H.; Wang, G.G. Understanding histone H3 lysine 36 methylation and its deregulation in disease. *Cell. Mol. Life Sci.* **2019**, *76*, 2899–2916. [CrossRef] [PubMed]
21. Huang, C.; Zhu, B. Roles of H3K36-specific histone methyltransferases in transcription: Antagonizing silencing and safeguarding transcription fidelity. *Biophys. Rep.* **2018**, *4*, 170–177. [CrossRef]
22. Edmunds, J.W.; Mahadevan, L.C.; Clayton, A.L. Dynamic histone H3 methylation during gene induction: HYPB/Setd2 mediates all H3K36 trimethylation. *EMBO J.* **2007**, *27*, 406–420. [CrossRef]
23. Qiao, Q.; Li, Y.; Chen, Z.; Wang, M.; Reinberg, D.; Xu, R.-M. The Structure of NSD1 Reveals an Autoregulatory Mechanism Underlying Histone H3K36 Methylation. *J. Biol. Chem.* **2011**, *286*, 8361–8368. [CrossRef]
24. Li, W.; Tian, W.; Yuan, G.; Deng, P.; Sengupta, D.; Cheng, Z.; Cao, Y.; Ren, J.; Qin, Y.; Zhou, Y.; et al. Molecular basis of nucleosomal H3K36 methylation by NSD methyltransferases. *Nat. Cell Biol.* **2021**, *590*, 498–503.
25. Yuan, G.; Ma, B.; Yuan, W.; Zhang, Z.; Chen, P.; Ding, X.; Feng, L.; Shen, X.; Chen, S.; Li, G.; et al. Histone H2A Ubiquitination Inhibits the Enzymatic Activity of H3 Lysine 36 Methyltransferases. *J. Biol. Chem.* **2013**, *288*, 30832–30842. [CrossRef] [PubMed]
26. Laugesen, A.; Højfeldt, J.W.; Helin, K. Molecular Mechanisms Directing PRC2 Recruitment and H3K27 Methylation. *Mol. Cell* **2019**, *74*, 8–18. [CrossRef]
27. Streubel, G.; Watson, A.; Jammula, S.G.; Scelfo, A.; Fitzpatrick, D.J.; Oliviero, G.; McCole, R.; Conway, E.; Glancy, E.; Negri, G.L.; et al. The H3K36me2 Methyltransferase Nsd1 Demarcates PRC2-Mediated H3K27me2 and H3K27me3 Domains in Embryonic Stem Cells. *Mol. Cell* **2018**, *70*, 371–379.e5. [CrossRef]
28. Hoang, N.-M.; Rui, L. DNA methyltransferases in hematological malignancies. *J. Genet. Genom.* **2020**, *47*, 361–372. [CrossRef] [PubMed]
29. Dhayalan, A.; Rajavelu, A.; Rathert, P.; Tamas, R.; Jurkowska, R.Z.; Ragozin, S.; Jeltsch, A. The Dnmt3a PWWP Domain Reads Histone 3 Lysine 36 Trimethylation and Guides DNA Methylation. *J. Biol. Chem.* **2010**, *285*, 26114–26120. [CrossRef]
30. Deevy, O.; Bracken, A.P. PRC2 functions in development and congenital disorders. *Development* **2019**, *146*, dev181354. [CrossRef]
31. Li, Y.; Chen, X.; Lu, C. The interplay between DNA and histone methylation: Molecular mechanisms and disease implications. *EMBO Rep.* **2021**, *22*, e51803. [CrossRef] [PubMed]
32. Weinberg, D.; Papillon-Cavanagh, S.; Chen, H.; Yue, Y.; Chen, X.; Rajagopalan, K.N.; Horth, C.; McGuire, J.T.; Xu, X.; Nikbakht, H.; et al. The histone mark H3K36me2 recruits DNMT3A and shapes the intergenic DNA methylation landscape. *Nat. Cell Biol.* **2019**, *573*, 281–286. [CrossRef]
33. Nielsen, A.L.; Jørgensen, P.; Lerouge, T.; Cerviño, M.; Chambon, P.; Losson, R. Nizp1, a Novel Multitype Zinc Finger Protein That Interacts with the NSD1 Histone Lysine Methyltransferase through a Unique C2HR Motif. *Mol. Cell. Biol.* **2004**, *24*, 5184–5196. [CrossRef]
34. Berardi, A.; Ghitti, M.; Quilici, G.; Musco, G. In silico derived small molecules targeting the finger-finger interaction between the histone lysine methyltransferase NSD1 and Nizp1 repressor. *Comput. Struct. Biotechnol. J.* **2020**, *18*, 4082–4092. [CrossRef]
35. Fang, Y.; Tang, Y.; Zhang, Y.; Pan, Y.; Jia, J.; Sun, Z.; Zeng, W.; Chen, J.; Yuan, Y.; Fang, D. The H3K36me2 methyltransferase NSD1 modulates H3K27ac at active enhancers to safeguard gene expression. *Nucleic Acids Res.* **2021**, *49*, 6281–6295. [CrossRef] [PubMed]
36. Yamada-Okabe, T.; Matsumoto, N. Decreased serum dependence in the growth of NIH3T3 cells from the overexpression of human nuclear receptor-binding SET-domain-containing protein 1 (NSD1) or fission yeast su(var)3-9, enhancer-of-zeste, trithorax 2 (SET2). *Cell Biochem. Funct.* **2007**, *26*, 146–150. [CrossRef] [PubMed]
37. Jeong, Y.; Kim, T.; Kim, S.; Hong, Y.-K.; Cho, K.S.; Lee, I.-S. Overexpression of histone methyltransferase NSD in Drosophila induces apoptotic cell death via the Jun-N-terminal kinase pathway. *Biochem. Biophys. Res. Commun.* **2018**, *496*, 1134–1140. [CrossRef] [PubMed]
38. Kim, S.; Kim, T.; Jeong, Y.; Choi, S.; Yamaguchi, M.; Lee, I.-S. The Drosophila histone methyltransferase NSD is positively regulated by the DRE/DREF system. *Genes Genom.* **2018**, *40*, 475–484. [CrossRef] [PubMed]
39. Kim, T.; Shin, H.; Song, B.; Won, C.; Yoshida, H.; Yamaguchi, M.; Cho, K.S.; Lee, I. Overexpression of H3K36 methyltransferase NSD in glial cells affects brain development in Drosophila. *Glia* **2020**, *68*, 2503–2516. [CrossRef]
40. Choi, S.; Song, B.; Shin, H.; Won, C.; Kim, T.; Yoshida, H.; Lee, D.; Chung, J.; Cho, K.S.; Lee, I.-S. Drosophila NSD deletion induces developmental anomalies similar to those seen in Sotos syndrome 1 patients. *Genes Genom.* **2021**, *43*, 737–748. [CrossRef]
41. Rayasam, G.V.; Wendling, O.; Angrand, P.-O.; Mark, M.; Niederreither, K.; Song, L.; Lerouge, T.; Hager, G.L.; Chambon, P.; Losson, R. NSD1 is essential for early post-implantation development and has a catalytically active SET domain. *EMBO J.* **2003**, *22*, 3153–3163. [CrossRef] [PubMed]

42. Kaneda, M.; Okano, M.; Hata, K.; Sado, T.; Tsujimoto, N.; Li, E.; Sasaki, H. Essential role for de novo DNA methyltransferase Dnmt3a in paternal and maternal imprinting. *Nat. Cell Biol.* **2004**, *429*, 900–903. [CrossRef]
43. Shirane, K.; Miura, F.; Ito, T.; Lorincz, M.C. NSD1-deposited H3K36me2 directs de novo methylation in the mouse male germline and counteracts Polycomb-associated silencing. *Nat. Genet.* **2020**, *52*, 1088–1098. [CrossRef]
44. Leonards, K.; Almosailleakh, M.; Tauchmann, S.; Bagger, F.O.; Thirant, C.; Juge, S.; Bock, T.; Méreau, H.; Bezerra, M.F.; Tzankov, A.; et al. Nuclear interacting SET domain protein 1 inactivation impairs GATA1-regulated erythroid differentiation and causes erythroleukemia. *Nat. Commun.* **2020**, *11*, 1–15. [CrossRef]
45. Amarilio, R.; Viukov, S.V.; Sharir, A.; Eshkar-Oren, I.; Johnson, R.; Zelzer, E. HIF1α regulation of Sox9 is necessary to maintain differentiation of hypoxic prechondrogenic cells during early skeletogenesis. *Development* **2007**, *134*, 3917–3928. [CrossRef]
46. Shao, R.; Zhang, Z.; Xu, Z.; Ouyang, H.; Wang, L.; Ouyang, H.; Zou, W. H3K36 methyltransferase NSD1 regulates chondrocyte differentiation for skeletal development and fracture repair. *Bone Res.* **2021**, *9*, 1–11.
47. Oishi, S.; Zalucki, O.; Vega, M.S.; Harkins, D.; Harvey, T.J.; Kasherman, M.; Davila, R.A.; Hale, L.; White, M.; Piltz, S.; et al. Investigating cortical features of Sotos syndrome using mice heterozygous for Nsd1. *Genes Brain Behav.* **2020**, *19*, e12637. [CrossRef] [PubMed]
48. Kurotaki, N.; Imaizumi, K.; Harada, N.; Masuno, M.; Kondoh, T.; Nagai, T.; Ohashi, H.; Naritomi, K.; Tsukahara, M.; Makita, Y.; et al. Haploinsufficiency of NSD1 causes Sotos syndrome. *Nat. Genet.* **2002**, *30*, 365–366. [CrossRef] [PubMed]
49. Tatton-Brown, K.; Rahman, N. Sotos syndrome. *Eur. J. Hum. Genet.* **2006**, *15*, 264–271. [CrossRef] [PubMed]
50. Dikow, N.; Maas, B.; Gaspar, H.; Kreiss-Nachtsheim, M.; Engels, H.; Kuechler, A.; Garbes, L.; Netzer, C.; Neuhann, T.M.; Koehler, U.; et al. The phenotypic spectrum of duplication 5q35.2-q35.3 encompassing NSD1: Is it really a reversed sotos syndrome? *Am. J. Med Genet. Part A* **2013**, *161*, 2158–2166. [CrossRef]
51. Tatton-Brown, K.; Loveday, C.; Yost, S.; Clarke, M.; Ramsay, E.; Zachariou, A.; Elliott, A.; Wylie, H.; Ardissone, A.; Rittinger, O.; et al. Mutations in Epigenetic Regulation Genes Are a Major Cause of Overgrowth with Intellectual Disability. *Am. J. Hum. Genet.* **2017**, *100*, 725–736. [CrossRef]
52. Madsen, R.R.; Knox, R.G.; Pearce, W.; Lopez, S.; Mahler-Araujo, B.; McGranahan, N.; Semple, R.K. Oncogenic PIK3CA promotes cellular stemness in an allele dose-dependent manner. *Proc. Natl. Acad. Sci. USA* **2019**, *116*, 8380–8389. [CrossRef]
53. Madsen, R.R. PI3K in stemness regulation: From development to cancer. *Biochem. Soc. Trans.* **2020**, *48*, 301–315. [CrossRef]
54. Choufani, S.; Cytrynbaum, C.; Chung, B.H.Y.; Turinsky, A.L.; Grafodatskaya, D.; Chen, Y.A.; Cohen, A.S.A.; Dupuis, L.; Butcher, D.T.; Siu, M.T.; et al. NSD1 mutations generate a genome-wide DNA methylation signature. *Nat. Commun.* **2015**, *6*, 10207. [CrossRef]
55. Martin-Herranz, D.E.; Aref-Eshghi, E.; Bonder, M.J.; Stubbs, T.M.; Choufani, S.; Weksberg, R.; Stegle, O.; Sadikovic, B.; Reik, W.; Thornton, J.M. Screening for genes that accelerate the epigenetic aging clock in humans reveals a role for the H3K36 methyltransferase NSD1. *Genome Biol.* **2019**, *20*, 1–19. [CrossRef] [PubMed]
56. Jaju, R.J. A novel gene, NSD1, is fused to NUP98 in the t(5;11)(q35;p15.5) in de novo childhood acute myeloid leukemia. *Blood* **2001**, *98*, 1264–1267. [CrossRef]
57. Hollink, I.H.I.M.; Heuvel-Eibrink, M.M.V.D.; Arentsen-Peters, S.T.C.J.M.; Pratcorona, M.; Abbas, S.; Kuipers, J.E.; Van Galen, J.F.; Beverloo, H.B.; Sonneveld, E.; Kaspers, G.-J.J.L.; et al. NUP98/NSD1 characterizes a novel poor prognostic group in acute myeloid leukemia with a distinct HOX gene expression pattern. *Blood* **2011**, *118*, 3645–3656. [CrossRef]
58. Wang, G.G.; Cai, L.; Pasillas, M.P.; Kamps, M.P. NUP98–NSD1 links H3K36 methylation to Hox-A gene activation and leukaemogenesis. *Nat. Cell Biol.* **2007**, *9*, 804–812. [CrossRef] [PubMed]
59. Thanasopoulou, A.; Tzankov, A.; Schwaller, J. Potent co-operation between the NUP98-NSD1 fusion and the FLT3-ITD mutation in acute myeloid leukemia induction. *Haematologica* **2014**, *99*, 1465–1471. [CrossRef]
60. Dolnik, A.; Engelmann, J.C.; Scharfenberger-Schmeer, M.; Mauch, J.; Kelkenberg-Schade, S.; Haldemann, B.; Fries, T.; Krönke, J.; Kühn, M.W.M.; Paschka, P.; et al. Commonly altered genomic regions in acute myeloid leukemia are enriched for somatic mutations involved in chromatin remodeling and splicing. *Blood* **2012**, *120*, e83–e92. [CrossRef] [PubMed]
61. Su, X.; Zhang, J.; Mouawad, R.; Compérat, E.; Rouprêt, M.; Allanic, F.; Malouf, G.G. NSD1 Inactivation and SETD2 Mutation Drive a Convergence toward Loss of Function of H3K36 Writers in Clear Cell Renal Cell Carcinomas. *Cancer Res.* **2017**, *77*, 4835–4845. [CrossRef]
62. Cancer Genome Atlas Network. Comprehensive genomic characterization of head and neck squamous cell carcinomas. *Nature* **2015**, *517*, 576–582. [CrossRef]
63. Mohammad, F.; Helin, K. Oncohistones: Drivers of pediatric cancers. *Genes Dev.* **2017**, *31*, 2313–2324. [CrossRef]
64. Schuhmacher, M.; Kusevic, D.; Kudithipudi, S.; Jeltsch, A. Kinetic Analysis of the Inhibition of the NSD1, NSD2 and SETD2 Protein Lysine Methyltransferases by a K36M Oncohistone Peptide. *ChemistrySelect* **2017**, *2*, 9532–9536. [CrossRef]
65. Farhangdoost, N.; Horth, C.; Hu, B.; Bareke, E.; Chen, X.; Li, Y.; Coradin, M.; Garcia, B.A.; Lu, C.; Majewski, J. Chromatin dysregulation associated with NSD1 mutation in head and neck squamous cell carcinoma. *Cell Rep.* **2021**, *34*, 108769. [CrossRef]
66. Rajagopalan, K.N.; Chen, X.; Weinberg, D.N.; Chen, H.; Majewski, J.; Allis, C.D.; Lu, C. Depletion of H3K36me2 recapitulates epigenomic and phenotypic changes induced by the H3.3K36M oncohistone mutation. *Proc. Natl. Acad. Sci. USA* **2021**, *118*. [CrossRef]

67. Brumbaugh, J.; Kim, I.S.; Ji, F.; Huebner, A.J.; Di Stefano, B.; Schwarz, B.A.; Charlton, J.; Coffey, A.; Choi, J.; Walsh, R.M.; et al. Inducible histone K-to-M mutations are dynamic tools to probe the physiological role of site-specific histone methylation in vitro and in vivo. *Nat. Cell Biol.* **2019**, *21*, 1449–1461. [CrossRef]
68. Bianco-Miotto, T.; Chiam, K.; Buchanan, G.; Jindal, S.; Day, T.K.; Thomas, M.; Pickering, M.A.; O'Loughlin, M.A.; Ryan, N.K.; Raymond, W.A.; et al. Global Levels of Specific Histone Modifications and an Epigenetic Gene Signature Predict Prostate Cancer Progression and Development. *Cancer Epidemiol. Biomark. Prev.* **2010**, *19*, 2611–2622. [CrossRef] [PubMed]
69. Bakardjieva-Mihaylova, V.; Kramarzova, K.S.; Slamova, M.; Svaton, M.; Rejlova, K.; Zaliova, M.; Dobiasova, A.; Fiser, K.; Stuchly, J.; Grega, M.; et al. Molecular Basis of Cisplatin Resistance in Testicular Germ Cell Tumors. *Cancers* **2019**, *11*, 1316. [CrossRef]
70. Lee, S.-T.; Wiemels, J.L. Genome-wide CpG island methylation and intergenic demethylation propensities vary among different tumor sites. *Nucleic Acids Res.* **2015**, *44*, 1105–1117. [CrossRef] [PubMed]
71. Park, S.; Supek, F.; Lehner, B. Systematic discovery of germline cancer predisposition genes through the identification of somatic second hits. *Nat. Commun.* **2018**, *9*, 1–13. [CrossRef]
72. Spurr, L.; Li, M.; Alomran, N.; Zhang, Q.; Restrepo, P.; Movassagh, M.; Trenkov, C.; Tunnessen, N.; Apanasovich, T.; Crandall, K.A.; et al. Systematic pan-cancer analysis of somatic allele frequency. *Sci. Rep.* **2018**, *8*, 7735. [CrossRef]
73. Chang, Y.; Zhang, X.; Horton, J.; Upadhyay, A.K.; Spannhoff, A.; Liu, J.; Snyder, J.P.; Bedford, M.T.; Cheng, X. Structural basis for G9a-like protein lysine methyltransferase inhibition by BIX-01294. *Nat. Struct. Mol. Biol.* **2009**, *16*, 312–317. [CrossRef] [PubMed]
74. Morishita, M.; Mevius, D.E.H.F.; Shen, Y.; Zhao, S.; di Luccio, E. BIX-01294 inhibits oncoproteins NSD1, NSD2 and NSD3. *Med. Chem. Res.* **2017**, *26*, 2038–2047. [CrossRef]
75. Drake, K.M.; Watson, V.G.; Kisielewski, A.; Glynn, R.; Napper, A.D. A Sensitive Luminescent Assay for the Histone Methyltransferase NSD1 and Other SAM-Dependent Enzymes. *ASSAY Drug Dev. Technol.* **2014**, *12*, 258–271. [CrossRef] [PubMed]
76. Rabal, O.; Castellar, A.; Oyarzabal, J. Novel pharmacological maps of protein lysine methyltransferases: Key for target deorphanization. *J. Chem.* **2018**, *10*, 1–19. [CrossRef] [PubMed]
77. Huang, H.; Howard, C.A.; Zari, S.; Cho, H.J.; Shukla, S.; Li, H.; Ndoj, J.; González-Alonso, P.; Nikolaidis, C.; Abbott, J.; et al. Covalent inhibition of NSD1 histone methyltransferase. *Nat. Chem. Biol.* **2020**, *16*, 1403–1410. [CrossRef]
78. Rahman, N. Mechanisms predisposing to childhood overgrowth and cancer. *Curr. Opin. Genet. Dev.* **2005**, *15*, 227–233. [CrossRef]
79. Cytrynbaum, C.; Choufani, S.; Weksberg, R. Epigenetic signatures in overgrowth syndromes: Translational opportunities. *Am. J. Med. Genet. Part C Semin. Med. Genet.* **2019**, *181*, 491–501. [CrossRef]
80. Ghandi, M.; Huang, F.W.; Jané-Valbuena, J.; Kryukov, G.; Lo, C.C.; McDonald, E.R., III; Barretina, J.; Gelfand, E.T.; Bielski, C.M.; Li, H.; et al. Next-generation characterization of the Cancer Cell Line Encyclopedia. *Nature* **2019**, *569*, 503–508. [CrossRef]
81. Barretina, J.; Caponigro, G.; Stransky, N.; Venkatesan, K.; Margolin, A.A.; Kim, S.; Garraway, L.A. The Cancer Cell Line Encyclopedia enables predictive modelling of anticancer drug sensitivity. *Nature* **2012**, *483*, 603–607. [CrossRef]
82. Brennan, K.; Shin, J.H.; Tay, J.K.; Prunello, M.; Gentles, A.J.; Sunwoo, J.B.; Gevaert, O. NSD1 inactivation defines an immune cold, DNA hypomethylated subtype in squamous cell carcinoma. *Sci. Rep.* **2017**, *7*, 1–12. [CrossRef]
83. Saloura, V.; Izumchenko, E.; Zuo, Z.; Bao, R.; Korzinkin, M.; Ozerov, I.; Zhavoronkov, A.; Sidransky, D.; Bedi, A.; Hoque, M.O.; et al. Immune profiles in primary squamous cell carcinoma of the head and neck. *Oral Oncol.* **2019**, *96*, 77–88. [CrossRef]
84. Han, J.; Jun, Y.; Kim, S.H.; Hoang, H.-H.; Jung, Y.; Kim, S.; Kim, J.; Austin, R.H.; Lee, S.; Park, S. Rapid emergence and mechanisms of resistance by U87 glioblastoma cells to doxorubicin in an in vitro tumor microfluidic ecology. *Proc. Natl. Acad. Sci. USA* **2016**, *113*, 14283–14288. [CrossRef] [PubMed]
85. Pereira, A.M.M.; Sims, D.; Dexter, T.; Fenwick, K.; Assiotis, I.; Kozarewa, I.; Mitsopoulos, C.; Hakas, J.; Zvelebil, M.; Lord, C.; et al. Genome-wide functional screen identifies a compendium of genes affecting sensitivity to tamoxifen. *Proc. Natl. Acad. Sci. USA* **2012**, *109*, 2730–2735. [CrossRef]
86. McNeer, N.; Philip, J.; Geiger, H.; Ries, R.E.; Lavallee, V.-P.; Walsh, M.; Shah, M.; Arora, K.; Emde, A.-K.; Robine, N.; et al. Abstract 2870: Genetic mechanisms of primary chemotherapy resistance in pediatric acute myeloid leukemia. *Tumor Biol.* **2019**, *33*, 1934–1943.
87. Barresi, V.; Di Bella, V.; Andriano, N.; Privitera, A.; Bonaccorso, P.; La Rosa, M.; Iachelli, V.; Spampinato, G.; Pulvirenti, G.; Scuderi, C.; et al. NUP-98 Rearrangements Led to the Identification of Candidate Biomarkers for Primary Induction Failure in Pediatric Acute Myeloid Leukemia. *Int. J. Mol. Sci.* **2021**, *22*, 4575. [CrossRef] [PubMed]
88. Franks, T.M.; McCloskey, A.; Shokhirev, M.N.; Benner, C.; Rathore, A.; Hetzer, M.W. Nup98 recruits the Wdr82-Set1A/COMPASS complex to promoters to regulate H3K4 trimethylation in hematopoietic progenitor cells. *Genes Dev.* **2017**, *31*, 2222–2234. [CrossRef] [PubMed]
89. Xu, H.; Valerio, D.G.; Eisold, M.E.; Sinha, A.; Koche, R.; Hu, W.; Chen, C.-W.; Chu, S.H.; Brien, G.; Park, C.Y.; et al. NUP98 Fusion Proteins Interact with the NSL and MLL1 Complexes to Drive Leukemogenesis. *Cancer Cell* **2016**, *30*, 863–878. [CrossRef]

Review

Structure, Activity and Function of the NSD3 Protein Lysine Methyltransferase

Philipp Rathert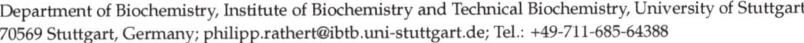

Department of Biochemistry, Institute of Biochemistry and Technical Biochemistry, University of Stuttgart, 70569 Stuttgart, Germany; philipp.rathert@ibtb.uni-stuttgart.de; Tel.: +49-711-685-64388

Abstract: NSD3 is one of six H3K36-specific lysine methyltransferases in metazoans, and the methylation of H3K36 is associated with active transcription. NSD3 is a member of the nuclear receptor-binding SET domain (NSD) family of histone methyltransferases together with NSD1 and NSD2, which generate mono- and dimethylated lysine on histone H3. NSD3 is mutated and hyperactive in some human cancers, but the biochemical mechanisms underlying such dysregulation are barely understood. In this review, the current knowledge of NSD3 is systematically reviewed. Finally, the molecular and functional characteristics of NSD3 in different tumor types according to the current research are summarized.

Keywords: NSD3; WHSC1L1; structure and function

1. Introduction

In eukaryotes, DNA is assembled into a higher order nucleoprotein structure called chromatin. Besides the condensation of the DNA, chromatin poses a variety of different functions centered around the regulation of transcription, replication, DNA repair and recombination. The main unit of chromatin is the nucleosome consisting of 147 base pairs (bp) of DNA, which is wrapped around the histone octamer comprising two molecules of each core histone: H2A, H2B, H3 and H4 [1]. The linker histone protein H1 is involved in packaging nucleosomes and proteins such as condensin, cohesin, CCCTC-binding factor (CTCF) or Yin Yang 1 (YY1) to organize the chromatin into higher order structures such as gene loops, topologically associated domains (TADs), chromosome territories, and chromosomes [2–4]. Chromatin adopts a highly condensed structure, called heterochromatin, where genes are less accessible and generally transcriptionally silent. In turn, decondensed chromatin, called euchromatin, is much more accessible and harbors the majority of actively transcribed genes [5].

In order to establish or maintain a cell-type-specific gene expression program, the chromatin structures need to be highly dynamic to allow access of transcription factors and other regulatory entities to the DNA at defined time points. These events are tightly regulated by post-translational modifications (PTMs) which are enriched at the unstructured and flexible N-terminal regions of the histone proteins. These histone tails protrude from the nucleosome core and are subject to a diverse array of PTMs, e.g., acetylation, phosphorylation, ubiquitination and methylation, often referred to as the "histone code" that extends the information potential of the genetic code [6–8]. The "histone code" hypothesis suggests that specific patterns of modifications function as a barcode and recruit distinct combinations of proteins or protein complexes to drive specific transcriptional programs [9,10].

Histone lysine methylation is among the best characterized PTM of the histone code and is attached to the basic side chains of lysine by a diverse set of sequence-specific lysine methyltransferases [11]. Histone lysine methylation mediates either an activating or repressive effect on gene transcription, which depends on the site, degree of methylation, genomic location, and the status of other coexisting PTMs [11]. The methylation of H3K36 is generally linked to the transcriptionally active state and introduced by six different methyltransferases, which can establish H3K36 methylation to various degrees [12]. The nuclear receptor-binding SET domain (NSD) family of histone methyltransferases is composed of three members of this family, namely NSD1, NSD2/MMSET/WHSC1, and NSD3/WHSC1L1 (referred to as NSD2 and NSD3 from here on) [13], which all generate mono and dimethylation of lysine 36 on histone H3 (H3K36me1/me2).

NSD3 was first characterized in 2001 as the third member of the NSD gene family [14,15]. Despite the physiologic importance of NSD family proteins, their mechanisms of action are only beginning to become elucidated. In the following review, the structural and functional features of NSD3 will be discussed in more detail with references to the other family members in case information is available.

2. Structural Features

The full-length (FL) members of the NSD family of histone methyltransferases are large multidomain proteins, which share most of the evolutionary conserved domains. They belong to the so-called SET domain-containing lysine-specific methyltransferases [16] and the domain involved in the catalytic activity is the SET domain, named after the Su(var)3-9, Enhancer-of-zeste and Trithorax (SET) proteins identified in Drosophila [17]. The SET domain is flanked by the associated with SET (AWS) and post-SET domains.

Besides the SET domain FL-NSD family members contain two PWWP domains named after its central core Pro-Trp-Trp-Pro motif, a five plant homeo domains (PHD) and a Cys-His-rich domain (C5HCH) domain (Figure 1). Crystal structures showed that the fifth PHD domain (PHD5) and the adjacent Cys-His-rich domain (C5HCH), located at the C terminus of NSD3, fold into a novel PHD-PHD-like module recognizing the unmodified H3K4 and trimethylated H3K9 by PHD5. This function is not conserved between members of the NSD family, with PHD5 of NSD2 showing stronger preference for unmethylated H3K9 (H3K9me0) than trimethylated H3K9 (H3K9me3), and the NSD1 PHD5-C5HCH showed no binding to histone peptides at all [18], but is in involved in binding to the transcription cofactor Nizp1 in NSD1 [19–21].

Not much information is available about the specific roles of the other domains of NSD3, and most functions can only be roughly implied from information published for NSD1 and 2. The first N-terminal PWWP domains of NSD1 and 2 were shown to bind to methylated H3K36 to stabilize NSD2, and probably NSD1, at chromatin, and the catalytic SET domain of NSD2 propagates this gene-activating mark to adjacent nucleosomes [22–27].

The PHD1-3 motifs of NSD2 were shown to be important for its H3K36me2 methylation activity. Specifically, the removal of PHD1 decreased H3K36me2 activity and PHD2 caused NSD2 localization into the cytoplasm, which resulted in a complete loss of activity [28]. More details are known for the PHD domains of NSD1. These were shown to mediate binding of NSD1 to methylated H3K4 and K9 with a preference for dimethylated lysines in vitro [21]. Only the PHD4, PHD5 and C4HCH domains show binding to both modifications, which is controversial as both methylation states are associated with opposite transcriptional states [29–31]. The binding of various states of H3K4 and H3K9 methylation would allow NSD1 to recognize genes in stages of transcriptional activation and repression. It was therefore hypothesized that the activities of NSD1 cofactors would ultimately lead to either the enforcement, or alternatively, to the reversal of repression mechanisms [21].

All three members of the NSD family of histone H3K36 methyltransferases share most of the common motifs except NSD2, which contains a so-called high mobility group (HMG) domain. The HMG domain of NSD2 was shown to interact with the DNA-binding domain of the androgen receptor (AR), thereby enhancing the nuclear translocation of both proteins [32]. Future studies are necessary to reveal whether the common corresponding domains of NSD3 have similar roles.

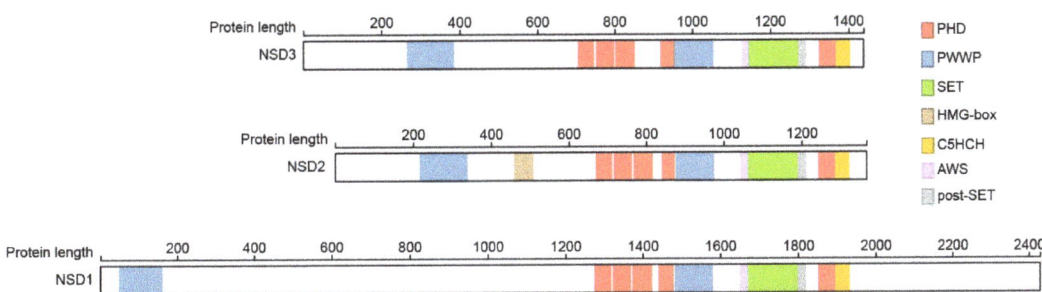

Figure 1. Structural relationship within the NSD family. The major domains of all three members of the NSD family of histone methyltransferases are highlighted. Numbers represent the number of amino acids in each full-length NSD protein. Proteins were extracted using ProteinPaint [33].

3. NSD3 Structure

The structure of the full-length NSD3 protein was never solved completely until now, due to its large protein size. An NSD3 construct containing amino acids 1054–1285, which spans the entire catalytic SET domain and additional residues on both sides without the reader domains, was crystallized in the presence of a histone H4 sequence flanking lysine 44 (H4K44), in which K44 was replaced by the unnatural amino acid norleucine (Nle) [34]. The catalytic part of NSD3 folds into a compact globular structure [34], which was confirmed later using cryo-electron microscopy (cryo-EM) studies on a larger version of NSD3 containing the C-terminal part of NSD3 starting from the first PHD domain (termed NSD3C) in complex with the nucleosome [35]. The histone peptide binds in a narrow groove and the lysine is occupying the substrate lysine channel. Interactions between the H3 tail and the SET domain are mainly mediated by hydrogen bonds, which tightly position the target lysine of H3 within the catalytic pocket. The hydrophobic side chain of the lysine points towards the methyl donor S-adenosylmethionine (SAM) through insertion into a hydrophobic pocket. [30]. The structures currently available for the NSD family show that a loop connecting the SET and post-SET domains can adopt multiple conformations, which are important for the regulation of the catalytic activity. This loop can extend over the H3 tail binding site of the SET domain, leading to autoinhibition [34,35] and significant reorganization of the autoinhibitory loop is observed in the structure of NSD3. In complex with the peptide, the autoinhibitory loop moves towards the C-terminus, which opens the substrate binding site for the peptide [34]. Similar to NSD3, the NSD1 and two post-SET domains are attached to the catalytic SET domain via an autoinhibitory loop region and inhibition is relieved upon nucleosome binding [13,36].

The recent cryo-EM studies provided a more detailed view on the importance of the nucleosome-bound DNA in the activation of NSD3 [35]. NSD3 forms several contacts with the nucleosomal DNA and inserts between the histone octamer and the DNA near the linker region leading to an unwrapped segment of DNA [35] (Figure 2a). The interactions between NSD3 and the unwrapped DNA are required for the full activity of NSD3 and several basic residues from the long N-terminal loop bind to the unwrapped segment of DNA. This interaction of NSD3 to the DNA is strengthened by additional salt bridges between lysine and arginine residues of the SET and post-Set domain and the phosphate backbone [35]. Interactions within this region of DNA not only stabilize the binding between NSD3 and the nucleosome core particle (NCP), but also enable the positioning of the H3 tail in the substrate-binding groove of the SET domain (Figure 2b). The interaction of NSD3 with the DNA at several positions, which leads to the partial unwrapping of the DNA, is essential for the correct positioning of K36 in the active center and is a key factor that determines NSD3 substrate specificity. Additionally, NSD3 makes extensive intermolecular contacts with a short section of the C terminus of histone H2A as well as a long fragment of H3 that contains the first α-helix and the N-terminal tail.

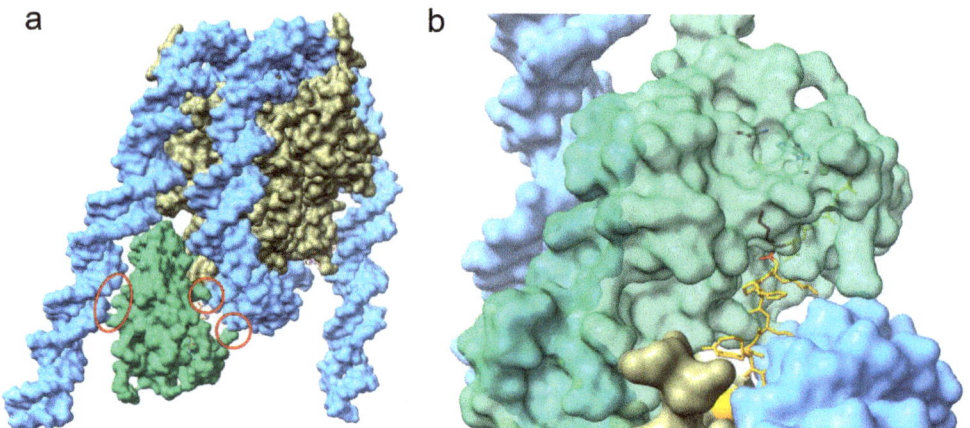

Figure 2. Surface representation of the structure of NSD3 bound to the nucleosome core particle (adopted from 7CRR PDB [35]). (**a**), NSD3 forms several contacts with the nucleosomal DNA (highlighted with red circles) and inserts between the DNA and the octamer. (**b**), Magnified view of the catalytic center of NSD3 showing the norleucine at position 36 is oriented in the catalytic center of NSD3. NSD3 is colored in green, the octamer in grey and the nucleosomal DNA in cyan. The bound H3 tail is shown in orange and the norleucine (inserted in the catalytic pocket of the SET domain) is depicted in red.

Furthermore, the AWS domain extends into the core histones and contacts the H2A C-terminal fragment through hydrophobic and electrostatic interactions. These contacts result in an extended conformation of NSD3, rendering NSD3 catalytically active and contributing to the precise positioning of NSD3 to specifically bind H3K36. However, it is possible that the conformational states observed differ with the full-length protein when compared to truncated constructs, which could influence the regulation of the enzyme activity by the autoinhibitory loop.

Additionally, the C-terminal part of NSD3 the crystal structure of the PWWP1 domain of NSD3 (residues 247–398) was solved and revealed a classical PWWP domain fold, as described previously [37,38]. An N-terminal β-barrel of 5 antiparallel β-strands (β1–β5), with a short helix insertion between β4 and β5 is followed by 3 α helices. The aromatic cage is formed by the aromatic amino acids Trp284, Tyr281, and Phe312, which are located at flexible loops connecting the different β-sheets. The aromatic cage could potentially accommodate an H3 peptide methylated at K36, indicated by the superimposition of the BRPF1-PWWP domain in complex with an H3K36me3 peptide [37,39].

4. Biochemical Features

The catalytic activity of the NSD family of histone H3K36 methyltransferases is restricted to a lower degree methylation of H3K36, and a specificity for mono and dimethylation is observed [12,40]. The substrate specificity of the NSD family of histone methyltransferases has long been debated and in vitro the catalytic domain (CTD) of NSD1, NSD2, and NSD3 were shown to recognize and methylate H3K4, H3K9, H3K27, H3K36, H3K79, and H4K20 peptides, with substantial differences in catalytic activities depending on the substrate [25]. NSD3 had previously been reported to specifically methylate H3K4 and H3K27 [41]. However, additional data with recombinant nucleosomes as substrate showed that the SET domains of all NSD family members specifically methylated K36 on histone H3. In contrast, when using recombinant histone octamers as substrate, the activity of NSD3 remained specific for H3 although with much lower activity, whereas the NSD2-SET domain mainly targeted H4 with very weak activity on H3 and the NSD1-SET domain methylated all components of the octamer, namely histone H3, H2A/H2B, and H4. Therefore, it was proposed that DNA acts as an allosteric effector of the NSD family proteins, such that H3K36 becomes the preferred target [42], which was recently confirmed through structural analysis [35].

Apart from the regulation of their enzymatic activity through binding of the nucleosome and the resulting clearance of the catalytic site from the autoregulatory loop, all members of the NSD family of histone methyltransferases are inhibited in their activity by different post translational modifications (PTMs) on histones. The ubiquitination of histone H2A at Lys119 [35,43] inhibits the activity of the whole NSD family of methyltransferases, which could be explained by the fact that they form extensive intermolecular contacts with the C terminus of histone H2A described for NSD3 [35]. Furthermore, the trimethylation of H3 at Lys4 also decreased the catalytic activity of NSD3, which correlates with the finding that the last PHD finger of NSD3 favors an unmodified Lys4 of H3 [18]. This suggests that binding of the unmodified H3 tail at lysine 4 contributes to some extent to the catalytic activity of NSD3. By contrast, the trimethylation of H3 at Lys27 did not alter the catalytic activity of NSD3 [35], which is intriguing because K27me3 rarely co-exists with K36me2 or K36me3 on the same histone. H3 polypeptide and PRC2 activity is greatly inhibited on nucleosomal substrates with preinstalled H3K36 methylation [44,45].

5. Cellular Features

NSD3 is ubiquitously expressed (Figure 3) and generates three major transcripts, a long (NSD3-long) isoform of 1437 amino acids, a short (NSD3-short) isoform containing 645 amino acids [14,15] and another short transcript called WHSC1-like 1 isoform 9 with methyltransferase activity to lysine (WHISTLE), which consists of 506 amino acids (Figure 4) [41].

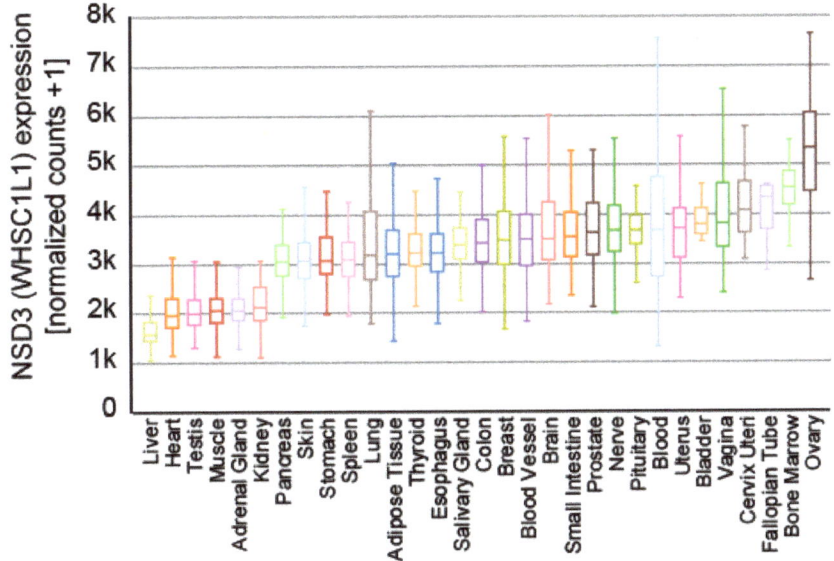

Figure 3. Expression of NSD3 across different human tissues. The data used for the analyses were obtained from the GTEx Portal (9783 samples). Visualization was generated with the UCSC Xena platform [46].

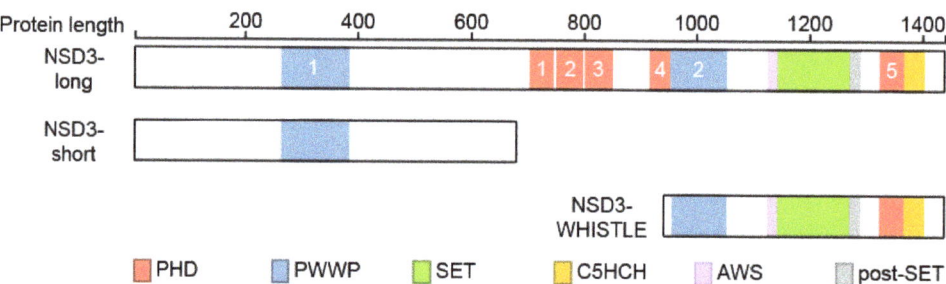

Figure 4. Isoforms of NSD3. Three isoforms of NSD3 are depicted and their domains were highlighted. Numbers of the different domains are indicated.

The NSD3-short protein lacks the catalytic SET domain and only contains the amino-terminal PWWP domain (Figure 2) [15] that binds to histone H3 when it is methylated on lysine 36 before [22]. NSD3-short was shown to interact with the bromodomain-containing protein 4 (BRD4) [47–49], which belongs to the bromodomain and extra-terminal domain (BET) protein family [50]. BRD4 plays an important role in controlling oncogene expression and genome stability and has sparked considerable interest as a drug target in multiple diseases in the past few years [51–53]. NSD3-short interacts with the extra terminal (ET) domain of BRD4 [48,49], which functions as an adaptor protein that links BRD4 to the chromatin remodeler CHD8 to enable transcriptional programs [48].

Both the NSD3-long and NSD3-short transcripts are co-expressed in many tissues [14,15], whereas WHISTLE was found to be mainly expressed in testis and in bone marrow mononuclear cells of AML and ALL patients [41]. In contrast to NSD3-long, WHISTLE only contains the second PWWP, SET, and post-SET domains (Figure 4) and was reported to facilitate transcriptional repression through its enzymatic activity and by recruiting HDACs [54], which is controversial to some extent, as all other reports connect NSD3 to transcriptional activation.

All NSD family proteins show methylation activity towards H3K36, which is restricted to mono and dimethylation [12,40]. Numerous studies in multiple systems support a role for H3K36 methylation in transcriptional activation [55,56]. While H3K36me3 exhibited characteristic enrichment within gene bodies, H3K36me2 shows a very distinctive genomic occupancy pattern and displays a significant enrichment in promoters and intergenic regions in various cell types [24,57,58] suggesting that H3K36me2 might play a role in enhancer regulation. Evidence for the function of H3K36me2 in the regulation of enhancer accessibility was provided recently through the investigation of Nsd1-mediated H3K36me2 distribution [45,58]. Interestingly, the simultaneous presence of H3K36me2 and H3K27me2, which is regulated through the activity of the polycomb repressive complex 2 (PRC2) [59], strongly correlate in embryonic stem cells (ESCs), whereas H3K36me3 and H3K27me3 are anticorrelated [45]. A switch from di- to trimethylation at K36 induces an increase in H3K27me3 [45], which results in the downregulation of the enhancer activity. In line with this observation, NSD2 was shown to regulate epithelial plasticity by altering enhancer activity. H3K27ac peaks residing within intergenic H3K36me2 domains are lost when H3K36me2 levels decrease, providing another indication that H3K36me2 mediates its effects by modulating enhancer activity [60]. Due to its comparable substrate specificity and structural similarity, an analogous function could be conceived for NSD3 as well, but this needs to be investigated experimentally.

Furthermore, H3K36me2 is required for recruitment of DNMT3A and maintenance of DNA methylation at intergenic regions [58]. Genome-wide analysis showed that the binding and activity of DNMT3A co-localize with H3K36me2 at non-coding regions of euchromatin [58]. Accordingly, the PWWP domain of DNMT3A shows dual recognition of H3K36me2/3 in vitro with a higher binding affinity towards H3K36me2 [58,61]. However, ChIP-seq experiments investigating different lysine methylation states should be taken with great care. Many antibodies which are raised against a specific methylation state can show high cross-reactivity to other states at the same lysine residue [62]. Until now, it was unclear whether NSD3 contributes to the above-mentioned deposition of H3K36me2 at intergenic regions in other cell types where its expression is dominant over NSD1 and NSD2 or if the activity of NSD3 is restricted to other regulatory genomic elements.

Analogous to other known lysine methyltransferases [63], members of the NSD family were shown to methylate non-histone proteins. Apart from histone substrates, NSD3 recently was reported to methylate the epidermal growth factor receptor (EGFR), leading to enhanced activation [64], and NSD1 was shown to mono- and dimethylate p65, an NF-κB family transcription factor, at K218 and K221, which stimulates the expression of p65-dependent tumorigenic genes [65]. Furthermore, NSD1 was shown to methylate histone H1 in a variant-specific manner [66].

6. The Role of NSD3 in Cancer

Knowledge about the function of NSD3 in individual diseases is sparse, and most of the information available is about its role in different tumors. NSD3 is located on chromosome 8p11.2, in a region which has been linked to various diseases and that is amplified in primary tumors and cell lines from breast carcinoma [14,15]. As well as NSD3, the 8p11.2 region contains a set of genes including TAM, FGFR1, and LETM2 [15,67].

Genomic alterations of NSD3 occur in multiple cancer types, implicating its cancer-promoting role [12,68]. In most cases, the fusion between the NUP98 and NSD3 genes was detected in patients with AML or myelodysplastic syndrome [69,70], which promotes hematopoietic transformation in the same fashion as already shown for the NUP98-NSD1 fusion protein, due to the structural similarity between the two [71]. Besides the fusion to NUP98, NSD3 fusion has been observed with NUTM1 in primary pulmonary NUT carcinoma [72–74], which is known to typically harbor the BRD4/3-NUT fusion oncoprotein [75].

In line with the function of NSD3-short as an adaptor protein of BRD4 and CHD8 [48], MLL-AF9 rearranged acute myeloid leukemia (AML) were proven to be dependent on NSD3 [48,51,76]. This was confirmed by the development of a chemical probe for the PWWP1 domain of NSD3, which leads to the reduced proliferation of AML cell lines through the downregulation of MYC mRNA [37].

In addition, the 8p11.2 region is amplified in many cancers [67], leading to the increased expression of NSD3 (Figure 5), and reports have described NSD3 to be essential for tumor maintenance and the suppression of NSD3 expression leads to reduced cell proliferation in lung cancer [77–79], breast cancer [80,81], and osteosarcoma [82]. Furthermore, the 8p11.2 region is amplified in breast cancer (BC) [14,80,81] and the overexpression of NSD3 is linked to overexpression of the estrogen receptor alpha (ERα) in breast cancer [80]. A similar scenario was described for colorectal cancer (CRC) [83]. Here, NSD3 was shown to be upregulated in CRC and the suppression of NSD3 expression resulted in a decrease in proliferation, migration, and EMT marker proteins such as E-cadherin and N-cadherin [83].

Thus far, only one non histone protein has been described, which is methylated by NSD3 [64]. The epidermal growth factor receptor (EGFR) was shown to be methylated by NSD3, leading to the enhanced activation of the associated ERK cascade without stimulation by EGF. In addition, nuclear EGFR was showed to enhance its interaction with proliferating-cell-nuclear-antigen (PCNA) resulted in enhanced proliferation in squamous cell carcinoma of the head and neck (SCCHN) [64].

Furthermore, over 260 mutations have been described within the NSD3 protein (Figure 6) and for most, the underlying change in protein function has not yet been described. Intermolecular contacts between NSD3 and nucleosomes are altered by several recurrent cancer-associated mutations. E1181K and T1232A substitution leads to enhanced enzymatic activity through preventing the autoinhibitory loop from blocking the active site, which improves the insertion of the target H3K36 into the catalytic pocket of NSD3 [35,79]. Both mutations were demonstrated to promote the proliferation of cancer cells and accelerated growth of xenograft tumors [35]. There is no specific information available on the effect of the other mutations observed in NSD3.

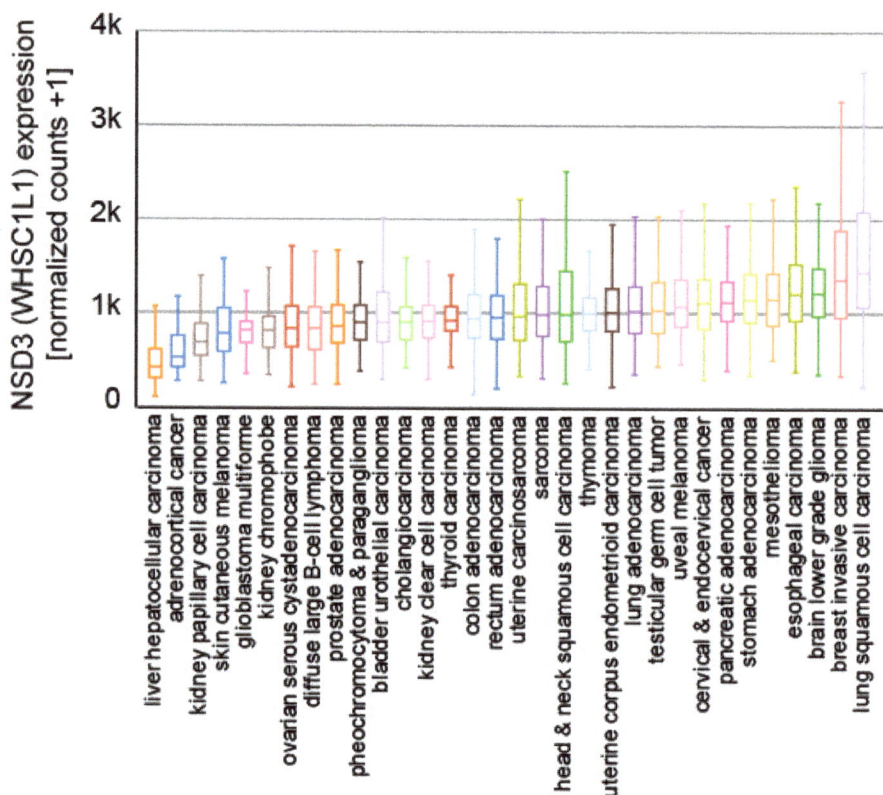

Figure 5. Expression of NSD3 in different indicated cancer subtypes. Samples from 9621 primary tumors of the TCGA Pan-Cancer set (https://www.cancer.gov/tcga, accessed on 21 July 2021) are presented and visualized with the UCSC Xena platform [46].

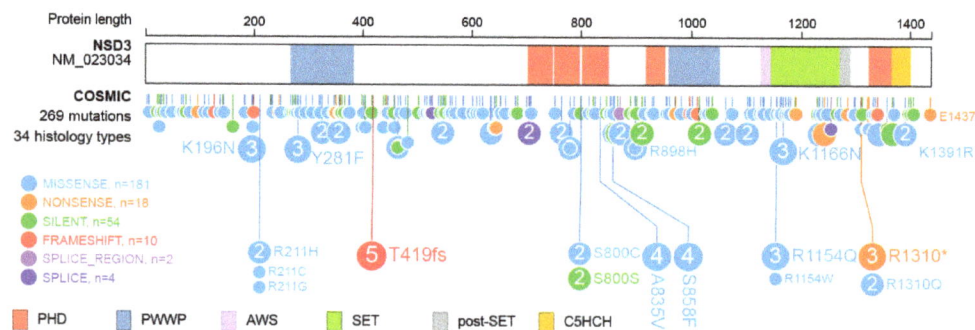

Figure 6. Mutations observed within NSD3. Visualization of NSD3 mutations listed in the Catalogue of Somatic Mutations In Cancer (COSMIC release 87) using ProteinPaint [33]. The major domains NSD3 are highlighted. Mutations are color coded as indicated. Numbers represent the number of amino acids.

Besides mutations in NSD3 itself, so-called onco-histones harboring mutations of the lysine at position 36 [84–87], lead to alterations of the function of NSD3. Given the importance of H3K36me2 in maintaining active enhancers to regulate epithelial-to-mesenchymal identity, tumor differentiation, and metastasis [45,60] it is inevitable that these onco-histones impose a strong negative impact on transcriptional maintenance. The incorporation of a lysine-to-methionine histone H3 mutant (H3K36M) led to a genome-wide reduction in H3K36me2 and H3K36me3 levels in different malignancies [60,86–89], which was attributed to a direct inhibitory effect of the H3.3K36M mutation on NSD2 and SET Domain Containing 2 (SETD2) [87]. Unfortunately, in these studies, the effect of the K36M mutation was not tested on NSD3 activity, but the comparable substrate specificities and structural similarities suggest a potential inhibitory effect on NSD3 as well.

Two recent publications shed more light on how altered NSD3 activity promotes tumor development and growth. These studies investigated the role of NSD3 in squamous cell lung cancer [79] and breast cancer [80]. Both showed that NSD3 acts as a factor that reprograms the chromatin landscape to promote oncogenic gene expression signatures. Elevated NSD3 expression [80] or hyperactivity [79] leads to an increase in H3K36me2 which inhibits the activity of the PRC2 complex [45]. This leads to the reexpression of developmental genes like MYC [79] or Notch3 [80], which promote stem cell like properties and in turn malignant transformation [79,80].

7. Outlook

Despite the recent achievements in the structural and biochemical analyses of NSD3 in complex with the nucleosome, which provided a molecular basis for the nucleosomal preference and activation mechanism of NSD proteins, not much information is available on cellular functions of NSD3 itself. Nevertheless, the fact that the methylation of H3K36 plays such an important role in regulating enhancer activity [45,60] and NSD3 is amplified in many cancers [14,67,73,77,79,80], suggests that NSD3 must play an important role in many different cellular processes. Epigenetic-based therapies are emerging as effective and valuable approaches in cancer and targeting NSD3 may indeed present a valuable approach [37,48,51,76]. However, the existence of at least six histone methyltransferases, which are capable of methylating H3K36, complicate the efforts in understanding the effects of NSD3 in cells, and further work will be needed to clarify these roles.

Funding: This research received no external funding.

Institutional Review Board Statement: Not applicable.

Informed Consent Statement: Not applicable.

Data Availability Statement: Not applicable.

Conflicts of Interest: The authors declare no conflict of interest.

References

1. Kornberg, R.D.; Lorch, Y.L. Twenty-Five Years of the Nucleosome, Fundamental Particle of the Eukaryote Chromosome. *Cell* **1999**, *98*, 285–294. [CrossRef]
2. Davis, L.; Onn, I.; Elliott, E. The emerging roles for the chromatin structure regulators CTCF and cohesin in neurodevelopment and behavior. *Cell. Mol. Life Sci.* **2018**, *75*, 1205–1214. [CrossRef]
3. Dekker, J.; Mirny, L. The 3D Genome as Moderator of Chromosomal Communication. *Cell* **2016**, *164*, 1110–1121. [CrossRef]
4. Atchison, M.L. Function of YY1 in Long-Distance DNA Interactions. *Front. Immunol.* **2014**, *5*, 45. [CrossRef] [PubMed]
5. Hildebrand, E.M.; Dekker, J. Mechanisms and Functions of Chromosome Compartmentalization. *Trends Biochem. Sci.* **2020**, *45*, 385–396. [CrossRef] [PubMed]
6. Turner, B.M. Histone acetylation and an epigenetic code. *Bioessays* **2000**, *22*, 836–845. [CrossRef]
7. Strahl, B.D.; Allis, C.D. The language of covalent histone modifications. *Nature* **2000**, *403*, 41–45. [CrossRef] [PubMed]
8. Fischle, W.; Wang, Y.M.; Allis, C.D. Histone and chromatin cross-talk. *Curr. Opin. Cell Biol.* **2003**, *15*, 172–183. [CrossRef]
9. Jenuwein, T.; Allis, C.D. Translating the Histone Code. *Science* **2001**, *293*, 1074–1080. [CrossRef] [PubMed]
10. Cosgrove, M.S.; Wolberger, C. How does the histone code work? *Biochem. Cell Biol.* **2005**, *83*, 468–476. [CrossRef]

11. Greer, E.L.; Shi, Y. Histone methylation: A dynamic mark in health, disease and inheritance. *Nat. Rev. Genet.* **2012**, *13*, 343–357. [CrossRef]
12. Li, J.; Ahn, J.H.; Wang, G.G. Understanding histone H3 lysine 36 methylation and its deregulation in disease. *Cell. Mol. Life Sci.* **2019**, *76*, 2899–2916. [CrossRef]
13. Bennett, R.L.; Swaroop, A.; Troche, C.; Licht, J.D. The Role of Nuclear Receptor–Binding SET Domain Family Histone Lysine Methyltransferases in Cancer. *Cold Spring Harb. Perspect. Med.* **2017**, *7*, a026708. [CrossRef]
14. Angrand, P.-O.; Apiou, F.; Stewart, A.F.; Dutrillaux, B.; Losson, R.; Chambon, P. NSD3, a New SET Domain-Containing Gene, Maps to 8p12 and Is Amplified in Human Breast Cancer Cell Lines. *Genomics* **2001**, *74*, 79–88. [CrossRef]
15. Stec, I.; Van Ommen, G.J.B.; den Dunnen, J.T. WHSC1L1, on Human Chromosome 8p11.2, Closely Resembles WHSC1 and Maps to a Duplicated Region Shared with 4p16.3. *Genomics* **2001**, *76*, 5–8. [CrossRef] [PubMed]
16. Wood, A.; Shilatifard, A. Posttranslational Modifications of Histones by Methylation. *Adv. Protein Chem.* **2004**, *67*, 201–222. [CrossRef]
17. Alvarez-Venegas, R.; Avramova, Z. SET-domain proteins of the Su(var)3-9, E(z) and Trithorax families. *Gene* **2002**, *285*, 25–37. [CrossRef]
18. He, C.; Li, F.D.; Zhang, J.H.; Wu, J.H.; Shi, Y.Y. The Methyltransferase NSD3 Has Chromatin-binding Motifs, PHD5-C5HCH, That Are Distinct from Other NSD (Nuclear Receptor SET Domain) Family Members in Their Histone H3 Recognition. *J. Biol. Chem.* **2013**, *288*, 4692–4703. [CrossRef]
19. Berardi, A.; Quilici, G.; Spiliotopoulos, D.; Corral-Rodriguez, M.A.; Martin-Garcia, F.; Degano, M.; Tonon, G.; Ghitti, M.; Musco, G. Structural basis for PHD(V)C5HCH(NSD1)-C2HR(Nizp1) interaction: Implications for Sotos syndrome. *Nucleic Acids Res.* **2016**, *44*, 3448–3463. [CrossRef]
20. Nielsen, A.L.; Jorgensen, P.; Lerouge, T.; Cerviño, M.; Chambon, P.; Losson, R. Nizp1, a Novel Multitype Zinc Finger Protein That Interacts with the NSD1 Histone Lysine Methyltransferase through a Unique C2HR Motif. *Mol. Cell. Biol.* **2004**, *24*, 5184–5196. [CrossRef]
21. Pasillas, M.P.; Shah, M.; Kamps, M.P. NSD1 PHD domains bind methylated H3K4 and H3K9 using interactions disrupted by point mutations in human sotos syndrome. *Hum. Mutat.* **2011**, *32*, 292–298. [CrossRef]
22. Vermeulen, M.; Eberl, H.C.; Matarese, F.; Marks, H.; Denissov, S.; Butter, F.; Lee, K.K.; Olsen, J.V.; Hyman, A.A.; Stunnenberg, H.G.; et al. Quantitative Interaction Proteomics and Genome-wide Profiling of Epigenetic Histone Marks and Their Readers. *Cell* **2010**, *142*, 967–980. [CrossRef]
23. Wu, H.; Zeng, H.; Lam, R.; Tempel, W.; Amaya, M.F.; Xu, C.; Dombrovski, L.; Qiu, W.; Wang, Y.; Min, J. Structural and Histone Binding Ability Characterizations of Human PWWP Domains. *PLoS ONE* **2011**, *6*, e18919. [CrossRef]
24. Kuo, A.J.; Cheung, P.; Chen, K.F.; Zee, B.M.; Kioi, M.; Lauring, J.; Xi, Y.X.; Park, B.H.; Shi, X.B.; Garcia, B.A.; et al. NSD2 Links Dimethylation of Histone H3 at Lysine 36 to Oncogenic Programming. *Mol. Cell* **2011**, *44*, 609–620. [CrossRef]
25. Morishita, M.; Mevius, D.; Di Luccio, E. In vitro histone lysine methylation by NSD1, NSD2/MMSET/WHSC1 and NSD3/WHSC1L. *BMC Struct. Biol.* **2014**, *14*, 1–13. [CrossRef]
26. Sankaran, S.M.; Wilkinson, A.W.; Elias, J.E.; Gozani, O. A PWWP Domain of Histone-Lysine N-Methyltransferase NSD2 Binds to Dimethylated Lys-36 of Histone H3 and Regulates NSD2 Function at Chromatin. *J. Biol. Chem.* **2016**, *291*, 8465–8474. [CrossRef] [PubMed]
27. Martinez-Garcia, E.; Popovic, R.; Min, D.-J.; Sweet, S.M.M.; Thomas, P.M.; Zamdborg, L.; Heffner, A.; Will, C.; Lamy, L.; Staudt, L.M.; et al. The MMSET histone methyl transferase switches global histone methylation and alters gene expression in t(4;14) multiple myeloma cells. *Blood* **2011**, *117*, 211–220. [CrossRef] [PubMed]
28. Huang, Z.; Wu, H.P.; Chuai, S.; Xu, F.N.; Yan, F.; Englund, N.; Wang, Z.F.; Zhang, H.L.; Fang, M.; Wang, Y.Z.; et al. NSD2 Is Recruited through Its PHD Domain to Oncogenic Gene Loci to Drive Multiple Myeloma. *Cancer Res.* **2013**, *73*, 6277–6288. [CrossRef]
29. Sims, R.J.; Reinberg, D. Histone H3 Lys 4 methylation: Caught in a bind? *Genes Dev.* **2006**, *20*, 2779–2786. [CrossRef]
30. Li, Y.L.; Chen, X.; Lu, C. The interplay between DNA and histone methylation: Molecular mechanisms and disease implications. *EMBO Rep.* **2021**, *22*, e51803. [CrossRef] [PubMed]
31. Lienert, F.; Mohn, F.; Tiwari, V.K.; Baubec, T.; Roloff, T.C.; Gaidatzis, D.; Stadler, M.B.; Schubeler, D. Genomic Prevalence of Heterochromatic H3K9me2 and Transcription Do Not Discriminate Pluripotent from Terminally Differentiated Cells. *PLoS Genet.* **2011**, *7*, e1002090. [CrossRef]
32. Kang, H.B.; Choi, Y.; Lee, J.M.; Choi, K.C.; Kim, H.C.; Yoo, J.Y.; Lee, Y.H.; Yoon, H.G. The histone methyltransferase, NSD2, enhances androgen receptor-mediated transcription. *FEBS Lett.* **2009**, *583*, 1880–1886. [CrossRef] [PubMed]
33. Zhou, X.; Edmonson, M.N.; Wilkinson, M.R.; Patel, A.; Wu, G.; Liu, Y.; Li, Y.J.; Zhang, Z.J.; Rusch, M.C.; Parker, M.; et al. Exploring genomic alteration in pediatric cancer using ProteinPaint. *Nat. Genet.* **2016**, *48*, 4–6. [CrossRef]
34. Morrison, M.J.; Boriack-Sjodin, P.A.; Swinger, K.K.; Wigle, T.J.; Sadalge, D.; Kuntz, K.W.; Scott, M.P.; Janzen, W.P.; Chesworth, R.; Duncan, K.W.; et al. Identification of a peptide inhibitor for the histone methyltransferase WHSC1. *PLoS ONE* **2018**, *13*, e0197082. [CrossRef] [PubMed]
35. Li, W.; Tian, W.; Yuan, G.; Deng, P.; Sengupta, D.; Cheng, Z.; Cao, Y.; Ren, J.; Qin, Y.; Zhou, Y.; et al. Molecular basis of nucleosomal H3K36 methylation by NSD methyltransferases. *Nature* **2021**, *590*, 498–503. [CrossRef]

36. Qiao, Q.; Li, Y.; Chen, Z.; Wang, M.Z.; Reinberg, D.; Xu, R.-M. The Structure of NSD1 Reveals an Autoregulatory Mechanism Underlying Histone H3K36 Methylation. *J. Biol. Chem.* **2011**, *286*, 8361–8368. [CrossRef]
37. Böttcher, J.; Dilworth, D.; Reiser, U.; Neumüller, R.A.; Schleicher, M.; Petronczki, M.; Zeeb, M.; Mischerikow, N.; Allali-Hassani, A.; Szewczyk, M.M.; et al. Fragment-based discovery of a chemical probe for the PWWP1 domain of NSD3. *Nat. Chem. Biol.* **2019**, *15*, 822–829. [CrossRef]
38. Qin, S.; Min, J.R. Structure and function of the nucleosome-binding PWWP domain. *Trends Biochem. Sci.* **2014**, *39*, 536–547. [CrossRef]
39. Vezzoli, A.; Bonadies, N.; Allen, M.D.; Freund, S.M.V.; Santiveri, C.M.; Kvinlaug, B.T.; Huntly, B.J.P.; Gottgens, B.; Bycroft, M. Molecular basis of histone H3K36me3 recognition by the PWWP domain of Brpf1. *Nat. Struct. Mol. Biol.* **2010**, *17*, 617–619. [CrossRef] [PubMed]
40. McDaniel, S.L.; Strahl, B.D. Shaping the cellular landscape with Set2/SETD2 methylation. *Cell. Mol. Life Sci.* **2017**, *74*, 3317–3334. [CrossRef]
41. Kim, S.M.; Kee, H.J.; Eom, G.H.; Choe, N.W.; Kim, J.Y.; Kim, Y.S.; Kim, S.K.; Kook, H.; Kook, H.; Seo, S.B. Characterization of a novel WHSC1-associated SET domain protein with H3K4 and H3K27 methyltransferase activity. *Biochem. Biophys. Res. Commun.* **2006**, *345*, 318–323. [CrossRef]
42. Li, Y.; Trojer, P.; Xu, C.F.; Cheung, P.; Kuo, A.; Drury, W.J.; Qiao, Q.; Neubert, T.A.; Xu, R.M.; Gozani, O.; et al. The Target of the NSD Family of Histone Lysine Methyltransferases Depends on the Nature of the Substrate. *J. Biol. Chem.* **2009**, *284*, 34283–34295. [CrossRef]
43. Yuan, G.; Ma, B.; Yuan, W.; Zhang, Z.Q.; Chen, P.; Ding, X.J.; Feng, L.; Shen, X.H.; Chen, S.; Li, G.H.; et al. Histone H2A Ubiquitination Inhibits the Enzymatic Activity of H3 Lysine 36 Methyltransferases. *J. Biol. Chem.* **2013**, *288*, 30832–30842. [CrossRef]
44. Yuan, W.; Xu, M.; Huang, C.; Liu, N.; Chen, S.; Zhu, B. H3K36 Methylation Antagonizes PRC2-mediated H3K27 Methylation. *J. Biol. Chem.* **2011**, *286*, 7983–7989. [CrossRef]
45. Streubel, G.; Watson, A.; Jammula, S.G.; Scelfo, A.; Fitzpatrick, D.J.; Oliviero, G.; McCole, R.; Conway, E.; Glancy, E.; Negri, G.L.; et al. The H3K36me2 Methyltransferase Nsd1 Demarcates PRC2-Mediated H3K27me2 and H3K27me3 Domains in Embryonic Stem Cells. *Mol. Cell* **2018**, *70*, 371–379.e5. [CrossRef]
46. Goldman, M.J.; Craft, B.; Hastie, M.; Repecka, K.; McDade, F.; Kamath, A.; Banerjee, A.; Luo, Y.H.; Rogers, D.; Brooks, A.N.; et al. Visualizing and interpreting cancer genomics data via the Xena platform. *Nat. Biotechnol.* **2020**, *38*, 675–678. [CrossRef] [PubMed]
47. Rahman, S.; Sowa, M.E.; Ottinger, M.; Smith, J.A.; Shi, Y.; Harper, J.W.; Howley, P.M. The Brd4 Extraterminal Domain Confers Transcription Activation Independent of pTEFb by Recruiting Multiple Proteins, Including NSD3. *Mol. Cell. Biol.* **2011**, *31*, 2641–2652. [CrossRef]
48. Shen, C.; Ipsaro, J.J.; Shi, J.; Milazzo, J.P.; Wang, E.; Roe, J.S.; Suzuki, Y.; Pappin, D.J.; Joshua-Tor, L.; Vakoc, C.R. NSD3-Short Is an Adaptor Protein that Couples BRD4 to the CHD8 Chromatin Remodeler. *Mol. Cell* **2015**, *60*, 847–859. [CrossRef]
49. Zhang, Q.; Zeng, L.; Shen, C.; Ju, Y.; Konuma, T.; Zhao, C.; Vakoc, C.R.; Zhou, M.M. Structural Mechanism of Transcriptional Regulator NSD3 Recognition by the ET Domain of BRD4. *Structure* **2016**, *24*, 1201–1208. [CrossRef]
50. Spriano, F.; Stathis, A.; Bertoni, F. Targeting BET bromodomain proteins in cancer: The example of lymphomas. *Pharmacol. Ther.* **2020**, *215*, 107631. [CrossRef]
51. Zuber, J.; Shi, J.W.; Wang, E.; Rappaport, A.R.; Herrmann, H.; Sison, E.A.; Magoon, D.; Qi, J.; Blatt, K.; Wunderlich, M.; et al. RNAi screen identifies Brd4 as a therapeutic target in acute myeloid leukaemia. *Nature* **2011**, *478*, 524–528. [CrossRef] [PubMed]
52. Dawson, M.A.; Prinjha, R.; Dittman, A.; Giotopoulos, G.; Bantscheff, M.; Chan, W.-I.; Robson, S.; Chung, C.-W.; Hopf, C.; Savitski, M.; et al. Inhibition of BET Recruitment to Chromatin As An Effective Treatment for MLL-Fusion Leukaemia. *Blood* **2011**, *118*, 55. [CrossRef]
53. Xu, Y.L.; Vakoc, C.R. Targeting Cancer Cells with BET Bromodomain Inhibitors. *Cold Spring Harb. Perspect. Med.* **2017**, *7*, a026674. [CrossRef] [PubMed]
54. Kim, S.M.; Kee, H.J.; Choe, N.; Kim, J.Y.; Kook, H.; Kook, H.; Seo, S.B. The histone methyltransferase activity of WHISTLE is important for the induction of apoptosis and HDAC1-mediated transcriptional repression. *Exp. Cell Res.* **2007**, *313*, 975–983. [CrossRef]
55. Bannister, A.J.; Schneider, R.; Myers, F.A.; Thorne, A.W.; Crane-Robinson, C.; Kouzarides, T. Spatial Distribution of Di- and Tri-methyl Lysine 36 of Histone H3 at Active Genes. *J. Biol. Chem.* **2005**, *280*, 17732–17736. [CrossRef] [PubMed]
56. Zhou, M.S.; Deng, L.W.; Lacoste, V.; Park, H.U.; Pumfery, A.; Kashanchi, F.; Brady, J.N.; Kumar, A. Coordination of Transcription Factor Phosphorylation and Histone Methylation by the P-TEFb Kinase during Human Immunodeficiency Virus Type 1 Transcription. *J. Virol.* **2004**, *78*, 13522–13533. [CrossRef]
57. Popovic, R.; Martine-Garcia, E.; Giannopoulou, E.G.; Zhang, Q.W.; Zhang, Q.Y.; Ezponda, T.; Shah, M.Y.; Zheng, Y.P.; Will, C.M.; Small, E.C.; et al. Histone Methyltransferase MMSET/NSD2 Alters EZH2 Binding and Reprograms the Myeloma Epigenome through Global and Focal Changes in H3K36 and H3K27 Methylation. *PLoS Genet.* **2014**, *10*, e1004566. [CrossRef]
58. Weinberg, D.N.; Papillon-Cavanagh, S.; Chen, H.F.; Yue, Y.; Chen, X.; Rajagopalan, K.N.; Horth, C.; McGuire, J.T.; Xu, X.J.; Nikbakht, H.; et al. The histone mark H3K36me2 recruits DNMT3A and shapes the intergenic DNA methylation landscape. *Nature* **2019**, *573*, 281–286. [CrossRef]

59. Piunti, A.; Shilatifard, A. The roles of Polycomb repressive complexes in mammalian development and cancer. *Nat. Rev. Mol. Cell Biol.* **2021**, *22*, 326–345. [CrossRef]
60. Yuan, S.; Natesan, R.; Sanchez-Rivera, F.J.; Li, J.Y.; Bhanu, N.V.; Yamazoe, T.; Lin, J.H.; Merrell, A.J.; Sela, Y.; Thomas, S.K.; et al. Global Regulation of the Histone Mark H3K36me2 Underlies Epithelial Plasticity and Metastatic Progression. *Cancer Discov.* **2020**, *10*, 854–871. [CrossRef]
61. Dhayalan, A.; Rajavelu, A.; Rathert, P.; Tamas, R.; Jurkowska, R.Z.; Ragozin, S.; Jeltsch, A. The Dnmt3a PWWP Domain Reads Histone 3 Lysine 36 Trimethylation and Guides DNA Methylation. *J. Biol. Chem.* **2010**, *285*, 26114–26120. [CrossRef] [PubMed]
62. Bock, I.; Dhayalan, A.; Kudithipudi, S.; Brandt, O.; Rathert, P.; Jeltsch, A. Detailed specificity analysis of antibodies binding to modified histone tails with peptide arrays. *Epigenetics* **2011**, *6*, 256–263. [CrossRef] [PubMed]
63. Carlson, S.M.; Gozani, O. Nonhistone Lysine Methylation in the Regulation of Cancer Pathways. *Cold Spring Harb. Perspect. Med.* **2016**, *6*, a026435. [CrossRef]
64. Saloura, V.; Vougiouklakis, T.; Zewde, M.; Deng, X.; Kiyotani, K.; Park, J.-H.; Matsuo, Y.; Lingen, M.; Suzuki, T.; Dohmae, N.; et al. WHSC1L1-mediated EGFR mono-methylation enhances the cytoplasmic and nuclear oncogenic activity of EGFR in head and neck cancer. *Sci. Rep.* **2017**, *7*, 40664. [CrossRef] [PubMed]
65. Lu, T.; Jackson, M.W.; Wang, B.L.; Yang, M.J.; Chance, M.R.; Miyagi, M.; Gudkov, A.V.; Stark, G.R. Regulation of NF-kappa B by NSD1/FBXL11-dependent reversible lysine methylation of p65. *Proc. Natl. Acad. Sci. USA* **2010**, *107*, 46–51. [CrossRef] [PubMed]
66. Kudithipudi, S.; Lungu, C.; Rathert, P.; Happel, N.; Jeltsch, A. Substrate Specificity Analysis and Novel Substrates of the Protein Lysine Methyltransferase NSD1. *Chem. Biol.* **2014**, *21*, 226–237. [CrossRef]
67. Voutsadakis, I.A. Amplification of 8p11.23 in cancers and the role of amplicon genes. *Life Sci.* **2021**, *264*, 118729. [CrossRef] [PubMed]
68. Han, X.; Piao, L.; Zhuang, Q.; Yuan, X.; Liu, Z.; He, X. The role of histone lysine methyltransferase NSD3 in cancer. *OncoTargets Ther.* **2018**, *11*, 3847–3852. [CrossRef]
69. Taketani, T.; Taki, T.; Nakamura, H.; Taniwaki, M.; Masuda, J.; Hayashi, Y. NUP98–NSD3 fusion gene in radiation-associated myelodysplastic syndrome with t(8;11)(p11;p15) and expression pattern of NSD family genes. *Cancer Genet. Cytogenet.* **2009**, *190*, 108–112. [CrossRef]
70. Rosati, R.; La Starza, R.; Veronese, A.; Aventin, A.; Schwienbacher, C.; Vallespi, T.; Negrini, M.; Martelli, M.F.; Mecucci, C. NUP98 is fused to the NSD3 gene in acute myeloid leukemia associated with t(8;11)(p11.2;p15). *Blood* **2002**, *99*, 3857–3860. [CrossRef]
71. Wang, G.G.; Cai, L.; Pasillas, M.P.; Kamps, M.P. NUP98–NSD1 links H3K36 methylation to Hox-A gene activation and leukaemogenesis. *Nat. Cell Biol.* **2007**, *9*, 804–812. [CrossRef] [PubMed]
72. Khan, J.; Whaley, R.; Cheng, L. Primary Pulmonary NUT Carcinoma with NSD3-NUTM1 Fusion. *Am. J. Clin. Pathol.* **2020**, *154*, S83–S84. [CrossRef]
73. Suzuki, S.; Kurabe, N.; Ohnishi, I.; Yasuda, K.; Aoshima, Y.; Naito, M.; Tanioka, F.; Sugimura, H. NSD3-NUT-expressing midline carcinoma of the lung: First characterization of primary cancer tissue. *Pathol. Res. Pract.* **2015**, *211*, 404–408. [CrossRef] [PubMed]
74. Kuroda, S.; Suzuki, S.; Kurita, A.; Muraki, M.; Aoshima, Y.; Tanioka, F.; Sugimura, H. Cytological Features of a Variant NUT Midline Carcinoma of the Lung Harboring theNSD3-NUTFusion Gene: A Case Report and Literature Review. *Case Rep. Pathol.* **2015**, *2015*, 1–5. [CrossRef] [PubMed]
75. French, C.A.; Rahman, S.; Walsh, E.M.; Kuhnle, S.; Grayson, A.R.; Lemieux, M.E.; Grunfeld, N.; Rubin, B.P.; Antonescu, C.R.; Zhang, S.L.; et al. NSD3–NUT Fusion Oncoprotein in NUT Midline Carcinoma: Implications for a Novel Oncogenic Mechanism. *Cancer Discov.* **2014**, *4*, 928–941. [CrossRef]
76. Rathert, P.; Roth, M.; Neumann, T.; Muerdter, F.; Roe, J.-S.; Muhar, M.; Deswal, S.; Cerny-Reiterer, S.; Peter, B.; Jude, J.; et al. Transcriptional plasticity promotes primary and acquired resistance to BET inhibition. *Nature* **2015**, *525*, 543–547. [CrossRef] [PubMed]
77. Mahmood, S.F.; Gruel, N.; Nicolle, R.; Chapeaublanc, E.; Delattre, O.; Radvanyi, F.; Bernard-Pierrot, I. PPAPDC1B and WHSC1L1 Are Common Drivers of the 8p11-12 Amplicon, Not Only in Breast Tumors But Also in Pancreatic Adenocarcinomas and Lung Tumors. *Am. J. Pathol.* **2013**, *183*, 1634–1644. [CrossRef]
78. Kang, D.; Cho, H.S.; Toyokawa, G.; Kogure, M.; Yamane, Y.; Iwai, Y.; Hayami, S.; Tsunoda, T.; Field, H.I.; Matsuda, K.; et al. The histone methyltransferase Wolf–Hirschhorn syndrome candidate 1-like 1 (WHSC1L1) is involved in human carcinogenesis. *Genes Chromosomes Cancer* **2013**, *52*, 126–139. [CrossRef]
79. Yuan, G.; Flores, N.M.; Hausmann, S.; Lofgren, S.M.; Kharchenko, V.; Angulo-Ibanez, M.; Sengupta, D.; Lu, X.; Czaban, I.; Azhibek, D.; et al. Elevated NSD3 histone methylation activity drives squamous cell lung cancer. *Nature* **2021**, *590*, 504–508. [CrossRef]
80. Irish, J.C.; Mills, J.N.; Turner-Ivey, B.; Wilson, R.C.; Guest, S.T.; Rutkovsky, A.; Dombkowski, A.; Kappler, C.S.; Hardiman, G.; Ethier, S.P. Amplification of WHSC1L1 regulates expression and estrogen-independent activation of ERα in SUM-44 breast cancer cells and is associated with ERα over-expression in breast cancer. *Mol. Oncol.* **2016**, *10*, 850–865. [CrossRef]
81. Rutkovsky, A.C.; Turner-Ivey, B.; Smith, E.L.; Spruill, L.S.; Mills, J.N.; Ethier, S.P. Development of mammary hyperplasia, dysplasia, and invasive ductal carcinoma in transgenic mice expressing the 8p11 amplicon oncogene NSD3 (WHSC1L1). *Cancer Res.* **2017**, *77*, 1835. [CrossRef]
82. Liu, Z.; Piao, L.; Zhuang, M.; Qiu, X.; Xu, X.; Zhang, D.; Liu, M.; Ren, D. Silencing of histone methyltransferase NSD3 reduces cell viability in osteosarcoma with induction of apoptosis. *Oncol. Rep.* **2017**, *38*, 2796–2802. [CrossRef] [PubMed]

83. Yi, L.; Yi, L.; Liu, Q.; Li, C. Downregulation of NSD3 (WHSC1L1) inhibits cell proliferation and migration via ERK1/2 deactivation and decreasing CAPG expression in colorectal cancer cells. *OncoTargets Ther.* **2019**, *12*, 3933–3943. [CrossRef]
84. Brumbaugh, J.; Kim, I.S.; Ji, F.; Huebner, A.J.; Di Stefano, B.; Schwarz, B.A.; Charlton, J.; Coffey, A.; Choi, J.; Walsh, R.M.; et al. Inducible histone K-to-M mutations are dynamic tools to probe the physiological role of site-specific histone methylation in vitro and in vivo. *Nat. Cell Biol.* **2019**, *21*, 1449–1461. [CrossRef]
85. Morgan, M.; Herz, H.M.; Gao, X.; Jackson, J.; Rickels, R.; Swanson, S.K.; Florens, L.; Washburn, M.P.; Eissenberg, J.C.; Shilatifard, A. Histone H3 lysine-to-methionine mutants as a paradigm to study chromatin signaling. *FEBS J.* **2015**, *282*, 406.
86. Fang, D.; Gan, H.Y.; Lee, J.H.; Han, J.; Wang, Z.Q.; Riester, S.M.; Jin, L.; Chen, J.J.; Zhou, H.; Wang, J.L.; et al. The histone H3.3K36M mutation reprograms the epigenome of chondroblastomas. *Science* **2016**, *352*, 1344–1348. [CrossRef] [PubMed]
87. Lu, C.; Jain, S.U.; Hoelper, D.; Bechet, D.; Molden, R.C.; Ran, L.L.; Murphy, D.; Venneti, S.; Hameed, M.; Pawel, B.R.; et al. Histone H3K36 mutations promote sarcomagenesis through altered histone methylation landscape. *Science* **2016**, *352*, 844–849. [CrossRef] [PubMed]
88. Zhang, Y.; Fang, D. The incorporation loci of H3.3K36M determine its preferential prevalence in chondroblastomas. *Cell Death Dis.* **2021**, *12*, 1–16. [CrossRef]
89. Zhang, Y.L.; Shan, C.M.; Wang, J.Y.; Bao, K.; Tong, L.; Jia, S.T. Molecular basis for the role of oncogenic histone mutations in modulating H3K36 methylation. *Sci. Rep.* **2017**, *7*, 43906. [CrossRef]

Review

The Role of Lysine Methyltransferase SET7/9 in Proliferation and Cell Stress Response

Alexandra Daks [1,†], Elena Vasileva [1,2,†], Olga Fedorova [1], Oleg Shuvalov [1] and Nickolai A. Barlev [1,*]

1. Institute of Cytology RAS, 194064 St. Petersburg, Russia; alexandra.daks@gmail.com (A.D.); slkd-k@mail.ru (E.V.); fedorovaolga0402@gmail.com (O.F.); oleg8988@mail.ru (O.S.)
2. Children's Hospital Los Angeles, University of Southern California, Los Angeles, CA 90027, USA
* Correspondence: nick.a.barlev@gmail.com
† These authors contributed equally to this work.

Abstract: Lysine-specific methyltransferase 7 (KMT7) SET7/9, aka Set7, Set9, or SetD7, or KMT5 was discovered 20 years ago, yet its biological role remains rather enigmatic. In this review, we analyze the particularities of SET7/9 enzymatic activity and substrate specificity with respect to its biological importance, mostly focusing on its two well-characterized biological functions: cellular proliferation and stress response.

Keywords: SET7/9; SETD7; lysine-specific methyltransferase (PKMT); cell proliferation; stress response; post-translational protein modification

1. Introduction

Methyltransferases are a compendium of diverse enzymes, most of which use S-adenosyl-methionine (Ado-Met) as a donor of methyl groups. The basic methyl group transfer reaction is the catalytic attack of a nucleophile (carbon, oxygen, nitrogen, or sulfur) on a methyl group to form methylated derivatives of proteins, lipids, polysaccharides, nucleic acids, and various small molecules. This methyl conjugation not only affects the bioconversion pathways of many drugs, but also affects the properties of endogenous neurotransmitters and hormones. Furthermore, methylation is fundamental to the regulation of gene expression. Unlike DNA methylation, which has been known since the middle of the last century, protein methylation was discovered relatively recently. Proteins can be methylated at different amino acids, however, for protein-protein interactions the most relevant and well-studied is methylation on lysine and arginine residues. Gene expression can be regulated by lysine methylation on two levels: methylation of histones and methylation of non-histone proteins that include transcription factors and chromatin modifiers.

2. The History of SET7/9 Discovery

SET7/9, a lysine methyltransferase (PKMT) encoded by the SETD7 gene ((su(var)3–9, enhancer of zeste, trithorax (SET) domain-containing protein 7) was discovered independently by two laboratories in 2001. Reinberg's lab named this enzyme Set9 and Ye Zhang's lab called it Set7 [1,2]. Later, these two names were unified as SET7/9. Studies from both groups identified SET7/9 as specific lysine 4 (K4) of histone H3 (H3K4) methyltransferase. Zhang's group indicated that Set7 was able to di-methylate H3K4, which led to transcriptional activation by counteracting SuVar39h1-mediated H3K9 methylation. However, the caveat in the interpretation of the in vitro methylation data was that these experiments were done on free histones, whereas it is well known that the basic unit of chromatin is the nucleosome that is formed by a histone octamer wrapped by 157 nucleotides of DNA. Moreover, the in vivo experiments relied on the modification-specific antibodies, which are notoriously famous for their off-target recognition. Numerous experiments from different groups, including ours, have clearly demonstrated that SET7/9 failed to methylate

nucleosomal histones. Furthermore, despite the early report, SET7/9 was convincingly shown to exert its functions as mono-methyltransferase but not as dimethyltransferase. For example, Dhayalan et al. showed that although SET7/9 was able to transfer two methyl groups to both histone- and non-histone targets in vitro, it did so with much lower efficacy (~10% of the mono-methylation rate) [3]. Extensive structural studies showed that the free-energy barrier for the transfer of the first methyl group by SET7/9 was 17–18 kcal/mol and the subsequent addition of the second methyl group imposed a 5 kcal/mol higher energy barrier for the transfer. Therefore, at least in vitro, SET7/9 acts preferentially as a monomethyltransferase. However, it should be noted that in cellulo studies in islet cells from R.G. Mirmira's group suggested that SET7/9 was associated with the di-methylation of H3K4 [4]. One can speculate that SET7/9-mediated mono-methylation can trigger the subsequent addition of a second or third methyl group by other yet unknown methyltransferases [3,5,6].

3. The Substrate Specificity of SET7/9

Another enigmatic and debatable feature of SET7/9 is its substrate recognition specificity. There is an obvious discrepancy between the predicted frequency of occurrence for the potential SET7/9 consensus motif to be found in its target proteins and the handful number of in vivo confirmed substrates of SET7/9 that have been reported to date. Using the sequence-based approach together with the comparison of the structures of SET7/9 bound to TAF10, histone H3, and p53, a conserved sequence K/R-S/T/A-K*-D/N/Q/K (K^* is the methylation site) was identified [7,8]. Later, it was clarified that the requirement for amino acid residues in +1 and +2 positions is not that stringent for the ability of SET7/9 to methylate the substrate. Thus, the majority of SET7/9 methylation targets share the G/R/H/K/P/S/T-K>R-S>K/Y/A/R/T/P/N-K* sequence motif [3]. However, in another study alternative, SET7/9-recognition amino acid sequences were reported. The p300/CBP-associated factor (PCAF), which is an acetyltransferase itself, was shown to be mono-methylated by SET7/9 methyltransferase at K78 and K89 in vitro forming the consensus motif A/F/I/V-K*-D/K (K^* is the methylation site) for SET7/9 modification [9]. Comparison of 45 known methylation consensus motifs of SET7/9 shows high complexity of the target sequence. There are two types of consensus sequences that could be identified as preferable sites for SET7/9 methylation. Most analyzed proteins contain the K/R-S/A-K-K/S/R (Type 1) consensus motif and display an enrichment in the positively charged amino acids situated in the flanking regions, while the Type 2 alternative consensus sequence is enriched in basic amino acids in the flanking region and often displays proline at -12 position [10]. Our unpublished results indicate that the recognition sequence motif is much longer than it is thought currently (Vasileva, Daks and Barlev, unpublished).

4. Cellular Localization of SET7/9

Another interesting feature of SET7/9 is that it can be found preferentially either in the nucleus or in the cytoplasm, depending on the cell line [6,11–15]. Notably, unlike other SET domain-containing KMTs, SET7/9 does not contain a defined nuclear localization signal in its sequence. Thus, it can be hypothesized that SET7/9 is imported into the nucleus via a direct interaction with importin 5a, or via protein-protein interactions with its target proteins. This hypothesis is supported by the observation that nuclear factor (NF)-kappa-B (NFκB) recruits SET7/9 to the promoters of NFκB-dependent genes [11]. In light of the fact that NFκB interacts with the actin-binding protein, ACTN4, it would be interesting to see whether SET7/9 also interacts with elements of the cytoskeleton [16]. The SET7/9 localization may also depend on the cell type. For example, in mouse embryonic fibroblasts (MEFs), SET7/9 retains Yes-associated protein (YAP) in the cytoplasm [17]. In contrast, in human monocytes, SET7/9 was observed with NFκB-p65, both in the cytoplasm and nucleus [11]. Likewise, in the human osteosarcoma cell line, U2-OS, SET7/9 was detected both in the cytoplasm and the nucleus. However, upon DNA damage SET7/9 accumulates in the nucleus [18]. Therefore, further research is required to elucidate how SET7/9 is transported into the nucleus.

5. The Structural Organization of SET7/9 Methyltransferase

SET7/9 is a member of the SET domain-containing methyltransferases family that transfers a methyl group on the target protein involving S-adenosyl-L-methionine (SAM) as a donor. The protein structure of SET7/9 includes three MORN (Membrane Occupation and Recognition Nexus) domains mediating protein-protein interactions with the substrates and one SET domain required for SET7/9 enzymatic activity (Figure 1). According to the study of H. Liu et al. the MORN domain repeats represent a concave structure which is enriched in negatively charged amino acids [19]. Thus, perhaps expectedly, it can bind a number of positively charged proteins, including the DNA binding domains of several transcription factors. Using the bioinformatic approach, it was revealed that the MORN repeat-containing proteins are expressed both in procaryotes and eucaryotes [20]. In addition to SET7/9, there is a number of MORN repeat-containing proteins including junctophilins (JPHs), ALS2 (Rho guanine nucleotide exchange factor, alsin), and MORN4 (retinophilin) that are expressed in mammals [21–23]. It is generally assumed that MORN repeats bind to lipids and are responsible for plasma membrane targeting. However, there is a growing volume of evidence suggesting that MORN repeats may mediate protein-protein interactions [24,25]. Indeed, by using a GST-pull-down assay coupled with mass spectrometry, we have demonstrated that MORN repeats are responsible for the majority of SET7/9 interactions [10].

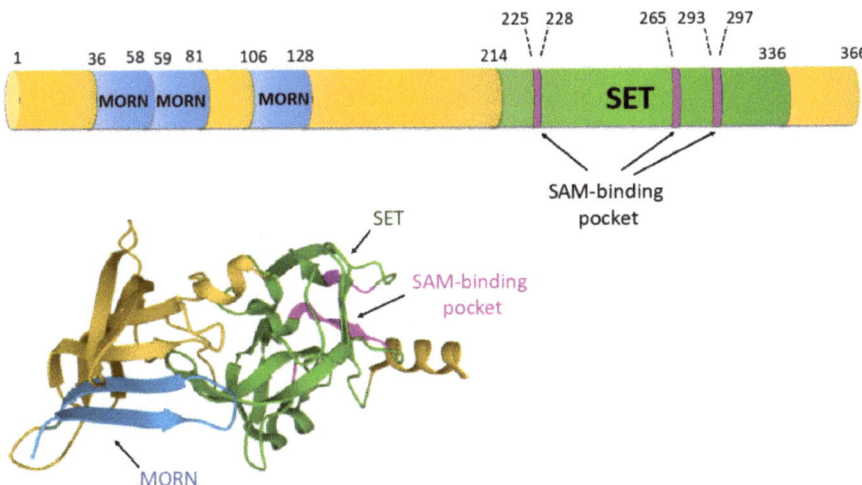

Figure 1. The domain organization and 3D-structure (PDB ID 1N6C, [26]) of the SET7/9 protein. The amino-terminal domain includes three MORN repeats. The SET domain is located in the C-terminus. Several amino acid residues of the SET domain form the SAM-binding pocket. SAM—S-adenosyl-L-methionine.

The resolved crystal structure of the SET domain revealed that the amino-terminal domain has a groove running across the extended beta sheet to the SET domain leading to a narrower channel running around the SET domain [27]. It was shown that this N-domain provides part of the binding site for basic histone tails as well as participates in determining the substrate specificity of the enzyme, while the C-terminus of the SET domain is important for the catalytic competence and contributes to the formation of the active site [27] (Figure 1).

6. Histone Targets of SET7/9

Posttranslational histone modifications such as methylation, phosphorylation, acetylation, ubiquitination and ADP-ribosylation define chromatin's dynamic structure and function. Histones as substrates for lysine methylation were first described in 1964 [28]. In particular, lysines H3K4, H3K9, H3K27, H3K36 and H4K20 are the preferred sites for methylation [28,29]. SET7/9 was initially identified as a methyltransferase that methylates H3K4 facilitating transcriptional activation by displacing the histone deacetylase NuRD complex (HDAC) [1], (Table 1). Since SET7/9-mediated methylation of H3K4 enhanced the following acetylation of histones and the latter correlates with gene activation, it was implied that H3K4 methylation by SET7/9 should positively regulate transcription. The Reinberg's group also demonstrated that the interplay between the Set9 and Suv39H1 histone methyltransferases was specific, as the methylation of H3K9 by another histone methyltransferase, G9a, was not affected by the Set9-mediated methylation of H3K4. Moreover, methylation of H3K4 was shown to reduce Suv39H1-mediated methylation at K9 of H3 histone (H3K9) [1]. In line with this notion is the fact that methylated H3K9 was shown to localize to a 20-kb silent heterochromatic region, whereas methylated H3K4 was detected exclusively in surrounding euchromatic regions [30].

Table 1. The substrates of SET7/9-dependent methylation playing roles in proliferation and cellular stress response.

SET7/9 Target Protein	Methylation Sites	Effect of the Modification	Reference
Histone H1	K12, K14, K17, K20, K21, K27, K111	Modulation of the affinity of histone H1 to chromatin during human pluripotent cells differentiation	[31]
Histone H1.4	K34, K127, K129, K130	Prevention of acetylation at the same sites, heterochromatin formation	[31]
Histone H2.A	K5, K13, K15	Unknown	[3]
Histone H2.B	K15	Unknown	[3]
Histone H3	K4	Activation of transcription	[1]
Suv39H1	K25, K123	Heterohromatin relaxation, genome instability	[32]
DNMT1	K142	Promotion of DNMT1 ubiquitination and proteasomal degradation	[33]
	K1094	Decrease of the DNMT1 level	[34]
TAF7	K5	Enhancement of TAF7 activity as co-factor of RNA polymerase II	[8]
TAF10	K189	Enhancement of TAF10 activity as co-factor of RNA polymerase II, activation of transcription of TAF10 target genes	[35]
YAP1	K494	Retention of YAP1 in the cytoplasm	[17]
β-catenin	K180	Promotion of β-catenin ubiquitination by (GSK)-3b and its subsequent proteasomal degradation	[36]
STAT3	K140	Dissociation of STAT3 from promoter elements, downregulation of STAT3-dependent genes expression	[37]
E2F1	K185	Promotion of E2F1 ubiquitination and subsequent proteasomal degradation	[38]
	Unknown	Enhancement of E2F transactivation of its target genes	[39]
pRb	K873	Enhancement of pRB-dependent repression of transcription	[40]
	K810	Promotion of p65/RelA ubiquitination and its subsequent proteasomal degradation	[41]

Table 1. *Cont.*

SET7/9 Target Protein	Methylation Sites	Effect of the Modification	Reference
YY1	K173, K411	Retention of YY1 in the cytoplasm	[42]
p65/RelA	K37	Translocation to the nucleus and transactivation of target genes	[43]
	K314, K315	Promotion of p65/RelA ubiquitination and subsequent proteasomal degradation	[44]
FOXO3	K270	Downregulation of FOXO3-dependent transactivation of BIM	[45]
	K271	Increase of the FOXO3 transactivation potential	[46]
Hif1α	K32	Suppression of Hif1α transactivation of its target genes	[47,48]
p53	K372	Stabilization, translocation to the nucleus and transactivation of target genes	[7]
SIRT1	K233, K235, K236, K238	Enhancement of SIRT1-dependent p53 acetylation and activation	[49]

Recently, it was shown that SET7/9 specifically methylates histone H1.4 at the K121, K129, K159, K171, K177 and K192 positions, competing for binding with the H3 histone protein. Methylation of H1.4 by SET7/9 upon binding to DNA tended to form less α-helix but more β-structure than unmethylated H1.4. There are two sites in H1.4 for methylation in vivo: K129 in the C-terminal domain and at K34 in the N-terminal domain. Methylation of H1.4 at K34 results in the reduction of the levels of acetylation by competition, contributing to the establishment of the proper heterochromatin patterns during differentiation [31].

Moreover, such modification as ADP-ribosylation of H3 by ARTD1 (PARP1) prevents H3 methylation by SET7/9, while poly(ADP-ribosyl)ation (PARylation) of histone H3 allowed subsequent methylation of H1 by SET7/9 [50]. Taken together, histone lysine methylation is a mark involved in the maintenance of genome expression and is dynamically regulated during the transcriptional activation.

In addition to histone H3 and H1, SET7/9 was also reported to methylate the free histones H2A and H2B [3,15]. Again, similar to H3, these histones were subject to SET7/9-mediated methylation only in a free state, and not as part of the nucleosomal core. The functional significance of these modifications is still unknown [3,15].

7. Non-Histone Targets of SET7/9

According to the PPI database, SET7/9 interacts with more than 120 different proteins, (BioGRID) and at present, more than 30 proteins are shown to be the targets of SET7/9. SET7/9 acts as regulator of such proteins as p53 [7], TAF10 [35], NFkB [43], YAP1 [17], PCAF [9], STAT3 [37], the nuclear receptors AR [51] and ERα [52], pRB [41] and many more. Perhaps it is not surprising that by regulating such crucial transcription factors SET7/9 participates in the orchestration of the cellular processes they are involved in. Here we focus on the effect of SET7/9 on cellular proliferation and stress response via methylation of the responsible factors.

General Effects of SET7/9 on Transcription

Since lysine methylation on H3K4 is commonly associated with transcriptional activation, while H3K4me1 signatures are closely connected with the location and activity of multiple enhancers, it is tempting to speculate that SET7/9 plays role in tissue-specific transcriptional regulation [53]. Surprisingly, RNAi-mediated SET7/9 knockdown as well as somatic SET7/9 knockout do not affect global nucleosomal H3K4me in vivo [54], while in another report SET7/9 knockdown of rat mesangial cells led to the global H3K4me1 depletion [55]. This discrepancy requires further experimental validation.

A large volume of experimental data published to date unequivocally points to SET7/9 as a transcriptional regulator. In addition to free histones, whose fate and biological significance remains to be addressed in the future, SET7/9 also methylates basal transcription factors, e.g., TAF10 and TAF7. SET7/9 mono-methylates the TBP-associated factor TAF10, a component of the general transcription factor complex TFIID at a single lysine residue located at the loop 2 region, thereby increasing the affinity for RNA polymerase II. SET7/9-mediated methylation of TAF10 enhances transcription of several TAF10-dependent genes [35]. The in vitro studies also showed that SET7/9 is able to methylate TAF7 at the lysine residue K5 [8], which points to SET7/9 being involved in the TAF7-dependent regulation of its target genes, particularly in response to heat shock [56] (Table 1).

Importantly, SET7/9 was also shown as a specific methyltransferase for Suv39H1, which methylates the latter at lysines 105 and 123. The SET7/9-methylated methylation of Suv39H1 results in heterochromatin relaxation and genome instability in response to DNA damage in cancer cells [32] (Table 1).

In addition to histone methylation, SET7/9 can modulate gene expression by methylating DNA methyltransferase (DNMT1). The knockdown of SET7/9 was shown to stabilize cellular DNMT1 levels in mammalian cells, while the overexpression of SET7/9 decreased the DNMT1 protein level. The methylation-promoted degradation of DNMT1 facilitated DNA demethylation resulting in the approximately 10% reduction of global DNA methylation [33]. There is interplay between monomethylation of DNMT1 lysine at position 142 by SET7/9 and phosphorylation of DNMT1 at Ser143 by AKT1 kinase. In mammalian cells, phosphorylated DNMT1 is more stable than methylated DNMT1 [57]. Depletion of AKT1 increased methylation of DNMT1, thereby attenuating the DNMT1 level in cells. Thus, it is prudent to say that SET7/9 is a regulator of DNMT1. However, given the low abundancy of DNMT1 methylation in vivo, additional research is required to establish the role of SET7/9 in the regulation of DNMT1.

8. SET7/9 and Cell Proliferation

8.1. SET7/9, β-Catenin and YAP1

β-catenin is a key mediator of the Wnt/β-catenin signaling pathway and plays an important role in cell fate determination, cell proliferation and tumorigenesis. SET7/9 was shown to monomethylate β-catenin at lysine residue 180 in vivo and in vitro, thereby providing a novel mechanism by which the Wnt/β-catenin signaling pathway is regulated in response to oxidative stress. The binding of Wnt to its receptor LRP5/6 induces dissociation of β-catenin and its negative regulator glycogen synthase kinase (GSK)-3β, resulting in the stabilization of the β-catenin protein, its translocation to the nucleus and subsequent transactivation of the target genes [36]. It was demonstrated that methylation of β-catenin by SET7/9 facilitates its phosphorylation by (GSK)-3β and subsequent β-catenin degradation. Expression of the Wnt/β-catenin target genes such as c-Myc and CyclinD1 were significantly enhanced by either the depletion of SET7/9 or the mutation in the methylation site (K180R) of the β-catenin protein to promote the growth of cancer cells [36] (Figure 2).

On the other hand, SET7/9 was reported as a regulator of the methylation-dependent checkpoint in the Hippo/YAP1/TAZ pathway. SET7/9 monomethylates the YAP1 protein leading to its cytoplasmic retention [17]. YAP1 and TAZ are integral components of the β-catenin destruction complex while the β-catenin/TCF4 complex binds enhancer elements of the YAP gene to drive YAP expression in colorectal cancer cells [58,59]. Taken together, these facts indicate that SET7/9 may be considered as one of the key regulators of the Wnt/β-catenin and Hippo signaling pathways.

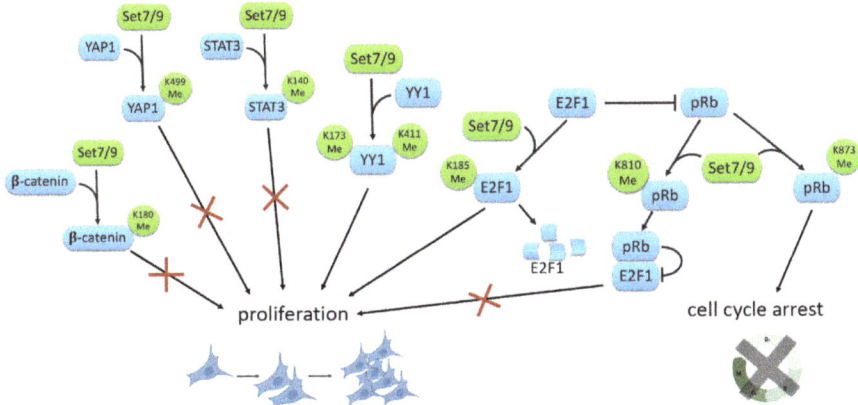

Figure 2. The scheme illustrating the participation of SET7/9 in regulation of proliferation.

8.2. SET7/9 and STAT3

Another example of SET7/9 involvement in the regulation of the cell cycle is presented by the SET7/9-mediated methylation of the STAT3 transcription factor, which results in harnessing the activity of the latter. STAT3 forms dimers through reciprocal phosphor-tyrosine–SH2 interactions after phosphorylation on tyrosine and serine residues in response to different cytokines and growth factors. This phosphorylated form of STAT3 binds to and activates the promoters of its target genes. STAT3 can be methylated at K140 by SET7/9 and demethylated by LSD1. Methylation of K140 decreases the steady-state level of activated STAT3 and hence the expression of many STAT3 target genes [37] (Figure 2).

8.3. SET7/9 and YY1

YY1 (Yin Yang1) is a multifunctional zinc-finger transcription factor involved in a variety of biological processes such as DNA repair, apoptosis, cell proliferation, differentiation and development. SET7/9 methylates YY1 at K173 and K411 positions and enhances the DNA-binding activity of YY1 both in vitro and in cellulo at specific genomic loci in cultured cells. Functionally, SET7/9-mediated methylation of YY1 augments its transcriptional function and hence cell proliferation [42].

8.4. SET7/9 and E2F1/Rb1

The retinoblastoma protein (pRb) is a tumor suppressor protein playing an important role in regulating progression through the early stages of the cell cycle. pRb negatively regulates entry into the S-phase, thereby affecting the early cell cycle control [60]. The retinoblastoma protein interacts and blunts transcriptional activity of the E2F (E2 promoter-binding factor) family of transcription factors [61]. In addition to cell cycle control, pRb activity is associated with other types of cell fate, such as differentiation, senescence and apoptosis [62,63]. Munro et al. demonstrated that SET7/9 regulates the pRb tumor suppressor activity by methylating it at the K873 position [40]. SET7/9-mediated methylation of the C-terminal region of pRb facilitates the interaction between methylated pRb and the heterochromatin protein HP1, resulting in pRb-dependent transcriptional repression, cell cycle arrest, and differentiation [40].

It should be mentioned that the interplay between methylation and phosphorylation was observed for histone and non-histone proteins. In line with this, the phosphorylation of pRb required for the release of E2F1 and hence cell cycle progression was attenuated by the methylation of pRb at K810 by SET7/9 [41,62,63]. Apparently, SET7/9 locks pRb in a hypophosphorylated, growth-arresting state, thereby limiting the E2F target gene expression. Thus, cell cycle control could be regulated by the methylation/phosphorylation switch.

Moreover, the methylation of E2F1 by SET7/9 at lysine 185 inhibits acetylation and phosphorylation at the nearest positions, stimulating ubiquitination-depended degradation of the E2F1 protein [38]. At the same time, SET7/9 was shown as a critical co-activator of E2F1-dependent transcription under conditions of DNA damage [39]. SET7/9 affected the activity of E2F1 by indirect modulation of histone modifications in the promoters of E2F1-dependent genes, thereby promoting cell proliferation and repressing apoptosis. However, SET7/9 differentially affected E2F1 transcription targets: it promoted the expression of the CCNE1 gene, thereby facilitating cell proliferation, and it repressed the TP73 gene, hence preventing apoptosis [39] (Figure 3). Additionally, it was demonstrated that LSD1 removes the methyl mark required for the E2F1 stabilization and function in apoptosis [38]. Collectively, SET7/9 seems to be a critical element for the regulation of the cell cycle upon stress.

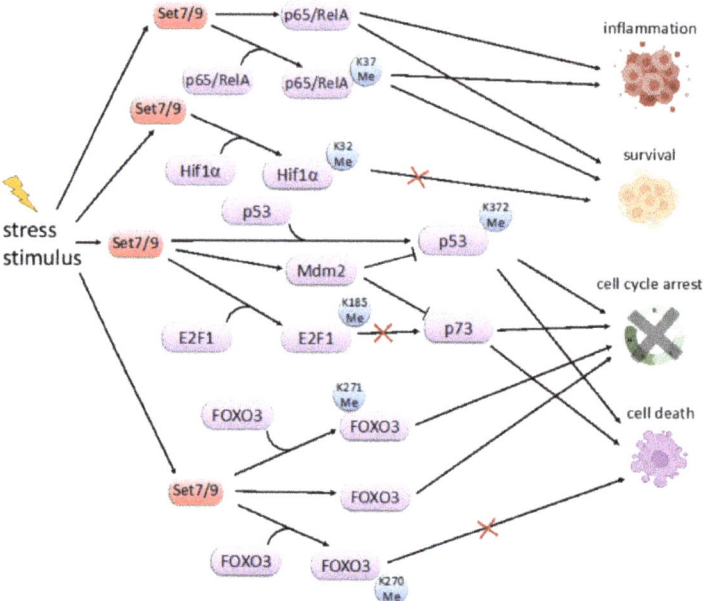

Figure 3. The scheme illustrating the key cellular stress effector pathways regulated by SET7/9. The functional outcomes are denoted on the right.

9. SET7/9 and Cell Stress Response

The tumor suppressor p53 was the first published non-histone methylation target for SET7/9. The p53 protein is the sequence-specific transcription factor that activates expression of its downstream transcription targets, whose products are involved in the regulation of cell cycle arrest and apoptosis [64]. The lysine-specific methylation of p53 at position K372 by SET7/9 is important for p53 transcriptional activation and stabilization mediated by its subsequent acetylation by p300/CBP [7]. Therefore, the cross talk between lysine methylation and acetylation is critical for p53 activation in response to DNA damage [18]. Importantly, both methylation and acetylation prevent p53 poly-ubiquitination mediated by an E3 ligase, Mdm2 [65]. The physical interaction between p53 and Mdm2 is critical for p53 ubiquitination and its subsequent degradation. Thus, small molecules that break this interaction were shown to stabilize p53 [66–68]. In line with this was the observation that SET7/9 physically interacts with Mdm2 and sequesters it away from p53 [69] (Figure 3). Both in vitro and in vivo experiments suggest that SET7/9 and Mdm2 have inverse expression. Accordingly, the unleashed expression of Mdm2 in cancer patients with diminished expression of SET7/9 correlated with poor survival outcomes [69].

It is worthy of note that SET7/9 was shown to be a new regulator of another p53-specific enzyme, Sirtulin 1 (SIRT1). The latter plays an important role during aging, metabolism and autophagy. SIRT1 interacts with SET7/9 mostly in response to DNA damage in human cells resulting in the dissociation of SIRT1 from p53 and the enhancement of p53 acetylation at K382. SET7/9 is able to both interact with and methylate SIRT1 at multiple sites [49]. The presence of SET7/9 attenuates the interaction between SIRT1 and p53, resulting in the transcriptional activation of p53 target genes and thereby inflicting cell cycle arrest and apoptosis [49].

The transcription factor FoxO3 of the Forkhead Box O (FoxO) family is involved in the regulation of the cellular response to ROS-induced DNA damage [70]. In response to oxidative stress, FoxO3 induces cell cycle arrest, apoptosis, autophagy, and hence can be considered as a tumor suppressor. It also affects metabolism and aging [46,70,71]. The phenotypic features of FoxO3-knockout mice support its involvement in the process of ageing and are exemplified by premature follicular activation, ovarian failure and early infertility [72].

FoxO3 inhibits transcription induced by ERα, thereby inhibiting the proliferation of breast cancer cells [73]. Accordingly, downregulation of the ERα activity is associated with poor prognosis in estrogen-dependent breast cancer and colorectal cancers [74]. SET7/9 methylates the FoxO3 protein at K271 [45]. Paradoxically, the methylation of FoxO3 destabilizes the protein but enhances its transcriptional activity towards the activation of pro-apoptotic genes. This paradoxical effect is similar to the one observed in the case of SET7/9-mediated methylation of E2F1. It should also be noted that SET7/9 apparently can methylate the additional FoxO3 lysine residue, K270 [46] (Figure 3). Surprisingly, this methylation has an opposite effect on the transcriptional activity of FoxO3, thereby preventing it from activation of the pro-apoptotic gene BIM and, hence, preventing cell death.

Hypoxia-inducible factor 1α (HIF-1α) is another target of SET7/9 methylation activity. HIF1a is a critical transcription factor for cellular hypoxic response. In response to oxygen deprivation, HIF1a is released from the ubiquitin-mediated degradation mediated by the von Hippel-Lindau disease tumor suppressor (VHL) E3 ligase [75]. SET7/9 methylates HIF-1α at K32, which competes with ubiquitination and its subsequent degradation. Thus, SET7/9 stabilizes the HIF-1α protein and stimulates the HIF-1α-dependent transcription of genes involved in the regulation of energy metabolism and angiogenesis to maintain tissue homeostasis [48]. However, several studies suggest that SET7/9-mediated methylation inhibits HIF1α transcriptional activity by preventing its DNA binding (Figure 3). Importantly, this effect was reversed by a SET7/9-specific inhibitor, (R)-PFI-2 [47,48]. Thus, additional experiments are required to elucidate the effect of SET7/9 on the function of HIF-1α.

10. Concluding Remarks

To assess the biological significance of SET7/9's role in proliferation and stress response in vivo, it is important to develop the relevant tools. In this respect, small-molecule inhibitors are promising tools that allow for the probing functions of methyltransferases in diseases. In this respect, Barsyte-Lovejoy et al. designed and synthesized a novel inhibitor, (R)-PFI-2, against SET7/9 [76]. This inhibitor has demonstrated low toxicity even at high concentrations in human cells [77]. Mori et al. has developed an inhibitor which is an amine analogue of adenosylmethionine, bearing various alkylamino groups for increasing the inhibitory activity [78].

Berberine is a naturally occurring isoquinoline alkaloid which is commonly used in traditional Chinese medicine and exhibits anti-oxidant, anti-inflammatory, and anti-cancer activities. Berberin was shown to augment the activity of SET7/9 towards NFκB by sensitizing human cancer cells to ionizing radiation or chemotherapy. Berberine negatively regulates NFκB through SET7/9-mediated lysine methylation. Such methylation leads to a decrease in miR-21 levels and Bcl-2 levels [79].

Since there is a significant correlation between SET7/9 and different types of cancer, the wide application of small molecules as experimental tools should significantly facilitate the experimental work directed towards the elucidation of SET7/9 in tumorigenesis and other diseases.

Author Contributions: Conceptualization, N.A.B.; writing—original draft preparation, A.D., E.V., O.F., O.S. and N.A.B.; writing—review and editing, N.A.B.; visualization, A.D., E.V., O.F. and O.S. All authors have read and agreed to the published version of the manuscript.

Funding: This research was funded by RUSSIAN SCIENCE FOUNDATION (RSF), #19-75-10059.

Institutional Review Board Statement: Not applicable.

Informed Consent Statement: Not applicable.

Conflicts of Interest: The authors declare that they have no conflict of interest.

References

1. Nishioka, K.; Chuikov, S.; Sarma, K.; Erdjument-Bromage, H.; Allis, C.D.; Tempst, P.; Reinberg, D. Set9, a Novel Histone H3 Methyltransferase that Facilitates Transcription by Precluding Histone Tail Modifications Required for Heterochromatin Formation. *Genes Dev.* **2002**, *16*, 479–489. [CrossRef] [PubMed]
2. Wang, H.; Cao, R.; Xia, L.; Erdjument-Bromage, H.; Borchers, C.; Tempst, P.; Zhang, Y. Purification and Functional Characterization of a Histone H3-Lysine 4-Specific Methyltransferase. *Mol. Cell* **2001**, *8*, 1207–1217. [CrossRef]
3. Dhayalan, A.; Kudithipudi, S.; Rathert, P.; Jeltsch, A. Specificity Analysis-Based Identification of New Methylation Targets of the SET7/9 Protein Lysine Methyltransferase. *Chem. Biol.* **2011**, *18*, 111–120. [CrossRef]
4. Chakrabarti, S.K.; Francis, J.; Ziesmann, S.M.; Garmey, J.C.; Mirmira, R.G. Covalent Histone Modifications Underlie the Developmental Regulation of Insulin Gene Transcription in Pancreatic β Cells. *J. Biol. Chem.* **2003**, *278*, 23617–23623. [CrossRef] [PubMed]
5. Evans-Molina, C.; Robbins, R.D.; Kono, T.; Tersey, S.A.; Vestermark, G.L.; Nunemaker, C.S.; Garmey, J.C.; Deering, T.G.; Keller, S.R.; Maier, B.; et al. Peroxisome Proliferator-Activated Receptor γ Activation Restores Islet Function in Diabetic Mice through Reduction of Endoplasmic Reticulum Stress and Maintenance of Euchromatin Structure. *Mol. Cell. Biol.* **2009**, *29*, 2053–2067. [CrossRef] [PubMed]
6. Pradhan, S.; Chin, H.G.; Estève, P.-O.; Jacobsen, S.E. SET7/9 Mediated Methylation of Non-Histone Proteins in Mammalian Cells. *Epigenetics* **2009**, *4*, 383–387. [CrossRef]
7. Chuikov, S.; Kurash, J.K.; Wilson, J.; Xiao, B.; Justin, N.; Ivanov, G.S.; McKinney, K.; Tempst, P.; Prives, C.; Gamblin, S.; et al. Regulation of p53 activity through lysine methylation. *Nature* **2004**, *432*, 353–360. [CrossRef]
8. Couture, J.-F.; Collazo, E.; Hauk, G.; Trievel, R.C. Structural Basis for the Methylation Site Specificity of SET7/9. *Nat. Struct. Mol. Biol.* **2006**, *13*, 140–146. [CrossRef]
9. Masatsugu, T.; Yamamoto, K. Multiple Lysine Methylation of PCAF by Set9 Methyltransferase. *Biochem. Biophys. Res. Commun.* **2009**, *381*, 22–26. [CrossRef]
10. Vasileva, E.; Shuvalov, O.; Petukhov, A.; Fedorova, O.; Daks, A.; Nader, R.; Barlev, N. KMT Set7/9 Is a New Regulator of Sam68 STAR-Protein. *Biochem. Biophys. Res. Commun.* **2020**, *525*, 1018–1024. [CrossRef]
11. Li, Y.; Reddy, M.A.; Miao, F.; Shanmugam, N.; Yee, J.-K.; Hawkins, D.; Ren, B.; Natarajan, R. Role of the Histone H3 Lysine 4 Methyltransferase, SET7/9, in the Regulation of NF-κB-Dependent Inflammatory Genes: Relevance to Diabetes and Inflammation. *J. Biol. Chem.* **2008**, *283*, 26771–26781. [CrossRef]
12. Del Rizzo, P.A.; Trievel, R.C. Substrate and Product Specificities of SET Domain Methyltransferases. *Epigenetics* **2011**, *6*, 1059–1067. [CrossRef]
13. Okabe, J.; Orlowski, C.; Balcerczyk, A.; Tikellis, C.; Thomas, M.; Cooper, M.E.; El-Osta, A. Distinguishing Hyperglycemic Changes by Set7 in Vascular Endothelial Cells. *Circ. Res.* **2012**, *110*, 1067–1076. [CrossRef] [PubMed]
14. Batista, I.; Helguero, L.A. Biological Processes and Signal Transduction Pathways Regulated by the Protein Methyltransferase SETD7 and Their Significance in Cancer. *Signal Transduct. Target. Ther.* **2018**, *3*, 19. [CrossRef] [PubMed]
15. Keating, S.T.; El-Osta, A. Transcriptional Regulation by the Set7 Lysine Methyltransferase. *Epigenetics* **2013**, *8*, 361–372. [CrossRef]
16. Aksenova, V.; Turoverova, L.; Khotin, M.; Magnusson, K.-E.; Tulchinsky, E.; Melino, G.; Pinaev, G.P.; Barlev, N.; Tentler, D. Actin-Binding Protein Alpha-Actinin 4 (ACTN4) Is a Transcriptional Co-Activator of RelA/p65 Sub-unit of NF-kB. *Oncotarget* **2013**, *4*, 362. [CrossRef] [PubMed]
17. Oudhoff, M.J.; Braam, M.J.; Freeman, S.A.; Wong, D.; Rattray, D.G.; Wang, J.; Antignano, F.; Snyder, K.; Refaeli, I.; Hughes, M.R.; et al. SETD7 Controls Intestinal Regeneration and Tumorigenesis by Regulating Wnt/β-Catenin and Hippo/YAP Signaling. *Dev. Cell* **2016**, *37*, 47–57. [CrossRef] [PubMed]
18. Ivanov, G.S.; Ivanova, T.; Kurash, J.; Ivanov, A.; Chuikov, S.; Gizatullin, F.; Herrera-Medina, E.M.; Rauscher, F., III; Reinberg, D.; Barlev, N.A. Methylation-Acetylation Interplay Activates p53 in Response to DNA Damage. *Mol. Cell. Biol.* **2007**, *27*, 6756–6769. [CrossRef]

19. Liu, H.; Li, Z.; Yang, Q.; Liu, W.; Wan, J.; Li, J.; Zhang, M. Substrate Docking–Mediated Specific and Efficient Lysine Methylation by the SET Domain–Containing Histone Methyltransferase SETD7. *J. Biol. Chem.* **2019**, *294*, 13355–13365. [CrossRef]
20. El-Gebali, S.; Mistry, J.; Bateman, A.; Eddy, S.R.; Luciani, A.; Potter, S.C.; Qureshi, M.; Richardson, L.J.; Salazar, G.A.; Smart, A.; et al. The Pfam Protein Families Database in 2019. *Nucleic Acids Res.* **2019**, *47*, D427–D432. [CrossRef]
21. Mecklenburg, K.L.; Freed, S.A.; Raval, M.; Quintero, O.A.; Yengo, C.M.; O'Tousa, J.E. Invertebrate and Vertebrate Class III Myosins Interact with MORN Repeat-Containing Adaptor Proteins. *PLoS ONE* **2015**, *10*, e0122502. [CrossRef] [PubMed]
22. Hadano, S.; Kunita, R.; Otomo, A.; Suzuki-Utsunomiya, K.; Ikeda, J.-E. Molecular and Cellular Function of ALS2/Alsin: Implication of Membrane Dynamics in Neuronal Development and Degeneration. *Neurochem. Int.* **2007**, *51*, 74–84. [CrossRef] [PubMed]
23. Landstrom, A.; Beavers, D.L.; Wehrens, X.H. The Junctophilin Family of Proteins: From Bench to Bedside. *Trends Mol. Med.* **2014**, *20*, 353–362. [CrossRef]
24. Guo, A.; Wang, Y.; Chen, B.; Wang, Y.; Yuan, J.; Zhang, L.; Hall, D.; Wu, J.; Shi, Y.; Zhu, Q.; et al. E-C Coupling Structural Protein Junctophilin-2 Encodes a Stress-Adaptive Transcription Regulator. *Science* **2018**, *362*, eaan3303. [CrossRef]
25. Sajko, S.; Grishkovskaya, I.; Kostan, J.; Graewert, M.; Setiawan, K.; Trübestein, L.; Niedermüller, K.; Gehin, C.; Sponga, A.; Puchinger, M. Structures of Three MORN Repeat Proteins and a Re-Evaluation of the Proposed Lipid-Binding Properties of MORN Repeats. *PLoS ONE* **2020**, *15*, e0242677. [CrossRef]
26. Sehnal, D.; Bittrich, S.; Deshpande, M.; Svobodová, R.; Berka, K.; Bazgier, V.; Velankar, S.; Burley, S.K.; Koča, J.; Rose, A.S. Mol* Viewer: Modern Web App for 3D Visualization and Analysis of Large Biomolecular Structures. *Nucleic Acids Res.* **2021**, *49*, W431–W437. [CrossRef] [PubMed]
27. Wilson, J.R.; Jing, C.; Walker, P.A.; Martin, S.R.; Howell, S.A.; Blackburn, G.M.; Gamblin, S.J.; Xiao, B. Crystal Structure and Functional Analysis of the Histone Methyltransferase SET7/9. *Cell* **2002**, *111*, 105–115. [CrossRef]
28. Vasileva, E.; Barlev, N. The World of SET-Containing Lysine Methyltransferases. *eLS* **2017**, 1–10. [CrossRef]
29. Kouzarides, T. Chromatin Modifications and Their Function. *Cell* **2007**, *128*, 693–705. [CrossRef]
30. Noma, K.-I.; Allis, C.D.; Grewal, S.I.S. Transitions in Distinct Histone H3 Methylation Patterns at the Heterochromatin Domain Boundaries. *Science* **2001**, *293*, 1150–1155. [CrossRef]
31. Castaño, J.; Morera, C.; Sesé, B.; Boue, S.; Bonet-Costa, C.; Martí, M.; Roque, A.; Jordan, A.; Barrero, M.J. SETD7 Regulates the Differentiation of Human Embryonic Stem Cells. *PLoS ONE* **2016**, *11*, e0149502. [CrossRef] [PubMed]
32. Wang, D.; Zhou, J.; Liu, X.; Lu, D.; Shen, C.; Du, Y.; Wei, F.-Z.; Song, B.; Lu, X.; Yu, Y. Methylation of SUV39H1 by SET7/9 Results in Heterochromatin Relaxation and Genome Instability. *Proc. Natl. Acad. Sci. USA* **2013**, *110*, 5516–5521. [CrossRef] [PubMed]
33. Estève, P.-O.; Chin, H.G.; Benner, J.; Feehery, G.R.; Samaranayake, M.; Horwitz, G.A.; Jacobsen, S.E.; Pradhan, S. Regulation of DNMT1 Stability through SET7-Mediated Lysine Methylation in Mammalian Cells. *Proc. Natl. Acad. Sci. USA* **2009**, *106*, 5076–5081. [CrossRef] [PubMed]
34. Wang, J.; Hevi, S.; Kurash, J.K.; Lei, H.; Gay, F.; Bajko, J.; Su, H.; Sun, W.; Chang, H.; Xu, G.; et al. The Lysine Demethylase LSD1 (KDM1) Is Required for Maintenance of Global DNA Methylation. *Nat. Genet.* **2008**, *41*, 125–129. [CrossRef] [PubMed]
35. Kouskouti, A.; Scheer, E.; Staub, A.; Tora, L.; Talianidis, I. Gene-Specific Modulation of TAF10 Function by SET9-Mediated Methylation. *Mol. Cell* **2004**, *14*, 175–182. [CrossRef]
36. Shen, C.; Wang, D.; Liu, X.; Gu, B.; Du, Y.; Wei, F.Z.; Cao, L.L.; Song, B.; Lu, X.; Yang, Q. SET7/9 Regulates Cancer Cell Proliferation by Influencing β-Catenin Stability. *FASEB J.* **2015**, *29*, 4313–4323. [CrossRef]
37. Yang, J.; Huang, J.; Dasgupta, M.; Sears, N.; Miyagi, M.; Wang, B.; Chance, M.R.; Chen, X.; Du, Y.; Wang, Y.; et al. Reversible Methylation of Promoter-Bound STAT3 by Histone-Modifying Enzymes. *Proc. Natl. Acad. Sci. USA* **2010**, *107*, 21499–21504. [CrossRef]
38. Kontaki, H.; Talianidis, I. Lysine Methylation Regulates E2F1-Induced Cell Death. *Mol. Cell* **2010**, *39*, 152–160. [CrossRef]
39. Lezina, L.; Aksenova, V.; Ivanova, T.; Purmessur, N.; Antonov, A.; Tentler, D.; Fedorova, O.; Garabadgiu, A.; Talianidis, I.; Melino, G. KMTase Set7/9 Is a Critical Regulator of E2F1 Activity upon Genotoxic Stress. *Cell Death Differ.* **2014**, *21*, 1889–1899. [CrossRef] [PubMed]
40. Munro, S.; Khaire, N.; Inche, A.; Carr, S.; La Thangue, N.B. Lysine Methylation Regulates the pRb Tumour Suppressor Protein. *Oncogene* **2010**, *29*, 2357–2367. [CrossRef]
41. Carr, S.M.; Munro, S.; Kessler, B.; Oppermann, U.; La Thangue, N.B. Interplay between Lysine Methylation and Cdk Phosphorylation in Growth Control by the Retinoblastoma Protein. *EMBO J.* **2010**, *30*, 317–327. [CrossRef]
42. Zhang, W.-J.; Wu, X.-N.; Shi, T.-T.; Xu, H.-T.; Yi, J.; Shen, H.-F.; Huang, M.-F.; Shu, X.-Y.; Wang, F.-F.; Peng, B.-L. Regulation of Transcription Factor Yin Yang 1 by SET7/9-Mediated Lysine Methylation. *Sci. Rep.* **2016**, *6*, 21718. [CrossRef]
43. Ea, C.-K.; Baltimore, D. Regulation of NF-κB Activity through Lysine Monomethylation of p65. *Proc. Natl. Acad. Sci. USA* **2009**, *106*, 18972–18977. [CrossRef] [PubMed]
44. Yang, X.-D.; Huang, B.; Li, M.; Lamb, A.; Kelleher, N.L.; Chen, L.-F. Negative Regulation of NF-κB Action by Set9-Mediated Lysine Methylation of the RelA Subunit. *EMBO J.* **2009**, *28*, 1055–1066. [CrossRef]
45. Xie, Q.; Hao, Y.; Tao, L.; Peng, S.; Rao, C.; Chen, H.; You, H.; Dong, M.; Yuan, Z. Lysine Methylation of FOXO3 Regulates Oxidative Stress-Induced Neuronal Cell Death. *EMBO Rep.* **2012**, *13*, 371–377. [CrossRef] [PubMed]
46. Calnan, D.R.; Webb, A.E.; White, J.L.; Stowe, T.R.; Goswami, T.; Shi, X.; Espejo, A.; Bedford, M.T.; Gozani, O.; Gygi, S.P.; et al. Methylation by Set9 Modulates FoxO3 Stability and Transcriptional Activity. *Aging* **2012**, *4*, 462–479. [CrossRef] [PubMed]

47. Kim, Y.; Nam, H.J.; Lee, J.; Park, D.Y.; Kim, C.; Yu, Y.S.; Kim, D.; Park, S.W.; Bhin, J.; Hwang, D.; et al. Methylation-Dependent Regulation of HIF-1α Stability Restricts Retinal and Tumour Angiogenesis. *Nat. Commun.* **2016**, *7*, 10347. [CrossRef] [PubMed]
48. Liu, X.; Chenxi, X.; Xu, C.; Leng, X.; Cao, H.; Ouyang, G.; Xiaoqian, L. Repression of Hypoxia-Inducible Factor α Signaling by Set7-Mediated Methylation. *Nucleic Acids Res.* **2015**, *43*, 5081–5098. [CrossRef] [PubMed]
49. Liu, X.; Wang, D.; Zhao, Y.; Tu, B.; Zheng, Z.; Wang, L.; Wang, H.; Gu, W.; Roeder, R.G.; Zhu, W.-G. Methyltransferase Set7/9 Regulates p53 Activity by Interacting with Sirtuin 1 (SIRT1). *Proc. Natl. Acad. Sci. USA* **2011**, *108*, 1925–1930. [CrossRef]
50. Kassner, I.; Barandun, M.; Fey, M.; Rosenthal, F.; Hottiger, M.O. Crosstalk between SET7/9-Dependent Methylation and ARTD1-Mediated ADP-Ribosylation of Histone H1. 4. *Epigenet. Chromatin* **2013**, *6*, 1. [CrossRef]
51. Gaughan, L.; Stockley, J.; Wang, N.; McCracken, S.R.; Treumann, A.; Armstrong, K.; Shaheen, F.; Watt, K.; McEwan, I.J.; Wang, C. Regulation of the Androgen Receptor by SET9-Mediated Methylation. *Nucleic Acids Res.* **2011**, *39*, 1266–1279. [CrossRef] [PubMed]
52. Subramanian, K.; Jia, D.; Kapoor-Vazirani, P.; Powell, D.R.; Collins, R.; Sharma, D.; Peng, J.; Cheng, X.; Vertino, P.M. Regulation of Estrogen Receptor α by the SET7 Lysine Methyltransferase. *Mol. Cell* **2008**, *30*, 336–347. [CrossRef] [PubMed]
53. Heintzman, N.D.; Stuart, R.K.; Hon, G.; Fu, Y.; Ching, C.W.; Hawkins, R.D.; Barrera, L.O.; Van Calcar, S.; Qu, C.; Ching, K.A.; et al. Distinct and Predictive Chromatin Signatures of Transcriptional Promoters and Enhancers in the Human Genome. *Nat. Genet.* **2007**, *39*, 311–318. [CrossRef]
54. Lehnertz, B.; Rogalski, J.C.; Schulze, F.M.; Yi, L.; Lin, S.; Kast, J.; Rossi, F.M. P53-Dependent Transcription and Tumor Suppression Are Not Affected in Set7/9-Deficient Mice. *Mol. Cell* **2011**, *43*, 673–680. [CrossRef]
55. Sun, G.; Reddy, M.A.; Yuan, H.; Lanting, L.; Kato, M.; Natarajan, R. Epigenetic Histone Methylation Modulates Fibrotic Gene Expression. *J. Am. Soc. Nephrol.* **2010**, *21*, 2069–2080. [CrossRef]
56. Nagashimada, M.; Ueda, T.; Ishita, Y.; Sakurai, H. TAF7 Is a Heat-Inducible Unstable Protein and Is Required for Sustained Expression of Heat Shock Protein Genes. *FEBS J.* **2018**, *285*, 3215–3224. [CrossRef] [PubMed]
57. Estève, P.-O.; Chang, Y.; Samaranayake, M.; Upadhyay, A.K.; Horton, J.R.; Feehery, G.R.; Cheng, X.; Pradhan, S. A Methylation and Phosphorylation Switch between an Adjacent Lysine and Serine Determines Human DNMT1 Stability. *Nat. Struct. Mol. Biol.* **2010**, *18*, 42–48. [CrossRef]
58. Konsavage, W.M.; Kyler, S.L.; Rennoll, S.A.; Jin, G.; Yochum, G.S. Wnt/β-Catenin Signaling Regulates Yes-associated Protein (YAP) Gene Expression in Colorectal Carcinoma Cells. *J. Biol. Chem.* **2012**, *287*, 11730–11739. [CrossRef]
59. Azzolin, L.; Panciera, T.; Soligo, S.; Enzo, E.; Bicciato, S.; Dupont, S.; Bresolin, S.; Frasson, C.; Basso, G.; Guzzardo, V.; et al. YAP/TAZ Incorporation in the β-Catenin Destruction Complex Orchestrates the Wnt Response. *Cell* **2014**, *158*, 157–170. [CrossRef]
60. Weinberg, R.A. The Retinoblastoma Protein and Cell Cycle Control. *Cell* **1995**, *81*, 323–330. [CrossRef]
61. Stevens, C.; La Thangue, N.B. E2F and Cell Cycle Control: A Double-Edged Sword. *Arch. Biochem. Biophys.* **2003**, *412*, 157–169. [CrossRef]
62. Classon, M.; Harlow, E. The Retinoblastoma Tumour Suppressor in Development and Cancer. *Nat. Cancer* **2002**, *2*, 910–917. [CrossRef]
63. Giacinti, C.; Giordano, A. RB and Cell Cycle Progression. *Oncogene* **2006**, *25*, 5220–5227. [CrossRef] [PubMed]
64. Kastan, M.B.; Onyekwere, O.; Sidransky, D.; Vogelstein, B.; Craig, R.W. Participation of p53 Protein in the Cellular Response to DNA Damage. *Cancer Res.* **1991**, *51*, 6304–6311. [CrossRef] [PubMed]
65. Daks, A.; Melino, D.; Barlev, N.A. The Role of Different E3 Ubiquitin Ligases in Regulation of the P53 Tumor Suppressor Protein. *Tsitologiia* **2013**, *55*, 673–687.
66. Fedorova, O.; Daks, A.; Petrova, V.; Petukhov, A.; Lezina, L.; Shuvalov, O.; Davidovich, P.; Kriger, D.; Lomert, E.; Tentler, D.; et al. Novel Isatin-Derived Molecules Activate p53 via Interference with Mdm2 to Promote Apoptosis. *Cell Cycle* **2018**, *17*, 1917–1930. [CrossRef] [PubMed]
67. Vassilev, L.T.; Vu, B.T.; Graves, B.; Carvajal, D.; Podlaski, F.; Filipovic, Z.; Kong, N.; Kammlott, U.; Lukacs, C.; Klein, C.; et al. In Vivo Activation of the p53 Pathway by Small-Molecule Antagonists of MDM2. *Science* **2004**, *303*, 844–848. [CrossRef]
68. Davidovich, P.B.; Aksenova, V.; Petrova, V.; Tentler, D.; Orlova, D.; Smirnov, S.V.; Gurzhiy, V.; Okorokov, A.; Garabadzhiu, A.V.; Melino, G.; et al. Discovery of Novel Isatin-Based p53 Inducers. *ACS Med. Chem. Lett.* **2015**, *6*, 856–860. [CrossRef]
69. Lezina, L.; Aksenova, V.; Fedorova, O.; Malikova, D.; Shuvalov, O.; Antonov, A.V.; Tentler, D.; Garabadgiu, A.V.; Melino, G.; Barlev, N.A. KMT Set7/9 Affects Genotoxic Stress Response via the Mdm2 Axis. *Oncotarget* **2015**, *6*, 25843. [CrossRef]
70. Greer, E.; Brunet, A. FOXO Transcription Factors at the Interface between Longevity and Tumor Suppression. *Oncogene* **2005**, *24*, 7410–7425. [CrossRef]
71. Hwangbo, D.S.; Gersham, B.; Tu, M.-P.; Palmer, M.; Tatar, M. Drosophila dFOXO Controls Lifespan and Regulates Insulin Signalling in Brain and Fat Body. *Nature* **2004**, *429*, 562–566. [CrossRef] [PubMed]
72. Castrillon, D.H.; Miao, L.; Kollipara, R.; Horner, J.W.; DePinho, R.A. Suppression of Ovarian Follicle Activation in Mice by the Transcription Factor Foxo3a. *Science* **2003**, *301*, 215–218. [CrossRef] [PubMed]
73. Zou, Y.; Tsai, W.-B.; Cheng, C.-J.; Hsu, C.; Chung, Y.M.; Li, P.-C.; Lin, S.-H.; Hu, M.C.-T. Forkhead Box Transcription Factor FOXO3a Suppresses Estrogen-Dependent Breast Cancer Cell Proliferation and Tumorigenesis. *Breast Cancer Res.* **2008**, *10*, R21. [CrossRef] [PubMed]

74. Bullock, M.D.; Bruce, A.; Sreekumar, R.; Curtis, N.; Cheung, T.; Reading, I.; Primrose, J.N.; Ottensmeier, C.; Packham, G.K.; Thomas, G.; et al. FOXO3 Expression during Colorectal Cancer Progression: Biomarker Potential Reflects a Tumour Suppressor Role. *Br. J. Cancer* **2013**, *109*, 387–394. [CrossRef]
75. Maxwell, P.H.; Wiesener, M.S.; Chang, G.-W.; Clifford, S.C.; Vaux, E.C.; Cockman, M.E.; Wykoff, C.C.; Pugh, C.W.; Maher, E.R.; Ratcliffe, P.J. The Tumour Suppressor Protein VHL Targets Hypoxia-Inducible Factors for Oxygen-Dependent Proteolysis. *Nature* **1999**, *399*, 271–275. [CrossRef]
76. Barsyte-Lovejoy, D.; Li, F.; Oudhoff, M.; Tatlock, J.H.; Dong, A.; Zeng, H.; Wu, H.; Freeman, S.A.; Schapira, M.; Senisterra, G.A.; et al. (R)-PFI-2 Is a Potent and Selective Inhibitor of SETD7 Methyltransferase Activity in Cells. *Proc. Natl. Acad. Sci. USA* **2014**, *111*, 12853–12858. [CrossRef]
77. Daks, A.; Mamontova, V.; Fedorova, O.; Petukhov, A.; Shuvalov, O.; Parfenyev, S.; Netsvetay, S.; Venina, A.; Kizenko, A.; Imyanitov, E. Set7/9 Controls Proliferation and Genotoxic Drug Resistance of NSCLC Cells. *Biochem. Biophys. Res. Commun.* **2021**, *572*, 41–48. [CrossRef]
78. Mori, S.; Iwase, K.; Iwanami, N.; Tanaka, Y.; Kagechika, H.; Hirano, T. Development of Novel Bisubstrate-Type Inhibitors of Histone Methyltransferase SET7/9. *Bioorg. Med. Chem.* **2010**, *18*, 8158–8166. [CrossRef]
79. Hu, H.-Y.; Li, K.-P.; Wang, X.-J.; Liu, Y.; Lu, Z.-G.; Dong, R.-H.; Guo, H.-B.; Zhang, M.-X. Set9, NF-κB, and MicroRNA-21 Mediate Berberine-Induced Apoptosis of Human Multiple Myeloma Cells. *Acta Pharmacol. Sin.* **2012**, *34*, 157–166. [CrossRef]

Review

Structure, Activity and Function of the Dual Protein Lysine and Protein N-Terminal Methyltransferase METTL13

Magnus E. Jakobsson

Department of Immunotechnology, Lund University, Medicon Village, 22100 Lund, Sweden; Magnus.Jakobsson@immun.lth.se

Abstract: METTL13 (also known as eEF1A-KNMT and FEAT) is a dual methyltransferase reported to target the N-terminus and Lys55 in the eukaryotic translation elongation factor 1 alpha (eEF1A). METTL13-mediated methylation of eEF1A has functional consequences related to translation dynamics and include altered rate of global protein synthesis and translation of specific codons. Aberrant regulation of METTL13 has been linked to several types of cancer but the precise mechanisms are not yet fully understood. In this article, the current literature related to the structure, activity, and function of METTL13 is systematically reviewed and put into context. The links between METTL13 and diseases, mainly different types of cancer, are also summarized. Finally, key challenges and opportunities for METTL13 research are pinpointed in a prospective outlook.

Keywords: post translational modification; lysine methylation; N-terminal methylation; translation; enzyme specificity; eEF1A; METTL13

1. Introduction

Cellular protein synthesis is guided and catalyzed by the ribosome, which uses messenger RNA as a template for protein synthesis in a process termed translation. Several elongation factors support the process of translation, and one prominent example is the eukaryotic elongation factor 1 alpha (eEF1A), which delivers aminoacyl-tRNA complexes to the ribosome acceptor (A)-site to provide substrate for protein synthesis. The function of a protein is often regulated by enzyme-mediated post-translational modification (PTM) [1]. Prominent examples include phosphorylation, glycosylation, acetylation, and methylation [2].

In cells, specific methyltransferase (MT) enzymes catalyze the transfer of a methyl group ($-CH_3$) from *S*-adenosylmethionine (AdoMet) to specific substrates to generate a methylated product and *S*-adenosylhomocysteine (AdoHcy) (Figure 1A,B). Protein methylation has been most extensively studied on lysine [3] and arginine [4], but emerging evidence suggests that also histidine methylation [5,6] is prevalent and important. In addition, methylation can occur on the side chains of glutamate, glutamine, asparagine, and cysteine as well as the protein N-terminus (Nt) and C-terminus [7]. Methylation of lysine and the protein Nt are biochemically similar. They both occur on primary amino groups corresponding to the α-amino group of the protein Nt and the ε-amino group of the lysine side chain (Figure 1C,D). Each amino group can accept up to three methyl groups yielding mono-, di-, and tri-methylated substrates (Figure 1C,D).

The protein Nt α-amino group and the ε-amino group of lysine are both chemical bases and exist in both possible protonation states; a neutral state and a positively charged state. The neutral, i.e., unprotonated state is characterized by a free electron pair capable of acting as a nucleophile in nucleophilic substitution reactions [8]. In cells, the differentially protonated forms are in equilibrium and their relative abundance is determined by the acid dissociation constant (pKa) and pH. Notably, the protein Nt has a pKa close to physiological pH whereas a lysine side chain typically has a pKa above 10 [9]. Consequently, the Nt is more chemically active under physiological conditions.

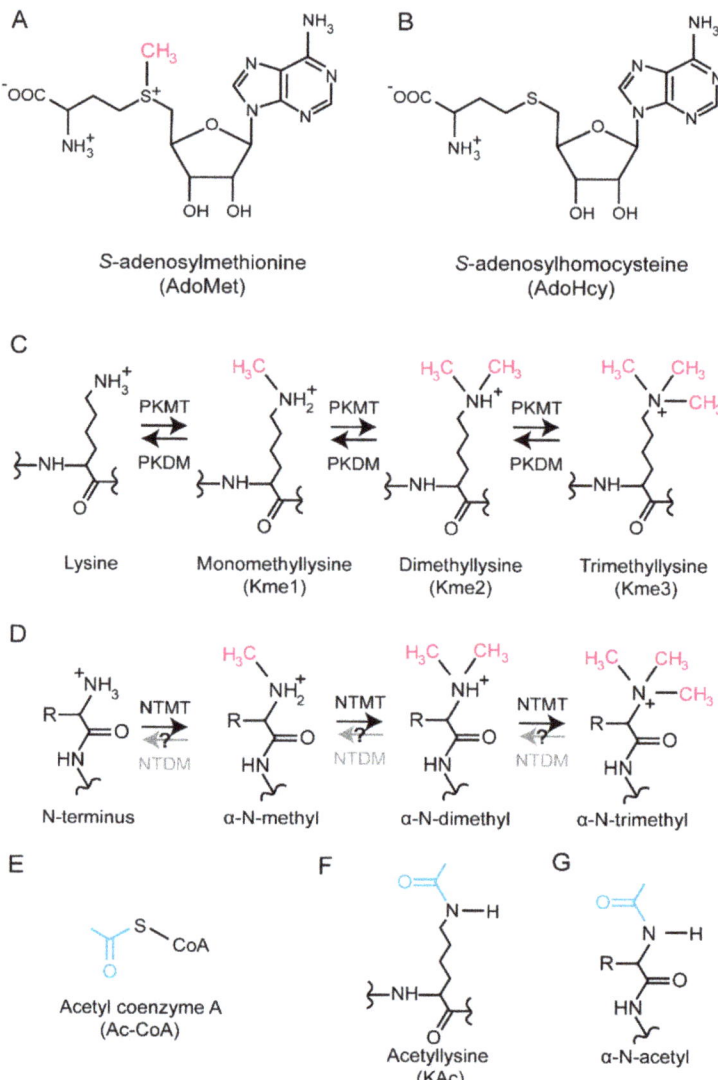

Figure 1. Biochemistry of protein lysine and N-terminal methylation. (**A**,**B**) Structures of AdoMet and AdoHcy. The chemical structures of the methyl donor (**A**) S-adenosylmethionine (AdoMet) and (**B**) the demethylated counterpart (AdoHcy) are shown. The transferred methyl group is highlighted (magenta). (**C**) Biochemistry of lysine methylation. Consecutive protein lysine methyltransferase (PKMT)-mediated methylation can introduce up to three methyl groups in a lysine side chain. The methyl groups can be enzymatically removed by protein lysine demethylase (PKDM) enzymes. (**D**) Biochemistry of protein N-terminal methylation. Consecutive protein N-terminal methyltransferase (NTMT)-mediated methylation can introduce up to three methyl groups on the α-amino group of proteins. There is yet no evidence of protein N-terminal demethylase (NTDM) enzymes, but their potential enzymatic activity is indicated (grey arrow, question mark). (**E**) Structure of acetyl-coenzyme A. The transferred acetyl group is highlighted (cyan). (**F**,**G**) Structures of (**F**) acetyl lysine and (**G**) α-N-acetylated protein terminus are shown.

Methylation of the Nt and lysine side chain have similar biochemical consequences. Firstly, methylation increases both the void occupancy and the hydrophobicity. Secondly, trimethylation renders a permanent positive charge and chemically saturates the amino group making it chemically inert. For reference, both sites can be acetylated by acetyl-CoA dependent acetyltransferases (Figure 1E). Nt and lysine acetylation also renders the amino groups chemically inert by occupying the "free" election pair but, in contrast to methylation, acetylation neutralizes the positive charge (Figure 1F,G).

It was recently reported that human methyltransferase-like protein 13 (METTL13) (also called eEF1A-KNMT or FEAT) trimethylates the Nt and dimethylates a specific lysine in position 55 in eEF1A (eEF1A-Lys55) to regulate mRNA translation and protein synthesis [10–12]. Here, we review the literature on METTL13 and discuss structural, biochemical, and cellular features as well as its links to disease. We end with a prospective outlook and propose directions for future research.

2. Structural Features

The human genome is predicted to encode over 200 enzymes with MT activity [13]. These are often categorized based on structural features and MT activity has been reported for five distinct protein folds [14]. The largest group of MTs corresponds to the so-called seven beta strand (7BS) domain containing enzymes that harbor a characteristic fold comprising 7BS and alternating alpha helices (Figure 2A).

METTL13 has a unique domain organization comprising two distinct 7BS domains, henceforth denoted MT13-N and MT13-C (Figure 2B–D). Although the domains belong to the same 7BS superfamily, they are not closely related [11,13]. The closest paralog for MT13-C is spermidine synthase, an enzyme that catalyzes the transfer of a propylamine group from S-adenosylmethioninamine to putrescine to generate spermidine [13,15] (Figure 2E). In contrast, the MT13-N domain has three close paralogs that are all established lysine-specific MTs, namely to CS-KMT [16,17] (also called METTL12), eEF1A-KMT4 (previously annotated as a splice variant of ECE2) [18] and eEF1A-KMT2 (also called METTL10) [19] (Figure 2E).

The evolutionary conservation of METTL13 has been explored through systematic BLAST searches throughout the eukaryotic kingdom [11]. This analysis revealed clear paralogs in several commonly used model organisms including *D. melanogaster*, *C. elegans*, and *A. thaliana* but not in *S. cerevisiae* [11]. In line with these observations, the *S. cerevisiae* eEF1A homolog lacks methylation at the site corresponding to human eEF1A-Lys55 [20]. However, *S. cerevisiae* eEF1A is Nt trimethylated and the responsible enzyme has been identified as YLR285Wp [21], a 7BS MT that bears no sequence homology to MT13-C [11,13]. This demonstrates that eEF1A Nt MT activity has arisen twice in evolution, underscoring the functional importance of the PTM.

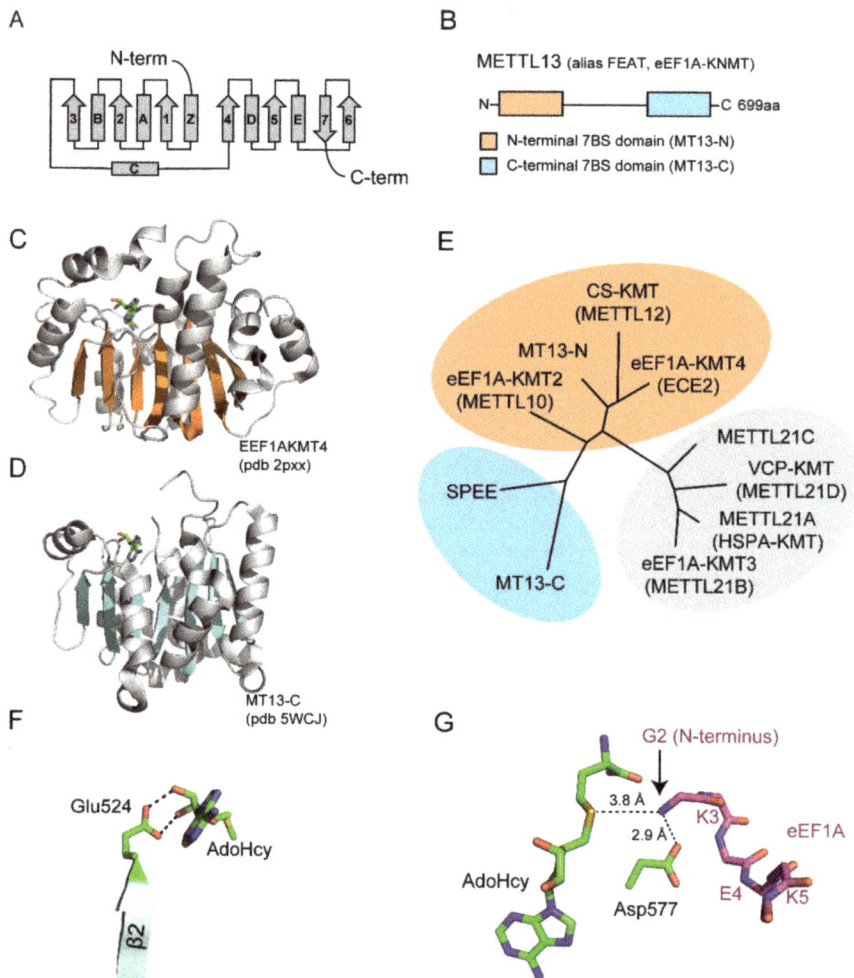

Figure 2. Domain organization and structure of METTL13. (**A**) Topology diagram of seven beta strand (7BS) methyltransferase fold. (**B**) Domain architecture of METTL13. (**C,D**) Structure of MT13-N-like eEF1A-KMT4 protein and the MT13-C domain. Ribbon representations are shown with beta strands highlighted in orange for (**C**) eEF1A-KMT4 (pdb # 2PXX) and blue for (**D**) MT13-C (pdb # 5WCJ). (**E**) Phylogenetic tree of METT13 domains and related methyltransferase enzymes. The tree was generated using the "Phylogeny.fr" platform [22] using METTL21A–D as an outgroup. (**F**) Structural model of MT13-C interaction with AdoHcy. Potential hydrogen bonds between Glu524 and the ribose moiety of AdoHcy are indicated (dashed lines). (**G**) Structural model of MT13-C and eEF1A N-terminal substrate peptide. Possible hydrogen bonds between METTL13-Asp577 and the eEF1A substrate peptide (stick representation, purple) are shown. The relative position and distance of the eEF1A N-terminus in relation to AdoHcy is indicated. The model was generated using the glide dock approach [23].

3. Biochemical Features

The catalytic activity of METTL13 is currently confined to the Nt and Lys55 of human eEF1A and the enzyme represents one out of five yet identified human eEF1A-KMTs (Figure 3). In vitro experiments with purified MT13-N and MT13-C domains have firmly demonstrated that MT13-N is responsible for dimethylation of Lys55 [10,11] and MT13-C is responsible for trimethylation of the protein Nt [11].

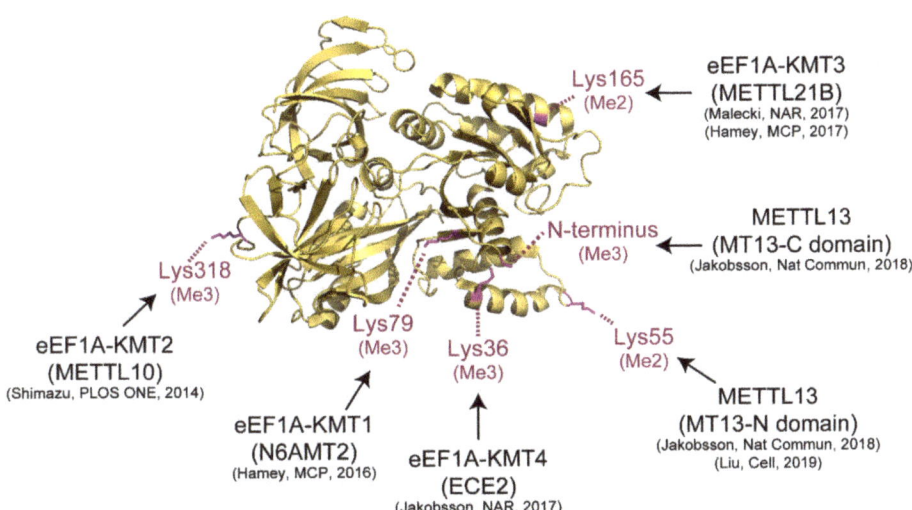

Figure 3. Methylation of human eukaryotic elongation factor 1 alpha. The structure of eEF1A (pdb # 1F60) is shown in ribbon representation (gold). Key methylation sites in human eEF1A with the predominant methylated forms (magenta) as well as the responsible MT enzymes are indicated.

3.1. The METTL13 C-Terminal MT Domain

The Nt of proteins is most frequently acetylated [24]; only in rare occasions subject to methylation [25]. Notably, the major Nt acetyltransferase A complex (NatA) is reported to acetylate substrates with a small amino acid such as Gly, Ala or Ser in the second position, and after excision of the initiator Met residue [26]. Therefore, the discovery of eEF1A Nt methylation was somewhat unexpected. Nt methylation is also a rare PTM and MT13-C is to date one out of three validated human Nt MTs. Aside from MT13-C, Nt MT activity has been reported for the closely related NTMT1 (also called METTL11A and NRMT1) and NTMT2 (also called METTL11B and NRMT2) enzymes that target the second residue in proteins Met-(Ala/Pro/Ser)-Pro-Lys-, after iMet excision [27–29]. Recent studies have further refined the NTMT1/2 consensus motif to X-Pro-Lys/Arg (X = Gly, Ser, Pro, or Ala) [30–32].

The specificity of MT13-C has been explored in depth. First, In vitro MT assays have demonstrated that MT13-C primarily methylates as 50-55 kDa protein in METTL13 KO cells extracts, corresponding to the molecular weight of eEF1A [11]. Second, Protein MTs can recognize substrates in different ways. Conceptually, they may recognize a folded substrate, or a linear sequence motif present in the substrate. MTs targeting the flexible histone tails often belong to the latter class [33,34] whereas some 7BS-MTs such as VCP-KMT [35,36] and METTL21A [37,38] require a folded substrate. To explore the mode of substrate recognition for MT13-C and to identify potential additional substrates, a peptide array harboring systematic mutations of the eEF1A Nt was utilized to define a general recognition motif for MT13-C. These experiments indicated that the domain is capable of methylating peptide sequences corresponding to the eEF1A Nt and that it can methylate a linear motif corresponding to [GAP]-[KRFYQH]-E-[KRQHIL] (amino acid in eEF1A is underlined) in In vitro settings. This degenerate motif was used to identify ~50 candidate substrates in the human proteome. Notably, none of these candidate substrates were efficiently methylated by MT13-C, suggesting that the enzyme is highly specific for the eEF1A Nt [11].

The extent and spread of human eEF1A Nt methylation have been explored in a set of cells and tissues. The stoichiometry of methylation has been assessed through quantitative mass spectrometry experiments and revealed the site to be primarily trimethylated in mouse liver, kidney, and intestine [11] as well as in human HAP-1 [11], HEK-293 [21], and HeLa [5] cells.

Taken together, the collective body of biochemical experiments suggest that MT13-C is a highly specific MT for the eEF1A Nt, trimethylation is the predominant form and that it is widespread across mammalian cells and tissues.

3.2. The METTL13 N-Terminal MT Domain

The MT13-N domain has not been characterized to the same depth as MT13-C, mainly due to lower solubility levels of recombinant forms of the domain (unpublished observation by the author). Nonetheless, MT13-N has been firmly demonstrated to possess eEF1A-Lys55 activity in independent studies [10,11].

Dimethylation of eEF1A-Lys55 has been known for long [39] and in a recent methylproteomic study, we reported identification of the PTM in wide range of human cells including A549, HCT116, HEK293, HeLa, MCF7, and SY5Y as well as human tissue biopsies from liver, colon, and prostate [5]. The relative abundance of the different methylated forms of eEF1A-Lys55 has also been explored in a set of mammalian cells and tissues corresponding to RPE1, 293T, NCI-H2170, NCIH520, PaTu8902, T3M4, and U2OS cells [10] as well as mouse liver, kidney, and intestine [11]. In all analyzed cells and tissues, the dimethylated form of Lys55 has been predominant.

The link between METTL13 and methylation of eEF1A-Lys55 has been reported and validated in independent methylproteomics studies. Liu and colleagues in the Gozani lab showed that methylation of Lys55 was the only methylationsite strikingly underrepresented in T3M4 METTL13 KO cells using a SILAC-based quantitative approach [10]. In similar experiments, we have reported both monomethylation of APOB-K1163 and dimethylation eEF1A-Lys55 as underrepresented in HAP-1 METTL13 KO cells [11]. Notably, the apparent significant under-representation of APOB-K1163me1 likely represents an experimental artefact and a remnant from bovine serum proteins added to the cell culture [40].

In summary, the body of data related to MT13-N suggests that the domain is highly specific for eEF1A-Lys55 and catalyzes dimethylation of the site in a broad range of mammalian cells and tissues.

4. Regulation

Proteins can be regulated at several different levels. Firstly, regulation can occur at the level of DNA and transcription. In addition, the stability of mRNA can be regulated. Finally, the function and stability of proteins can be regulated, for example by PTMs. Notably, METTL13 has been reported as regulated at all three levels.

At the level of transcription, HNL1 has been reported to upregulate METTL13 [41]. METTL13 protein levels are regulated at the mRNA level through the micro RNA miR-16, which targets the 3' UTR of the METTL13 mRNA and mediates its degradation [42,43]. Notably, miR-16 has been linked to OvCa and the micro RNA is underrepresented in both ovarian cancer OvCa cell lines and primary ovarian tissues [44]. Specific implications of METTL13 and OvCa are further detailed below.

PTMs are key determinants of protein function and they can act as both positive and negative regulators of protein stability. For example, phosphorylation can both promote the stability and mediate degradation of proteins [45]. Moreover, distinct branches of poly-ubiquitination are linked to proteasomal degradation and autophagic clearance [46]. Intriguingly, METTL13 is reported to be both ubiquitinated and phosphorylated on multiple sites, whereof some are located in the active MT domains (Figure 4). However, the potential role of these PTMs in regulating the function and stability of METTL13 protein has not yet been explored.

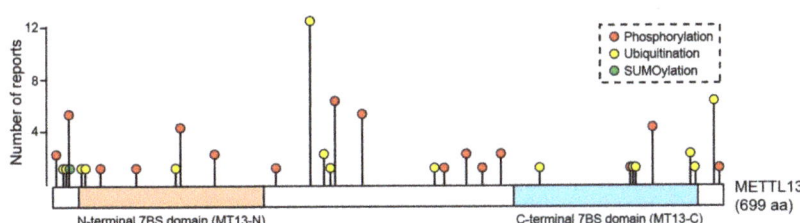

Figure 4. The PTM landscape of METTL13. The data were retrieved from the "PhosphoSitePlus" database [47] (www.phosphosite.org; (accessed on 15 September 2021)).

In summary, little is known about the regulation of METTL13 but evidence suggests that it might be regulated at multiple levels including DNA, mRNA, and protein.

5. Cellular Features

Global transcriptomic data indicate that the METTL13 gene is ubiquitously expressed in human cells and tissues (Figure 5A). Moreover, recent exhaustive proteomics datasets also indicate that METTL13 protein is present in most, if not all, cells and tissues (Figure 5B). Notably, across multiple proteomics datasets, METTL13 is one of the more abundant METTL-proteins, and invariably the most abundant eEF1A-KMT (Figure 5B).

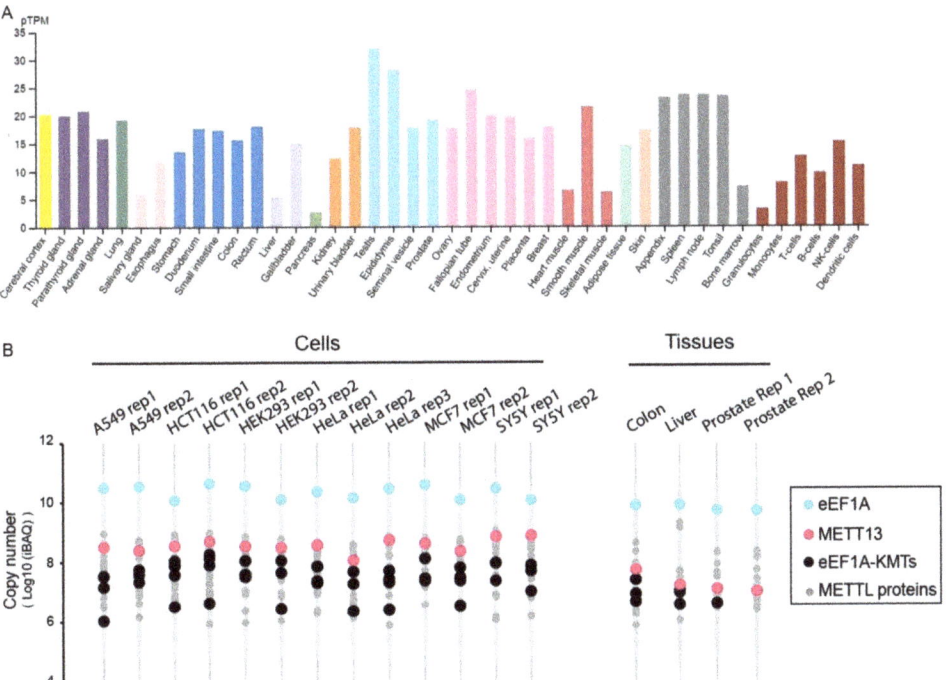

Figure 5. METTL13 expression in cells and tissues. (**A**) RNA data showing protein-transcripts per million in a range of human organs and cell types. The data were retrieved from the human protein atlas (HPA, https://www.proteinatlas.org/ (accessed on 15 September 2021)) and color coded according to tissue or cell type. (**B**) Proteome data showing METTL13 protein levels in cultured human cells and primary tissue biopsies. The data were retrieved from ProteomeXchange (dataset PXD004452) [48]. eEF1A and METTL13 are highlighted and other eEF1A-KMTs (METTL10, EEF1AKMT4, N6AMT2, METTL21B) and METTL- proteins are indicated.

Human eEF1A represents the hitherto only validated METTL13 substrate and the molecular effects of METTL13 have mainly been linked to translation. Liu et al have shown that METTL13-mediated dimethylation of eEF1A-Lys55 increases the GTPase activity of the elongation factor and thereby increases the overall rate of translation and protein synthesis [10] (Figure 6). We have instead used a ribosome foot printing approach to assess the role of METTL13 mediated methylation in the context of translation dynamics. Our method relies on the well-established notion that the ribosome samples cellular eEF1A-GTP-aminoacyl-tRNA complexes in its acceptor (A)-site. Consequently, the frequency of a codon in the A-site can be used as a proxy for its rate of translation [49]. The analysis revealed that cells lacking METTL13-mediated eEF1A methylation displayed a faster translation of histidine codons and a slower translation of alanine codons [11] (Figure 6).

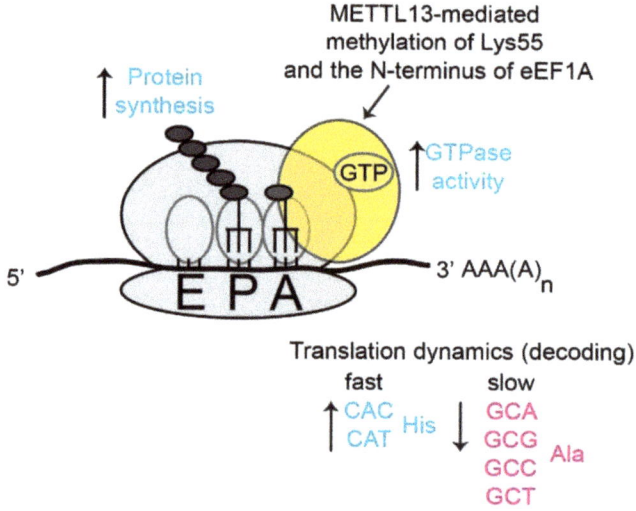

Figure 6. Molecular effects of METTL13-mediated eEF1A methylation on translation.

Notably, the deletion of other human eEF1A-KMTs [18,50] as well as MTs targeting ribosomal proteins [51] and rRNA [52] have led to related phenotypes with altered translation of specific codons. Collectively, these observations corroborate that methylation of the translational apparatus is frequent and represents a mechanism to regulate proteins synthesis [7,53]. In analogy to the "histone code" in epigenetics [54,55], it has been proposed that the multiple reported PTMs on eEF1A may constitute an "eEF1A code" [56,57] that dynamically regulates gene activity at the level of translation.

6. Connection to Diseases

METTL13 has been evaluated in the context of disease biology, mainly cancer, but the reports are somewhat scattered and do not point in a unified direction. This can be a consequence of cell-type specific functions of METTL13 or pleiotropic phenotypes related to its general role as a regulator of protein synthesis.

A seminal study by Takahashi et al first linked METTL13 to cancer [58]. They generated a mouse model that over-expressed METTL13 in a tissue-specific manner and observed that mice developed tumors in the organs where the gene was overexpressed, suggesting the enzyme is a general driver of tumorigenesis. In subsequent studies from the Takahashi lab, METTL13 plasma levels were reported as elevated in several cancer types, and specifically in OvCa [59], suggesting the protein may have clinical utility as a biomarker.

The abundance of METTL13 has been linked to both favorable and poor cancer prognosis. For clear cell renal cell carcinoma, high levels of METTL13 has been linked to favorable prognosis and the enzyme has been reported to inhibit growth and metastasis [60]. In bladder cancer, METTL13 has been reported to negatively regulate key cancer hallmarks including proliferation, migration and invasion [61].

In contrast, METTL13 levels have been reported as elevated and linked to unfavorable prognosis for other cancer types. In head and neck squamous cell carcinoma, METTL13 expression is increased at both the transcript and protein level and high levels are associated with poor prognosis [62]. In hepatocellular carcinoma, METTL13 has been implied in mediating tumor growth and metastasis [41]. Finally, a recent study revealed that both METTL13 and eEF1A-K55me2 levels are upregulated in pancreatic and lung cancer and high levels of both these markers were linked to low patient survival [10]. The authors also convincingly demonstrated that METTL13 depletion sensitized cancer cells to PI3K and mTOR pathway inhibition [10]. Importantly, independent studies have shown that downregulation of METTL13 levels by miR-16 induces apoptosis [42]. Taken together, this indicates that inhibition of METTL13 may represent an effective strategy in combinatorial cancer therapy approaches.

METTL13 has 157 mutations annotated in the Catalog of Somatic Mutations In Cancer (COSMIC) database (Figure 7) and it has been highlighted as the most mutated METTL protein in comprehensive transcriptomics cancer datasets [63]. However, as METTL13 is a dual 7BS domain MT it is also one of the larger 7BS MTs which can represent an explanation for the high number of annotated mutations.

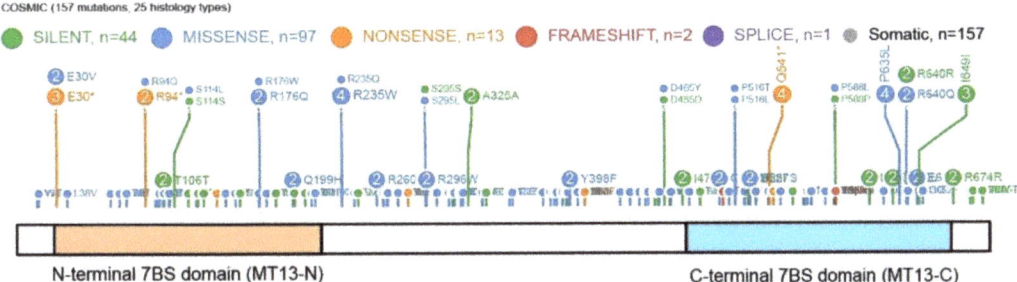

Figure 7. The mutational landscape of METTL13. The data was retrieved from the COSMIC database and the visualization is modified from ProteinPaint [64].

In addition to the numerous links to different types of cancer, METTL13 has also been associated with hearing loss. In detail, a dominant mutation corresponding to Arg544Gln in the METTL13 protein has been linked to deafness [65].

In summary, aberrant expression of METTL13 has been linked to a wide range of cancers and it has been suggested to function as both oncogene and tumor suppressor.

7. Conclusions and Outlook

During the last decade, significant discoveries have been made to increase the understanding of METTL13 enzymatic activity, cellular features, and links to disease. Future research efforts will likely extend on the current knowledge status, especially how aberrantly regulated METTL13 relates to cancer etiology and progression.

Biochemically, future focus will likely be devoted to comprehending the potential dynamic nature of METTL13-mediated eEF1A methylation. While methylation of histone proteins [66] and eEF1A-Lys165 [50] have been reported as dynamic, there are yet no reports of potential demethylases targeting the eEF1A Nt or Lys55. Here a combination of heavy methyl [67] and dynamic SILAC [68] can be used to assess cellular turnover of both bulk and methylation modified species of eEF1A, to uncover potential methylation dynamics.

From a clinical perspective, METTL13 is clearly linked to key cancer hallmarks and recent evidence suggests that combined targeting of METTL13 and key cellular signaling pathways may represent an effective therapeutic strategy in cancer management. Here, large-scale synthetic lethality studies using genome-wide CRISPR KO libraries can globally uncover co-dependencies of METTL13 and other genes, particularly key signaling hubs. Such experiments have potential to uncover novel strategies for combination cancer therapy.

Funding: Research in the Jakobsson lab is funded by The Crafoord Foundation [ref 20200526], Stiftelsen Cancera, Mats Paulssons Stiftelse for Research, Innovation and Community Development, EUMSCA-COFUND [CanFaster: 847583], Fru Berta Kamprads Stiftelse [FBKS-2021-31-(330)], Gunnar Nilssons Cancerstiftelse [GN-2021-9 - (247)], and CREATE Health Cancer Center.

Institutional Review Board Statement: Not applicable.

Informed Consent Statement: Not applicable.

Data Availability Statement: The proteomics data used to generate Figure 5B is available through ProteomeXchange (dataset PXD004452).

Acknowledgments: The author thanks Levon Halabelian for the structural model presented as Figure 2G and Jana Hladílková for providing input on the manuscript text.

Conflicts of Interest: The author declares no conflict of interest.

References

1. Walsh, C.T.; Garneau-Tsodikova, S.; Gatto, G.J., Jr. Protein posttranslational modifications: The chemistry of proteome diversifications. *Angew. Chem. Int. Ed.* **2005**, *44*, 7342–7372. [CrossRef]
2. Olsen, J.V.; Mann, M. Status of large-scale analysis of post-translational modifications by mass spectrometry. *Mol. Cell. Proteom.* **2013**, *12*, 3444–3452. [CrossRef] [PubMed]
3. Moore, K.E.; Carlson, S.M.; Camp, N.D.; Cheung, P.; James, R.G.; Chua, K.F.; Wolf-Yadlin, A.; Gozani, O. A general molecular affinity strategy for global detection and proteomic analysis of lysine methylation. *Mol. Cell* **2013**, *50*, 444–456. [CrossRef] [PubMed]
4. Larsen, S.C.; Sylvestersen, K.B.; Mund, A.; Lyon, D.; Mullari, M.; Madsen, M.V.; Daniel, J.A.; Jensen, L.J.; Nielsen, M.L. Proteome-wide analysis of arginine monomethylation reveals widespread occurrence in human cells. *Sci. Signal.* **2016**, *9*, rs9. [CrossRef] [PubMed]
5. Kapell, S.; Jakobsson, M.E. Large-scale identification of protein histidine methylation in human cells. *NAR Genom. Bioinform.* **2021**, *3*, lqab045. [CrossRef] [PubMed]
6. Jakobsson, M.E. Enzymology and significance of protein histidine methylation. *J. Biol. Chem.* **2021**, *297*, 101130. [CrossRef] [PubMed]
7. Clarke, S.G. Protein methylation at the surface and buried deep: Thinking outside the histone box. *Trends Biochem. Sci.* **2013**, *38*, 243–252. [CrossRef] [PubMed]
8. Lewis, J.C.A.; Wolfenden, R. The burden borne by protein methyltransferases: Rates and Equilibria of non-enzymatic methylation of amino acid side chains by SAM in water. *Biochemistry* **2021**, *60*, 854–858. [CrossRef]
9. Grimsley, G.R.; Scholtz, J.M.; Pace, C.N. A summary of the measured pK values of the ionizable groups in folded proteins. *Protein Sci.* **2008**, *18*, 247–251. [CrossRef]
10. Liu, S.; Hausmann, S.; Carlson, S.M.; Fuentes, M.E.; Francis, J.; Pillai, R.; Lofgren, S.M.; Hulea, L.; Tandoc, K.; Lu, J.; et al. METTL13 methylation of eEF1A increases translational output to promote tumorigenesis. *Cell* **2019**, *176*, 491–504.e21. [CrossRef]
11. Jakobsson, M.E.; Małecki, J.; Halabelian, L.; Nilges, B.S.; Pinto, R.; Kudithipudi, S.; Munk, S.; Davydova, E.; Zuhairi, F.R.; Arrowsmith, C.; et al. The dual methyltransferase METTL13 targets N terminus and Lys55 of eEF1A and modulates codon-specific translation rates. *Nat. Commun.* **2018**, *9*, 1–15. [CrossRef] [PubMed]
12. Miura, G. METTLing with translation. *Nat. Chem. Biol.* **2019**, *15*, 207. [CrossRef] [PubMed]
13. Petrossian, T.; Clarke, S.G. Uncovering the human methyltransferasome. *Mol. Cell. Proteom.* **2011**, *10*, M110.000976. [CrossRef] [PubMed]
14. Schubert, H.L.; Blumenthal, R.M.; Cheng, X. Many paths to methyltransfer: A chronicle of convergence. *Trends Biochem. Sci.* **2003**, *28*, 329–335. [CrossRef]
15. Wu, H.; Min, J.; Ikeguchi, Y.; Zeng, H.; Dong, A.; Loppnau, P.; Pegg, A.E.; Plotnikov, A.N. Structure and mechanism of spermidine synthases. *Biochemistry* **2007**, *46*, 8331–8339. [CrossRef]
16. Rhein, V.F.; Carroll, J.; Ding, S.; Fearnley, I.M.; Walker, J.E. Human METTL12 is a mitochondrial methyltransferase that modifies citrate synthase. *FEBS Lett.* **2017**, *591*, 1641–1652. [CrossRef] [PubMed]
17. Małecki, J.; Jakobsson, M.E.; Ho, A.Y.; Moen, A.; Rustan, A.C.; Falnes, P.Ø. Uncovering human METTL12 as a mitochondrial methyltransferase that modulates citrate synthase activity through metabolite-sensitive lysine methylation. *J. Biol. Chem.* **2017**, *292*, 17950–17962. [CrossRef] [PubMed]

18. Jakobsson, M.E.; Małecki, J.; Nilges, B.S.; Moen, A.; Leidel, S.; Falnes, P.Ø. Methylation of human eukaryotic elongation factor alpha (eEF1A) by a member of a novel protein lysine methyltransferase family modulates mRNA translation. *Nucleic Acids Res.* **2017**, *45*, 8239–8254. [CrossRef]
19. Shimazu, T.; Barjau, J.; Sohtome, Y.; Sodeoka, M.; Shinkai, Y. Selenium-based S-adenosylmethionine analog reveals the mammalian seven-beta-strand methyltransferase METTL10 to be an EF1A1 lysine methyltransferase. *PLoS ONE* **2014**, *9*, e105394. [CrossRef]
20. White, J.T.; Cato, T.; Deramchi, N.; Gabunilas, J.; Roy, K.R.; Wang, C.; Chanfreau, G.F.; Clarke, S.G. Protein methylation and translation: Role of lysine modification on the function of yeast elongation factor 1A. *Biochemistry* **2019**, *58*, 4997–5010. [CrossRef] [PubMed]
21. Hamey, J.; Winter, D.; Yagoub, D.; Overall, C.M.; Hart-Smith, G.; Wilkins, M.R. Novel N-terminal and lysine methyltransferases that target translation elongation factor 1A in yeast and human. *Mol. Cell. Proteom.* **2016**, *15*, 164–176. [CrossRef] [PubMed]
22. Dereeper, A.; Guignon, V.; Blanc, G.; Audic, S.; Buffet, S.; Chevenet, F.; Dufayard, J.-F.; Guindon, S.; Lefort, V.; Lescot, M.; et al. Phylogeny.fr: Robust phylogenetic analysis for the non-specialist. *Nucleic Acids Res.* **2008**, *36*, W465–W469. [CrossRef]
23. Friesner, R.A.; Murphy, R.B.; Repasky, M.P.; Frye, L.L.; Greenwood, J.R.; Halgren, T.A.; Sanschagrin, P.C.; Mainz, D.T. Extra precision glide: Docking and scoring incorporating a model of hydrophobic enclosure for protein−ligand complexes. *J. Med. Chem.* **2006**, *49*, 6177–6196. [CrossRef]
24. Aksnes, H.; Ree, R.; Arnesen, T. Co-translational, post-translational, and non-catalytic roles of N-terminal acetyltransferases. *Mol. Cell* **2019**, *73*, 1097–1114. [CrossRef] [PubMed]
25. Chen, P.; Sobreira, T.J.P.; Hall, M.C.; Hazbun, T.R. Discovering the N-terminal methylome by repurposing of proteomic datasets. *J. Proteome Res.* **2021**, *20*, 4231–4247. [CrossRef]
26. Arnesen, T.; van Damme, P.; Polevoda, B.; Helsens, K.; Evjenth, R.; Colaert, N.; Varhaug, J.E.; Vandekerckhove, J.; Lillehaug, J.R.; Sherman, F.; et al. Proteomics analyses reveal the evolutionary conservation and divergence of N-terminal acetyltransferases from yeast and humans. *Proc. Natl. Acad. Sci. USA* **2009**, *106*, 8157–8162. [CrossRef] [PubMed]
27. Tooley, C.E.S.; Petkowski, J.J.; Muratore-Schroeder, T.L.; Balsbaugh, J.L.; Shabanowitz, J.; Sabat, M.; Minor, W.; Hunt, D.F.; Macara, I.G. NRMT is an α-N-methyltransferase that methylates RCC1 and retinoblastoma protein. *Nature* **2010**, *466*, 1125–1128. [CrossRef] [PubMed]
28. Chen, T.; Muratore, T.L.; Schaner-Tooley, C.E.; Shabanowitz, J.; Hunt, N.F.; Macara, I.G. N-terminal alpha-methylation of RCC1 is necessary for stable chromatin association and normal mitosis. *Nat. Cell Biol.* **2007**, *9*, 596–603. [CrossRef] [PubMed]
29. Petkowski, J.; Bonsignore, L.A.; Tooley, J.G.; Wilkey, D.W.; Merchant, M.L.; Macara, I.G.; Tooley, C.E.S. NRMT2 is an N-terminal monomethylase that primes for its homologue NRMT1. *Biochem. J.* **2013**, *456*, 453–462. [CrossRef] [PubMed]
30. Diaz, K.; Meng, Y.; Huang, R. Past, present, and perspectives of protein N-terminal methylation. *Curr. Opin. Chem. Biol.* **2021**, *63*, 115–122. [CrossRef] [PubMed]
31. Dong, C.; Mao, Y.; Tempel, W.; Qin, S.; Li, L.; Loppnau, P.; Huang, R.; Min, J. Structural basis for substrate recognition by the human N-terminal methyltransferase 1. *Genes Dev.* **2015**, *29*, 2343–2348. [CrossRef]
32. Dong, C.; Dong, G.; Li, L.; Zhu, L.; Tempel, W.; Liu, Y.; Huang, R.; Min, J. An asparagine/glycine switch governs product specificity of human N-terminal methyltransferase NTMT2. *Commun. Biol.* **2018**, *1*, 183. [CrossRef] [PubMed]
33. Rathert, P.; Dhayalan, A.; Murakami, M.; Zhang, X.; Tamas, R.; Jurkowska, R.; Komatsu, Y.; Shinkai, Y.; Cheng, X.; Jeltsch, A. Protein lysine methyltransferase G9a acts on non-histone targets. *Nat. Chem. Biol.* **2008**, *4*, 344–346. [CrossRef] [PubMed]
34. Dhayalan, A.; Kudithipudi, S.; Rathert, P.; Jeltsch, A. Specificity analysis-based identification of new methylation targets of the SET7/9 protein lysine methyltransferase. *Chem. Biol.* **2011**, *18*, 111–120. [CrossRef]
35. Kernstock, S.; Davydova, E.; Jakobsson, M.; Moen, A.; Pettersen, S.J.; Mælandsmo, G.M.; Egge-Jacobsen, W.; Falnes, P.Ø. Lysine methylation of VCP by a member of a novel human protein methyltransferase family. *Nat. Commun.* **2012**, *3*, 1038. [CrossRef] [PubMed]
36. Cloutier, P.; Lavallee-Adam, M.; Faubert, D.; Blanchette, M.; Coulombe, B. A newly uncovered group of distantly related lysine methyltransferases preferentially interact with molecular chaperones to regulate their activity. *PLoS Genet.* **2013**, *9*, e1003210. [CrossRef] [PubMed]
37. Jakobsson, M.E.; Moen, A.; Bousset, L.; Egge-Jacobsen, W.; Kernstock, S.; Melki, R.; Falnes, P.Ø. Identification and characterization of a novel human methyltransferase modulating Hsp70 protein function through lysine methylation. *J. Biol. Chem.* **2013**, *288*, 27752–27763. [CrossRef] [PubMed]
38. Jakobsson, M.E.; Moen, A.; Falnes, P.Ø. Correspondence: On the enzymology and significance of HSPA1 lysine methylation. *Nat. Commun.* **2016**, *7*, 11464. [CrossRef] [PubMed]
39. Dever, T.; Costello, C.; Owens, C.; Rosenberry, T.; Merrick, W. Location of seven post-translational modifications in rabbit elongation factor 1α including dimethyllysine, trimethyllysine, and glycerylphosphorylethanolamine. *J. Biol. Chem.* **1989**, *264*, 20518–20525. [CrossRef]
40. Marcos, E.; Mazur, A.; Cardot, P.; Rayssiguier, Y. Quantitative determination of apolipoprotein B in bovine serum by radial immunodiffusion. *Comp. Biochem. Physiol. Part B Comp. Biochem.* **1989**, *94*, 171–173. [CrossRef]
41. Li, L.; Zheng, Y.-L.; Jiang, C.; Fang, S.; Zeng, T.-T.; Zhu, Y.-H.; Li, Y.; Xie, D.; Guan, X.-Y. HN1L-mediated transcriptional axis AP-2γ/METTL13/TCF3-ZEB1 drives tumor growth and metastasis in hepatocellular carcinoma. *Cell Death Differ.* **2019**, *26*, 2268–2283. [CrossRef]

42. Liang, H.; Fu, Z.; Jiang, X.; Wang, N.; Wang, F.; Wang, X.; Zhang, S.; Wang, Y.; Yan, X.; Guan, W.-X.; et al. miR-16 promotes the apoptosis of human cancer cells by targeting FEAT. *BMC Cancer* **2015**, *15*, 1–9. [CrossRef] [PubMed]
43. Su, X.F.; Li, N.; Meng, F.L.; Chu, Y.L.; Li, T.; Gao, X.Z. MiR-16 inhibits hepatocellular carcinoma progression by targeting FEAT through NF-κB signaling pathway. *Eur. Rev. Med. Pharmacol. Sci.* **2019**, *23*, 10274–10282. [CrossRef]
44. Bhattacharya, R.; Nicoloso, M.; Arvizo, R.; Wang, E.; Cortez, A.; Rossi, S.; Calin, G.; Mukherjee, P. MiR-15a and MiR-16 control Bmi-1 expression in ovarian cancer. *Cancer Res.* **2009**, *69*, 9090–9095. [CrossRef]
45. Wu, C.; Ba, Q.; Lu, D.; Li, W.; Salovska, B.; Hou, P.; Mueller, T.; Rosenberger, G.; Gao, E.; Di, Y.; et al. Global and site-specific effect of phosphorylation on protein turnover. *Dev. Cell* **2021**, *56*, 111–124. [CrossRef] [PubMed]
46. Pohl, C.; Dikic, I. Cellular quality control by the ubiquitin-proteasome system and autophagy. *Science* **2019**, *366*, 818–822. [CrossRef] [PubMed]
47. Hornbeck, P.V.; Kornhauser, J.M.; Tkachev, S.; Zhang, B.; Skrzypek, E.; Murray, B.; Latham, V.; Sullivan, M. PhosphoSitePlus: A comprehensive resource for investigating the structure and function of experimentally determined post-translational modifications in man and mouse. *Nucleic Acids Res.* **2011**, *40*, D261–D270. [CrossRef] [PubMed]
48. Bekker-Jensen, D.B.; Kelstrup, C.D.; Batth, T.S.; Larsen, S.C.; Haldrup, C.; Bramsen, J.B.; Sørensen, K.D.; Høyer, S.; Ørntoft, T.F.; Andersen, C.L.; et al. An optimized shotgun strategy for the rapid generation of comprehensive human proteomes. *Cell Syst.* **2017**, *4*, 587–599. [CrossRef]
49. Nedialkova, D.D.; Leidel, S.A. Optimization of codon translation rates via tRNA modifications maintains proteome integrity. *Cell* **2015**, *161*, 1606–1618. [CrossRef]
50. Małecki, J.; Aileni, V.K.; Ho, A.Y.; Schwarz, J.; Moen, A.; Sørensen, V.; Nilges, B.S.; Jakobsson, M.E.; Leidel, S.; Falnes, P.Ø. The novel lysine specific methyltransferase METTL21B affects mRNA translation through inducible and dynamic methylation of Lys-165 in human eukaryotic elongation factor 1 alpha (eEF1A). *Nucleic Acids Res.* **2017**, *45*, 4370–4389. [CrossRef]
51. Małecki, J.M.; Odonohue, M.-F.; Kim, Y.; Jakobsson, M.E.; Gessa, L.; Pinto, R.; Wu, J.; Davydova, E.; Moen, A.; Olsen, J.V.; et al. Human METTL18 is a histidine-specific methyltransferase that targets RPL3 and affects ribosome biogenesis and function. *Nucleic Acids Res.* **2021**, *49*, 3185–3203. [CrossRef] [PubMed]
52. Pinto, R.; Vågbø, C.B.; Jakobsson, M.E.; Kim, Y.; Baltissen, M.P.; O'Donohue, M.-F.; Guzmán, U.H.; Małecki, J.M.; Wu, J.; Kirpekar, F.; et al. The human methyltransferase ZCCHC4 catalyses N6-methyladenosine modification of 28S ribosomal RNA. *Nucleic Acids Res.* **2020**, *48*, 830–846. [CrossRef] [PubMed]
53. Clarke, S.G. The ribosome: A hot spot for the identification of new types of protein methyltransferases. *J. Biol. Chem.* **2018**, *293*, 10438–10446. [CrossRef] [PubMed]
54. Jenuwein, T.; Allis, C.D. Translating the histone code. *Science* **2001**, *293*, 1074–1080. [CrossRef]
55. Strahl, B.D.; Allis, C.D. The language of covalent histone modifications. *Nature* **2000**, *403*, 41–45. [CrossRef] [PubMed]
56. Hamey, J.; Wilkins, M.R. Methylation of elongation factor 1A: Where, who, and why? *Trends Biochem. Sci.* **2018**, *43*, 211–223. [CrossRef]
57. Jakobsson, M.E.; Małecki, J.; Falnes, P.Ø. Regulation of eukaryotic elongation factor 1 alpha (eEF1A) by dynamic lysine methylation. *RNA Biol.* **2018**, *15*, 314–319. [CrossRef]
58. Takahashi, A.; Tokita, H.; Takahashi, K.; Takeoka, T.; Murayama, K.; Tomotsune, D.; Ohira, M.; Iwamatsu, A.; Ohara, K.; Yazaki, K.; et al. A novel potent tumour promoter aberrantly overexpressed in most human cancers. *Sci. Rep.* **2011**, *1*, 15. [CrossRef]
59. Li, Y.; Kobayashi, K.; Mona, M.M.; Satomi, C.; Okano, S.; Inoue, H.; Tani, K.; Takahashi, A. Immunogenic FEAT protein circulates in the bloodstream of cancer patients. *J. Transl. Med.* **2016**, *14*, 275. [CrossRef] [PubMed]
60. Liu, Z.; Sun, T.; Piao, C.; Zhang, Z.; Kong, C. METTL13 inhibits progression of clear cell renal cell carcinoma with repression on PI3K/AKT/mTOR/HIF-1α pathway and c-Myc expression. *J. Transl. Med.* **2021**, *19*, 1–15. [CrossRef] [PubMed]
61. Zhang, Z.; Zhang, G.; Kong, C.; Zhan, B.; Dong, X.; Man, X. METTL13 is downregulated in bladder carcinoma and suppresses cell proliferation, migration and invasion. *Sci. Rep.* **2016**, *6*, 19261. [CrossRef] [PubMed]
62. Wang, X.; Li, K.; Wan, Y.; Chen, F.; Cheng, M.; Xiong, G.; Wang, G.; Chen, S.; Chen, Z.; Chen, J.; et al. Methyltransferase like 13 mediates the translation of snail in head and neck squamous cell carcinoma. *Int. J. Oral Sci.* **2021**, *13*, 1–11. [CrossRef] [PubMed]
63. Campeanu, I.J.; Jiang, Y.; Liu, L.; Pilecki, M.; Najor, A.; Cobani, E.; Manning, M.; Zhang, X.M.; Yang, Z.-Q. Multi-omics integration of methyltransferase-like protein family reveals clinical outcomes and functional signatures in human cancer. *Sci. Rep.* **2021**, *11*, 1–14. [CrossRef] [PubMed]
64. Zhou, X.; Edmonson, M.N.; Wilkinson, M.R.; Patel, A.; Wu, G.; Liu, Y.; Li, Y.; Zhang, Z.; Rusch, M.C.; Parker, M.; et al. Exploring genomic alteration in pediatric cancer using ProteinPaint. *Nat. Genet.* **2016**, *48*, 4–6. [CrossRef] [PubMed]
65. Yousaf, R.; Ahmed, Z.M.; Giese, A.P.; Morell, R.J.; Lagziel, A.; Dabdoub, A.; Wilcox, E.R.; Riazuddin, S.; Friedman, T.B.; Riazuddin, S. Modifier variant of METTL13 suppresses human GAB1–associated profound deafness. *J. Clin. Investig.* **2018**, *128*, 1509–1522. [CrossRef] [PubMed]
66. Black, J.; van Rechem, C.; Whetstine, J.R. Histone lysine methylation dynamics: Establishment, regulation, and biological impact. *Mol. Cell* **2012**, *48*, 491–507. [CrossRef]
67. Ong, S.-E.; Mittler, G.; Mann, M. Identifying and quantifying in vivo methylation sites by heavy methyl SILAC. *Nat. Methods* **2004**, *1*, 119–126. [CrossRef] [PubMed]
68. Schwanhäusser, B.; Gossen, M.; Dittmar, G.; Selbach, M. Global analysis of cellular protein translation by pulsed SILAC. *Proteomics* **2009**, *9*, 205–209. [CrossRef] [PubMed]

Review

Structure, Activity, and Function of SETMAR Protein Lysine Methyltransferase

Michael Tellier

Sir William Dunn School of Pathology, University of Oxford, South Parks Road, Oxford OX1 3RE, UK; michael.tellier@path.ox.ac.uk; Tel.: +44-1865-275583

Abstract: SETMAR is a protein lysine methyltransferase that is involved in several DNA processes, including DNA repair via the non-homologous end joining (NHEJ) pathway, regulation of gene expression, illegitimate DNA integration, and DNA decatenation. However, SETMAR is an atypical protein lysine methyltransferase since in anthropoid primates, the SET domain is fused to an inactive DNA transposase. The presence of the DNA transposase domain confers to SETMAR a DNA binding activity towards the remnants of its transposable element, which has resulted in the emergence of a gene regulatory function. Both the SET and the DNA transposase domains are involved in the different cellular roles of SETMAR, indicating the presence of novel and specific functions in anthropoid primates. In addition, SETMAR is dysregulated in different types of cancer, indicating a potential pathological role. While some light has been shed on SETMAR functions, more research and new tools are needed to better understand the cellular activities of SETMAR and to investigate the therapeutic potential of SETMAR.

Keywords: SETMAR; Metnase; H3K36me2; Hsmar1; non-homologous end joining repair; NHEJ; transposase; transposable elements; histone; methyltransferase

1. Introduction

In eukaryotes, DNA is wrapped around proteins called histone to form the chromatin, a nucleoprotein structure. The basic unit of chromatin is the nucleosome, which is composed of 147 base pairs (bp) and eight histone proteins: two histones H2A, two H2B, two H3, and two H4 [1]. An additional histone, H1, is positioned between nucleosomes to regulate their packaging. DNA processes, such as transcription, replication, or DNA repair, are therefore dependent on chromatin remodeling [2]. A key factor in chromatin regulation is the addition, removal, and reading of post-translational modifications (PTMs) that can be deposited on the tails of histones, especially H3 and H4. Frequent histone PTMs include methylation, acetylation, phosphorylation, or ubiquitination but numerous other PTMs have also been described, referred as a whole as the histone code [3]. The histone code hypothesis implies that specific patterns of histone PTMs are associated and involved in specific DNA processes. For example, euchromatin, also known as open chromatin, is associated with active transcription and histone marks such as histone H3 lysine 4 trimethylation (H3K4me3), H3K36me3, H3K79 mono- di-, and trimethylation (me1, me2, me3), and acetylation of H3K9 or H3K27 (H3K9ac or H3K27ac). In contrast, heterochromatin, or close chromatin, is linked to inactive transcription and the histone marks H3K9me3 and H3K27me3 [4,5].

Dimethylation of H3K36 (H3K36me2) is associated with regulation of gene expression and DNA damage repair [6]. In humans, H3K36me2 is catalyzed by several histone methyltransferases, including ASH1L, NSD1-3, SETD3, SETMAR, and SMYD2 [7]. The interest in SETMAR has started to grow following the sequencing of the human genome as *SETMAR* was found to be one of the 47 genes containing or derived from a domesticated transposable element [8]. Later works have shown that in mammals, SETMAR exists in two different versions, as a single SET histone methyltransferase domain in most mammals while in anthropoid primates, SETMAR is a fusion between the SET domain and an inactive domesticated DNA transposase, Hsmar1 [8–11]. SETMAR is expressed ubiquitously and has been found to dimethylate H3K4 and H3K36 [12]. Since its discovery, the human SETMAR has been associated with numerous cellular processes, including DNA damage repair via the non-homologous end joining (NHEJ) pathway, illegitimate DNA integration, restart of stalled replication forks, chromosomal decatenation, suppression of chromosomal translocations, and regulation of gene expression [12–21]. In addition, SETMAR is dysregulated in several cancers, such as glioblastoma, leukemia, hematologic neoplasms, breast and colon cancer, and mantle cell lymphoma [16,22–30].

2. Structural Features

2.1. Domain Architecture

SETMAR is composed of two domains, the protein lysine methyltransferase (SET) domain and the DNA transposase (MAR) domain (Figure 1A). The protein lysine methyltransferase domain is constituted by the SET subdomain, named after the Su(var)3-9, enhancer-of-zeste and Trithorax (SET) proteins originally identified in Drosophila [31–33]. Two other subdomains, the pre-SET and the post-SET flank the SET subdomain. The SET domain contains binding sites for the lysine ligand and the co-factor S-adenosylmethionine (SAM), which provides the methyl groups.

The DNA transposase domain of SETMAR is catalytically inactive, as several inactivating mutations in the catalytic subdomain have neutralized the DNA transposition activity of SETMAR [10,11,34]. In contrast, the DNA binding subdomain is under strong purifying selection [9] and retains its binding activity to the inverted-terminal repeats (ITRs) of Hsmar1 transposon remnants both in vitro [9] and in SETMAR chromatin immunoprecipitation followed by sequencing (ChIP-seq) experiments [21,35,36]. The DNA binding specificity of SETMAR is regulated via an interaction with PRPF19, with the SETMAR-PRPF19 complex able to bind non-ITRs sequences [37]. Similarly to the ancestral Hsmar1 DNA transposase, the DNA transposase domain of SETMAR also forms a homodimer [38]. While SETMAR cannot perform DNA transposition, it was shown to perform in vitro 5'end nicking on Hsmar1 transposon ends, in presence of DMSO and Mn^{2+} [10], and to act as an endonuclease, such as Artemis, trimming DNA overhangs [39–41].

2.2. Isoforms

The *SETMAR* gene encodes eight different mRNA isoforms in human tissues with variable combinations of domains and subdomains (Figure 1B). Surprisingly, only one isoform encodes for a SETMAR protein containing both the SET domain and the DNA transposase domain (ENST00000358065.4). The remaining isoforms encode proteins containing only the DNA transposase domain (three isoforms), only an active SET domain (one isoform), or no complete domain (three isoforms). Out of the eight isoforms, four isoforms, encoding full-length SETMAR, SET domain only, or the DNA transposase domain only, are dominantly expressed in tissues. Interestingly, the most expressed isoform encodes a SETMAR protein containing only the DNA transposase domain. Two splicing factors have been found to regulate SETMAR alternative splicing, NONO and SFPQ, in bladder cancer [30]. As these two splicing factors are ubiquitously expressed, NONO and SFPQ are likely to be general regulators of SETMAR alternative splicing.

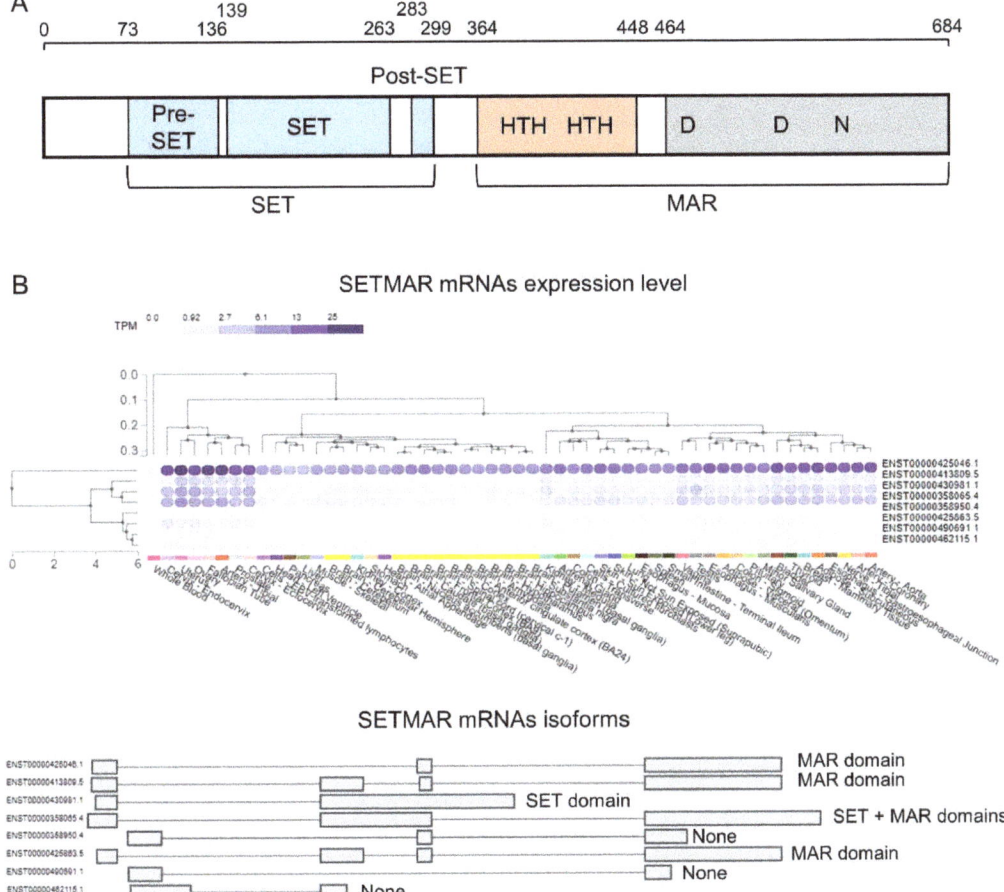

Figure 1. Protein organization and expression level of SETMAR in human. (**A**) SETMAR is composed of two domains, the protein methyltransferase domain (SET) that includes the pre-SET, the SET- and the post-SET subdomains, and the DNA transposase domain (MAR) that contains the DNA binding (orange) and the catalytic (gray) subdomains. (**B**) *SETMAR* encodes for eight different isoforms. One isoform encodes for the full-length SETMAR, one isoform includes only the SET domain, and three isoforms contain only the DNA transposase domain. Out of the eight mRNA isoforms, only four are expressed at a high level in human tissues (from GTEx).

2.3. Structure

The complete structure of SETMAR has not been determined yet. Currently, only structures of the SET domain and of the DNA transposase catalytic subdomain have been resolved (Figure 2A,B) [38,42]. In agreement with other eukaryotic DNA transposases, such as Mos1 [43], the MAR domain of SETMAR forms a homodimer (Figure 2B). A potential model of a SETMAR full-length isoform is provided by AlphaFold (Figure 2C) [44]. However, the model represents only a monomer while SETMAR is known to form a dimer via the DNA transposase domain. The structures of full-length SETMAR homodimer and of combination between the different isoforms (full-length/SET-depleted isoform) remain to be obtained to understand better how SETMAR methyltransferase activity is regulated.

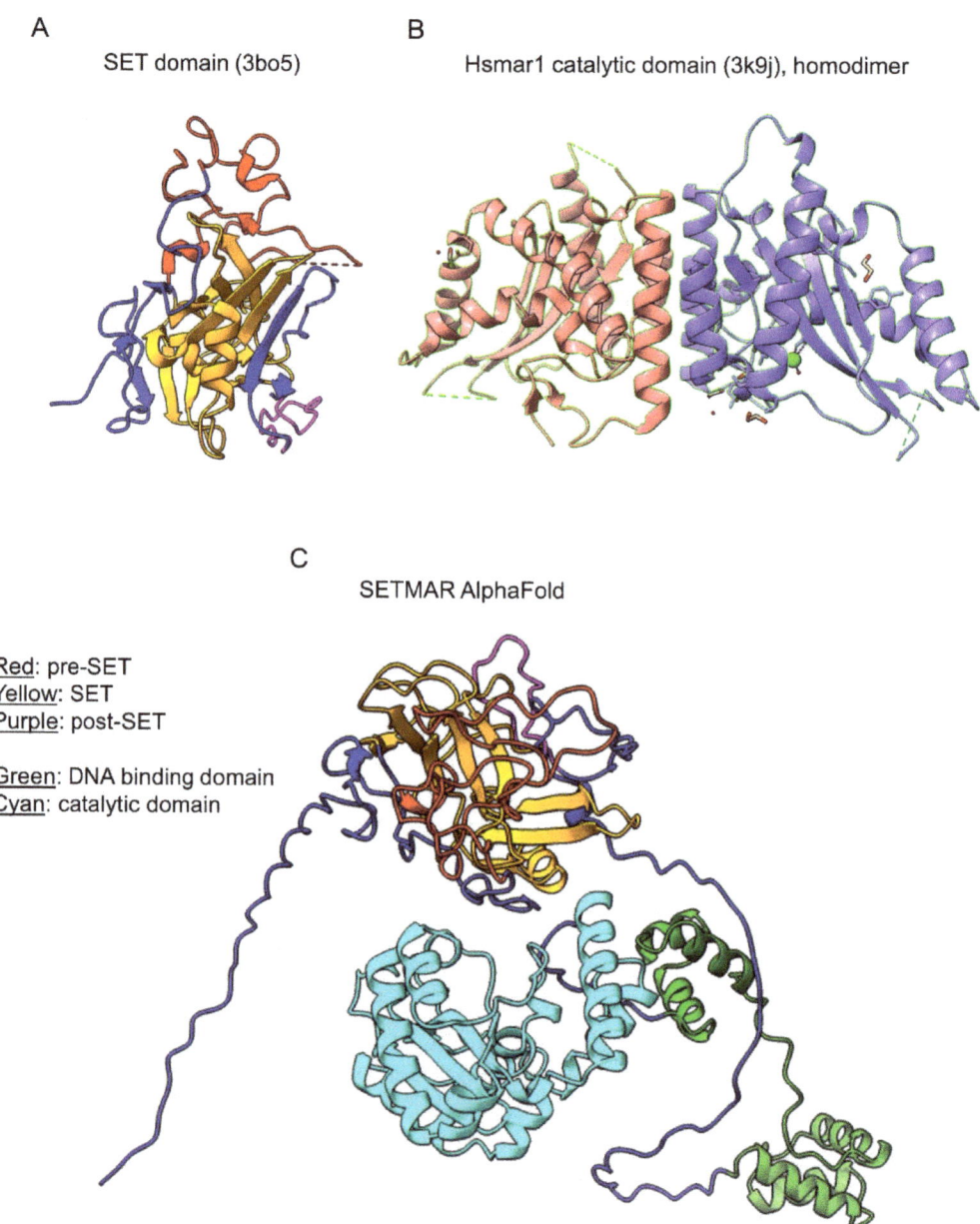

Figure 2. Crystal structures of SETMAR. (**A**) Structure of the SET domain of SETMAR with the pre-SET, SET, and post-SET subdomains shown in red, yellow, and purple, respectively. (**B**) Structure of the catalytic subdomain of the DNA transposase domain of SETMAR, shown as a homodimer. Each monomer is represented by a different color. (**B,C**) Proposed structure of a SETMAR monomer by AlphaFold with the pre-SET, SET, and post-SET subdomains shown in red, yellow, and purple, respectively, and the DNA binding and catalytic subdomains shown in green and cyan, respectively.

3. Biological Roles of SETMAR

3.1. Substrates

In addition to its own automethylation [15,45], SETMAR has been found to methylate snRNP70, a pre-mRNA splicing factor, and SPTBN2, a spectrin [45], and to catalyze H3K4me2 [12], H3K27me3 [30], and H3K36me2 [12,19,21,28].

The H3K36me2 histone methyltransferase activity of SETMAR has been the most intensively investigated. The initial study found that SETMAR could dimethylate H3K36, and H3K4 to a lesser extent, in vitro [12]. In a later study investigating the link between H3K36me2 and NHEJ in HT1904 cells, the authors have shown that overexpression or knockdown of SETMAR is associated with an increased or decreased H3K36me2 level by ChIP around DNA double strand breaks (DSBs), respectively [19]. Furthermore, overexpression of a methyltransferase-deficient SETMAR, D261S, compared to WT SETMAR fails to induce H3K36me2 around DSBs [19]. However, another study could not replicate SETMAR dimethylation of H3K36 in vitro by LC-MS/MS proteomics and Western blot [45]. Carlson et al. found that SETMAR could only weakly methylate histone H3 in vitro, potentially on H3K115, a residue located in the DNA-histone dyad interface [45,46]. A study in U2OS cells revealed that while overexpression of WT SETMAR did not increase H3K36me2 level, overexpression of a methyltransferase-deficient SETMAR, N223A, was associated with a decreased H3K36me2 level by Western blot and ChIP [21]. Another study in glioblastoma cell lines reported that knockdown of SETMAR by shRNA or siRNA is associated with a decreased H3K36me2 level [28]. A recent study in bladder cancer found that SETMAR could also mediate H3K27me3 [30], a mark associated with heterochromatin and gene repression.

As non-histone targets, SETMAR was found to perform automethylation on two different residues, lysine 335 and lysine 498 [15,45]. However, each residue has been found to be methylated in only one study. Lysine 335 methylation did not affect SETMAR methyltransferase activity but might be involved in the regulation of protein-protein interactions [45]. In contrast, methylation of lysine 498 was shown to inhibit SETMAR activity in chromosome decatenation [15]. Carlson et al. found that SETMAR could methylate two other non-histone targets, the splicing factor snRNP70 on the lysine 170 and SPTBN2, a spectrin (residue not provided) [45]. However, it remains unknown whether the SETMAR mediated methylation affects the activities of these proteins.

3.2. Regulation

The expression of SETMAR has been found to be regulated by SOX11 [23], a transcription factor involved in development, including neurogenesis and skeletogenesis, and disease, such as neurodevelopmental disorders, osteoarthiritis, and cancers [47]. The Notch signaling pathway, involved in developmental and homeostatic processes [48], also regulates SETMAR expression as knockdown of each of the four Notch receptors decreases SETMAR expression in colon cancer stem cells [24].

SETMAR activities have been proposed to be regulated by different mechanisms including automethylation, phopshorylation, the presence of an autoinhibitory loop from the post-SET domain, and by alternative splicing. Automethylation of SETMAR has been observed on two residues, lysine 335 and lysine 498 [15,45]. While only lysine 498 was found to regulate SETMAR activity in DNA decatenation, the mechanism behind this automethylation regulation remains unclear. Following DNA damage, SETMAR is phosphorylated on its Ser508 residue, located in the catalytic subdomain of the transposase domain, by Chk1 and is dephosphorylated by PP2A [49]. Overexpression of a Ser508 to alanine mutant compared to overexpression of wild-type SETMAR results in a decreased association of SETMAR to DSBs, a reduced DSB repair in vivo, and a higher nuclease activity in vitro [49]. Of note, phosphorylation of the equivalent residue in another transposase, Ser170 in Mos1, is performed by the cAMP-dependent protein kinase (PKA) and prevents the active transport of the transposase to the nucleus and also interferes with the formation of the paired-end complex (a transposase dimer bound to two ITRs) [50]. SETMAR binding to ITRs might therefore be also regulated by phosphorylation on its Ser508 residue.

Another regulation of SETMAR histone methyltransferase activity is provided by the loop connecting the SET and post-SET domains that can take an autoinhibitory conformation that will interfere with the accession of the histone tail substrate. The presence of an autoinhibitory loop is not specific to SETMAR as it has also been found in other H3K36 methyltransferases, such as NSD1, SETD2, or ASH1L [51–54]. Interaction with a nucleosome is thought to stabilize an active conformation of the post-SET loop, promoting methylation of H3K36 [51]. However, it remains to be determined whether this autoinhibitory loop also regulates the methylation of non-histone targets.

SETMAR is also regulated by alternative splicing, which can produce methyltransferase deficient isoforms without a complete SET domain (Figure 1B). Two splicing factors were found to regulate SETMAR alternative splicing, NONO and SFPQ, with a knockdown of NONO promoting the production of the methyltransferase deficient SETMAR isoform over the full-length isoform [30]. In addition, the use of an alternative TSS, which adds 13 amino acid, has been associated with an increased protein stability [27].

3.3. Sequence Specificity

The sequence specificity of SETMAR methyltransferase activity remains unclear due to the low number of substrates that have been found. However, based on the methylation proteomics performed by Carlson et al., it has been proposed that SETMAR could recognize the following consensus sequence on proteins: KR(I/L) [45].

3.4. Connection to Cell Signaling Pathways

Excluding the regulation of SETMAR expression by the Notch signaling pathway in colon cancer stem cells [24], SETMAR has not been directly connected to any other specific cell signaling pathway. However, SETMAR is phosphorylated on the residue Ser508 by Chk1, a serine/threonine kinase that coordinates DNA damage response and cell cycle checkpoint response [55]. Phosphorylation of the Ser508 residue is associated with a better recruitment of SETMAR to DSBs and an increased DNA repair [49]. In turn, phosphorylated SETMAR increases Chk1 stability as it interferes with the interaction between DDB1 and Chk1, which is required for ubiquitination of Chk1 and its degradation via the proteasome [56]. In addition, overexpression of the wild-type full-length SETMAR in the U2OS cell line resulted in transcriptional changes that are enriched for five signaling pathways: Rap1, PI3K-Akt, calcium, cAMP, and Hippo [21]. Further work is required but SETMAR might be involved in the regulation of some cellular signaling pathways.

3.5. Connection to Chromatin Regulation

Dimethylation of H3K36 is associated with several chromatin processes, including regulation of gene expression through controlling the distribution of H3K27me3 and DNA methylation [57], and DNA damage repair [6]. SETMAR has been associated with both regulation of gene expression and repair of DSB by the NHEJ pathway [19,21]. A recent paper has associated expression of the full-length SETMAR, which contains the active SET domain, with an increased H3K27me3 level [30]. However, it remains unclear whether SETMAR could trimethylate H3K27 or the change in H3K27me3 level is a consequence of transcriptional change due to the expression of the full-length SETMAR, as H3K36me2/me3 and H3K27me3 are known to be anti-correlated [58]. In addition, H3K36me2 has been shown to recruit the DNA methyltransferase DNMT3A to maintain DNA methylation at non-coding regions of euchromatin [57]. It will, therefore, be of interest to investigate whether SETMAR mediated H3K36me2 could also regulate DNA methylation maintenance, especially around Hsmar1 ITRs bound by SETMAR.

3.6. Cellular Roles and Function

In mouse, where SETMAR contains only the SET domain, homozygous deletion is associated with several phenotypic defects in vision/eye, behavior/neurological phenotype, metabolism, pigmentation, skeleton phenotype, and immune system, indicating a developmental role for the SET domain [59]. In humans, SETMAR is involved in several cellular functions, including DNA damage repair via the NHEJ, DNA integration, cell cycle and DNA replication, and the regulation of gene expression [12–21]. SETMAR activities are regulated via protein-protein interactions, such as the splicing factors PRPF19 and SPF27 [37], the DNA damage repair factors PRPF19, DNA ligase IV, and XRCC4 [13], and the DNA replication factors topoisomerase 2α (TOP2A), PCNA, and RAD9 [17].

The NHEJ pathway is one of the four pathways used by a cell to resolve DSBs but, in comparison to homologous recombination, can result in mutations and insertions-deletions [60]. Illegitimate DNA integration, such as genomic insertion of a plasmid, is also mediated by the NHEJ [61]. SETMAR has been proposed to act in most steps of NHEJ (Figure 3). Following DNA damage, SETMAR is phosphorylated on Ser508 by Chk1 [49] and recruited by PRPF19 to DSBs [37] where it dimethylates H3K36 on the surrounding nucleosomes [19]. Increased level of H3K36me2 helps to recruit and stabilize Ku70 and NBS1 to the DNA free ends, promoting DNA repair [19]. In addition to the dimethylation of H3K36, the catalytic subdomain of the transposase domain of SETMAR has also been proposed to act as an endonuclease, such as Artemis, to trim DNA overhangs [39,40,62]. Processing of non-compatible ends is required before ligation by the DNA ligase IV and XRCC4 complex can happen. As SETMAR can interact with DNA ligase IV and XRCC4 [13], it is thought that SETMAR could also help to recruit the ligation complex. A recent work with an in vivo system to follow NHEJ repair and cell lines overexpressing either SETMAR or its domains alone found that, while each SETMAR domain has an effect on NHEJ, no change was observed with the full-length SETMAR [20]. More works with better in vivo approaches are still required to properly understand the roles of SETMAR in NHEJ (see Part 5).

SETMAR has also been proposed to act in DNA replication and cell cycle regulation (Figure 4A). Knockdown, knockout, or overexpression of SETMAR has more or less of an effect on cell cycle depending on the cell type, with generally a higher SETMAR expression correlating with an increased growth rate [17,20,22,45,63]. While phosphorylation of SETMAR on the Ser508 residue by Chk1 is associated with DNA damage repair via NHEJ, the unphosphorylated form of SETMAR is involved in the response to replication stress [49]. The involvement of SETMAR in replication fork restart has been shown via interactions between SETMAR and PCNA and RAD9, and by a higher sensitivity of cells, following knockdown of SETMAR, to hydroxyurea (HU), a treatment that decreases the production of nucleotides and therefore induces replication stress [17,41,64,65]. In contrast, a recent CRISPR/Cas9 screening against DNA-damaging agents did not find SETMAR as a gene increasing the sensitivity of cells to HU [66]. The endonuclease activity of SETMAR is thought to be involved in the cleavage of branched DNA structures resulting from stress replication forks, producing DSBs that can be resolved by DNA damage repair pathways and thus allowing the restart of replication forks [17,41,64,65]. A recent work has proposed that SETMAR role in the response to replication stress was mediated by the dimethylation of H3K36 at stalled replication forks, facilitating the recruitment of DNA repair factors, rather than a cleavage of stalled forks via SETMAR endonuclease activity [63]. It has been recently suggested that SETMAR could bind to 12 bp motifs that are enriched in the regions of replication origins but more work is required to determine whether SETMAR binding is occurring simultaneously to DNA replication and whether SETMAR binding plays a role in DNA replication [35]. In addition, SETMAR interaction with TOP2A has been found to enhance TOP2A function in chromosome decatenation and to promote resistance to topoisomerase II inhibitors in breast cancer cells (Figure 4B) [15,16]. SETMAR mediated enhancement of TOP2A function in chromosome decatenation is negatively regulated by the automethylation of SETMAR on its residue lysine 498 [15].

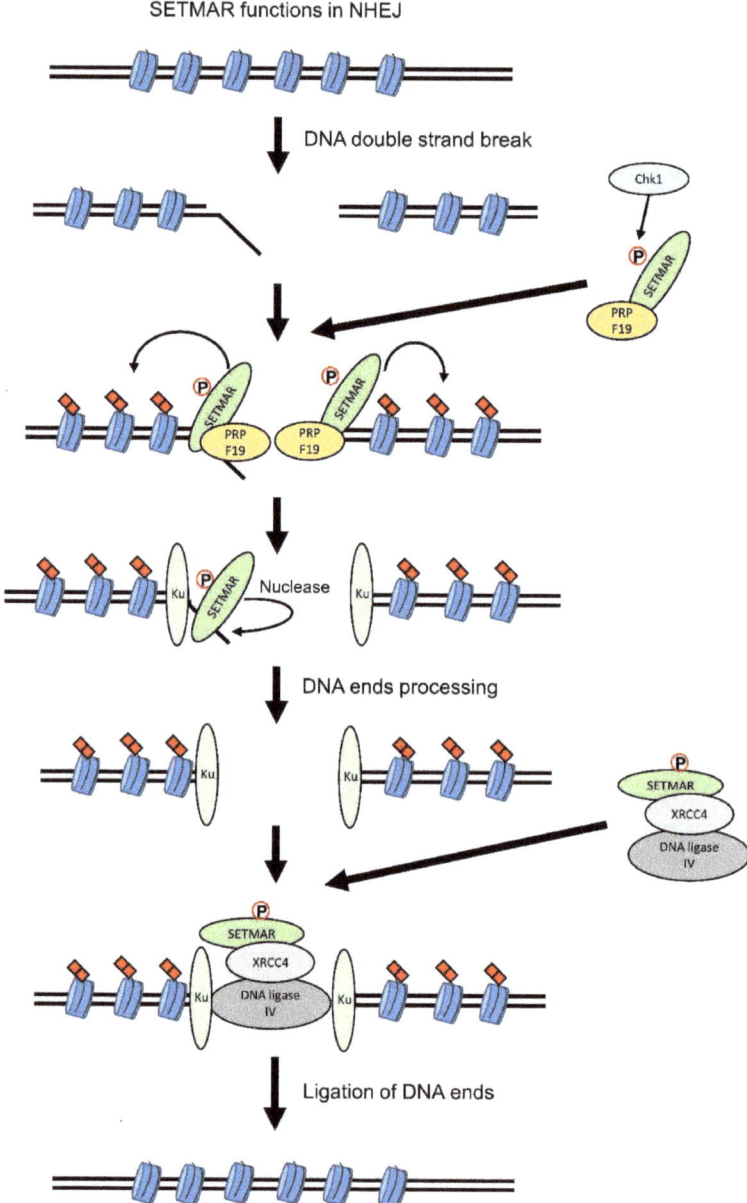

Figure 3. SETMAR functions in non-homologous end joining (NHEJ). SETMAR has been found to act on several aspects of the NHEJ pathway. Following DNA double-strand breaks (DSBs), SETMAR is phosphorylated by Chk1 on its Ser508 residue and interacts with PRPF19. The SETMAR-PRPF19 complex binds to the DSB site and SETMAR dimethylates H3K36 on the neighboring nucleosomes. H3K36me2 stabilizes the association of Ku70 and NBS1 to DSB sites. SETMAR is also involved in the processing of the DNA ends via its endonuclease activity, similarly to Artemis. Following DNA ends processing, SETMAR is also involved in the ligation of the two DNA ends as it interacts with the DNA ligase IV and XRCC4 complex. For the readability of the figure, the proteins are shown as monomer.

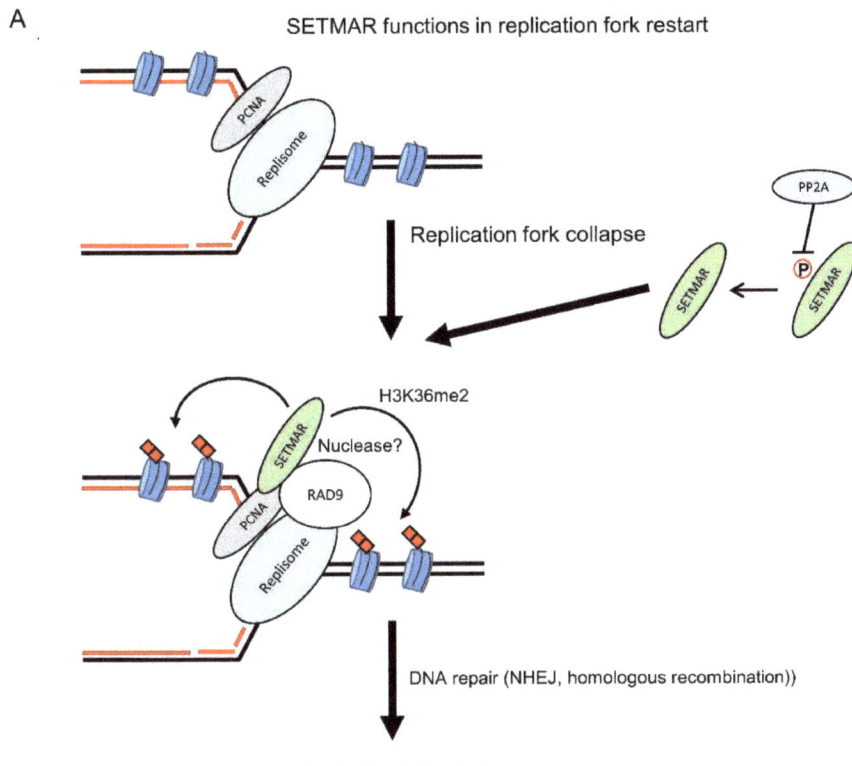

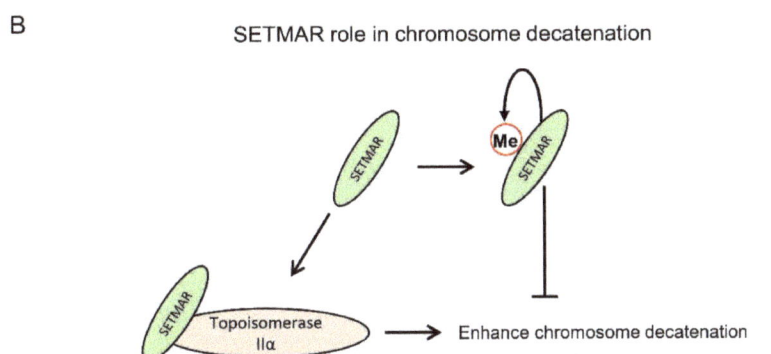

Figure 4. SETMAR functions in replication fork restart and chromosome decatenation. (**A**) Following replication fork collapse, the unphosphorylated form of SETMAR, mediated by the protein phosphatase PP2A, can interact with PCNA and RAD9 to promote replication fork restart via the dimethylation of H3K36, which helps recruiting DNA repair factors, and its endonuclease activity. (**B**) SETMAR can also enhance chromosome decatenation via its interaction with TOP2A. SETMAR activity in chromosome decatenation is regulated through its automethylation on the lysine 498 residue. For the readability of the figure, the proteins are shown as monomer.

Cordaux et al. proposed in one of the first papers published on SETMAR that the remnants of the Hsmar1 transposable element, including SETMAR and the Hsmar1 binding sites scattered across the genome, could represent a case of a new gene regulatory network with SETMAR binding Hsmar1 ITRs to regulate gene expression via histone methylation [9,67]. This hypothesis was confirmed later with three papers showing that SETMAR binds Hsmar1 ITRs in vivo [21,35,36], and that the overexpression of full-length SETMAR in U2OS cells upregulated the expression of 960 genes, that are enriched with genes containing Hsmar1 binding sites, while overexpression of a methyltransferase dead SETMAR hardly affected gene expression [21]. However, the mechanism behind SETMAR function in gene expression remains unclear as no consistent changes in H3K36me2 level was observed across upregulated genes [21] and SETMAR does not generally bind to promoter regions [21,35,36]. Regulation of gene expression could be mediated via binding of SETMAR to enhancers (Figure 5A), which are non-coding RNA genes regulating promoter activity via enhancer-promoter loops, and/or via regulation of the 3D genome organization (Figure 5B), as SETMAR could bind two Hsmar1 remnants located apart on the same chromosome or on different chromosomes. It has been recently shown that some enhancers contain SETMAR binding sites [35] but more work is required to determine whether SETMAR binding could affect the activity of these enhancers and the expression of their associated protein-coding genes.

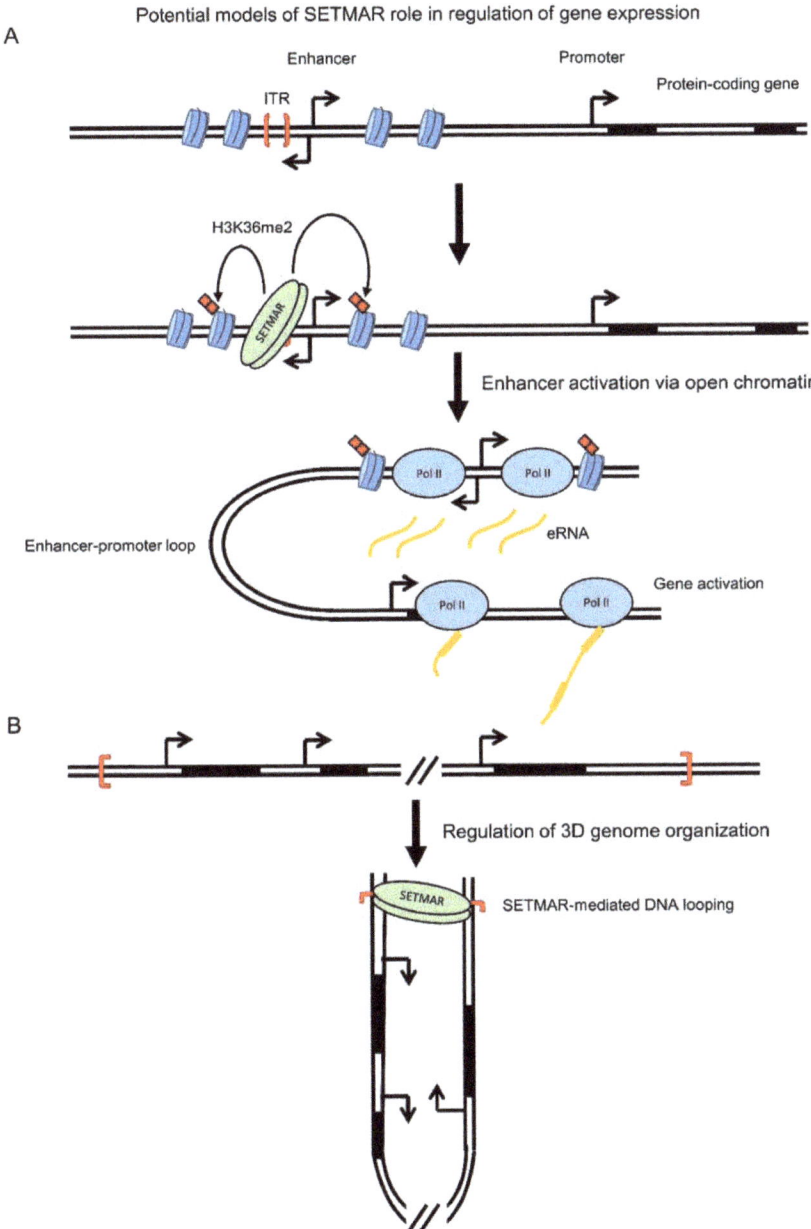

Figure 5. Potential models of SETMAR role in regulation of gene expression. (**A**) SETMAR could regulate gene expression via its binding to enhancer regions. Following SETMAR binding, dimethylation of H3K36 of the neighboring nucleosomes can create an open chromatin environment, via a decrease in H3K27me3, allowing the binding of transcription factors. Activation of the enhancer will then promote transcription of its associated protein-coding gene. (**B**) SETMAR could also regulate gene expression via an involvement in the 3D genome. Binding of SETMAR on two single ITRs can mediate the formation of a DNA loop that could regulate the expression of the genes present in the loop. Additionally, SETMAR could participate in 3D genome organization via the ITRs located near cohesin-CTCF-anchored loops.

4. Connection to Diseases

While some mutations have been found in *SETMAR* in different cancers (Figure 6A), no recurrent mutations have been observed. A recent study on 100 colon cancer samples with high microsatellite instability found only one sample with a single frameshift mutation, c.1409delA, in *SETMAR* [29]. Of note is the C226S mutation (Figure 6A) that has been found in five ovarian cancer samples as the mutation is located within the conserved NHSC motif of the SET domain. In cell lines, the mutation N223A, located in the same motif, decreases H3K36me2 level and affects gene expression and DNA repair [12,20,21], which might be mimicked by the C226S mutation.

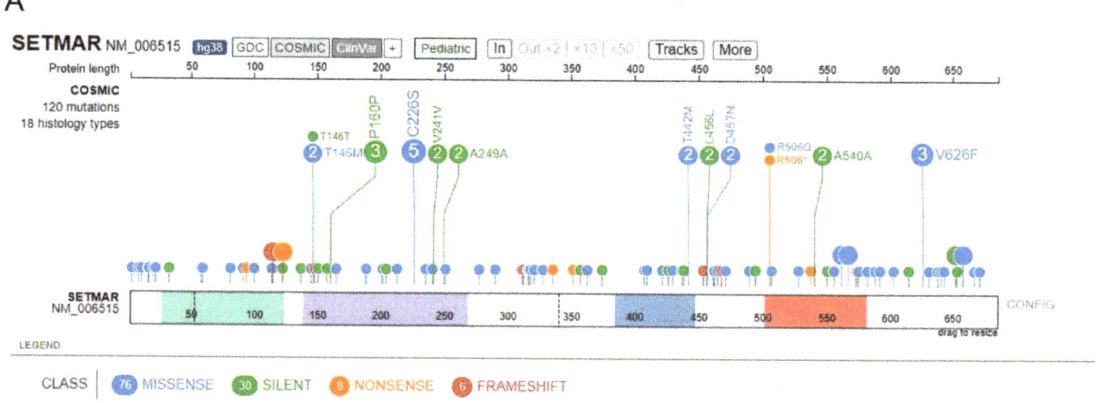

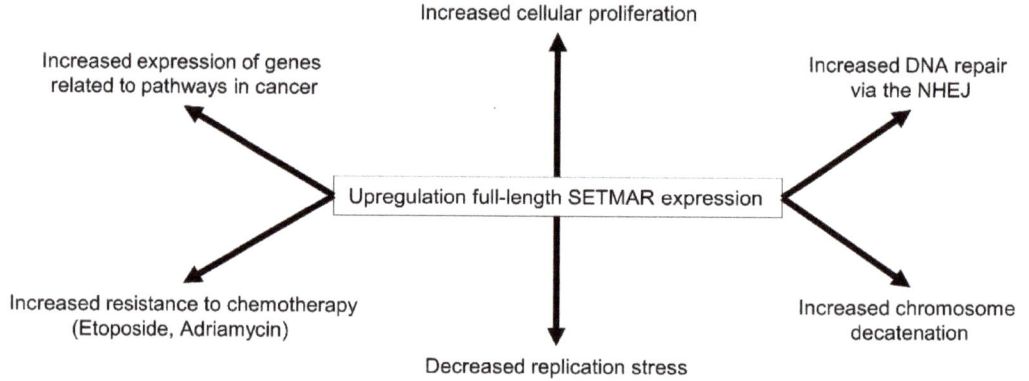

Figure 6. SETMAR roles in disease. (**A**) Visualization of SETMAR known mutations in cancer using ProteinPaint [68]. SETMAR major domains are shown in color at the bottom. The amino acid mutated is shown next to the circle while the number of observations of each mutation is shown within the circle. (**B**) Upregulation of SETMAR expression in cancer is associated with several phenotypes that can promote the tumor's growth and resistance to treatment.

Dysregulation of SETMAR expression has been associated with several cancers, including glioblastoma, leukemia, hematologic neoplasms, breast and colon cancer, and mantle cell lymphoma (Figure 6B) [16,22–30]. Overexpression of the full-length wild-type SETMAR in the osteosarcoma U2OS cell line is associated with broad changes in the transcriptome that are enriched in pathways connected to cancer, such as angiogenesis or response to hypoxia, and signaling pathways linked to cellular proliferation, such as the PI3K-AKT and the Hippo pathways [21,69]. In bladder cancer cells, expression of the full-length SETMAR has been associated with an inhibition of lymph node metastasis via an increase in H3K27me3 at the promoters of metastatic oncogenes, inhibiting their transcription [30]. In addition, SETMAR, via its role in the NHEJ pathway, is important in the survival of glioblastoma cancer cells to radiation therapy, which can cause cancer relapse [28].

On the treatment side, knockdown of SETMAR increases the sensitivity of leukemia cell lines to etoposide [22] and of breast cancer cell lines to the anthracycline Adriamycin [16]. Ciprofloxacin, a Quinolone drug acting on bacterial DNA gyrase, inhibits the NHEJ activity of the transposase domain of SETMAR [70]. Combination of ciprofloxacin and cisplatin shows a higher efficacy against the A549 cancer cell line, both in tissue culture and in a mouse A549 xenograft model, compared to ciprofloxacin or cisplatin alone [70]. However, the cytotoxic activity of ciprofloxacin against cancerous cells requires a high, non-pharmacological concentration, which currently restricts its use as a potential cancer treatment [71–73].

5. Directions for Future Research

After ~15 years of investigation on SETMAR functions, many questions remain unanswered. While specific antibodies against SETMAR are available [27,74], a major limitation remains the current lack of tools to study SETMAR in vivo. Current in vivo works have used knockdown, knockout, and overexpression approaches, which limit the interpretation of the primary functions of SETMAR as secondary/indirect effects are present. It will, therefore, be important in future studies to use, for example, endogenous targeted degradation approaches, such as auxin-inducible degradation [75] or the dTAG system [76], to obtain a quick and specific degradation of SETMAR. In addition, these targeted degradation systems could also help to investigate the functions of SETMAR isoforms that are expressed from alternative TSSs and poly(A) sites.

Current efforts in targeting SETMAR has found ciprofloxacin as an inhibitor of SETMAR transposase domain [70], but the cytotoxicity of this small molecule remains an issue [71–73]. The development of less cytotoxic small molecules targeting SETMAR DNA transposase domain is still needed before potential clinical use. Another important need will be the development of a specific inhibitor of SETMAR methyltransferase activity, which will be critical to determine the targets of SETMAR. In addition, a specific inhibitor could potentially be clinically relevant, by itself or in combination, as SETMAR is dysregulated in several cancers. Another possibility could be the development of small molecules promoting the ubiquitination and degradation of SETMAR by the proteasome, such as PROTACs or molecular glues [77].

As SETMAR is known to form a dimer and that out of the four most expressed isoforms, three are able to dimerize, a better understanding of the isoforms combination that are present in SETMAR dimers is required. Another significant question is why two out of the four expressed isoforms encodes only for the DNA transposase domain? With the exception of the viral vSET [78], human histone methyltransferases exists usually as monomers [79]. However, the DNA transposase domain of SETMAR enforces the formation of a dimer, which could result in the presence of two SET domains, potentially affecting the methyltransferase activity because of steric clashes. Interestingly, the phylogenetic analysis of a SETMAR isoform encoding a deleted SET domain and an active DNA binding domain (ENST00000413809.5) shows that this isoform is specific to anthropoid primates that contains the domesticated DNA transposase domain [20]. Therefore, the presence of multiple SETMAR isoforms encoding for a deleted SET domain could represent a way to decrease the possibility of a SETMAR dimer to contain two SET domains. However, a global phylogenetic analysis of the different SETMAR isoforms across mammals is required to determine whether all the SET deleted isoforms are specific to the species that contain the domesticated DNA Hsmar1 transposase. In addition, it will be important to determine which SETMAR dimers are active, i.e., SETMAR dimers with two SET domains, with only one SET domain, or without the SET domain.

Another question is whether SETMAR could be involved in pre-mRNA splicing and/or alternative splicing. In addition to methylating snRNP70, which is part of the U1 snRNP that recognizes the 5' splice site [45], SETMAR also interacts with two pre-mRNA splicing factors, PRPF19 and SPF27, that are part of the NineTeen Complex (NPC) [37]. While overexpression of wild-type SETMAR or of a methyltransferase-deficient SETMAR does not affect the exon inclusion/exclusion near SETMAR binding sites [21], it remains to be determined if the methylation of snRNP70 or the modulation of SETMAR expression level or methyltransferase activity affect pre-mRNA splicing.

More research is also needed to determine whether SETMAR could be involved in development. In mice, SETMAR homozygous knockout results in developmental defects [59]. In humans cells, SETMAR expression is regulated by SOX11 [23], a transcription factor involved in neurogenesis and skeletogenesis [47], while overexpression of SETMAR in U2OS cells results in transcriptional changes in numerous genes involved in development, especially brain development [21].

Funding: This research received no external funding.

Institutional Review Board Statement: Not applicable.

Informed Consent Statement: Not applicable.

Data Availability Statement: Not applicable.

Acknowledgments: The Genotype-Tissue Expression (GTEx) Project was supported by the Common Fund of the Office of the Director of the National Institutes of Health, and by NCI, NHGRI, NHLBI, NIDA, NIMH, and NINDS. The data used for the analyses described in this manuscript were obtained from the GTEx Portal on 15/10/21.

Conflicts of Interest: The authors declare no conflict of interest.

References

1. Strahl, B.D.; Allis, C.D. The language of covalent histone modifications. *Nature* **2000**, *403*, 41–45. [CrossRef]
2. Vignali, M.; Hassan, A.H.; Neely, K.E.; Workman, J.L. ATP-dependent chromatin-remodeling complexes. *Mol. Cell Biol.* **2000**, *20*, 1899–1910. [CrossRef] [PubMed]
3. Jenuwein, T.; Allis, C.D. Translating the histone code. *Science* **2001**, *293*, 1074–1080. [CrossRef]
4. Kouzarides, T. Chromatin modifications and their function. *Cell* **2007**, *128*, 693–705. [CrossRef]
5. Chi, P.; Allis, C.D.; Wang, G.G. Covalent histone modifications-miswritten, misinterpreted and mis-erased in human cancers. *Nat. Rev. Cancer* **2010**, *10*, 457–469. [CrossRef] [PubMed]
6. Li, J.; Ahn, J.H.; Wang, G.G. Understanding histone H3 lysine 36 methylation and its deregulation in disease. *Cell Mol. Life Sci.* **2019**, *76*, 2899–2916. [CrossRef]
7. Wagner, E.J.; Carpenter, P.B. Understanding the language of Lys36 methylation at histone H3. *Nat. Rev. Mol. Cell Biol.* **2012**, *13*, 115–126. [CrossRef]
8. Lander, E.S.; Linton, L.M.; Birren, B.; Nusbaum, C.; Zody, M.C.; Baldwin, J.; Devon, K.; Dewar, K.; Doyle, M.; FitzHugh, W.; et al. Initial sequencing and analysis of the human genome. *Nature* **2001**, *409*, 860–921. [CrossRef] [PubMed]
9. Cordaux, R.; Udit, S.; Batzer, M.A.; Feschotte, C. Birth of a chimeric primate gene by capture of the transposase gene from a mobile element. *Proc. Natl. Acad. Sci. USA* **2006**, *103*, 8101–8106. [CrossRef] [PubMed]
10. Liu, D.; Bischerour, J.; Siddique, A.; Buisine, N.; Bigot, Y.; Chalmers, R. The human SETMAR protein preserves most of the activities of the ancestral Hsmar1 transposase. *Mol. Cell Biol.* **2007**, *27*, 1125–1132. [CrossRef]
11. Miskey, C.; Papp, B.; Mates, L.; Sinzelle, L.; Keller, H.; Izsvak, Z.; Ivics, Z. The ancient mariner sails again: Transposition of the human Hsmar1 element by a reconstructed transposase and activities of the SETMAR protein on transposon ends. *Mol. Cell Biol.* **2007**, *27*, 4589–4600. [CrossRef] [PubMed]
12. Lee, S.H.; Oshige, M.; Durant, S.T.; Rasila, K.K.; Williamson, E.A.; Ramsey, H.; Kwan, L.; Nickoloff, J.A.; Hromas, R. The SET domain protein Metnase mediates foreign DNA integration and links integration to nonhomologous end-joining repair. *Proc. Natl. Acad. Sci. USA* **2005**, *102*, 18075–18080. [CrossRef]

13. Hromas, R.; Wray, J.; Lee, S.H.; Martinez, L.; Farrington, J.; Corwin, L.K.; Ramsey, H.; Nickoloff, J.A.; Williamson, E.A. The human set and transposase domain protein Metnase interacts with DNA Ligase IV and enhances the efficiency and accuracy of non-homologous end-joining. *DNA Repair* **2008**, *7*, 1927–1937. [CrossRef]
14. Williamson, E.A.; Farrington, J.; Martinez, L.; Ness, S.; O'Rourke, J.; Lee, S.H.; Nickoloff, J.; Hromas, R. Expression levels of the human DNA repair protein metnase influence lentiviral genomic integration. *Biochimie* **2008**, *90*, 1422–1426. [CrossRef]
15. Williamson, E.A.; Rasila, K.K.; Corwin, L.K.; Wray, J.; Beck, B.D.; Severns, V.; Mobarak, C.; Lee, S.H.; Nickoloff, J.A.; Hromas, R. The SET and transposase domain protein Metnase enhances chromosome decatenation: Regulation by automethylation. *Nucleic Acids Res.* **2008**, *36*, 5822–5831. [CrossRef] [PubMed]
16. Wray, J.; Williamson, E.A.; Royce, M.; Shaheen, M.; Beck, B.D.; Lee, S.H.; Nickoloff, J.A.; Hromas, R. Metnase mediates resistance to topoisomerase II inhibitors in breast cancer cells. *PLoS ONE* **2009**, *4*, e5323. [CrossRef]
17. De Haro, L.P.; Wray, J.; Williamson, E.A.; Durant, S.T.; Corwin, L.; Gentry, A.C.; Osheroff, N.; Lee, S.H.; Hromas, R.; Nickoloff, J.A. Metnase promotes restart and repair of stalled and collapsed replication forks. *Nucleic Acids Res.* **2010**, *38*, 5681–5691. [CrossRef]
18. Wray, J.; Williamson, E.A.; Chester, S.; Farrington, J.; Sterk, R.; Weinstock, D.M.; Jasin, M.; Lee, S.H.; Nickoloff, J.A.; Hromas, R. The transposase domain protein Metnase/SETMAR suppresses chromosomal translocations. *Cancer Genet. Cytogenet.* **2010**, *200*, 184–190. [CrossRef] [PubMed]
19. Fnu, S.; Williamson, E.A.; De Haro, L.P.; Brenneman, M.; Wray, J.; Shaheen, M.; Radhakrishnan, K.; Lee, S.H.; Nickoloff, J.A.; Hromas, R. Methylation of histone H3 lysine 36 enhances DNA repair by nonhomologous end-joining. *Proc. Natl. Acad. Sci. USA* **2011**, *108*, 540–545. [CrossRef]
20. Tellier, M.; Chalmers, R. The roles of the human SETMAR (Metnase) protein in illegitimate DNA recombination and non-homologous end joining repair. *DNA Repair* **2019**, *80*, 26–35. [CrossRef]
21. Tellier, M.; Chalmers, R. Human SETMAR is a DNA sequence-specific histone-methylase with a broad effect on the transcriptome. *Nucleic Acids Res.* **2019**, *47*, 122–133. [CrossRef] [PubMed]
22. Wray, J.; Williamson, E.A.; Sheema, S.; Lee, S.H.; Libby, E.; Willman, C.L.; Nickoloff, J.A.; Hromas, R. Metnase mediates chromosome decatenation in acute leukemia cells. *Blood* **2009**, *114*, 1852–1858. [CrossRef]
23. Wang, X.; Bjorklund, S.; Wasik, A.M.; Grandien, A.; Andersson, P.; Kimby, E.; Dahlman-Wright, K.; Zhao, C.; Christensson, B.; Sander, B. Gene expression profiling and chromatin immunoprecipitation identify DBN1, SETMAR and HIG2 as direct targets of SOX11 in mantle cell lymphoma. *PLoS ONE* **2010**, *5*, e14085. [CrossRef] [PubMed]
24. Apostolou, P.; Toloudi, M.; Ioannou, E.; Kourtidou, E.; Chatziioannou, M.; Kopic, A.; Komiotis, D.; Kiritsis, C.; Manta, S.; Papasotiriou, I. Study of the interaction among Notch pathway receptors, correlation with stemness, as well as their interaction with CD44, dipeptidyl peptidase-IV, hepatocyte growth factor receptor and the SETMAR transferase, in colon cancer stem cells. *J Recept Signal Transduct. Res.* **2013**, *33*, 353–358. [CrossRef]
25. Apostolou, P.; Toloudi, M.; Kourtidou, E.; Mimikakou, G.; Vlachou, I.; Chatziioannou, M.; Kipourou, V.; Papasotiriou, I. Potential role for the Metnase transposase fusion gene in colon cancer through the regulation of key genes. *PLoS ONE* **2014**, *9*, e109741. [CrossRef]
26. Jeyaratnam, D.C.; Baduin, B.S.; Hansen, M.C.; Hansen, M.; Jorgensen, J.M.; Aggerholm, A.; Ommen, H.B.; Hokland, P.; Nyvold, C.G. Delineation of known and new transcript variants of the SETMAR (Metnase) gene and the expression profile in hematologic neoplasms. *Exp. Hematol.* **2014**, *42*, 448–456. [CrossRef] [PubMed]
27. Dussaussois-Montagne, A.; Jaillet, J.; Babin, L.; Verrelle, P.; Karayan-Tapon, L.; Renault, S.; Rousselot-Denis, C.; Zemmoura, I.; Auge-Gouillou, C. SETMAR isoforms in glioblastoma: A matter of protein stability. *Oncotarget* **2017**, *8*, 9835–9848. [CrossRef]
28. Kaur, E.; Nair, J.; Ghorai, A.; Mishra, S.V.; Achareker, A.; Ketkar, M.; Sarkar, D.; Salunkhe, S.; Rajendra, J.; Gardi, N.; et al. Inhibition of SETMAR-H3K36me2-NHEJ repair axis in residual disease cells prevents glioblastoma recurrence. *Neuro Oncol.* **2020**, *22*, 1785–1796. [CrossRef] [PubMed]
29. Moon, S.W.; Son, H.J.; Mo, H.Y.; Choi, E.J.; Yoo, N.J.; Lee, S.H. Mutation and expression alterations of histone methylation-related NSD2, KDM2B and SETMAR genes in colon cancers. *Pathol. Res. Pract.* **2021**, *219*, 153354. [CrossRef] [PubMed]
30. Xie, R.; Chen, X.; Cheng, L.; Huang, M.; Zhou, Q.; Zhang, J.; Chen, Y.; Peng, S.; Chen, Z.; Dong, W.; et al. NONO Inhibits Lymphatic Metastasis of Bladder Cancer via Alternative Splicing of SETMAR. *Mol. Ther.* **2021**, *29*, 291–307. [CrossRef] [PubMed]
31. Jones, R.S.; Gelbart, W.M. The Drosophila Polycomb-group gene Enhancer of zeste contains a region with sequence similarity to trithorax. *Mol. Cell Biol.* **1993**, *13*, 6357–6366. [CrossRef]
32. Tschiersch, B.; Hofmann, A.; Krauss, V.; Dorn, R.; Korge, G.; Reuter, G. The protein encoded by the Drosophila position-effect variegation suppressor gene Su(var)3-9 combines domains of antagonistic regulators of homeotic gene complexes. *EMBO J.* **1994**, *13*, 3822–3831. [CrossRef]
33. Stassen, M.J.; Bailey, D.; Nelson, S.; Chinwalla, V.; Harte, P.J. The Drosophila trithorax proteins contain a novel variant of the nuclear receptor type DNA binding domain and an ancient conserved motif found in other chromosomal proteins. *Mech. Dev.* **1995**, *52*, 209–223. [CrossRef]
34. Tellier, M.; Chalmers, R. Compensating for over-production inhibition of the Hsmar1 transposon in Escherichia coli using a series of constitutive promoters. *Mob. DNA* **2020**, *11*, 5. [CrossRef]
35. Antoine-Lorquin, A.; Arensburger, P.; Arnaoty, A.; Asgari, S.; Batailler, M.; Beauclair, L.; Belleannee, C.; Buisine, N.; Coustham, V.; Guyetant, S.; et al. Two repeated motifs enriched within some enhancers and origins of replication are bound by SETMAR isoforms in human colon cells. *Genomics* **2021**, *113*, 1589–1604. [CrossRef] [PubMed]

36. Miskei, M.; Horvath, A.; Viola, L.; Varga, L.; Nagy, E.; Fero, O.; Karanyi, Z.; Roszik, J.; Miskey, C.; Ivics, Z.; et al. Genome-wide mapping of binding sites of the transposase-derived SETMAR protein in the human genome. *Comput. Struct. Biotechnol. J.* **2021**, *19*, 4032–4041. [CrossRef] [PubMed]
37. Beck, B.D.; Park, S.J.; Lee, Y.J.; Roman, Y.; Hromas, R.A.; Lee, S.H. Human Pso4 is a metnase (SETMAR)-binding partner that regulates metnase function in DNA repair. *J. Biol. Chem.* **2008**, *283*, 9023–9030. [CrossRef]
38. Goodwin, K.D.; He, H.; Imasaki, T.; Lee, S.H.; Georgiadis, M.M. Crystal structure of the human Hsmar1-derived transposase domain in the DNA repair enzyme Metnase. *Biochemistry* **2010**, *49*, 5705–5713. [CrossRef]
39. Beck, B.D.; Lee, S.S.; Williamson, E.; Hromas, R.A.; Lee, S.H. Biochemical characterization of metnase's endonuclease activity and its role in NHEJ repair. *Biochemistry* **2011**, *50*, 4360–4370. [CrossRef]
40. Mohapatra, S.; Yannone, S.M.; Lee, S.H.; Hromas, R.A.; Akopiants, K.; Menon, V.; Ramsden, D.A.; Povirk, L.F. Trimming of damaged 3′ overhangs of DNA double-strand breaks by the Metnase and Artemis endonucleases. *DNA Repair* **2013**, *12*, 422–432. [CrossRef]
41. Kim, H.S.; Chen, Q.; Kim, S.K.; Nickoloff, J.A.; Hromas, R.; Georgiadis, M.M.; Lee, S.H. The DDN catalytic motif is required for Metnase functions in non-homologous end joining (NHEJ) repair and replication restart. *J. Biol. Chem.* **2014**, *289*, 10930–10938. [CrossRef]
42. Chen, Q.; Georgiadis, M. Crystallization of and selenomethionine phasing strategy for a SETMAR-DNA complex. *Acta Cryst. F Struct. Biol. Commun.* **2016**, *72*, 713–719. [CrossRef] [PubMed]
43. Richardson, J.M.; Colloms, S.D.; Finnegan, D.J.; Walkinshaw, M.D. Molecular architecture of the Mos1 paired-end complex: The structural basis of DNA transposition in a eukaryote. *Cell* **2009**, *138*, 1096–1108. [CrossRef]
44. Jumper, J.; Evans, R.; Pritzel, A.; Green, T.; Figurnov, M.; Ronneberger, O.; Tunyasuvunakool, K.; Bates, R.; Zidek, A.; Potapenko, A.; et al. Highly accurate protein structure prediction with AlphaFold. *Nature* **2021**, *596*, 583–589. [CrossRef] [PubMed]
45. Carlson, S.M.; Moore, K.E.; Sankaran, S.M.; Reynoird, N.; Elias, J.E.; Gozani, O. A Proteomic Strategy Identifies Lysine Methylation of Splicing Factor snRNP70 by the SETMAR Enzyme. *J. Biol. Chem.* **2015**, *290*, 12040–12047. [CrossRef] [PubMed]
46. Manohar, M.; Mooney, A.M.; North, J.A.; Nakkula, R.J.; Picking, J.W.; Edon, A.; Fishel, R.; Poirier, M.G.; Ottesen, J.J. Acetylation of histone H3 at the nucleosome dyad alters DNA-histone binding. *J. Biol. Chem.* **2009**, *284*, 23312–23321. [CrossRef] [PubMed]
47. Tsang, S.M.; Oliemuller, E.; Howard, B.A. Regulatory roles for SOX11 in development, stem cells and cancer. *Semin. Cancer Biol.* **2020**, *67*, 3–11. [CrossRef]
48. Bray, S.J. Notch signalling in context. *Nat. Rev. Mol. Cell Biol.* **2016**, *17*, 722–735. [CrossRef]
49. Hromas, R.; Williamson, E.A.; Fnu, S.; Lee, Y.J.; Park, S.J.; Beck, B.D.; You, J.S.; Leitao, A.; Nickoloff, J.A.; Lee, S.H. Chk1 phosphorylation of Metnase enhances DNA repair but inhibits replication fork restart. *Oncogene* **2012**, *31*, 4245–4254. [CrossRef]
50. Bouchet, N.; Jaillet, J.; Gabant, G.; Brillet, B.; Briseno-Roa, L.; Cadene, M.; Auge-Gouillou, C. cAMP protein kinase phosphorylates the Mos1 transposase and regulates its activity: Evidences from mass spectrometry and biochemical analyses. *Nucleic Acids Res.* **2014**, *42*, 1117–1128. [CrossRef] [PubMed]
51. Qiao, Q.; Li, Y.; Chen, Z.; Wang, M.; Reinberg, D.; Xu, R.M. The structure of NSD1 reveals an autoregulatory mechanism underlying histone H3K36 methylation. *J. Biol. Chem.* **2011**, *286*, 8361–8368. [CrossRef] [PubMed]
52. An, S.; Yeo, K.J.; Jeon, Y.H.; Song, J.J. Crystal structure of the human histone methyltransferase ASH1L catalytic domain and its implications for the regulatory mechanism. *J. Biol. Chem.* **2011**, *286*, 8369–8374. [CrossRef] [PubMed]
53. Zheng, W.; Ibanez, G.; Wu, H.; Blum, G.; Zeng, H.; Dong, A.; Li, F.; Hajian, T.; Allali-Hassani, A.; Amaya, M.F.; et al. Sinefungin derivatives as inhibitors and structure probes of protein lysine methyltransferase SETD *J. Am. Chem. Soc.* **2012**, *134*, 18004–18014. [CrossRef]
54. Rogawski, D.S.; Ndoj, J.; Cho, H.J.; Maillard, I.; Grembecka, J.; Cierpicki, T. Two Loops Undergoing Concerted Dynamics Regulate the Activity of the ASH1L Histone Methyltransferase. *Biochemistry* **2015**, *54*, 5401–5413. [CrossRef] [PubMed]
55. Liu, Q.; Guntuku, S.; Cui, X.S.; Matsuoka, S.; Cortez, D.; Tamai, K.; Luo, G.; Carattini-Rivera, S.; DeMayo, F.; Bradley, A.; et al. Chk1 is an essential kinase that is regulated by Atr and required for the G(2)/M DNA damage checkpoint. *Genes Dev.* **2000**, *14*, 1448–1459. [CrossRef]
56. Williamson, E.A.; Wu, Y.; Singh, S.; Byrne, M.; Wray, J.; Lee, S.H.; Nickoloff, J.A.; Hromas, R. The DNA repair component Metnase regulates Chk1 stability. *Cell Div.* **2014**, *9*, 1. [CrossRef] [PubMed]
57. Weinberg, D.N.; Papillon-Cavanagh, S.; Chen, H.; Yue, Y.; Chen, X.; Rajagopalan, K.N.; Horth, C.; McGuire, J.T.; Xu, X.; Nikbakht, H.; et al. The histone mark H3K36me2 recruits DNMT3A and shapes the intergenic DNA methylation landscape. *Nature* **2019**, *573*, 281–286. [CrossRef] [PubMed]
58. Nojima, T.; Tellier, M.; Foxwell, J.; Ribeiro de Almeida, C.; Tan-Wong, S.M.; Dhir, S.; Dujardin, G.; Dhir, A.; Murphy, S.; Proudfoot, N.J. Deregulated Expression of Mammalian lncRNA through Loss of SPT6 Induces R-Loop Formation, Replication Stress, and Cellular Senescence. *Mol. Cell* **2018**, *72*, 970–984. [CrossRef] [PubMed]
59. Dickinson, M.E.; Flenniken, A.M.; Ji, X.; Teboul, L.; Wong, M.D.; White, J.K.; Meehan, T.F.; Weninger, W.J.; Westerberg, H.; Adissu, H.; et al. High-throughput discovery of novel developmental phenotypes. *Nature* **2016**, *537*, 508–514. [CrossRef]
60. Chang, H.H.Y.; Pannunzio, N.R.; Adachi, N.; Lieber, M.R. Non-homologous DNA end joining and alternative pathways to double-strand break repair. *Nat. Rev. Mol. Cell Biol.* **2017**, *18*, 495–506. [CrossRef]
61. Wurtele, H.; Little, K.C.; Chartrand, P. Illegitimate DNA integration in mammalian cells. *Gene. Ther.* **2003**, *10*, 1791–1799. [CrossRef]

62. Rath, A.; Hromas, R.; De Benedetti, A. Fidelity of end joining in mammalian episomes and the impact of Metnase on joint processing. *BMC Mol. Biol.* **2014**, *15*, 6. [CrossRef]
63. Sharma, N.; Speed, M.C.; Allen, C.P.; Maranon, D.G.; Williamson, E.; Singh, S.; Hromas, R.; Nickoloff, J.A. Distinct roles of structure-specific endonucleases EEPD1 and Metnase in replication stress responses. *NAR Cancer* **2020**, *2*, zcaa008. [CrossRef]
64. Kim, H.S.; Kim, S.K.; Hromas, R.; Lee, S.H. The SET Domain Is Essential for Metnase Functions in Replication Restart and the 5′ End of SS-Overhang Cleavage. *PLoS ONE* **2015**, *10*, e0139418. [CrossRef] [PubMed]
65. Kim, H.S.; Williamson, E.A.; Nickoloff, J.A.; Hromas, R.A.; Lee, S.H. Metnase Mediates Loading of Exonuclease 1 onto Single Strand Overhang DNA for End Resection at Stalled Replication Forks. *J. Biol. Chem.* **2017**, *292*, 1414–1425. [CrossRef]
66. Olivieri, M.; Cho, T.; Alvarez-Quilon, A.; Li, K.; Schellenberg, M.J.; Zimmermann, M.; Hustedt, N.; Rossi, S.E.; Adam, S.; Melo, H.; et al. A Genetic Map of the Response to DNA Damage in Human Cells. *Cell* **2020**, *182*, 481–496. [CrossRef]
67. Feschotte, C. Transposable elements and the evolution of regulatory networks. *Nat. Rev. Genet.* **2008**, *9*, 397–405. [CrossRef]
68. Zhou, X.; Edmonson, M.N.; Wilkinson, M.R.; Patel, A.; Wu, G.; Liu, Y.; Li, Y.; Zhang, Z.; Rusch, M.C.; Parker, M.; et al. Exploring genomic alteration in pediatric cancer using ProteinPaint. *Nat. Genet.* **2016**, *48*, 4–6. [CrossRef] [PubMed]
69. Duronio, R.J.; Xiong, Y. Signaling pathways that control cell proliferation. *Cold Spring Harb. Perspect. Biol.* **2013**, *5*, a008904. [CrossRef]
70. Williamson, E.A.; Damiani, L.; Leitao, A.; Hu, C.; Hathaway, H.; Oprea, T.; Sklar, L.; Shaheen, M.; Bauman, J.; Wang, W.; et al. Targeting the transposase domain of the DNA repair component Metnase to enhance chemotherapy. *Cancer Res.* **2012**, *72*, 6200–6208. [CrossRef]
71. El-Rayes, B.F.; Grignon, R.; Aslam, N.; Aranha, O.; Sarkar, F.H. Ciprofloxacin inhibits cell growth and synergises the effect of etoposide in hormone resistant prostate cancer cells. *Int. J. Oncol.* **2002**, *21*, 207–211. [CrossRef]
72. Herold, C.; Ocker, M.; Ganslmayer, M.; Gerauer, H.; Hahn, E.G.; Schuppan, D. Ciprofloxacin induces apoptosis and inhibits proliferation of human colorectal carcinoma cells. *Br. J. Cancer* **2002**, *86*, 443–448. [CrossRef]
73. Aranha, O.; Grignon, R.; Fernandes, N.; McDonnell, T.J.; Wood, D.P., Jr.; Sarkar, F.H. Suppression of human prostate cancer cell growth by ciprofloxacin is associated with cell cycle arrest and apoptosis. *Int. J. Oncol.* **2003**, *22*, 787–794. [CrossRef]
74. Arnaoty, A.; Gouilleux-Gruart, V.; Casteret, S.; Pitard, B.; Bigot, Y.; Lecomte, T. Reliability of the nanopheres-DNA immunization technology to produce polyclonal antibodies directed against human neogenic proteins. *Mol. Genet. Genom.* **2013**, *288*, 347–363. [CrossRef]
75. Natsume, T.; Kiyomitsu, T.; Saga, Y.; Kanemaki, M.T. Rapid Protein Depletion in Human Cells by Auxin-Inducible Degron Tagging with Short Homology Donors. *Cell Rep.* **2016**, *15*, 210–218. [CrossRef]
76. Nabet, B.; Roberts, J.M.; Buckley, D.L.; Paulk, J.; Dastjerdi, S.; Yang, A.; Leggett, A.L.; Erb, M.A.; Lawlor, M.A.; Souza, A.; et al. The dTAG system for immediate and target-specific protein degradation. *Nat. Chem. Biol.* **2018**, *14*, 431–441. [CrossRef]
77. Alabi, S.B.; Crews, C.M. Major advances in targeted protein degradation: PROTACs, LYTACs, and MADTACs. *J. Biol. Chem.* **2021**, *296*, 100647. [CrossRef]
78. Wei, H.; Zhou, M.M. Dimerization of a viral SET protein endows its function. *Proc. Natl. Acad. Sci. USA* **2010**, *107*, 18433–18438. [CrossRef]
79. Dou, Y.; Milne, T.A.; Ruthenburg, A.J.; Lee, S.; Lee, J.W.; Verdine, G.L.; Allis, C.D.; Roeder, R.G. Regulation of MLL1 H3K4 methyltransferase activity by its core components. *Nat. Struct. Mol. Biol.* **2006**, *13*, 713–719. [CrossRef]

Review

Structure, Activity, and Function of PRMT1

Charlène Thiebaut [1,2,3], Louisane Eve [1,2,3], Coralie Poulard [1,2,3] and Muriel Le Romancer [1,2,3,*]

1 Université de Lyon, F-69000 Lyon, France; charlene.thiebaut@inserm.fr (T.C.);
louisane.eve@lyon.unicancer.fr (E.L.); coralie.poulard@lyon.unicancer.fr (P.C.)
2 Inserm U1052, Centre de Recherche en Cancérologie de Lyon, F-69000 Lyon, France
3 CNRS UMR5286, Centre de Recherche en Cancérologie de Lyon, F-69000 Lyon, France
* Correspondence: muriel.leromancer@lyon.unicancer.fr; Tel.: +33-4-78-78-28-22

Abstract: PRMT1, the major protein arginine methyltransferase in mammals, catalyzes monomethylation and asymmetric dimethylation of arginine side chains in proteins. Initially described as a regulator of chromatin dynamics through the methylation of histone H4 at arginine 3 (H4R3), numerous non-histone substrates have since been identified. The variety of these substrates underlines the essential role played by PRMT1 in a large number of biological processes such as transcriptional regulation, signal transduction or DNA repair. This review will provide an overview of the structural, biochemical and cellular features of PRMT1. After a description of the genomic organization and protein structure of PRMT1, special consideration was given to the regulation of PRMT1 enzymatic activity. Finally, we discuss the involvement of PRMT1 in embryonic development, DNA damage repair, as well as its participation in the initiation and progression of several types of cancers.

Keywords: PRMT1; arginine methylation; H4R3 methylation; transcriptional regulation; cell signaling; DNA damage repair; cancer

1. Introduction

Arginine methylation is a common and widespread post-translational modification (PTM) in eukaryotes that regulates numerous biological processes. Currently, nine protein arginine methyltransferases (PRMTs) have been described which are divided into three families according to the type of methylarginine produced. Type I PRMTs (PRMT-1, 2, 3, 4, 6 and 8) generate ω-N^G-monomethylarginine (MMA) and ω-N^G, N^G-asymmetric dimethylarginine (ADMA), Type II PRMTs (PRMT-5 and 9) generate MMA and ω-N^G, N'^G-symmetric dimethylarginine (SDMA) and finally the unique Type III PRMT, PRMT7, generates MMA. Mechanistically, all PRMTs catalyze the transfer of a methyl group from S-adenosyl methionine (AdoMet) to the guanidino nitrogen atom of arginine [1]. Though considered for a long time as a stable mark, it is now well-known that arginine methylation is a dynamic PTM that can be removed by arginine demethylases [2].

PRMT1, which is the major type I PRMT, is responsible for 85% of the activity attributed to type I PRMTs in mammals [3]. Moreover, it plays key roles in various cellular processes such as transcriptional regulation, signal transduction or DNA damage repair, owing to the diversity of its histone and non-histone substrates [1].

The aim of this review is to provide an overview of the literature concerning PRMT1 structure, activities and functions. After a detailed description of the genomic organization and the protein structures of the different PRMT1 isoforms, the substrate specificity and the regulatory mechanisms of PRMT1 itself will be discussed. Finally, the cellular roles and functions of PRMT1, as well as its involvement in cancer, will be addressed.

2. Structural Features

2.1. Genomic Organization

Human PRMT1 is encoded by the *PRMT1* gene located on chromosome 19 (19q13.3) and composed of 12 exons and 11 introns. At the 5′ end of this genomic locus of

11.3 kilobases (kb) are four alternative exons (e1a–e1d) involved in the synthesis of at least seven splice variants of PRMT1 (v1–v7) [4,5] (Figure 1A,B). More recently, next-generation sequencing led to the identification of a novel exon located between exons 11 and 12, and 58 additional alternative splice variants of the *PRMT1* gene. Among them, 34 are speculated to encode additional protein isoforms of PRMT1 but remain to be characterized [6].

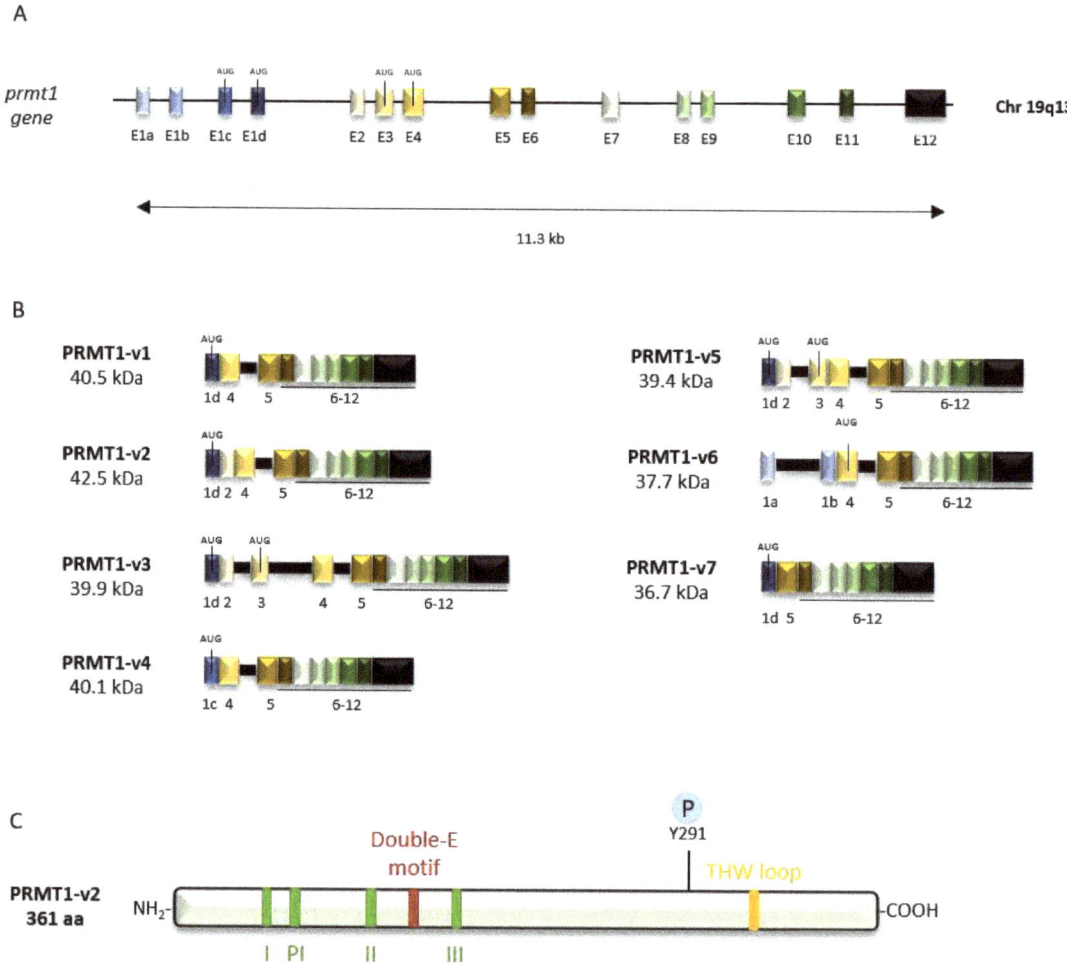

Figure 1. Genomic organization and protein structure of PRMT1. (**A**) Genomic organization of the *PRMT1* gene which spans 11.3 kilobases (kb) and possesses 12 constitutive exons including exon 1 subdivided into 4 alternative exons, represented by a scale of blue-colored boxes; (**B**) Exon composition of the different PRMT-1 isoforms (v1 to v7). The sequences of intron boundaries are represented by the black boxes. Molecular weight of each protein isoform is indicated in kilodaltons (kDa); (**C**) Protein structure of PRMT1-v2. The PRMT signature motifs (I, Post-I (PI), II, III) as well as double-E motif, the THW loop and the phosphorylation site Y291 are represented (adapted from [5,6]).

2.2. Protein Structure

At the protein level, human PRMT1 shares a high degree of homology with the different members of the PRMT family that is conserved in eukaryotes. Phylogenetic studies based on the methyltransferase domain highlighted that PRMT1 is closely related to PRMT8 [7]. The canonical structure of PRMT1 includes three functional domains: (i) the

N-terminal methyltransferase domain characterized by the Rossmann fold constituting the AdoMet binding pocket, (ii) the C-terminal β-barrel domain which forms a cylindrical structure corresponding to the arginine-substrate binding site and (iii) the α-helical dimerization arm which originates from the N-terminal part of β-barrel domain and connects to the Rossmann fold of a second monomer [8].

The catalytic core of PRMT1 is composed of 6 highly conserved peptide motifs essential for the methyltransferase activity. Motif I (VLDVGSGTG) delimits the AdoMet-binding site and is stabilized by motifs II (VDI) and III (LAPDG). The binding of the AdoMet in this pocket is favored by the formation of hydrogen bonds with the glutamic acid residue of the post-motif I (VIGIE). In addition, the double-E motif (SEWMGYCLFYESM) and the THW loop (YTHWK) define the peptidyl arginine-substrate pocket (Figure 1C). The double-E motif is composed of two negatively charged glutamic acid residues (E144 and E153) that neutralize the positively charged guanidium group of the target arginine, whereas the THW loop stabilizes three dynamic α-helices (αX, αY, αZ) located at the N-terminus of the Rossmann fold that participates in peptidyl arginine recognition [9–11]. To illustrate the organization of the catalytic core of PRMT1, an extensive study of the crystal structure of rat PRMT1 which shares 96% identity with the amino acid sequence of human PRMT1 was performed by Zhang and Cheng [10].

Dimerization of PRMTs is a conserved process, crucial for their methyltransferase activity. This mechanism is mediated by the dimerization arm that interacts with the outer surface of the AdoMet binding site through hydrophobic interactions and hydrogen bonds [12]. PRMT1 mutants displaying a mutation or a deletion of the dimerization arm were key to demonstrating the importance of dimerization for AdoMet binding, substrate specificity and the processivity of the methyltransferase activity [10]. As previously described for the yeast PRMT1 counterpart, Hmt1, rat PRMT1 dimers can be assembled into oligomers through hydrophilic interactions [13,14]. This oligomerization is notably associated with a stimulation of the PRMT1 methyltransferase activity [14].

2.3. PRMT1 Isoforms

To date, seven PRMT1 isoforms, PRMT1-v1 to PRMT1-v7, that differ in length and sequence of their N-terminal region have been identified (Figure 1B). These variations of the N-terminal sequence can impact enzymatic activity and substrate specificity. Unlike PRMT1-v7 which is catalytically inactive, variants PRMT1-v1 to PRMT1-v6 exhibit a methyltransferase activity in vitro on different previously described PRMT1 substrates. However, PRMT1-v3 and PRMT1-v4 display a lower methylation efficiency compared to the others. Studies of Goulet et al. also showed that each substrate can be preferentially methylated by a particular isoform. For example, Sam68 and SmB are mainly methylated by PRMT1-v1 and PRMT1-v2 [5]. Currently, studies describing the functionality of the PRMT1-v7 variant are lacking. Although it has retained the ability to heterodimerize with other isoforms, it does not seem to be involved in the regulation of their activity [5].

Differences in enzymatic activities observed among the different PRMT1 isoforms can be partly explained by their subcellular localization. Using a GFP-PRMT1 isoform reporter system, Goulet et al. showed that PRMT1-v1 and -v7 are mainly nuclear, whereas PRMT1-v2 is primarily cytoplasmic [5]. The nucleocytoplasmic shuttling of PRMT1-v2 depends on a leucin-rich nuclear export sequence (NES) encoded by the retained exon 2, but also on its enzymatic activity [15]. Interestingly, there is also a tissue-specific expression pattern of the different PRMT1 isoforms. PRMT1-v1, -v2 and -v3 are ubiquitously expressed in human tissues [4], whereas PRMT1-v4 to -v7 are tissue-specific. More precisely, expression of PRMT1-v4 and -v5 is restricted to the heart and pancreas, respectively; yet, PRMT1-v7 is detectable in the heart and skeletal muscle. PRMT1-v6 expression has so far not been detected in any normal human tissues but was detected in certain breast cancer cell lines [5].

3. Biochemical Features

3.1. Sequence Specificity

PRMT1, like the other type I PRMTs, except PRMT4, catalyzes the asymmetric dimethylation of arginine residues localized in glycine/arginine rich regions and more particularly within RGG or RXR motifs [10]. "RGG" sequences that are often located in regions rich in "RG" dinucleotides are also described as "RGG/RG" motifs that can be subdivided into 4 categories according to the number of repeats: "Tri-RGG", "Di-RGG", "Tri-RG" or "Di-RG" motifs [16]. Many substrates of PRMT1 contain a combination of these different motifs such as TAF15 (3 Tri-RGG, 1 Di-RGG) or Sam68 (1 Di-RGG, 1 Tri-RG, 1 Di-RG). Structurally, the presence of glycine residues near the target arginine induces a conformational flexibility that facilitates substrate recognition [17].

The modification of a single residue in conserved motifs like "RGG" can abolish the activity of PRMT1 towards the mutated substrate. For instance, the helicase eIF4A1 that contains an "RGG" motif is methylated by PRMT1, whereas the eIF4A3 isoform in which "RGG" is replaced by an "RSG" sequence is not a substrate for PRMT1. However, it was shown in the same study that PRMT1 is able to methylate synthetic peptides that contain a "RSG" sequence [18]. This suggests that other residues located at a long distance from the target arginine can also be involved in its recognition. This hypothesis was substantiated by a study of Osborne et al., which showed that the affinity of PRMT1 for its arginine substrate relies on long-range interactions involving an acidic residue located away from the PRMT1 active site and probably a positively-charged residue on the substrate [19].

3.2. Product Specificity

Understanding mechanisms that regulate the degree (mono- or dimethylation) and the type (symmetric or asymmetric) of methylation catalyzed by each member of the PRMT family is a major challenge. Indeed, MMA, ADMA and SDMA induce distinct and sometimes antagonistic biological effects as notably described for mono- and dimethylated H3R2 [20,21].

Studies conducted by Gui et al. on rat PRMT1 that shares 96% sequence identity with its human counterpart, identified two conserved methionine residues, M48 and M155, located in the active site that position the target arginine in a favorable configuration for asymmetric dimethylation. Interestingly, M48 also participates in the specific recognition of the target arginine in multi-arginine protein substrates [20]. Mutations in M48L and M155A induce an imbalance in the proportion of MMAs and ADMAs, but do not allow SDMA generation [20]. However, when M48 is mutated to phenylalanine (M48F), a switch in PRMT1 activity occurs, enabling it to induce symmetrical dimethylation. This is consistent with the fact that product specificity of PRMT5 which catalyzes SDMA formation is controlled by the conserved F379 residue in its active site [22]. More recently, mutagenesis studies showed that H293S mutation of the PRMT1 active site does not affect the production of MMA and ADMA by itself, but leads to a predominant formation of SDMA when it is associated with the M48F mutation [23].

The product specificity of PRMT1 which is non-stochastic and regioselective can also be guided by the substrate itself. It seems that the N-terminal arginyl-groups of substrates constitute the main targets for PRMT1 methylation, whereas positively-charged C-terminal residues (including arginines) participate in long-range interactions with acidic residues of PRMT1. This strengthens the affinity of PRMT1 for its arginine substrates [19,24].

Interestingly, the amino acid sequence of the substrate can also direct the degree of methylation (mono- or dimethylation) by regulating PRMT1 processivity [24,25]. Whether PRMT1 dimethylates its substrates in a distributive or processive manner is a matter of debate in the literature. While numerous studies support that PRMT1 acts distributively by transiently releasing MMA and replacing the methyl donor between the two methyl-group transfers [26–28], Obianyo and co-workers described a semi-processive activity of PRMT1. In this model the mono-methylated intermediate remained associated with the enzyme but the product S-adenosylhomocysteine (AdoHcy) was replaced by a novel AdoMet to allow the second reaction [19,29,30]. Studies on the catalytic activity and processivity of PRMT1 are ongoing, and the latest data indicate that the degree of processivity of PRMT1 depends on its dimerization but is also dependent on cofactor or enzyme concentrations [10,25].

3.3. Regulation of PRMT1 Expression and Enzymatic Activity

Many studies have sought to decipher the different levels of regulation of PRMT1 expression and enzymatic activity. Indeed, substrate methylation by PRMT1 is a highly regulated and dynamic phenomenon, occurring directly through PRMT1 PTMs or through its association with co-regulators. In addition, crosstalk between different PTMs on the same substrate can influence arginine methylation by PRMT1. Finally, methyl marks on arginine can be removed by PAD4 which demethylates histones by converting MMA to citrulline [31] or by JMJD6 which directly removes the methyl group to convert methylarginine into arginine [32]. More recently, JMJD1B, a well-known lysine demethylase for H3K9me2, has also been described as effective in demethylating H4R3me1 and H4R3me2a [33].

3.3.1. Regulation of PRMT1 Expression

PRMT1 can be regulated at the level of its expression. Indeed, a very recent study discovered that the serine/threonine kinase mTOR is involved in the regulation of PRMT1 expression in a fasting context. Forty-eight hours of experimental fasting was shown to induce a decrease in STAT1 phosphorylation mediated by mTOR, leading to the inhibition of STAT1 binding to the PRMT1 promoter. In this fasting condition, the decrease in PRMT1 expression induced a decrease in mitochondrial mass and thus a decrease in cellular energy availability [34]. Moreover, the expression level of PRMT1 can also be regulated by microRNAs (miR). This is the case for example for miR-503 that has a tumor suppressor role and whose expression is low in several types of cancers. In hepatocellular carcinoma cells, miR-503 directly targets PRMT1 and reduces its expression level. Consequently a decrease in cell invasion, migration and epithelial-mesenchymal transition are observed [35].

3.3.2. Post-Translational Modification of PRMT1

Unlike other PRMTs, few PTMs of PRMT1 have been described to date. A first study in 2004 conducted using mass spectrometry found that PRMT1 is phosphorylated on Y291. Using non-natural amino acid mutagenesis, the authors showed that phosphorylation of PRMT1 on Y291 alters protein-protein interactions and substrate specificity. Indeed, Y291 phosphorylation of PRMT1 decreases its interaction with hnRNP, and enzymatic activity on hnRNP in vitro. This is due to the negative charge of the phosphate group that modifies the tertiary structure of the enzyme and in particular of the THW loop [36]. Following this first finding, another study in keratinocytes revealed that PRMT1 is a substrate of the kinase CSNK1a1. Although phosphorylation of PRMT1 by CSNK1a1 does not affect the methylation efficiency of PRMT1 on several known substrates, it seems that it modulates its transcriptional activity on some target genes. Indeed, phosphorylated PRMT1 seems to induce the transcription of genes involved in proliferation and repress the expression of genes involved in keratinocyte differentiation [37]. More recently, in ovarian cancer cells, it was shown that PRMT1 can be phosphorylated by DNA-PK in response to cisplatin, thus inducing its recruitment on chromatin and its enzymatic activity towards H4R3 [38].

PRMT1 activity is also modulated by its degradation mediated by the proteasome pathway. In this context, a study in human embryonic kidney cells showed that PRMT1 is polyubiquitinated by the E3 ubiquitin ligase, TRIM48. Thus, the polyubiquitination of PRMT1 decreases the level of methylation of the substrate ASK1, a kinase involved in the cellular stress response. Downregulation of PRMT1 thus promotes cell death induced by ASK1-mediated oxidative stress. Polyubiquination of PRMT1 also negatively impacts FOXO1 methylation and its transcriptional activity [39]. Another in vivo study used an engineered ubiquitin transfer method called "orthogonal UB transfer" to profile E3 substrate specificity. This method showed that PRMT1 is polyubiquitinated by two other E3 ubiquitin ligases, CHIP and E4B, leading to its proteasome-mediated degradation.

Nevertheless, the physiological consequences of this polyubiquitination were not investigated in this study [40]. Given the importance of PRMT1, it probably undergoes many other PTMs including methylation, such as PRMT5 which is methylated by PRMT4 [41], or OGT-glycosylation [42]. Although other modifications (i.e., acetylation and sumoylation) have not been described in the PRMT family, it is likely that these events exist.

3.3.3. PRMT1 Association with Co-Regulators

PRMT1 activity can also be regulated through its interaction with non-substrate proteins that modulate its methyltransferase activity. The first regulators were described in 1996, with the BTG1 (B-cell translocation gene 1) and BTG2. This study showed in vitro that the interaction of BTG1 and BTG2 with PRMT1 positively modulates its enzymatic activity towards a substrate, hnRNPA1 [43]. Several years later, our team discovered a new regulator of PRMT1, hCAF1. We showed by in vitro methylation assay that hCAF1 inhibits PRMT1-mediated methylation of histone H4 on arginine 3 (H4R3) by PRMT1. This observation was confirmed in breast cancer cells where depletion of hCAF1 induces a strong reduction in the overall level of asymmetric arginine methylation, indicating that hCAF1 modulates PRMT1 activity towards several substrates [44]. Interestingly, a study in HeLa cells revealed a crosstalk between PRMT1 and PRMT2. Indeed, PRMT2 binds to PRMT1 without methylating it and potentiates its enzymatic activity towards H4R3. Surprisingly, PRMT2-mediated activation of PRMT1 also induces an increase in SDMA levels in vivo, implying possible further crosstalk between the different enzymes of the PRMT family [45].

PRMT1 activity can also be modulated by exogenous regulators. For instance, the serine/threonine phosphatase PP2A has been described to regulate PRMT1 activity. PRMT1 methylate hepatitis C virus NS3 protein and inhibits its helicase activity. PP2A binds to PRMT1 and inhibits its enzymatic activity towards a NS3 protein, which affects inhibitory role of PRMT1 on the helicase activity of NS3. Interestingly, the hepatitis C virus upregulates PP2A expression, thus counteracting the downregulation of NS3 by PRMT1. This study highlights the complexity of the pathways regulating PRMT1 enzymatic activity [46].

In addition, other regulators have been identified, such as RALY [47], TR3 [48], PDGF-BB [49], or GFI1 [50]. Moreover, other mechanisms of regulation of PRMT1 have been uncovered, such as oxidative stress [12] or iron deficiency [51].

3.3.4. PTMs Influencing PRMT1 Activity

In parallel to the direct regulation of PRMT1 by PTMs or by the binding of coregulators, a crosstalk between arginine methylation and different PTMs deposited by other enzymes on the same substrate has been described. For example, a 2006 study showed that methylation of H4R3 by PRMT1 at the pS2 promoter is required to activate its expression. Interestingly, this study showed that histone hypoacetylation is necessary for the recruitment of PRMT1 to the promoter and for the deposition of the H4R3 methylation mark. The patient SE translocation (SET) protein, which is part of the INHAT complex, prevents the acetylation of the histone at the pS2 promoter [52]. Another study investigated the effect of histone H4 phosphorylation on serine 1 (H4S1). The authors showed by in vitro methylation assays that H4S1 phosphorylation leads to a 3-fold decrease in PRMT1-mediated H4R3 methylation. Interestingly, mass spectrometry analysis revealed MMA as a PRMT1 major product. Indeed, further in vitro methylation assays revealed a 3-fold decrease in ADMA, due to an approximate 11-fold reduction in PRMT1 catalytic efficiency. Moreover, H4S1 phosphorylation also leads to a 8-, 5-, and 3-fold decrease in PRMT3, PRMT8 and PRMT5 activity, respectively [53].

These in vitro studies highlighted the complex crosstalk between the different PTMs in the histone code and the tight regulation of the activity of each enzyme. Although this phenomenon has only been described on H4R3 for PRMT1, this is probably because PRMT1 was first described as a histone methyltransferase catalyzing H4R3 methylation [54]. Many non-histone substrates have since been described, and likely display similar crosstalk that remains to be depicted.

3.4. Substrates

Arginine 3 of histone H4 was the first substrate described for PRMT1 [54,55]. The asymmetric dimethylation of H4R3 constitutes an activating mark of transcription [56]. It was also demonstrated that PRMT1 methylates histone H2A at R3, R11 and R29, although the latter two are not localized within a consensus motif recognized by PRMT1 [57]. Further studies are expected to clarify the impact of these two histone marks on transcriptional activity. In addition to the activity of PRMT1 as a chromatin modifying enzyme, a plethora of non-histone substrates of PRMT1 have been identified and can be classified according to their cellular functions: transcriptional and translational regulation, RNA-processing, DNA damage repair and signal transduction. A list of the currently identified substrates of PRMT1 is available in Table 1.

It is important to note that some substrates are common to different types of PRMTs and that competitive mechanisms may exist. This hypothesis is supported by the observations of Dhar et al. who showed that inhibition of PRMT1 induces a decrease in the level of ADMA concomitant with an increase in MMA and SDMA levels [58].

Table 1. List of non-histone substrates of PRMT1 classified according to their cellular functions.

Biological Function	Substrate	Methylation Site	Biological Outcome	Reference
Transcriptional Regulation Transcriptional regulation	BRCA1	Within the 504–802 region	Promotes BRCA1 recruitment to specific promoters	[59]
	C/EBPα	R35, R156, R165	Prevents C/EBPα interaction with the corepressor HDAC3	[60]
	c-Myc	R299, R346	Promotes c-Myc interaction with p300	[61]
	EZH2	R342	Prevents EZH2 target gene expression	[62]
	FOXO1	R248, R250	Prevents FOXO1 phosphorylation by Akt	[63]
	FOXP3	R48, R51	Enhances FOXP3 transcriptional activity	[64]
	GLI1	R597	Enhances GLI1 binding to target gene promoters	[65]
	MyoD	R121	Promotes MyoD DNA-binding and transcriptional activity	[66]
	Nrf2	R437	Promotes Nrf2 DNA-binding and transcriptional activity	[67]
	PR	R637	Accelerates PR recycling and transcriptional activity	[68]
	RACO-1	R98, R109	Promotes c-Jun/AP1 activation	[69]
	RelA	R30	Prevents RelA DNA-binding and represses NF-κB target genes	[70]
	RIP40	R240, R650, R948	Favors RIP140 nuclear export and prevents the recruitment of HDAC3	[71]
	RunX1	R206, R210	Prevents Sin3a binding and promotes RUNX1 transcriptional activity	[72]
	STAT1	R31	Prevents STAT1 association with PIAS1 and enhances IFNα/β induced transcription	[73]
	TAF15	R203	Affects the subcellular localization of TAF-15 and enhances its transcriptional activity	[74]

Table 1. Cont.

Biological Function	Substrate	Methylation Site	Biological Outcome	Reference
	FUS/TLS	R216, R218, R242, R394	Participates in the nuclear cytoplasmic shuttling of FUS/TLS and enhances its transcriptional activity	[75,76]
	TOP3B	R833, R835	Promotes TOP3B interaction with TDRD3, stress granule localization and topoisomerase activity	[77]
	Twist1	R34	Regulates the nuclear import of Twist1 and represses E-cadherin expression	[78]
	CNBP	R25, R27	Prevents its RNA binding activity	[79]
	G3BP1	R435, R447	Prevents stress granule formation during oxidative stress	[80]
	hnRNPA1	R214, R226, R223, R240	Prevents hnRNPA1 ITAF activity and RNA-binding ability	[81]
	HSP70	R416, R447	Enhances HSP70 RNA-binding and -stabilization abilities	[82]
RNA- processing	NS3	R1493	Affects NS3 RNA-binding and helicase activity	[46,83]
	RBM15	R578	Promotes RBM15 degradation by CNOT4 (RNA splicing)	[84]
	Sam68	Within the 276–343 region	Prevents Sam68 poly(U) RNA-binding activity	[85,86]
	SF2/ASF	R93, R97, R109	Affects SF2/ASF nucleocytoplasmic distribution and modulates the alternative splicing of target genes	[87,88]
	eIF4A1	R362	Prevents eIF4A1 interaction with eIF4G1 and inhibits ATPase activity	[18,89]
Translational Regulation	eIF4G1	R689, R698	Regulates eIF4G1 stability and the assembly of the translation initiation complex	[90]
	rpS3	R64, R65, R67	Promotes rpS3 import into the nucleolus and ribosome assembly	[91]
	53BP1	Within the 1319–1480 region	Promotes 53BP1 recruitment to DNA-damage sites	[92]
	APE1	R301	Promotes APE1 mitochondrial translocation (translocase Tom20) and protects mitochondrial DNA from oxidative damage	[93]
	DNA pol β	R137	Prevents DNA pol β interaction with PCNA in BER pathway	[94]
DNA damage repair	E2F-1	R109	Promotes E2F-1-dependent apoptosis in DNA-damaged cells	[95]
	FEN1	Not determined	Stabilizes FEN1 and upregulates its DNA damage repair activities	[96]
	hnRNPK	R296, R299	Prevents PKCδ-dependent apoptosis during DNA damage	[97]
	hnRNPUL1	R584, R618, R620, R645, R656	Promotes hnRNPUL1 association with NBS1 and recruitment to DNA-damage sites	[98]
	MRE11	GAR domain	Promotes MRE11 recruitment to DNA-damage sites and favors its exonuclease activity	[99,100]

Table 1. Cont.

Biological Function	Substrate	Methylation Site	Biological Outcome	Reference
Signal transduction	RunX1	R233, R237	Confers resistance to apoptosis under stress condition and DNA damage accumulation	[101]
	ASK1	R78, R80	Prevents the stress-induced ASK1-JNK1 signaling	[102]
	Axin	R378	Favors Axin stability and consequently prevents Wnt/β-catenin signaling	[103]
	BAD	R94, R96	Prevents BAD phosphorylation by Akt and subsequent survival signaling	[104]
	CaMKII	R9, R275	Prevents CaMKII-dependent signaling in cardiomyocytes	[105]
	CDK4	R55, R73, R82, R163	Prevents the formation of a CDK4/Cyc D3 complex and subsequent cell cycle progression	[106]
	cTnI	R146, R148	Induces cardiac cell hypertrophy	[107]
	EGFR	R198, R200	Upregulates EGFR signaling	[108]
	ERα	R260	Promotes the formation of the ERα/PI3K/Src/FAK complex and subsequent activation of downstream kinase cascades	[109]
	INCENP	R887	Enhances INCENP binding-affinity to AURKB and promotes cell division	[110]
	KCNQ	R333, R345, R353, R435	Promotes PIP2 binding and subsequent KCNQ channel activity	[111]
	MYCN	R65	Enhances MYCN stability through CDK-dependent phosphorylation	[112]
	NONO	R251	Favors NONO oncogenic function	[113]
	p38 MAPK	R49, R149	Promotes p38 MAPK phosphorylation by MKK3 and the subsequent activation of MAPKAK2 involved in erythroid differentiation	[114]
	Smad4	R272	Promotes Smad4 phosphorylation by GSK3 and support the activation of the canonical Wnt signaling	[115]
	Smad6	R74, R81	Participates in BMP signaling and prevents NF-κB activation	[116,117]
	Smad7	R57, R67	Facilitates TGF-β signaling	[118]
	TRAF6	R88, R125	Prevents TRAF6 ubiquitin ligase activity and regulates Toll-like receptor signaling	[119]
	TSC2	R1457, R1459	Blocks the Akt-dependent phosphorylation of TSC2 and regulates mTORC1 activity	[120]

4. Cellular Features

4.1. Connection with Chromatin Dynamics and Transcriptional Regulation

Arginine methylation was first described as a PTM of histones that regulates reader protein recruitment and therefore chromatin dynamics. The main target of PRMT1 at the chromatin level is the arginine 3 of histone H4 (H4R3) [54,55]. Asymmetrically dimethylated H4R3, H4R3me2a, is associated with an active form of the chromatin and recognized by different Tudor domain-containing proteins, such as TDRD3 [121]. This protein, with no intrinsic activity, serves as a scaffold coregulator for the assembly of protein complexes at the transcription start sites of target genes. More precisely, TDRD3 can recruit, through its OB-fold domain, the DNA Topoisomerase IIIβ [122] and can directly interact with the RNA Polymerase II, previously methylated at R1810 by PRMT4 also known as CARM1 [123]. Therefore, this complex assembled through TDRD3 and likely involving other actors promotes transcription at H4R3me2a loci (Figure 2).

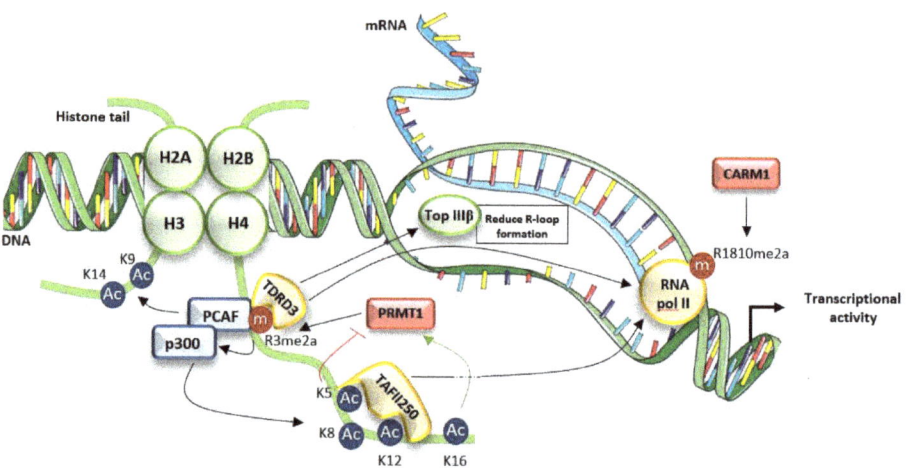

Figure 2. PRMT1 regulates chromatin dynamics. PRMT1-dependent H4R3 methylation (R3me2a) allows the recruitment of the Tudor domain-containing protein, TDRD3, which in turn associates with topoisomerase IIIβ (Top IIIβ) to reduce R-loop formation and RNA polymerase II (RNA pol II) to promote transcriptional activity. Concomitantly, H4R3me2a-dependent activation of histone acetyltransferases p300 and pCAF induces acetylation of H4K5, H4K8, H4K12 but also of H3K9 and H3K14. H4K5 and H4K12 are involved in the recruitment of TAFII250 that associates with RNA pol II. H4K5ac and H4K16ac are also involved in PRMT1-activity regulation. Ac = Acetylation, m = methylation.

Interestingly, H4R3me2a can also recruit chromatin modifying enzymes involved in transcriptional regulation by depositing other histone marks on chromatin. Indeed, methylation of H4R3 by PRMT1 promotes the subsequent acetylation of H4K8 and H4K12 by the histone acetyltransferase p300 [56]. An H4R3me2a-dependent induction of H4K5 and H4K12 acetylation, allowing the recruitment of the transcription initiation factor TAFII250 and therefore contributing to chromatin opening, was also suggested using the chicken β-globin locus as a model [124]. Finally, the ability of H3R4me2a to act in trans to promote the acetylation of histone H3K9 and H3K14 by the histone acetyltransferases p300 and PCAF was demonstrated within the β-major globin promoter in murine erythroleukemia cells [124,125]. It is worth noting that PCAF directly interacts with H4R3me2a and this could explain how PRMT1-dependent methylation potentiates H3K9 and H3K14 acetylation [125] (Figure 2).

Conversely, the activity of PRMT1 on H4R3 is inhibited by the presence of acetylation, propionylation, crotonylation, butyrylation or 2-hydroxyisobutyrylation of H4K5 [126]. Moreover, H4K5ac combined with H4K8ac or H4K12ac increases its repressive effect on PRMT1 activity. There is currently one known exception, as acetylated H4K16 is associated with an increase in PRMT1 activity. Interestingly, the inducing effect of H4K16ac dominates the repressive effect of H4K5ac when the 2 histone marks co-exist [53,127] (Figure 2).

Aside from chromatin regulation, a large number of transcription factors whose activity can be regulated by PTMs are known PRMT1 substrates (Table 1). PRMT1-dependent methylation can notably increase their stability and thus promote their transactivation function. This type of mechanism has been described for FOXO1 whose methylation by PRMT1 prevents its proteosomal degradation and favors its nuclear localization [63]. The methyltransferase activity of PRMT1 can also impact interactions between transcription factors and their corepressors. For example, PRMT1 was shown to act as a coactivator of RUNX1 by inducing its methylation at R206 and R210, and thereby preventing its interaction with the transcriptional corepressor SIN3A [72]. Similarly, C/EBPα methylation at R35, R156 and R165 blocks its interaction with the corepressor HDAC3 [60].

4.2. Connection to Cell Signaling Pathways

4.2.1. Steroid Receptors

To date, PRMT1 has been shown to methylate two steroid receptors; estrogen receptor (ERα) and progesterone receptor (PR). These arginine methylation events control different signaling pathways involved in breast tumorigenesis.

Estrogen Receptor (ERα)

ERα regulates many physiological processes, notably the growth and survival of breast tumor cells, acting as a ligand-dependent transcription factor. Aside from the well described transcriptional effects, estrogen also mediates extranuclear events called nongenomic signaling via its receptor [128]. Our group showed that ERα is methylated on the residue R260 (met260ERα) by PRMT1 in response to estrogen or IGF-1 [109,129]. This event is a prerequisite for the formation of a signaling complex containing met260ERα, Src and PI3K, which orchestrates cell proliferation and survival. The involvement of this complex in breast carcinogenesis will be addressed in Section 5.1 of this review. Met260ERα is a transient event downregulated by the arginine demethylase JMJD6 [130].

Progesterone Receptor (PR)

Our group also demonstrated that PRMT1 methylates PR on the residue R637, within a RGG consensus site. This methylation event decreases PR stability in order to accelerate its recycling and its transcriptional activity. In addition, PRMT1 depletion decreases the expression of a specific subset of progesterone-target genes, involved in breast cancer cell proliferation and migration [68].

4.2.2. Akt Signaling Pathway

Several reports demonstrated that specific arginine methylation, catalyzed by PRMT1 within the Akt consensus phosphorylation motif, works as an inhibitor of Akt-dependent survival signaling.

FOXO

Forkhead box O (FOXO) is a family of transcription factors controlling a large variety of biological processes including cell survival [131]. Several studies revealed that FOXO proteins are phosphorylated by Akt, resulting (i) in the export of FOXO proteins from the nucleus to the cytoplasm [132,133] and (ii) in FOXO proteasomal degradation through polyubiquitination [134,135]. Interestingly, a member of the FOXO family, FOXO1 was shown to be methylated by PRMT1 on R248 and R250, in the consensus Akt phosphorylation site, impeding Akt phosphorylation on S253 [63]. This methylation event results in a decrease in its cytoplasmic localization and its subsequent degradation. PRMT1 depletion decreases oxidative-stress-induced apoptosis regulated by the Akt-FOXO1 pathway. These results indicated that PRMT1 arginine methylation can act as a modulator of Akt-phosphorylation by regulating responses to oxidative stress in mammalian cells.

BAD

Similarly, PRMT1 binds and methylates the proapoptotic protein BCL-2 antagonist of cell death (BAD) on R94 and R96, in the Akt consensus site. PRMT1 methylation on these two residues inhibits Akt phosphorylation on S99, a modification that is necessary for its interaction and sequestration with 14-3-3 proteins, resulting in cell survival [104].

4.2.3. NF-κB Signaling

NF-κB plays an important role in the transcriptional regulation of genes involved in inflammation and cell survival. Toll-like receptor (TLR), when activated by lipopolysaccharides, triggers the recruitment of the adaptor protein Myd88 and the subsequent activation of the transcription factor NF-κB. TGFβ inhibits TLR signaling through the methylation of SMAD6 by PRMT1. Indeed, the binding of methylated SMAD6 to Myd88 results in its degradation, impeding TLR signaling to NF-κB [117]. Moreover, PRMT1 serves as a coactivator of NF-κB, synergistically with CARM1, although the underlying mechanisms are not fully elucidated [136]. More recently, the methylation of the RelA subunit of NF-κB by PRMT1 was identified as a repressive mark modulating TNFα/NF-κB response [70].

4.2.4. Wnt Signaling

Wnt signaling plays important roles in embryonic development and cell proliferation. Aberrant Wnt signaling leads to several human diseases including cancer. Axin is a negative regulator of the Wnt pathway, as it is a key scaffold protein for the β-catenin destruction complex. PRMT1-induced methylation of axin enhances its interaction with GSK3β, leading to a decrease in axin ubiquitination and degradation [103]. Therefore, PRMT1 seems to be a new modulator of Wnt/β-catenin signaling. Moreover, PRMT1 also regulates this pathway by methylating substrates prior to their phosphorylation by GSK3β and its sequestration in endolysosomes, a key event in Wnt signaling [115]. Altogether PRMT1 appears as an important modulator of the Wnt pathway at the interface of protein phosphorylation and trafficking.

4.3. Cellular Role and Functions

4.3.1. Embryogenesis and Development

The critical role of PRMT1 in embryogenesis and development was first suggested by the study of Pawlak et al. which showed that PRMT1 knockout mouse embryos, generated by insertion of a gene trap retrovirus in the second intron of the *PRMT1* gene, failed to develop beyond embryonic day 6.5, which would coincide with the exhaustion of the maternal stock of PRMT1 enzymes and methylated substrates [137]. It is worth noting that homozygous PRMT1 mutant embryonic stem (ES) cells isolated from mutant preimplantation blastocysts at day 3.5 are viable and retained the morphology and the same doubling time as wild-type ES cells. Moreover, in these cells, loss of PRMT1 activity is not balanced by the activation of other methyltransferases. Therefore, PRMT1 activity does not seem to be required for cell viability [137].

Early lethality of homozygous PRMT1 KO mouse embryos, as well as their uterus-enclosed localization, makes it difficult to study the epigenetic regulation of vertebrate development and emphasizes the importance to develop other models. Among them, Zebra fish embryos constitute a promising model as they are suitable for genetic manipulation approaches and express a highly conserved PRMT1 protein (90% identity with human PRMT1) at different stages of embryogenesis. A study conducted by Tsai et al. showed that PRMT1 knockdown, by antisense morpholino oligo injection into one-cell stage zebra fish embryos, induces developmental defects at gastrulation notably including a shortened body-length. This highlighted the importance of the methyltransferase activity of PRMT1 in early embryogenesis [138]. More recently, Shibata et al. used the TALEN genome editing technology to knockout PRMT1 in the diploid anuran *Xenopus tropicalis* that undergoes an external and biphasic development (embryogenesis and metamorphosis). They observed that H4R3me2a methylation by PRMT1 is not required for early embryogenesis but is essential for the growth and development of various organs including the brain, liver and intestine during late embryonic developmental stages, occurring prior to metamorphosis. This effect is directly related to the drastic inhibition of cell proliferation associated with

PRMT1 KO in this model [139]. Interestingly, *Xenopus* embryos were already used to demonstrate the involvement of the xPRMT1b gene in early neural determination [140].

Another interesting aspect is the potential involvement of PRMT1 in placental development. A study of Sato et al. showed that murine placental expression of two PRMT1 isoforms is differentially regulated during the gestational period. More precisely, while PRMT1-v1 expression reaches a maximum at embryonic day E11 before decreasing, PRMT1-v2 expression increases from E13. This balance between the two isoforms explains the change in subcellular localization of PRMT1 observed between early and late stages of gestation; though further studies are required to determine the exact role played by PRMT1 in the placenta [141].

4.3.2. DNA Damage Repair

The conditional knockout of PRMT1 in mouse embryonic fibroblasts is associated with a severe genetic instability characterized by the occurrence of spontaneous DNA damage, chromosome copy number variations and defective mitotic checkpoint [142]. The relevance of PRMT1 in the maintenance of genome integrity is based on the methylation and subsequent regulation of key factors involved in the major DNA repair pathways.

The first substrate of PRMT1, involved in DNA damage repair, to be identified was MRE11 (Meiotic recombination 11). This component of the MRN complex (MRE11/RAD50/NBS1), recruited early upon DNA double-strand break (DSB), participates in the initiation of DNA repair pathways by homologous recombination (HR) or by non-homologous end joining (NHEJ). Methylation of the C-terminal GAR motif of MRE11 at R587 by PRMT1 does not seem to participate in the formation of the MRN complex but it promotes the relocalization of MRE11 from PML nuclear bodies to DNA-damage sites and it favors its exonuclease activity [92,99,100]. These events are essential to allow the recruitment of RAD51 and the subsequent activation of HR [100]. By using a model of knock-in mice that express the mutated MRE11RK protein devoid of methylarginines, Yu et al. also demonstrated that MRE11 methylation participates in the activation of the ATR/CHK1 checkpoint signaling [143]. Finally, methylated MRE11 is involved in telomere maintenance and regulates DNA replication by controlling the intra-S phase checkpoint in response to DNA damage [99,144].

The choice of pathways between NHEJ or HR is directly influenced by the DNA-end structure of DNA DSBs. Among the actors that play a pivotal role to orient this choice are the tumor suppressor protein BRCA1, which promotes HR repair by activating DNA-end resection, and p53-Binding Protein 1 (53BP1) that inversely activates NHEJ repair by inhibiting the recruitment of BRCA1 to DNA DSBs [145]. Interestingly, these two proteins are methylated by PRMT1, suggesting that arginine methylation may play an important role in directing the switch from HR to NHEJ repair pathways. More precisely, 53BP1 is methylated by PRMT1 at a canonical GAR motif localized in its kinetochore-binding domain and this methylation is essential for its DNA-binding activities [92,146]. Concerning BRCA1, the methylation status of the 504–802 protein region, that encompasses the DNA-binding domain, directly influences its interaction with transcription factors such as Sp1 or STAT1 and its subsequent recruitment to specific promoters [59].

The base excision repair mechanism (BER) that can correct single-stranded DNA breaks and oxidative or alkylation damage is also regulated by PRMT1, which methylates two major players in this pathway, namely the Flap endonuclease 1 (FEN1) and the DNA polymerase β (DNA Pol β). Methylation of FEN1 by PRMT1, at an arginine residue that remains to be determined, stabilizes the protein without disturbing its localization [96]. Moreover, unlike PRMT5-dependent methylation at residue R192 which strengthens the interaction between FEN1 and the DNA polymerase processivity factor PCNA necessary for a faithful and efficient BER, PRMT1-dependent methylation of FEN1 does not seem to impact this interaction [96,147]. Interestingly, methylation of the DNA Pol β by PRMT1 on R137 abolishes its binding with PCNA without affecting its enzymatic activities (polymerase and dRA-lyase) [94]. This suggests that methylation could regulate the sequential interaction of FEN1 and DNA Pol β with PCNA during BER.

5. PRMT1 in Cancer

Since the substrates methylated by most PRMTs regulate various biological functions essential for cellular homeostasis, it is not surprising that a dysregulation of arginine methylation may contribute to cancer initiation and progression. The involvement of PRMT1 in carcinogenesis is no longer questioned due to its overexpression or aberrant splicing observed in numerous types of cancers.

5.1. Breast Cancer

Various studies have shown that *PRMT1* gene expression is higher in breast tumor samples than in healthy tissue suggesting the involvement of PRMT1 in breast carcinogenesis [5,148]. Despite the detection of PRMT1-v1, v2 and v3 isoforms in breast tumor tissue, it seems that only the predominant PRMT1-v1 variant is correlated with clinical parameters such as histological grade [148].

ERα is an important PRMT1 substrate whose methylation can be associated with the development of breast cancer. Our group highlighted that a PRMT1-dependent hypermethylation of ERα at R260, induced in response to estrogen or IGF-1, is observed in different subtypes of breast cancers and regulates cell proliferation and survival [109,129]. We notably showed that the signaling complex containing met260ERα, Src and PI3K (described in Section 4.2.1 of this review) is expressed at low levels in the cytoplasm of normal mammary epithelial cells but highly expressed in 55% of breast tumors [149]. Moreover, its overexpression is correlated with the activation of Akt (pAkt), the main effector of the pathway, showing that this signaling pathway exists in vivo. In addition, a high expression of the complex is an independent marker of poor prognostic [149] and has been linked with resistance to tamoxifen [150,151].

Another interesting aspect is the key role of PRMT1 in the maintenance of stem-cell-like properties of breast cancer cells. PRMT1-dependent EGFR methylation on R198 and R200 upregulates different signaling cascades, notably those involving Akt, ERK or STAT3 in triple-negative breast cancer (TNBC) cells, MDA-MB-468. EGFR/ERK-dependent activation of ZEB1, a transcription factor that regulates epithelial-mesenchymal transition, may be implicated in cancer stem cell maintenance [152]. Interestingly, asymmetric dimethylation of H4R3 by PRMT1 at the ZEB1 promoter is another mechanism described to activate this factor and therefore promotes migration, invasion and acquisition of stem cell characteristics. It is worth noting that ZEB1 may simultaneously contribute to the PRMT1-dependent inhibition of senescence in breast cancer cells [153].

PRMT1-dependent methylation also inhibits the tumor suppressive function of some substrates. For example, methylation of C/EBPα at R35, R156 and R165 by PRMT1 prevents its interaction with the corepressor HDAC3, thus promoting the expression of cell-cycle genes such as cyclin D1 and the subsequent growth of breast cancer cells [60]. In the same line, BRCA1 methylation by PRMT1 affects its recruitment to responsive promoters but also its ability to interact with certain partners such as Sp1 or STAT1. As a result, this can significantly affect the tumor suppressive activity of BRCA1 [59].

5.2. Colorectal Cancer

Two clinical reports demonstrated the unfavorable prognosis associated with PRMT1 expression in colorectal cancer (CRC) patients by discussing the respective involvement of PRMT1-v1 and PRMT-v2 isoforms [154,155]. Mechanistically, it was described that H4R3me2a can recruit SMARCA4, an ATPase subunit of the SWI/SNF complex, to the promoter of certain target genes including EGFR to promote their expression. PRMT1-dependent enhancing of EGFR signaling is associated with a significant increase in the proliferative and migratory abilities of human CRC cells [156]. Moreover, methylation of EGFR at R198 and R200 by PRMT1 leads to an EGF-dependent hyperactivation of

EGFR signaling and confers cells with resistance to the anti-EGFR monoclonal antibody, cetuximab. Indeed, in CRC patients, the rate of EGFR methylation is directly correlated with a higher recurrence rate after cetuximab treatment and a poorer overall patient survival [108].

Recently, the non-POU domain-containing octamer-binding protein (NONO), which is overexpressed in CRC tissue, was described as a substrate of PRMT1. Methylation of NONO at R251 is required to promote its oncogenic function including the induction of CRC cell proliferation, migration and invasion [113].

5.3. Lung Cancer

As described for other cancers, PRMT1 expression is significantly increased in lung cancer tissue compared to non-neoplastic ones though very little data are available in the literature to explain its role in lung carcinogenesis [157]. A study by Avasarala et al. highlighted that PRMT1 participates in non-small cell lung cancer progression and metastasis through the methylation of the EMT-associated transcription factor Twist1 at R34. PRMT1-dependent Twist1 methylation is associated with inhibition of E-cadherin expression [78]. Moreover, PRMT1 can methylate the inner centromere protein (INCENP) at R887 to favor its interaction and the subsequent activation of aurora kinase B in A549 non-small cell lung cancer cells. This mechanism regulates the alignment and segregation of chromosomes during cell division to promote the growth of cancer cells [110].

5.4. Other Cancers

Dysregulation of PRMT1 expression has been reported in several other types of cancers, albeit the molecular mechanisms that drive the initiation and progression of these cancers remain incompletely understood. The limited data available in the literature indicate that PRMT1 is particularly dysregulated in bladder cancer, esophageal squamous cell carcinoma, as well as in acute myeloid leukemia [157–159]. Interestingly, in ovarian carcinomas, upregulation of PRMT1 expression is associated with an increased methylation of the apoptosis signal-regulated kinase 1 (ASK1), which confers tumor cells with resistance to platinum-based chemotherapeutic agents [160]. Moreover, in prostate cancer, the methylation status of H4R3 is significantly correlated with clinical features, such as tumor grade or the risk of prostate cancer recurrence. This study highlighted the fact that histone modifications can also serve as a prognostic marker [161].

5.5. PRMT1 Inhibitors

In 2004, the symmetrical sulfonated urea salt named arginine methylation inhibitor-1 (AMI-1) was the first PRMT inhibitor characterized [162]. Since then, two substrate competitive inhibitors, MS023 and GSK3368715, that broadly target type I PRMTs (Table 2), were developed and displayed antitumor activities notably on xenograft mouse models of acute myeloid leukemia or breast cancer, respectively [163–165]. Promisingly, the GSK3368715 inhibitor is currently undergoing a first-time clinical trial (NCT03666988) for patients with solid tumors and diffuse large B-cell lymphoma. However, high affinity of these inhibitors for other type I PRMTs, renders the identification and characterization of specific PRMT1-dependent effects difficult.

Currently, two PRMT1-specific inhibitors, TC-E-5003 and C7280948, are mentioned in the literature (Table 2). TC-E-5003 displays significant antitumor activity in vitro on breast or lung cancer cell lines and inhibits the growth of xenografted A549 lung cancer cells in mice [166]. Concerning C7280948, a study of Yin et al. showed that it suppresses colorectal cancer cell proliferation, migration and invasion [113]. Additionally, a structure-based virtual screening of different libraries of compounds allowed the identification of several potential PRMT1-specific inhibitors, the properties of which were detailed by Hu et al. [167]. Although these inhibitors are promising, more studies are needed to characterize and consider their clinical potential.

Table 2. List of PRMT inhibitors targeting PRMT1. ND: Not defined in literature.

Name	Mechanism of Action	Target(s)	IC50	Reference
AMI-1	Substrate competitive SAM uncompetitive	PRMT1	8.81 µM	[162]
MS023	Substrate competitive SAM uncompetitive	PRMT1	30 nM	[163]
		PRMT3	119 nM	
		PRMT4/CARM1	83 nM	
		PRMT6	4 nM	
		PRMT8	5 nM	
GSK3368715	Substrate competitive SAM uncompetitive Reversible	PRMT1	33.1 nM	[165]
		PRMT3	162 nM	
		PRMT4/CARM1	38 nM	
		PRMT6	4.7 nM	
		PRMT8	3.9 nM	
TC-E-5003	ND	PRMT1	1.5 µM	[166]
C7280948	Interaction with the substrate-binding pocket	PRMT1	12.8 µM	[113]

6. Outlook

Over the last twenty years since the discovery of PRMT1, the number of studies conducted on this enzyme has constantly increased. This interest, which persists today, has improved our knowledge on the diversity of its substrates and the numerous biological functions regulated by PRMT1. Its key role in cancer initiation and progression makes PRMT1 an interesting target for the development of new anti-cancer therapeutic strategies. Therefore, the development of inhibitors that target PRMT1 activity is an ongoing challenge that may offer new therapeutic opportunities for various pathologies in the coming years.

Author Contributions: T.C., E.L., P.C. and L.R.M. wrote the manuscript and revised it. All authors have read and agreed to the published version of the manuscript.

Funding: T.C., E.L., P.C. and L.R.M.'s laboratory is funded with grants from "La Ligue contre le Cancer", "Fondation ARC Cancer" and "Fondation de France". C.T. was supported from a fellowship from "Fondation de France" and L.E. was supported from a fellowship from "La Ligue contre le Cancer".

Acknowledgments: We thank B. Manship for proofreading the manuscript. The illustrations were created by using Servier Medical Art (SERVIER SAS, Suresnes, France).

Conflicts of Interest: The authors declare no conflict of interest.

References

1. Yang, Y.; Bedford, M.T. Protein arginine methyltransferases and cancer. *Nat. Rev. Cancer* **2012**, *13*, 37–50. [CrossRef]
2. Zhang, J.; Jing, L.; Li, M.; He, L.; Guo, Z. Regulation of histone arginine methylation/demethylation by methylase and demethylase. *Mol. Med. Rep.* **2019**, *19*, 3963. [CrossRef]
3. Tang, J.; Frankel, A.; Cook, R.; Kim, S.; Paik, W.; Williams, K.; Clarke, S.; Herschman, H. PRMT1 is the predominant type I protein arginine methyltransferase in mammalian cells. *J. Biol. Chem.* **2000**, *275*, 7723–7730. [CrossRef]
4. Scorilas, A.; Black, M.H.; Talieri, M.; Diamandis, E.P. Genomic Organization, Physical Mapping, and Expression Analysis of the Human Protein Arginine Methyltransferase 1 Gene. *Biochem. Biophys. Res. Commun.* **2000**, *278*, 349–359. [CrossRef]
5. Goulet, I.; Gauvin, G.; Boisvenue, S.; Côté, J. Alternative splicing yields protein arginine methyltransferase 1 isoforms with distinct activity, substrate specificity, and subcellular localization. *J. Biol. Chem.* **2007**, *282*, 33009–33021. [CrossRef]
6. Bedford, M.T.; Richard, S. Arginine methylation: An emerging regulator of protein function. *Mol. Cell* **2005**, *18*, 263–272. [CrossRef]

7. Wu, H.; Min, J.; Lunin, V.V.; Antoshenko, T.; Dombrovski, L. Structural Biology of Human H3K9 Methyltransferases. *PLoS ONE* **2010**, *5*, e8570. [CrossRef]
8. Tewary, S.K.; Zheng, Y.G.; Ho, M.C. Protein arginine methyltransferases: Insights into the enzyme structure and mechanism at the atomic level. *Cell. Mol. Life Sci.* **2019**, *76*, 2917–2932. [CrossRef]
9. Jain, K.; Warmack, R.A.; Debler, E.W.; Hadjikyriacou, A.; Stavropoulos, P.; Clarke, S.G. Protein arginine methyltransferase product specificity is mediated by distinct active-site architectures. *J. Biol. Chem.* **2016**, *291*, 18299–18308. [CrossRef]
10. Zhang, X.; Cheng, X. Structure of the predominant protein arginine methyltransferase PRMT1 and analysis of its binding to substrate peptides. *Structure* **2003**, *11*, 509–520. [CrossRef]
11. Fuhrmann, J.; Clancy, K.; Thompson, P. Chemical biology of protein arginine modifications in epigenetic regulation. *Chem. Rev.* **2015**, *115*, 5413–5461. [CrossRef]
12. Morales, Y.; Nitzel, D.V.; Price, O.M.; Gui, S.; Li, J.; Qu, J.; Hevel, J.M. Redox Control of Protein Arginine Methyltransferase 1 (PRMT1) Activity. *J. Biol. Chem.* **2015**, *290*, 14915–14926. [CrossRef] [PubMed]
13. Weiss, V.; McBride, A.; Soriano, M.; Filman, D.; Silver, P.; Hogle, J. The structure and oligomerization of the yeast arginine methyltransferase, Hmt1. *Nat. Struct. Biol.* **2000**, *7*, 1165–1171. [CrossRef]
14. Feng, Y.; Xie, N.; Jin, M.; Stahley, M.; Stivers, J.; Zheng, Y. A transient kinetic analysis of PRMT1 catalysis. *Biochemistry* **2011**, *50*, 7033–7044. [CrossRef]
15. Herrmann, F.; Fackelmayer, F.O. Nucleo-cytoplasmic shuttling of protein arginine methyltransferase 1 (PRMT1) requires enzymatic activity. *Genes Cells* **2009**, *14*, 309–317. [CrossRef]
16. Thandapani, P.; O'Connor, T.; Bailey, T.; Richard, S. Defining the RGG/RG motif. *Mol. Cell* **2013**, *50*, 613–623. [CrossRef]
17. Morales, Y.; Cáceres, T.; May, K.; Hevel, J. Biochemistry and regulation of the protein arginine methyltransferases (PRMTs). *Arch. Biochem. Biophys.* **2016**, *590*, 138–152. [CrossRef]
18. Wooderchak, W.; Zang, T.Z.; Zhou, Z.S.; Acuna, M.; Taharam, S.M.; Hevel, J. Substrate profiling of PRMT1 reveals amino acid sequences that extend beyond the "RGG" paradigm. *Biochemistry* **2008**, *47*, 9456–9466. [CrossRef] [PubMed]
19. Osborne, T.C.; Obianyo, O.; Zhang, X.; Cheng, X.; Thompson, P.R. Protein Arginine Methyltransferase 1: Positively Charged Residues in Substrate Peptides Distal to the Site of Methylation Are Important for Substrate Binding and Catalysis. *Biochemistry* **2007**, *46*, 13370. [CrossRef]
20. Gui, S.; Wooderchak, W.; Daly, M.; Porter, P.; Johnson, S.; Hevel, J. Investigation of the molecular origins of protein-arginine methyltransferase I (PRMT1) product specificity reveals a role for two conserved methionine residues. *J. Biol. Chem.* **2011**, *286*, 29118–29126. [CrossRef]
21. Kirmizis, A.; Santos-Rosa, H.; Penkett, C.J.; Singer, M.A.; Vermeulen, M.; Mann, M.; Bähler, J.; Green, R.D.; Kouzarides, T. Arginine methylation at histone H3R2 controls deposition of H3K4 trimethylation. *Nature* **2007**, *449*, 928. [CrossRef]
22. Gui, S.; Gathiaka, S.; Li, J.; Qu, J.; Acevedo, O.; Hevel, J.M. A remodeled protein arginine methyltransferase 1 (PRMT1) generates symmetric dimethylarginine. *J. Biol. Chem.* **2014**, *289*, 9320–9327. [CrossRef] [PubMed]
23. Gathiaka, S.; Boykin, B.; Cáceres, T.; Hevel, J.M.; Acevedo, O. Understanding protein arginine methyltransferase 1 (PRMT1) product specificity from molecular dynamics. *Bioorg. Med. Chem.* **2016**, *24*, 4949–4960. [CrossRef]
24. Gui, S.; WL, W.-D.; Zang, T.; Chen, D.; Daly, M.; Zhou, Z.; Hevel, J. Substrate-induced control of product formation by protein arginine methyltransferase 1. *Biochemistry* **2013**, *52*, 199–209. [CrossRef]
25. Brown, J.I.; Koopmans, T.; van Strien, J.; Martin, N.I.; Frankel, A. Kinetic Analysis of PRMT1 Reveals Multifactorial Processivity and a Sequential Ordered Mechanism. *ChemBioChem* **2018**, *19*, 85–99. [CrossRef]
26. Kölbel, K.; Ihling, C.; Bellmann-Sickert, K.; Neundorf, I.; Beck-Sickinger, A.G.; Sinz, A.; Kühn, U.; Wahle, E. Type I arginine methyltransferases PRMT1 and PRMT-3 act distributively. *J. Biol. Chem.* **2009**, *284*, 8274–8282. [CrossRef]
27. Lakowski, T.M.; Frankel, A. Kinetic analysis of human protein arginine N-methyltransferase 2: Formation of monomethyl- and asymmetric dimethyl-arginine residues on histone H4. *Biochem. J.* **2009**, *421*, 253–261. [CrossRef]
28. Hu, H.; Luo, C.; Zheng, Y.G. Transient kinetics define a complete kinetic model for protein arginine methyltransferase 1. *J. Biol. Chem.* **2016**, *291*, 26722–26738. [CrossRef]
29. Obianyo, O.; Osborne, T.C.; Thompson, P.R. Kinetic mechanism of protein arginine methyltransferase. *Biochemistry* **2008**, *47*, 10420–10427. [CrossRef]
30. Obianyo, O.; Causey, C.P.; Jones, J.E.; Thompson, P.R. Activity-Based Protein Profiling of Protein Arginine Methyltransferase 1. *ACS Chem. Biol.* **2011**, *6*, 1127–1135. [CrossRef]
31. Wang, Y.; Wysocka, J.; Sayegh, J.; Lee, Y.; Perlin, J.; Leonelli, L.; Sonbuchner, L.; McDonald, C.; Cook, R.; Dou, Y.; et al. Human PAD4 regulates histone arginine methylation levels via demethylimination. *Science* **2004**, *306*, 279–283. [CrossRef]
32. Chang, B.; Chen, Y.; Zhao, Y.; Bruick, R. JMJD6 is a histone arginine demethylase. *Science* **2007**, *318*, 444–447. [CrossRef]
33. Li, S.; Ali, S.; Duan, X.; Liu, S.; Du, J.; Liu, C.; Dai, H.; Zhou, M.; Zhou, L.; Yang, L.; et al. JMJD1B Demethylates H4R3me2s and H3K9me2 to Facilitate Gene Expression for Development of Hematopoietic Stem and Progenitor Cells. *Cell Rep.* **2018**, *23*, 389–403. [CrossRef]
34. Zhang, X.; Li, L.; Li, Y.; Li, Z.; Zhai, W.; Sun, Q.; Yang, X.; Roth, M.; Lu, S. mTOR regulates PRMT1 expression and mitochondrial mass through STAT1 phosphorylation in hepatic cell. *Biochim. Biophys. Acta Mol. Cell Res.* **2021**, *1868*, 119017. [CrossRef]
35. Li, B.; Liu, L.; Li, X.; Wu, L. miR-503 suppresses metastasis of hepatocellular carcinoma cell by targeting PRMT1. *Biochem. Biophys. Res. Commun.* **2015**, *464*, 982–987. [CrossRef]

36. Rust, H.L.; Subramanian, V.; West, G.M.; Young, D.D.; Schultz, P.G.; Thompson, P.R. Using unnatural amino acid mutagenesis to probe the regulation of PRMT1. *ACS Chem. Biol.* **2014**, *9*, 649–655. [CrossRef]
37. Bao, X.; Siprashvili, Z.; Zarnegar, B.J.; Shenoy, R.M.; Rios, E.J.; Nady, N.; Qu, K.; Mah, A.; Webster, D.E.; Rubin, A.J.; et al. CSNK1a1 Regulates PRMT1 to Maintain the Progenitor State in Self-Renewing Somatic Tissue. *Dev. Cell* **2017**, *43*, 227–239.e5. [CrossRef]
38. Musiani, D.; Giambruno, R.; Massignani, E.; Ippolito, M.R.; Maniaci, M.; Jammula, S.; Manganaro, D.; Cuomo, A.; Nicosia, L.; Pasini, D.; et al. PRMT1 Is Recruited via DNA-PK to Chromatin Where It Sustains the Senescence-Associated Secretory Phenotype in Response to Cisplatin. *Cell Rep.* **2020**, *30*, 1208–1222.e9. [CrossRef]
39. Hirata, Y.; Katagiri, K.; Nagaoka, K.; Morishita, T.; Kudoh, Y.; Hatta, T.; Naguro, I.; Kano, K.; Udagawa, T.; Natsume, T.; et al. TRIM48 Promotes ASK1 Activation and Cell Death through Ubiquitination-Dependent Degradation of the ASK1-Negative Regulator PRMT1. *Cell Rep.* **2017**, *21*, 2447–2457. [CrossRef]
40. Bhuripanyo, K.; Wang, Y.; Liu, X.; Zhou, L.; Liu, R.; Duong, D.; Zhao, B.; Bi, Y.; Zhou, H.; Chen, G.; et al. Identifying the substrate proteins of U-box E3s E4B and CHIP by orthogonal ubiquitin transfer. *Sci. Adv.* **2018**, *4*, e1701393. [CrossRef]
41. Nie, M.; Wang, Y.; Guo, C.; Li, X.; Wang, Y.; Deng, Y.; Yao, B.; Gui, T.; Ma, C.; Liu, M.; et al. CARM1-mediated methylation of protein arginine methyltransferase 5 represses human γ-globin gene expression in erythroleukemia cells. *J. Biol. Chem.* **2018**, *293*, 17454–17463. [CrossRef]
42. Sakabe, K.; Hart, G.W. O-GlcNAc Transferase Regulates Mitotic Chromatin Dynamics. *J. Biol. Chem.* **2010**, *285*, 34460–34468. [CrossRef]
43. Lin, W.J.; Gary, J.D.; Yang, M.C.; Clarke, S.; Herschman, H.R. The mammalian immediate-early TIS21 protein and the leukemia-associated BTG1 protein interact with a protein-arginine N-methyltransferase. *J. Biol. Chem.* **1996**, *271*, 15034–15044. [CrossRef]
44. Robin-Lespinasse, Y.; Sentis, S.; Kolytcheff, C.; Rostan, M.C.; Corbo, L.; Le Romancer, M. hCAF1, a new regulator of PRMT1-dependent arginine methylation. *J. Cell Sci.* **2007**, *120*, 638–647. [CrossRef]
45. Pak, M.L.; Lakowski, T.M.; Thomas, D.; Vhuiyan, M.I.; Hüsecken, K.; Frankel, A. A protein arginine N-methyltransferase 1 (PRMT1) and 2 heteromeric interaction increases PRMT1 enzymatic activity. *Biochemistry* **2011**, *50*, 8226–8240. [CrossRef]
46. Duong, F.H.T.; Christen, V.; Berke, J.M.; Penna, S.H.; Moradpour, D.; Heim, M.H. Upregulation of protein phosphatase 2Ac by hepatitis C virus modulates NS3 helicase activity through inhibition of protein arginine methyltransferase 1. *J. Virol.* **2005**, *79*, 15342–15350. [CrossRef]
47. Gasperini, L.; Rossi, A.; Cornella, N.; Peroni, D.; Zuccotti, P.; Potrich, V.; Quattrone, A.; Macchi, P. The hnRNP RALY regulates PRMT1 expression and interacts with the ALS-linked protein FUS: Implication for reciprocal cellular localization. *Mol. Biol. Cell* **2018**, *29*, 3067–3081. [CrossRef]
48. Lei, N.; Zhang, X.; Chen, H.; Wang, Y.; Zhan, Y.; Zheng, Z.; Shen, Y.; Wu, Q. A feedback regulatory loop between methyltransferase PRMT1 and orphan receptor TR3. *Nucleic Acids Res.* **2009**, *37*, 832–848. [CrossRef]
49. Sun, Q.; Liu, L.; Mandal, J.; Molino, A.; Stolz, D.; Tamm, M.; Lu, S.; Roth, M. PDGF-BB induces PRMT1 expression through ERK1/2 dependent STAT1 activation and regulates remodeling in primary human lung fibroblasts. *Cell. Signal.* **2016**, *28*, 307–315. [CrossRef]
50. Vadnais, C.; Chen, R.; Fraszczak, J.; Yu, Z.; Boulais, J.; Pinder, J.; Frank, D.; Khandanpour, C.; Hébert, J.; Dellaire, G.; et al. GFI1 facilitates efficient DNA repair by regulating PRMT1 dependent methylation of MRE11 and 53BP1. *Nat. Commun.* **2018**, *9*, 1418. [CrossRef]
51. Inoue, H.; Hanawa, N.; Katsumata, S.-I.; Aizawa, Y.; Katsumata-Tsuboi, R.; Tanaka, M.; Takahashi, N.; Uehara, M. Iron deficiency negatively regulates protein methylation via the downregulation of protein arginine methyltransferase. *Heliyon* **2020**, *6*, e05059. [CrossRef]
52. Wagner, S.; Weber, S.; Kleinschmidt, M.A.; Nagata, K.; Bauer, U.-M. SET-mediated promoter hypoacetylation is a prerequisite for coactivation of the estrogen-responsive pS2 gene by PRMT1. *J. Biol. Chem.* **2006**, *281*, 27242–27250. [CrossRef] [PubMed]
53. Fulton, M.D.; Dang, T.; Brown, T.; Zheng, Y.G. Effects of substrate modifications on the arginine dimethylation activities of PRMT1 and PRMT5. *Epigenetics* **2020**, *31*, 1–18. [CrossRef] [PubMed]
54. Strahl, B.D.; Briggs, S.D.; Brame, C.J.; Caldwell, J.A.; Koh, S.S.; Ma, H.; Cook, R.G.; Shabanowitz, J.; Hunt, D.F.; Stallcup, M.R.; et al. Methylation of histone H4 at arginine 3 occurs in vivo and is mediated by the nuclear receptor coactivator PRMT1. *Curr. Biol.* **2001**, *11*, 996–1000. [CrossRef]
55. Stallcup, M.R. Role of protein methylation in chromatin remodeling and transcriptional regulation. *Oncogene* **2001**, *20*, 3014–3020. [CrossRef]
56. Wang, H.; Huang, Z.; Xia, L.; Feng, Q.; Erdjument-Bromage, H.; Strahl, B.; Briggs, S.; Allis, C.; Wong, J.; Tempst, P.; et al. Methylation of histone H4 at arginine 3 facilitating transcriptional activation by nuclear hormone receptor. *Science* **2001**, *293*, 853–857. [CrossRef] [PubMed]
57. Waldmann, T.; Izzo, A.; Kamieniarz, K.; Richter, F.; Vogler, C.; Sarg, B.; Lindner, H.; Young, N.L.; Mittler, G.; Garcia, B.A.; et al. Methylation of H2AR29 is a novel repressive PRMT6 target. *Epigenetics Chromatin* **2011**, *4*, 11. [CrossRef]
58. Dhar, S.; Vemulapalli, V.; Patananan, A.N.; Huang, G.L.; Di Lorenzo, A.; Richard, S.; Comb, M.J.; Guo, A.; Clarke, S.G.; Bedford, M.T. Loss of the major type i arginine methyltransferase PRMT1 causes substrate scavenging by other PRMTs. *Sci. Rep.* **2013**, *3*, 138–152. [CrossRef]

59. Guendel, I.; Carpio, L.; Pedati, C.; Schwartz, A.; Teal, C.; Kashanchi, F.; Kehn-Hall, K. Methylation of the Tumor Suppressor Protein, BRCA1, Influences Its Transcriptional Cofactor Function. *PLoS ONE* **2010**, *5*, e11379. [CrossRef]
60. Liu, L.-M.; Sun, W.-Z.; Fan, X.-Z.; Xu, Y.-L.; Cheng, M.-B.; Zhang, Y. Molecular Cell Biology Methylation of C/EBPa by PRMT1 Inhibits Its Tumor-Suppressive Function in Breast Cancer. *Cancer Res.* **2019**, *79*, 2865–2877. [CrossRef]
61. Tikhanovich, I.; Zhao, J.; Bridges, B.; Kumer, S.; Roberts, B.; Weinman, S.A. Arginine methylation regulates c-Myc–dependent transcription by altering promoter recruitment of the acetyltransferase p300. *J. Biol. Chem.* **2017**, *292*, 13333–13344. [CrossRef]
62. Li, Z.; Wang, D.; Lu, J.; Huang, B.; Wang, Y.; Dong, M.; Fan, D.; Li, H.; Gao, Y.; Hou, P.; et al. Methylation of EZH2 by PRMT1 regulates its stability and promotes breast cancer metastasis. *Cell Death Differ.* **2020**, *27*, 3226–3242. [CrossRef]
63. Yamagata, K.; Daitoku, H.; Takahashi, Y.; Namiki, K.; Hisatake, K.; Kako, K.; Mukai, H.; Kasuya, Y.; Fukamizu, A. Arginine methylation of FOXO transcription factors inhibits their phosphorylation by Akt. *Mol. Cell* **2008**, *32*, 221–231. [CrossRef] [PubMed]
64. Kagoya, Y.; Saijo, H.; Matsunaga, Y.; Guo, T.; Saso, K.; Anczurowski, M.; Wang, C.H.; Sugata, K.; Murata, K.; Butler, M.O.; et al. Arginine methylation of FOXP3 is crucial for the suppressive function of regulatory T cells. *J. Autoimmun.* **2019**, *97*, 10–21. [CrossRef] [PubMed]
65. Wang, Y.; Hsu, J.-M.; Kang, Y.; Wei, Y.; Lee, P.-C.; Chang, S.-J.; Hsu, Y.-H.; Hsu, J.L.; Wang, H.-L.; Chang, W.-C.; et al. Oncogenic functions of Gli in pancreatic adenocarcinoma are supported by its PRMT1-mediated methylation. *Cancer Res.* **2016**, *76*, 7049. [CrossRef] [PubMed]
66. Liu, Q.; Zhang, X.; Cheng, M.; Zhang, Y. PRMT1 activates myogenin transcription via MyoD arginine methylation at R121. *Biochim. Biophys. Acta Gene Regul. Mech.* **2019**, *1862*, 194442. [CrossRef] [PubMed]
67. Liu, X.; Li, H.; Liu, L.; Lu, Y.; Gao, Y.; Geng, P.; Li, X.; Huang, B.; Zhang, Y.; Lu, J. Methylation of arginine by PRMT1 regulates Nrf2 transcriptional activity during the antioxidative response. *Biochim. Biophys. Acta* **2016**, *1863*, 2093–2103. [CrossRef]
68. Malbeteau, L.; Poulard, C.; Languilaire, C.; Mikaelian, I.; Flamant, F.; Le Romancer, M.; Corbo, L. PRMT1 Is Critical for the Transcriptional Activity and the Stability of the Progesterone Receptor. *IScience* **2020**, *23*, 101236. [CrossRef]
69. Davies, C.C.; Chakraborty, A.; Diefenbacher, M.E.; Skehel, M.; Behrens, A. Arginine methylation of the c-Jun coactivator RACO-1 is required for c-Jun/AP-1 activation. *EMBO J.* **2013**, *32*, 1556. [CrossRef]
70. Reintjes, A.; Fuchs, J.E.; Kremser, L.; Lindner, H.H.; Liedl, K.R.; Huber, L.A.; Valovka, T. Asymmetric arginine dimethylation of RelA provides a repressive mark to modulate TNFα/NF-κB response. *Proc. Natl. Acad. Sci. USA* **2016**, *113*, 4326–4331. [CrossRef]
71. Huq, M.D.M.; Gupta, P.; Tsai, N.-P.; White, R.; Parker, M.G.; Wei, L.-N. Suppression of receptor interacting protein 140 repressive activity by protein arginine methylation. *EMBO J.* **2006**, *25*, 5094–5104. [CrossRef]
72. Zhao, X.; Jankovic, V.; Gural, A.; Huang, G.; Pardanani, A.; Menendez, S.; Zhang, J.; Dunne, R.; Xiao, A.; Erdjument-Bromage, H.; et al. Methylation of RUNX1 by PRMT1 abrogates SIN3A binding and potentiates its transcriptional activity. *Genes Dev.* **2008**, *22*, 640. [CrossRef]
73. Mowen, K.A.; Tang, J.; Zhu, W.; Schurter, B.T.; Shuai, K.; Herschman, H.R.; David, M. Arginine Methylation of STAT1 Modulates IFNα/β-Induced Transcription. *Cell* **2001**, *104*, 731–741. [CrossRef]
74. Jobert, L.; Argentini, M.; Tora, L. PRMT1 mediated methylation of TAF15 is required for its positive gene regulatory function. *Exp. Cell Res.* **2009**, *315*, 1273–1286. [CrossRef] [PubMed]
75. Tradewell, M.; Yu, Z.; Tibshirani, M.; Boulanger, M.; Durham, H.; Richard, S. Arginine methylation by PRMT1 regulates nuclear-cytoplasmic localization and toxicity of FUS/TLS harbouring ALS-linked mutations. *Hum. Mol. Genet.* **2012**, *21*, 136–149. [CrossRef]
76. Du, K.; Arai, S.; Kawamura, T.; Matsushita, A.; Kurokawa, R. TLS and PRMT1 synergistically coactivate transcription at the survivin promoter through TLS arginine methylation. *Biochem. Biophys. Res. Commun.* **2011**, *404*, 991–996. [CrossRef] [PubMed]
77. Huang, L.; Wang, Z.; Narayanan, N.; Yang, Y. Arginine methylation of the C-terminus RGG motif promotes TOP3B topoisomerase activity and stress granule localization. *Nucleic Acids Res.* **2018**, *46*, 3061–3074. [CrossRef]
78. Avasarala, S.; Van Scoyk, M.; Kumar, M.; Rathinam, K.; Zerayesus, S.; Zhao, X.; Zhang, W.; Pergande, M.R.; Borgia, J.A.; Degregori, J.; et al. PRMT1 Is a Novel Regulator of Epithelial-Mesenchymal-Transition in Non-small Cell Lung Cancer. *J. Biol. Chem.* **2015**, *290*, 13479–13489. [CrossRef]
79. Wei, H.-M.; Hu, H.-H.; Chang, G.-Y.; Lee, Y.-J.; Li, Y.-C.; Chang, H.-H.; Li, C. Arginine methylation of the cellular nucleic acid binding protein does not affect its subcellular localization but impedes RNA binding. *FEBS Lett.* **2014**, *588*, 1542–1548. [CrossRef] [PubMed]
80. Tsai, W.-C.; Gayatri, S.; Reineke, L.C.; Sbardella, G.; Bedford, M.T.; Lloyd, R.E. Arginine Demethylation of G3BP1 Promotes Stress Granule Assembly. *J. Biol. Chem.* **2016**, *291*, 22671. [CrossRef]
81. Wall, M.L.; Lewis, S.M. Methylarginines within the RGG-Motif Region of hnRNP A1 Affect Its IRES Trans-Acting Factor Activity and Are Required for hnRNP A1 Stress Granule Localization and Formation. *J. Mol. Biol.* **2017**, *429*, 295–307. [CrossRef]
82. Wang, L.; Jia, Z.; Xie, D.; Zhao, T.; Tan, Z.; Zhang, S.; Kong, F.; Wei, D.; Xie, K. Methylation of HSP70 orchestrates its binding to and stabilization of BCL2 mRNA and renders pancreatic cancer cells resistant to therapeutics. *Cancer Res.* **2021**, *80*, 4500–4513. [CrossRef] [PubMed]
83. Rho, J.; Choi, S.; Seong, Y.R.; Choi, J.; Im, D.-S. The Arginine-1493 Residue in QRRGRTGR1493G Motif IV of the Hepatitis C Virus NS3 Helicase Domain Is Essential for NS3 Protein Methylation by the Protein Arginine Methyltransferase 1. *J. Virol.* **2001**, *75*, 8031. [CrossRef] [PubMed]

84. Zhang, L.; Tran, N.-T.; Su, H.; Wang, R.; Lu, Y.; Tang, H.; Aoyagi, S.; Guo, A.; Khodadadi-Jamayran, A.; Zhou, D.; et al. Cross-talk between PRMT1-mediated methylation and ubiquitylation on RBM15 controls RNA splicing. *eLife* **2015**, *4*, e60742. [CrossRef] [PubMed]
85. Côté, J.; Boisvert, F.-M.; Boulanger, M.-C.; Bedford, M.T.; Richard, S. Sam68 RNA Binding Protein Is an In Vivo Substrate for Protein Arginine N-Methyltransferase 1. *Mol. Biol. Cell* **2003**, *14*, 274. [CrossRef]
86. Rho, J.; Choi, S.; Jung, C.R.; Im, D.S. Arginine methylation of Sam68 and SLM proteins negatively regulates their poly(U) RNA binding activity. *Arch. Biochem. Biophys.* **2007**, *466*, 49–57. [CrossRef]
87. Sinha, R.; Allemand, E.; Zhang, Z.; Karni, R.; Myers, M.P.; Krainer, A.R. Arginine Methylation Controls the Subcellular Localization and Functions of the Oncoprotein Splicing Factor SF2/ASF. *Mol. Cell. Biol.* **2010**, *30*, 2762–2774. [CrossRef]
88. Jia, H.; Du, C.H.; Bao, S.L.; Zheng, H.Y. Protein arginine methyltransferase 1 methylates SF2/ASF at arginine. *Chin. J. Cancer Biother.* **2009**, *16*, 216–220. [CrossRef]
89. Tahara, S.M.; Acuna, M. Discrimination of eIF4A isoforms by protein arginine methyltransferase 1 (PRMT1). *FASEB J.* **2006**, *20*, A109–A110. [CrossRef]
90. Hsu, J.H.-R.; Hubbell-Engler, B.; Adelmant, G.; Huang, J.; Joyce, C.E.; Vazquez, F.; Weir, B.A.; Montgomery, P.; Tsherniak, A.; Giacomelli, A.O.; et al. Prmt1-mediated translation regulation is a crucial vulnerability of cancer. *Cancer Res.* **2017**, *77*, 4613. [CrossRef]
91. Shin, H.-S.; Jang, C.-Y.; Kim, H.D.; Kim, T.-S.; Kim, S.; Kim, J. Arginine methylation of ribosomal protein S3 affects ribosome assembly. *Biochem. Biophys. Res. Commun.* **2009**, *385*, 273–278. [CrossRef]
92. Boisvert, F.-M.; Rhie, A.; Richard, S.; Doherty, A.J. The GAR Motif of 53BP1 is Arginine Methylated by PRMT1 and is Necessary for 53BP1 DNA Binding Activity. *Cell Cycle* **2005**, *4*, 1834–1841. [CrossRef]
93. Zhang, Y.; Zhang, Q.; Li, L.L.; Mu, D.; Hua, K.; Ci, S.; Shen, L.; Zheng, L.; Shen, B.; Guo, Z. Arginine methylation of APE1 promotes its mitochondrial translocation to protect cells from oxidative damage. *Free Radic. Biol. Med.* **2020**, *158*, 60–73. [CrossRef]
94. El-Andaloussi, N.; Valovka, T.; Toueille, M.; Hassa, P.O.; Gehrig, P.; Covic, M.; Hübscher, U.; Hottiger, M.O. Methylation of DNA polymerase ß by protein arginine methyltransferase 1 regulates its binding to proliferating cell nuclear antigen. *FASEB J.* **2007**, *21*, 26–34. [CrossRef] [PubMed]
95. Zheng, S.; Moehlenbrink, J.; Lu, Y.; Zalmas, L.; Sagum, C.; Carr, S.; McGouran, J.; Alexander, L.; Fedorov, O.; Munro, S.; et al. Arginine methylation-dependent reader-writer interplay governs growth control by E2F-1. *Mol. Cell* **2013**, *52*, 37–51. [CrossRef] [PubMed]
96. He, L.; Hu, Z.; Sun, Y.; Zhang, M.; Zhu, H.; Jiang, L.; Zhang, Q.; Mu, D.; Zhang, J.; Gu, L.; et al. PRMT1 is critical to FEN1 expression and drug resistance in lung cancer cells. *DNA Repair* **2020**, *95*, 102953. [CrossRef] [PubMed]
97. Yang, J.; Chiou, Y.; Fu, S.; Shih, I.; Weng, T.; Lin, W.; Lin, C. Arginine methylation of hnRNPK negatively modulates apoptosis upon DNA damage through local regulation of phosphorylation. *Nucleic Acids Res.* **2014**, *42*, 9908–9924. [CrossRef]
98. Gurunathan, G.; Yu, Z.; Coulombe, Y.; Masson, J.-Y.; Richard, S. Arginine methylation of hnRNPUL1 regulates interaction with NBS1 and recruitment to sites of DNA damage. *Sci. Rep.* **2015**, *5*, 1–9. [CrossRef]
99. Boisvert, F.M.; Déry, U.; Masson, J.Y.; Richard, S. Arginine methylation of MRE11 by PRMT1 is required for DNA damage checkpoint control. *Genes Dev.* **2005**, *19*, 671–676. [CrossRef]
100. Déry, U.; Coulombe, Y.; Rodrigue, A.; Stasiak, A.; Richard, S.; Masson, J.-Y. A Glycine-Arginine Domain in Control of the Human MRE11 DNA Repair Protein. *Mol. Cell. Biol.* **2008**, *28*, 3058. [CrossRef]
101. Matsumura, T.; Nakamura-Ishizu, A.; Anurag Muddineni, S.S.N.; Tan, D.Q.; Wang, C.Q.; Tokunaga, K.; Tirado-Magallanes, R.; Sian, S.; Benoukraf, T.; Okuda, T.; et al. Hematopoietic stem cells acquire survival advantage by loss of RUNX1 methylation identified in familial leukemia. *Blood* **2020**, *136*, 1919–1932. [CrossRef]
102. Cho, J.-H.; Lee, M.-K.; Yoon, K.W.; Lee, J.; Cho, S.-G.; Choi, E.-J. Arginine methylation-dependent regulation of ASK1 signaling by PRMT1. *Cell Death Differ.* **2012**, *19*, 859. [CrossRef]
103. Cha, B.; Kim, W.; Kim, Y.K.; Hwang, B.N.; Park, S.Y.; Yoon, J.W.; Park, W.S.; Cho, J.W.; Bedford, M.T.; Jho, E.H. Methylation by protein arginine methyltransferase 1 increases stability of Axin, a negative regulator of Wnt signaling. *Oncogene* **2011**, *30*, 2379–2389. [CrossRef]
104. Sakamaki, J.; Daitoku, H.; Ueno, K.; Hagiwara, A.; Yamagata, K.; Fukamizu, A. Arginine methylation of BCL-2 antagonist of cell death (BAD) counteracts its phosphorylation and inactivation by Akt. *Proc. Natl. Acad. Sci. USA* **2011**, *108*, 6085–6090. [CrossRef] [PubMed]
105. Pyun, J.; Kim, H.; Jeong, M.; Ahn, B.; Vuong, T.; Lee, D.; Choi, S.; Koo, S.; Cho, H.; Kang, J. Cardiac specific PRMT1 ablation causes heart failure through CaMKII dysregulation. *Nat. Commun.* **2018**, *9*, 1–15. [CrossRef]
106. Dolezal, E.; Infantino, S.; Drepper, F.; Börsig, T.; Singh, A.; Wossning, T.; Fiala, G.J.; Minguet, S.; Warscheid, B.; Tarlinton, D.M.; et al. The BTG2-PRMT1 module limits pre-B cell expansion by regulating the CDK4-Cyclin-D3 complex. *Nat. Immunol.* **2017**, *18*, 911–920. [CrossRef] [PubMed]
107. Onwuli, D.O.; Samuel, S.F.; Sfyri, P.; Welham, K.; Goddard, M.; Abu-Omar, Y.; Loubani, M.; Rivero, F.; Matsakas, A.; Benoit, D.M.; et al. The inhibitory subunit of cardiac troponin (cTnI) is modified by arginine methylation in the human heart. *Int. J. Cardiol.* **2019**, *282*, 76–80. [CrossRef]

108. Liao, H.-W.; Hsu, J.-M.; Xia, W.; Wang, H.-L.; Wang, Y.-N.; Chang, W.-C.; Arold, S.T.; Chou, C.-K.; Tsou, P.-H.; Yamaguchi, H.; et al. PRMT1-mediated methylation of the EGF receptor regulates signaling and cetuximab response. *J. Clin. Investig.* **2015**, *125*, 4529–4543. [CrossRef] [PubMed]
109. Le Romancer, M.; Treilleux, I.; Leconte, N.; Robin-Lespinasse, Y.; Sentis, S.; Bouchekioua-Bouzaghou, K.; Goddard, S.; Gobert-Gosse, S.; Corbo, L. Regulation of Estrogen Rapid Signaling through Arginine Methylation by PRMT1. *Mol. Cell* **2008**, *31*, 212–221. [CrossRef]
110. Deng, X.; Keudell, G. Von Suzuki, T.; Dohmae, N.; Nakakido, M.; Piao, L.; Yoshioka, Y.; Nakamura, Y.; Hamamoto, R. PRMT1 promotes mitosis of cancer cells through arginine methylation of INCENP. *Oncotarget* **2015**, *6*, 35173. [CrossRef] [PubMed]
111. Kim, H.J.; Jeong, M.H.; Kim, K.R.; Jung, C.Y.; Lee, S.Y.; Kim, H.; Koh, J.; Vuong, T.A.; Jung, S.; Yang, H.; et al. Protein arginine methylation facilitates KCNQ channel-PIP2 interaction leading to seizure suppression. *eLife* **2016**, *5*, e17159. [CrossRef] [PubMed]
112. Eberhardt, A.; Hansen, J.N.; Koster, J.; Lotta, L.T., Jr.; Wang, S.; Livingstone, E.; Qian, K.; Valentijn, L.J.; Zheng, Y.G.; Schor, N.F.; et al. Protein arginine methyltransferase 1 is a novel regulator of MYCN in neuroblastoma. *Oncotarget* **2016**, *7*, 63629. [CrossRef] [PubMed]
113. Yin, X.-K.; Wang, Y.-L.; Wang, F.; Feng, W.-X.; Bai, S.-M.; Zhao, W.-W.; Feng, L.-L.; Wei, M.-B.; Qin, C.-L.; Wang, F.; et al. PRMT1 enhances oncogenic arginine methylation of NONO in colorectal cancer. *Oncogene* **2021**, *40*, 1375. [CrossRef] [PubMed]
114. Liu, M.; Hua, W.; Chen, C.; Lin, W. The MKK-Dependent Phosphorylation of p38α Is Augmented by Arginine Methylation on Arg49/Arg149 during Erythroid Differentiation. *Int. J. Mol. Sci.* **2020**, *21*, 3546. [CrossRef]
115. Albrecht, L.V.; Ploper, D.; Tejeda-Muñoz, N.; De Robertis, E.M. Arginine methylation is required for canonical Wnt signaling and endolysosomal trafficking. *Proc. Natl. Acad. Sci. USA* **2018**, *115*, E5317–E5325. [CrossRef]
116. Xu, J.; Wang, A.H.; Oses-Prieto, J.; Makhijani, K.; Katsuno, Y.; Pei, M.; Yan, L.; Zheng, Y.G.; Burlingame, A.; Brückner, K.; et al. Arginine Methylation Initiates BMP-Induced Smad Signaling. *Mol. Cell* **2013**, *51*, 5–19. [CrossRef]
117. Zhang, T.; Wu, J.; Ungvijanpunya, N.; Jackson-Weaver, O.; Gou, Y.; Feng, J.; Ho, T.; Shen, Y.; Liu, J.; Richard, S.; et al. Smad6 Methylation Represses NFκB Activation and Periodontal Inflammation. *J. Dent. Res.* **2018**, *97*, 810–819. [CrossRef]
118. Katsuno, Y.; Qin, J.; Oses-Prieto, J.; Wang, H.; Jackson-Weaver, O.; Zhang, T.; Lamouille, S.; Wu, J.; Burlingame, A.; Xu, J.; et al. Arginine methylation of SMAD7 by PRMT1 in TGF-β-induced epithelial-mesenchymal transition and epithelial stem-cell generation. *J. Biol. Chem.* **2018**, *293*, 13059–137072. [CrossRef]
119. Tikhanovich, I.; Kuravi, S.; Artigues, A.; Villar, M.T.; Dorko, K.; Nawabi, A.; Roberts, B.; Weinman, S.A. Dynamic Arginine Methylation of Tumor Necrosis Factor (TNF) Receptor-associated Factor 6 Regulates Toll-like Receptor Signaling. *J. Biol. Chem.* **2015**, *290*, 22236–22249. [CrossRef]
120. Gen, S.; Matsumoto, Y.; Kobayashi, K.-I.; Suzuki, T.; Inoue, J.; Yamamoto, Y. Stability of tuberous sclerosis complex 2 is controlled by methylation at R1457 and R1459. *Sci. Rep.* **2020**, *10*, 1–9. [CrossRef]
121. Yang, Y.; Lu, Y.; Espejo, A.; Wu, J.; Xu, W.; Liang, S.; Bedford, M.T. TDRD3 is an Effector Molecule for Arginine Methylated Histone Marks. *Mol. Cell* **2010**, *40*, 1016. [CrossRef] [PubMed]
122. Yang, Y.; McBride, K.M.; Hensley, S.; Lu, Y.; Chedin, F.; Bedford, M.T. Arginine methylation facilitates the recruitment of TOP3B to chromatin to prevent R-loop accumulation. *Mol. Cell* **2014**, *53*, 484. [CrossRef]
123. Sims, R.J., III; Rojas, L.A.; Beck, D.B.; Bonasio, R.; Schüller, R.; Drury, W.J., III; Eick, D.; Reinberg, D. The C-Terminal Domain of RNA Polymerase II Is Modified by Site-Specific Methylation. *Science* **2011**, *332*, 99. [CrossRef]
124. Huang, S.; Litt, M.; Felsenfeld, G. Methylation of histone H4 by arginine methyltransferase PRMT1 is essential in vivo for many subsequent histone modifications. *Genes Dev.* **2005**, *19*, 1885–1893. [CrossRef]
125. Li, X.; Hu, X.; Patel, B.; Zhou, Z.; Liang, S.; Ybarra, R.; Qiu, Y.; Felsenfeld, G.; Bungert, J.; Huang, S. H4R3 methylation facilitates β-globin transcription by regulating histone acetyltransferase binding and H3 acetylation. *Blood* **2010**, *115*, 2028–2037. [CrossRef]
126. Fulton, M.D.; Zhang, J.; He, M.; Ho, M.-C.; Zheng, Y.G. Intricate Effects of α-Amino and Lysine Modifications on Arginine Methylation of the N-Terminal Tail of Histone H4. *Biochemistry* **2017**, *56*, 3539–3548. [CrossRef] [PubMed]
127. Feng, Y.; Wang, J.; Asher, S.; Hoang, L.; Guardiani, C.; Ivanov, I.; Zheng, Y.G. Histone H4 acetylation differentially modulates arginine methylation by an in cis mechanism. *J. Biol. Chem.* **2011**, *286*, 20323–20334. [CrossRef]
128. Thiebaut, C.; Vlaeminck-Guillem, V.; Trédan, O.; Poulard, C.; Le Romancer, M. Non-genomic signaling of steroid receptors in cancer. *Mol. Cell. Endocrinol.* **2021**, *538*, 111453. [CrossRef]
129. Choucair, A.; Pham, T.H.; Omarjee, S.; Jacquemetton, J.; Kassem, L.; Trédan, O.; Rambaud, J.; Marangoni, E.; Corbo, L.; Treilleux, I.; et al. The arginine methyltransferase PRMT1 regulates IGF-1 signaling in breast cancer. *Oncogene* **2019**, *38*, 4015–4027. [CrossRef]
130. Poulard, C.; Rambaud, J.; Hussein, N.; Corbo, L.; Le Romancer, M. JMJD6 regulates ERα methylation on arginine. *PLoS ONE* **2014**, *9*, e87982. [CrossRef] [PubMed]
131. Greer, E.L.; Brunet, A. FOXO transcription factors at the interface between longevity and tumor suppression. *Oncogene* **2005**, *24*, 7410–7425. [CrossRef] [PubMed]
132. Brunet, A.; Bonni, A.; Zigmond, M.J.; Lin, M.Z.; Juo, P.; Hu, L.S.; Anderson, M.J.; Arden, K.C.; Blenis, J.; Greenberg, M.E. Akt Promotes Cell Survival by Phosphorylating and Inhibiting a Forkhead Transcription Factor. *Cell* **1999**, *96*, 857–868. [CrossRef]
133. Kops, G.; Burgering, B. Forkhead transcription factors: New insights into protein kinase B (c-akt) signaling. *J. Mol. Med.* **1999**, *77*, 656–665. [CrossRef] [PubMed]
134. Huang, H.; Tindall, D.J. Regulation of FoxO protein stability via ubiquitination and proteasome degradation. *Biochim. Biophys. Acta* **2011**, *1813*, 1961. [CrossRef] [PubMed]

135. Matsuzaki, H.; Daitoku, H.; Hatta, M.; Tanaka, K.; Fukamizu, A. Insulin-induced phosphorylation of FKHR (Foxo1) targets to proteasomal degradation. *Proc. Natl. Acad. Sci. USA* **2003**, *100*, 11285–11290. [CrossRef]
136. Hassa, P.O.; Covic, M.; Bedford, M.T.; Hottiger, M.O. Protein Arginine Methyltransferase 1 Coactivates NF-κB-Dependent Gene Expression Synergistically with CARM1 and PARP1. *J. Mol. Biol.* **2008**, *377*, 668–678. [CrossRef]
137. Pawlak, M.R.; Scherer, C.A.; Chen, J.; Roshon, M.J.; Ruley, H.E. Arginine N-Methyltransferase 1 Is Required for Early Postimplantation Mouse Development, but Cells Deficient in the Enzyme Are Viable. *Mol. Cell. Biol.* **2000**, *20*, 4859. [CrossRef]
138. Tsai, Y.; Pan, H.; Hung, C.; Hou, P.; Li, Y.; Lee, Y.; Shen, Y.; Wu, T.; Li, C. The predominant protein arginine methyltransferase PRMT1 is critical for zebrafish convergence and extension during gastrulation. *FEBS J.* **2011**, *278*, 905–917. [CrossRef]
139. Shibata, Y.; Okada, M.; Miller, T.C.; Shi, Y.-B. Knocking out histone methyltransferase PRMT1 leads to stalled tadpole development and lethality in Xenopus tropicalis. *Biochim. Biophys. Acta Gen. Subj.* **2020**, *1864*, 129482. [CrossRef]
140. Batut, J.; Vandel, L.; Leclerc, C.; Daguzan, C.; Moreau, M.; Néant, I. The Ca2+-induced methyltransferase xPRMT1b controls neural fate in amphibian embryo. *Proc. Natl. Acad. Sci. USA* **2005**, *102*, 15128–15133. [CrossRef]
141. Sato, A.; Kim, J.D.; Mizukami, H.; Nakashima, M.; Kako, K.; Ishida, J.; Itakura, A.; Takeda, S.; Fukamizu, A. Gestational changes in PRMT1 expression of murine placentas. *Placenta* **2018**, *65*, 47–54. [CrossRef]
142. Yu, Z.; Chen, T.; Hébert, J.; Li, E.; Richard, S. A Mouse PRMT1 Null Allele Defines an Essential Role for Arginine Methylation in Genome Maintenance and Cell Proliferation. *Mol. Cell. Biol.* **2009**, *29*, 2982. [CrossRef] [PubMed]
143. Yu, Z.; Vogel, G.; Coulombe, Y.; Dubeau, D.; Spehalski, E.; Hébert, J.; Ferguson, D.O.; Masson, J.Y.; Richard, S. The MRE11 GAR motif regulates DNA double-strand break processing and ATR activation. *Cell Res.* **2012**, *22*, 305. [CrossRef] [PubMed]
144. Mitchell, T.R.H.; Glenfield, K.; Jeyanthan, K.; Zhu, X.-D. Arginine Methylation Regulates Telomere Length and Stability. *Mol. Cell. Biol.* **2009**, *29*, 4918–4934. [CrossRef]
145. Zhao, F.; Kim, W.; Kloeber, J.A.; Lou, Z. DNA end resection and its role in DNA replication and DSB repair choice in mammalian cells. *Exp. Mol. Med.* **2020**, *52*, 1705–1714. [CrossRef]
146. Adams, M.M.; Wang, B.; Xia, Z.; Morales, J.C.; Lu, X.; Donehower, L.A.; Bochar, D.A.; Elledge, S.J.; Carpenter, P.B. 53BP1 Oligomerization is Independent of its Methylation by PRMT1. *Cell Cycle* **2005**, *4*, 12. [CrossRef] [PubMed]
147. Guo, Z.; Zheng, L.; Xu, H.; Dai, H.; Zhou, M.; Pascua, M.; Chen, Q.; Shen, B. Methylation of FEN1 suppresses nearby phosphorylation and facilitates PCNA binding. *Nat. Chem. Biol.* **2010**, *6*, 766–773. [CrossRef]
148. Mathioudaki, K.; Scorilas, A.; Ardavanis, A.; Lymberi, P.; Tsiambas, E.; Devetzi, M.; Apostolaki, A.; Talieri, M. Clinical evaluation of PRMT1 gene expression in breast cancer. *Tumor Biol.* **2011**, *32*, 575–582. [CrossRef]
149. Poulard, C.; Treilleux, I.; Lavergne, E.; Bouchekioua-Bouzaghou, K.; Goddard-Léon, S.; Chabaud, S.; Trédan, O.; Corbo, L.; Le Romancer, M. Activation of rapid oestrogen signalling in aggressive human breast cancers. *EMBO Mol. Med.* **2012**, *4*, 1200–1213. [CrossRef]
150. Poulard, C.; Jacquemetton, J.; Trédan, O.; Cohen, P.A.; Vendrell, J.; Ghayad, S.E.; Treilleux, I.; Marangoni, E.; Le Romancer, M. Oestrogen non-genomic signalling is activated in tamoxifen-resistant breast cancer. *Int. J. Mol. Sci.* **2019**, *20*, 2773. [CrossRef] [PubMed]
151. Jacquemetton, J.; Kassem, L.; Poulard, C.; Dahmani, A.; De Plater, L.; Montaudon, E.; Sourd, L.; Morisset, L.; El Botty, R.; Chateau-Joubert, S.; et al. Analysis of genomic and non-genomic signaling of estrogen receptor in PDX models of breast cancer treated with a combination of the PI3K inhibitor Alpelisib (BYL719) and fulvestrant. *Breast Cancer Res.* 2021, in press.
152. Nakai, K.; Xia, W.; Liao, H.; Saito, M.; Hung, M.; Yamaguchi, H. The role of PRMT1 in EGFR methylation and signaling in MDA-MB-468 triple-negative breast cancer cells. *Breast Cancer* **2018**, *25*, 74–80. [CrossRef]
153. Gao, Y.; Zhao, Y.; Zhang, J.; Lu, Y.; Liu, X.; Geng, P.; Huang, B.; Zhang, Y.; Lu, J. The dual function of PRMT1 in modulating epithelial-mesenchymal transition and cellular senescence in breast cancer cells through regulation of ZEB1. *Sci. Rep.* **2016**, *6*, 1–13. [CrossRef] [PubMed]
154. Papadokostopoulou, A.; Mathioudaki, K.; Scorilas, A.; Xynopoulos, D.; Ardavanis, A.; Kouroumalis, E.; Talieri, M. Colon cancer and protein arginine methyltransferase 1 gene expression. *Anticancer Res.* **2009**, *29*, 1361–1366. [PubMed]
155. Mathioudaki, K.; Papadokostopoulou, A.; Scorilas, A.; Xynopoulos, D.; Agnanti, N.; Talieri, M. The PRMT1 gene expression pattern in colon cancer. *Br. J. Cancer* **2008**, *99*, 2094. [CrossRef]
156. Yao, B.; Gui, T.; Zeng, X.; Deng, Y.; Wang, Z.; Wang, Y.; Yang, D.; Li, Q.; Xu, P.; Hu, R.; et al. PRMT1-mediated H4R3me2a recruits SMARCA4 to promote colorectal cancer progression by enhancing EGFR signaling. *Genome Med.* **2021**, *13*, 1–21. [CrossRef]
157. Yoshimatsu, M.; Toyokawa, G.; Hayami, S.; Unoki, M.; Tsunoda, T.; Field, H.I.; Kelly, J.D.; Neal, D.E.; Maehara, Y.; Ponder, B.A.J.; et al. Dysregulation of PRMT1 and PRMT6, Type I arginine methyltransferases, is involved in various types of human cancers. *Int. J. Cancer* **2011**, *128*, 562–573. [CrossRef]
158. Zhao, Y.; Lu, Q.; Li, C.; Wang, X.; Jiang, L.; Huang, L.; Wang, C.; Chen, H. PRMT1 regulates the tumour-initiating properties of esophageal squamous cell carcinoma through histone H4 arginine methylation coupled with transcriptional activation. *Cell Death Dis.* **2019**, *10*, 1–17. [CrossRef]
159. He, X.; Zhu, Y.; Lin, Y.; Li, M.; Du, J.; Dong, H.; Sun, J.; Zhu, L.; Wang, H.; Ding, Z.; et al. PRMT1-mediated FLT3 arginine methylation promotes maintenance of FLT3-ITD + acute myeloid leukemia. *Blood* **2019**, *134*, 548–560. [CrossRef]
160. Matsubara, H.; Fukuda, T.; Awazu, Y.; Nanno, S.; Shimomura, M.; Inoue, Y.; Yamauchi, M.; Yasui, T.; Sumi, T. PRMT1 expression predicts sensitivity to platinum-based chemotherapy in patients with ovarian serous carcinoma. *Oncol. Lett.* **2021**, *21*, 1. [CrossRef]

161. Seligson, D.B.; Horvath, S.; Shi, T.; Yu, H.; Tze, S.; Grunstein, M.; Kurdistani, S.K. Global histone modification patterns predict risk of prostate cancer recurrence. *Nature* **2005**, *435*, 1262–1266. [CrossRef]
162. Cheng, D.; Yadav, N.; King, R.W.; Swanson, M.S.; Weinstein, E.J.; Bedford, M.T. Small Molecule Regulators of Protein Arginine Methyltransferases. *J. Biol. Chem.* **2004**, *279*, 23892–23899. [CrossRef]
163. Eram, M.S.; Shen, Y.; Szewczyk, M.; Wu, H.; Senisterra, G.; Li, F.; Butler, K.V.; Kaniskan, H.Ü.; Speed, B.A.; Seña, C.; et al. A Potent, Selective and Cell-active Inhibitor of Human Type I Protein Arginine Methyltransferases. *ACS Chem. Biol.* **2016**, *11*, 772. [CrossRef]
164. Fong, J.Y.; Pignata, L.; Goy, P.-A.; Kawabata, K.C.; Lee, S.C.-W.; Koh, C.M.; Musiani, D.; Massignani, E.; Kotini, A.G.; Penson, A.; et al. Therapeutic Targeting of RNA Splicing Catalysis through Inhibition of Protein Arginine Methylation. *Cancer Cell* **2019**, *36*, 194. [CrossRef] [PubMed]
165. Fedoriw, A.; Rajapurkar, S.R.; O'Brien, S.; Gerhart, S.V.; Mitchell, L.H.; Adams, N.D.; Rioux, N.; Lingaraj, T.; Ribich, S.A.; Pappalardi, M.B.; et al. Anti-tumor Activity of the Type I PRMT Inhibitor, GSK3368715, Synergizes with PRMT5 Inhibition through MTAP Loss. *Cancer Cell* **2019**, *36*, 100–114.e25. [CrossRef]
166. Zhang, P.; Tao, H.; Yu, L.; Zhou, L.; Zhu, C. Developing protein arginine methyltransferase 1 (PRMT1) inhibitor TC-E-5003 as an antitumor drug using INEI drug delivery systems. *Drug Deliv.* **2020**, *27*, 491–501. [CrossRef] [PubMed]
167. Hu, H.; Qian, K.; Ho, M.-C.; Zheng, Y.G. Small Molecule Inhibitors of Protein Arginine Methyltransferases. *Expert Opin. Investig. Drugs* **2016**, *25*, 335. [CrossRef] [PubMed]

Review

Structure, Activity and Function of the PRMT2 Protein Arginine Methyltransferase

Vincent Cura [1,2,3,4] and Jean Cavarelli [1,2,3,4,*]

1. Institut de Génétique et de Biologie Moléculaire et Cellulaire, 67404 Illkirch, France; cura@igbmc.fr
2. Centre National de la Recherche Scientifique, UMR 7104, 67404 Illkirch, France
3. Institut National de la Santé et de la Recherche Médicale, U1258, 67404 Illkirch, France
4. Université de Strasbourg, 67000 Strasbourg, France
* Correspondence: jean.cavarelli@igbmc.fr; Tel.: +33-(0)3-6948-5274

Abstract: PRMT2 belongs to the protein arginine methyltransferase (PRMT) family, which catalyzes the arginine methylation of target proteins. As a type I enzyme, PRMT2 produces asymmetric dimethyl arginine and has been shown to have weak methyltransferase activity on histone substrates in vitro, suggesting that its authentic substrates have not yet been found. PRMT2 contains the canonical PRMT methylation core and a unique Src homology 3 domain. Studies have demonstrated its clear implication in many different cellular processes. PRMT2 acts as a coactivator of several nuclear hormone receptors and is known to interact with a multitude of splicing-related proteins. Furthermore, PRMT2 is aberrantly expressed in several cancer types, including breast cancer and glioblastoma. These reports highlight the crucial role played by PRMT2 and the need for a better characterization of its activity and cellular functions.

Keywords: protein arginine methylation; PRMT2; epigenetics; SH3; cancer

1. Introduction

Arginine methylation is a widespread posttranslational modification in eukaryotes catalyzed by protein arginine methyltransferases (PRMTs), a class of enzymes that transfers methyl groups from S-adenosyl-L-methionine (SAM) to guanidine nitrogen atoms in arginine residues of target proteins.

Methylation makes arginine bulkier and more hydrophobic as well as reducing its H-bonding potential, thereby altering interactions with other proteins or nucleic acids [1,2]. Arginine methylation is involved in different cellular processes, including transcriptional regulation, RNA metabolism, DNA repair and signal transduction (see [2,3] for recent reviews). The nine PRMTs identified in mammals have been classified into three types. Type I PRMTs, including PRMT1, 2, 3, 4, 6 and 8, catalyze the formation of asymmetric dimethylarginine, while type II PRMTs (PRMT5 and PRMT9) produce symmetric dimethylarginine. PRMT7, the only type III PRMT, generates mono-methylarginine.

This review is focused on PRMT2, one of the least functionally characterized PRMTs. The difficulty in detecting its importance in cellular processes was initially attributed to its low methyl transferase activity on classical PRMT substrates, namely, histone tails. However, various studies have since demonstrated the implication of PRMT2 in transcriptional regulation independently of its catalytic activity and, therefore, in cancer. Furthermore, recent results in the systematic analysis of PRMT interactomes shed new light on PRMT2 interactants and potential substrates [4]. The interaction of PRMT2 with RNA binding proteins and splicing factors is discussed.

2. Structural features
2.1. Sequence

PRMT2 (or HRMT1L1) was first identified in the human genome through sequence homology with PRMT1 [5]. Phylogenetic analysis [6] and sequence comparisons have established that the PRMT2 methylation module is closely related to all type I PRMTs (35% to 39% sequence identity between PRMT1, PRMT3, PRMT6, PRMT8 and PRMT4 (CARM1) from mouse) (Figure 1). PRMT2 is present in all vertebrates, except in reptiles and birds, and has also been found in cnidaria, echinoderms and cephalochordates [7]. It is mainly localized in the nucleus, excluded from nucleolus, but is also found at low levels in the cytoplasm [8,9].

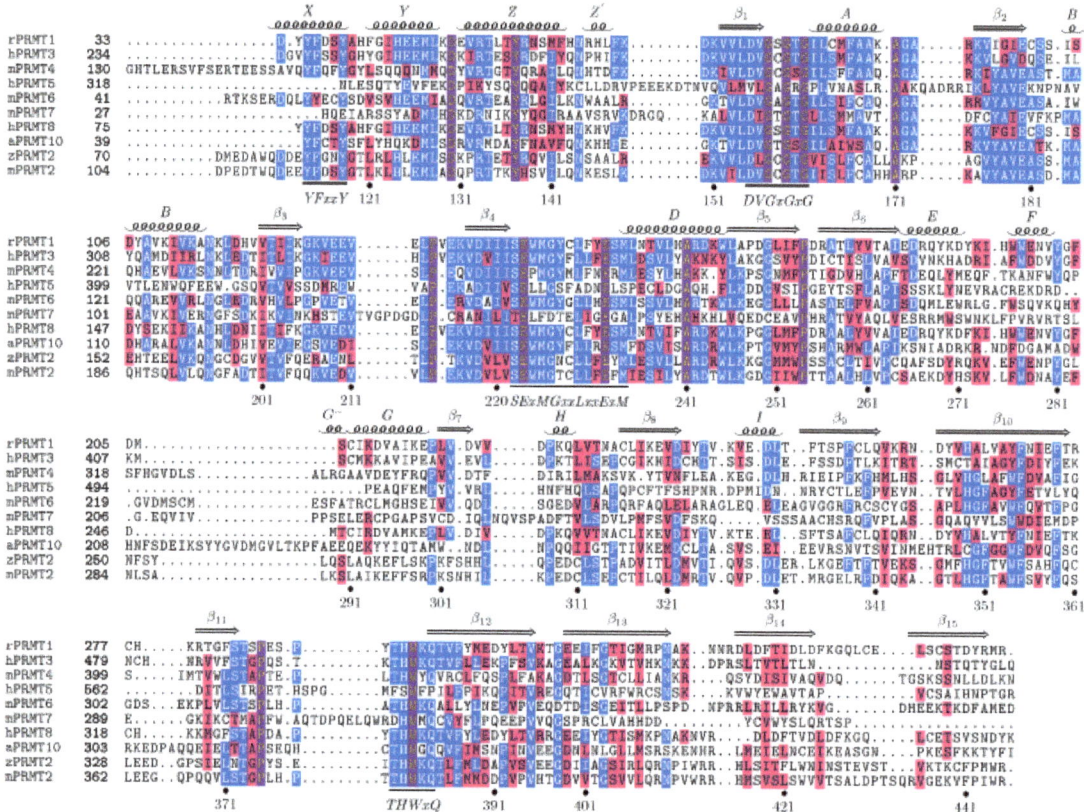

Figure 1. Structure-based sequence alignment of selected PRMTs. Ten PRMT sequences are aligned based on their crystal structures. The alignment is restricted to the catalytic core. The secondary structure of mPRMT2 is drawn above the alignment. The SAM-binding domain, the β-barrel domains and the dimerization arm are colored green, yellow and blue, respectively. The mPRMT2 residue numbering is shown below the sequences. The four signature sequences are localized, and their consensus is written below. Amino acids are shaded according to similarity to the consensus sequence. Amino acids highlighted are either invariant (violet) or similar (blue) as defined by the following grouping: F, Y and W; I, L, M and V; R and K; D and E; and G and A; S, T, N and Q. Abbreviations are as follows: m/Mus musculus, h/Homo sapiens, r/Rattus norvegicus, a/Arabidopsis thaliana and z/zebrafish (*D. rerio*). This figure, adapted from [10], was produced with the program TEXSHADE [11].

The PRMT2 sequence contains the canonical PRMT methylation core composed of two domains: An SAM-binding domain adopting a Rossmann fold followed by a β-barrel interrupted by a protruding helix–coil (Figure 2). PRMT2 exhibits all the conserved motifs involved in the SAM and peptide binding characteristics of PRMTs. However, these enzymes mainly differ in terms of the potential presence of additional domains. PRMT2 is characterized by an N-terminal extension containing a 50 residue Src homology 3 (SH3) domain located downstream of an unfolded N-terminal extremity that varies slightly in size, depending on the species.

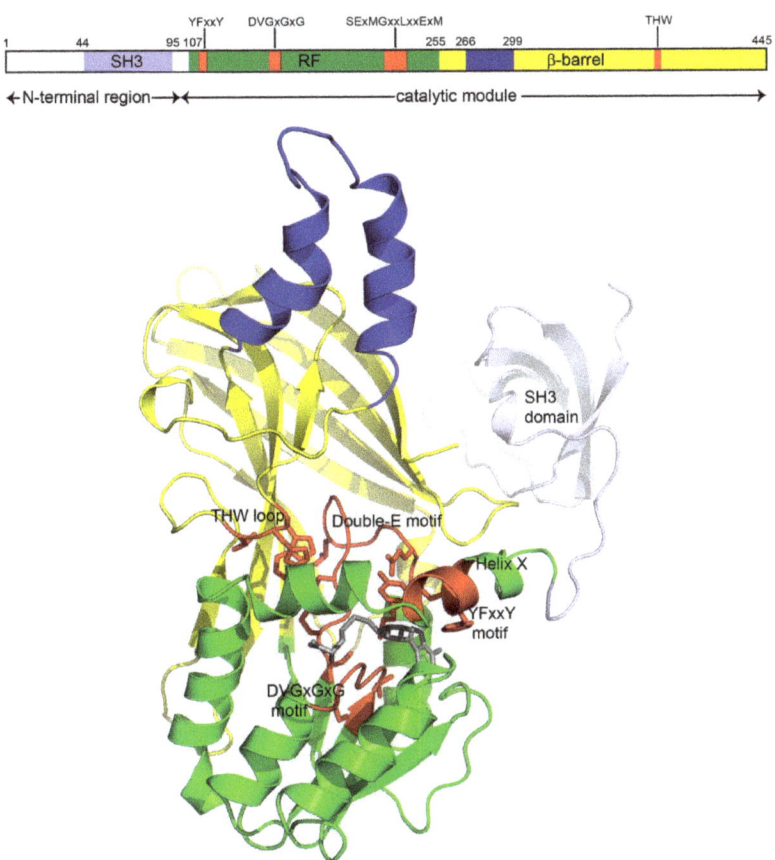

Figure 2. Full-length mPRMT2. Top: Scheme showing the modular organization of PRMT2. SH3 domain is indicated in light blue. The Rossmann fold is shown in green, the β barrel in yellow and the dimerization arm in blue. Motif YFxxY, motif DVGxGxG, double-E loop and motif THW are shown in red. Bottom: Three-dimensional model of full-length monomeric mouse PRMT2 generated with AlphaFold [12]. The modelized methylation module was replaced by the X-ray structure (PDB 5FUL) after superimposition. S-adenosyl-L-homocysteine (SAH) is displayed as a gray stick model. The 3D cartoon was generated with PyMol (http://www.pymol.org).

Different isoforms resulting from alternative mRNA splicing have been found in several organisms. Nevertheless, according to sequence conservation, only one isoform is common to every species and is considered as the canonical form. In humans, in addition to the full-length PRMT2 expressed from a gene of eleven exons [5], six alternatively spliced PRMT2 isoforms have been detected (UniProtKB–P55345) and four of them (PRMT2L2, PRMT2α, β, and γ) have been isolated from breast cancer cells [13,14] (Figure 3). In all of these variants, the β barrel domain containing the dimerization helix–coil and the THW loop required for a fully active enzyme is missing. The sequences restricted to the SH3- and a major part of the Rossman-fold domain appear dramatically modified compared to the full-length PRMT2, leading to catalytically inactive proteins.

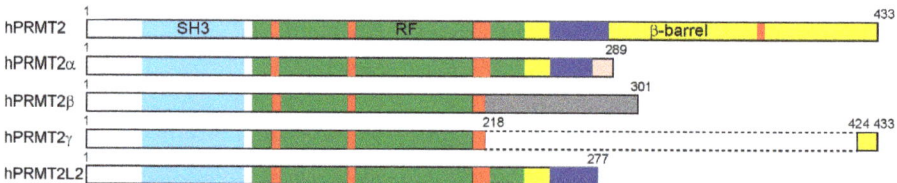

Figure 3. Human PRMT2 isoforms. PRMT2α is obtained after deletion of exons 8 to 10, including a modification of the 12 last residues due to a frameshift (in salmon). PRMT2β lacks exons 7 to 9, leading to a 301 amino acid sequence with a specific C-terminal sequence resulting from a frame-shift (in gray). The PRMT2γ isoform is produced by the removal of exons 7 to 10 corresponding to an in-frame deletion of 205 amino acids (indicated by dotted lines). PRMT2L2 being slightly smaller than PRMT2α results from alternative polyadenylation. Motif YFxxY, motif DVGxGxG, double-E loop and motif THW are represented in red. Dimerization helices are shown in blue.

2.2. Structure

Several X-ray structures of the PRMT2 methylation core from two different organisms have been determined [10]. The structure of PRMT2 from *D. rerio* was solved with the co-factor product of the reaction S-adenosyl-L-homocysteine (SAH) (PDB: 5FUB) and sinefungin (PDB: 5G02), while the structure of *M. musculus* (mPRMT2) was obtained in complex with SAH (PDB: 5FUL) and three inhibitors (PDB: 5FWA, 5FWD and 5JMQ).

As expected, the monomeric structure of the PRMT2 catalytic module is very similar to that of all the known PRMT structures, especially of type I PRMTs. It consists of an SAM-binding domain (residues 107-254 and mPRMT2 numbering) adopting a Rossmann fold and a β-barrel domain composed of eight strands (residues 255–265 and residues 299–445) (Figure 2). The helical dimerization arm encompasses residues 266–298. The two domains are connected by the strictly conserved cis-proline 254 [15]. Unfortunately, the N-terminal module of mPRMT2 is missing in the electron density map despite being present in the crystal [10]. Similar wobbly domain behavior has previously been observed for the PH N-terminal domain of CARM1 [15]. However, a structure of the isolated SH3 domain of human PRMT2 was determined by NMR in 2005 (PDB: 1X2P). It displays a classical SH3 fold containing 50 residues, which form five antiparallel β-strands folded into a barrel structure. SH3 domains are known to bind to target proteins through sequences containing proline and hydrophobic amino acids and are usually involved in protein–protein interactions [16,17].

2.3. Co-Factor Binding Site

The SAM binding pocket is formed by the motif sequence DVGCGTG (Figure 1). Hydrogen bonds involving the residue E209 carboxylate and the S237 Oγ maintain the SAM adenine amino group. N1 interacts with the V208 main chain carbonyl, and the E180 carboxylate forms two hydrogen bonds with the ribose hydroxyl oxygens. For the homocysteine moiety, the carboxylate group binds to R133, the amino group interacts with the C158 carbonyl, and M127 makes van der Waals contact with the S atom. Helix X, which harbors the conserved YFxxY motif, closes the SAM binding pocket (Figures 2 and 4a).

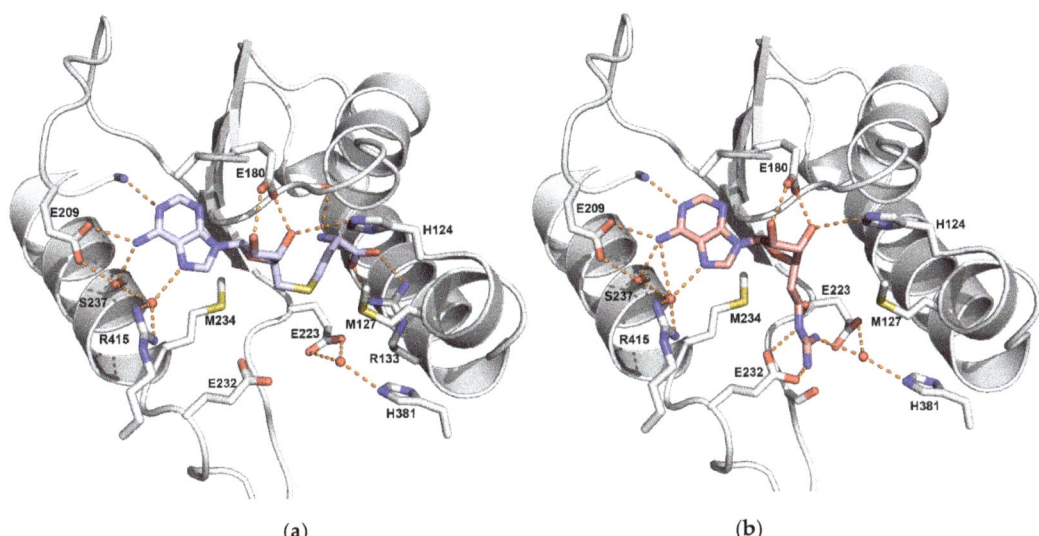

Figure 4. Interactions in the active site of mouse PRMT2 in complex with SAH (in blue) (PDB 5FUL) (**a**) and with Cp1 (in salmon) (PDB 5FWA) (**b**). The guanidinium group of Cp1 is held between the carboxylates of the double-E motif residues E223 and E232 by a set of H-bonds and salt bridges and mimics the arginine substrate guanidinium group. Water molecules are shown as red spheres, and hydrogen bonds are indicated by dotted lines. This figure was generated with PyMol (http://www.pymol.org).

2.4. Substrate Binding Pocket

The strictly conserved glutamate residues E223 and E232 of the double-E motif form a pair of salt bridges with the positively charged guanidinium group of the substrate arginine. These two glutamates are involved in the generally agreed PRMT catalytic mechanism by positioning the guanidinium group and modulating its nucleophilicity to favor methyl transfer [18,19].

PRMT2 structures obtained with SAM/arginine-like inhibitors indicate that the guanidinium group is positioned in the arginine pocket between the E223 and E232 carboxylates [10]. Helix X interacts via Y118 and Y114 with catalytic E232, together with the THW loop, allowing the formation of the substrate arginine pocket required for catalysis (Figures 2 and 4).

2.5. Dimerization

Homodimerization is a feature conserved in all type I PRMTs and is essential for catalytic activity (see [3] for a recent review). In PRMT2, the dimer formation involves the dimerization arm from one monomer and helices Y, Z, A, and B from the other monomer, leading to the classical doughnut-shaped structure with a central hole, common to all type I PRMTs (Figure 5). Monomers are related to each other by a twofold rotational symmetry and are both able to bind the substrates. Small angle X-ray scattering (SAXS) experiments confirmed that PRMT2 behaves as a dimer in solution [10].

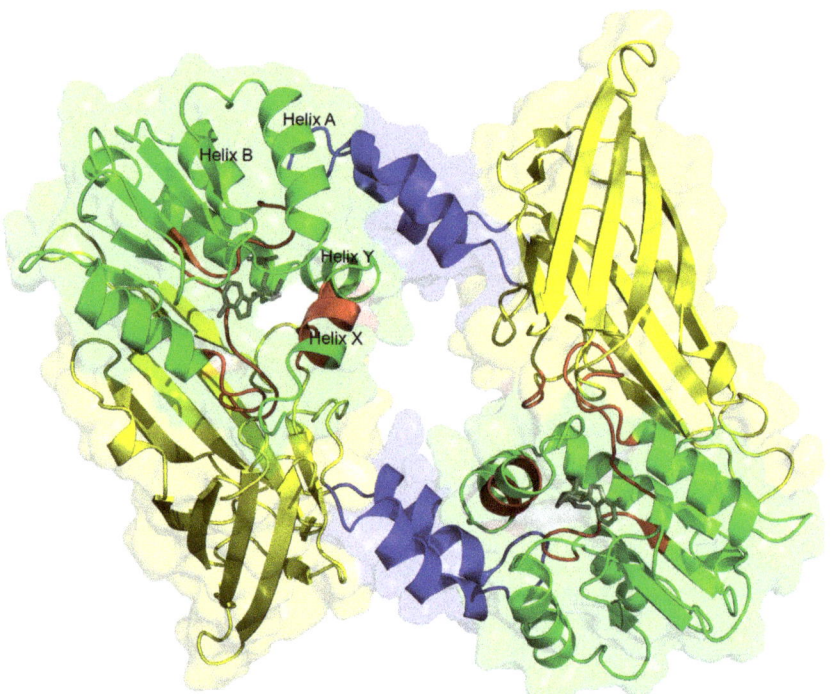

Figure 5. Crystal structure of dimeric mouse PRMT2 (PDB 5FUL). The N terminal part is not visible in the structure. The Rossmann fold domain is shown in green, the β barrel in yellow and the dimerization arm in blue. Helices A, B, X and Y interacting with dimerization helix–coil are indicated. SAH is displayed as a gray stick model, and SAM/SAH-binding motifs are in red. Structure cartoon was generated with PyMol (http://www.pymol.org).

3. Activity

The first attempt to detect human PRMT2 (hPRMT2) activity proved unsuccessful. No activity could be detected using a recombinant hPRMT2, suggesting that the enzyme could be inactive [20]. Indeed, the in vitro methyltransferase activity of PRMT2 on histones is described as being weak compared to that of PRMT1, CARM1 and PRMT6 [21]. Initially, a small, but significant, activity was described on histone H4 [22]. Further experiments demonstrated that PRMT2 catalyzes asymmetric dimethylation of histone H3 arginine 8 in cells and that the presence of H3R8me2a at promoters is required to regulate target gene expression [23–26].

We detected a low methyltransferase activity signal using purified mPRMT2 and either H3 or H4 histone-tail peptides as substrates in vitro [10]. However, a stronger signal corresponding to PRMT2 automethylation could be revealed, suggesting that the enzyme is potentially fully active but that the optimal conditions, in terms of substrate or interacting partner, were not met. It is noteworthy that automethylation has been detected in several PRMTs [27–30]. A systematic analysis of protein methylation in mouse tissues revealed that R84 is methylated in mouse and human PRMT2 [31,32]. This arginine localizes in the SH3 domain and could, therefore, correspond to an automethylation site.

We discovered that an SAM-based compound, Cp1, reported as an inhibitor for PRMT1, PRMT6 and CARM [33], was also able to inhibit the activity of PRMT2 in vitro slightly more efficiently than SAH [10]. The IC$_{50}$ values were 16.3 ± 3.8 µM for Cp1 and 18.3 ± 2.0 µM for SAH. However, thermal shift assays showed that binding of SAH increased PRMT2's melting temperature (Tm) by 5.3 °C ± 1.2 °C with respect to the apo protein. The Tm shift reached 15 °C ± 1.9 °C with Cp1, indicating a stronger affinity of PRMT2 for Cp1 than for SAH. The X-ray structure revealed the key interactions occurring in the active site required for its recognition and specificity [10] (Figure 4b).

4. Interactions

4.1. Coactivation

It has been shown that PRMT2 acts as a coactivator of several nuclear hormone receptors. Using a yeast two-hybrid system, Qi and coworkers showed that PRMT2 interacts directly with estrogen receptor alpha (ERα) [34]. Three ERα regions, namely, AF-1, the DNA binding domain, and the hormone binding domain, were identified as interaction areas. The ER-interacting region on PRMT2, encompassing amino acids 133–275, is localized in the Rossmann fold domain. PRMT2 is able to enhance ER transcriptional activity. In another study, Meyer et al. [9] pointed out the interaction with ERα was strongly dependent on the cellular background, suggesting the involvement of differentially expressed coregulators.

The same authors identified PRMT2 as an AR-associated protein binding directly to the receptor via the C-terminal part (residues 271-433) of PRMT2 [9]. They demonstrated that PRMT2 acts as a strong coactivator of the androgen receptor (AR) in the presence of AR agonists. The coactivation function seems to depend on the methyltransferase activity of PRMT2. Furthermore, AR and PRMT2 colocalize and translocate from the cytoplasm into the nucleus when androgens are present.

PRMT2 promotes apoptosis by inhibiting NF-κB-dependent transcription [35]. The SAM-binding domain interacts with IκB-α by its ankyrin domain, which mediates the interaction with NF-κB. PRMT2 blocks nuclear export of IκB-α, causing increased levels of IκB-α in the nucleus and preventing NF-κB from binding DNA in mouse fibroblasts. The regulation role of PRMT2 on NF-κB was pointed out by Dalloneau and co-workers in the pulmonary inflammatory and airway distress syndrome induced by lipopolysaccharide (LPS) [36]. After LPS treatment, PRMT2 is downregulated in lungs and in macrophages, which allows the binding of NF-κB to the promoters of its target genes, such as cytokines IL-6 and TNF-α, leading to the inflammatory response.

In addition to its role as a transcriptional co-activator, PRMT2 has also been found to be involved in diverse cellular processes, such as energy homeostasis. The observation that PRMT2 null mice are leaner than wildtype animals, associated with perturbed energy metabolism, resistance to obesity and enhanced leptin sensitivity, suggested an involvement of PRMT2 in the regulation of feeding via a leptin-dependent pathway [37]. The authors showed that PRMT2 colocalizes with the transcription factor STAT3 in hypothalamic nuclei, where it binds and methylates STAT3 at the R31 residue. These results revealed that PRMT2 is a pivotal modulator of hypothalamic leptin–STAT3 signaling and energy homeostasis.

In 2015, Hussein and coworkers found that PRMT2 expression was reduced in diabetes-relevant high glucose conditions in macrophages. PRMT2 enhances ATP-binding cassette transporter A1 (ABCA1) expression induced by the liver X receptor (LXR) [38]. Thus, PRMT2 represents a glucose-sensitive factor that controls ABCA1-dependent cholesterol efflux and could provide a potential explanation behind the atherosclerosis development in diabetic patients. Although the mechanism is not known, this effect may be related to the discovery made by Li et al., who demonstrated that PRMT2 inhibits macrophage-derived foam cell formation [39].

PRMT2 interacts directly with PRMT1 to increase PRMT1 activity and influences the substrate specificity of the resulting complex both in vitro and in HeLa cells [40]. The binding requires the dimerization arm and catalytic activity of PRMT1. A study revealed that the SH3 domain regulates the interaction between PRMT1 and PRMT2 in a methylation-dependent manner. PRMT2 interacts with the retinoblastoma protein (RB) to regulate E2F transcriptional activity [41]. In contrast to other PRMTs, PRMT2 binds directly to RB through its SAM-binding domain, forming a ternary complex with E2F1.

The authors of this study showed that PRMT2 repressed E2F1 transcriptional activity in an RB-dependent manner, delaying cell cycle progression from G1 to the S phase.

Blythe and coworkers showed that PRMT2 is directly recruited by β-catenin to target gene promoters during dorsal development in Xenopus, leading to histone H3 dimethylation on arginine 8 [23]. Associated with H3K4 trimethylation, H3R8me2 activates the Spindlin1-Wnt/β-catenin signaling pathway implicating the activity of PRMT2 in the expression of Wnt target genes [24].

Hou and coworkers showed that PRMT2 regulates the function of the actin nucleator Cobl by arginine methylation [42]. This posttranslational modification is crucial for proper Cobl association with G-actin. Both catalytic and SH3 domains are required for PRMT2–Cobl interaction and activity. The two methylated arginine residues are located in the second WH2 domain of Cobl, which is known to bind strongly to actin [43]. Thus, through Cobl methylation, PRMT2 plays a role in neuronal morphogenesis and dendritic arborization regulation in the central nervous system.

Additionally, PRMT2 expression was found to be selectively upregulated in alveolar epithelial cells of mouse lungs in response to chronic hypoxia. These results demonstrate that PRMT2 expression may be linked to asymmetric dimethylarginine metabolism [44].

4.2. Splicing

Protein arginine methylation is a posttranslational modification occurring on many proteins implicated in RNA processing [31]. In a systematic analysis of PRMT interactome, Wei and coworkers [4] found a significant enrichment for RNA binding domains in proteins interacting with PRMTs and revealed their importance in RNA splicing as well as in the assembly and function of ribosomes. These RNA-binding factors include heterogeneous nuclear ribonucleoproteins (hnRNPs) and serine/arginine-rich (SR) proteins that play a crucial role in pre-mRNA splicing. In these cases, arginine methylation constitutes a regulatory process controlling subcellular localization and protein–protein and RNA–protein interactions (see for reviews [45–47]). Thus, examples of the implication of PRMTs in RNA splicing have already been described: Sm proteins SmB/B0, SmD1, and SmD3 are methylated by PRMT5 [48,49]. RBM15, which regulates RNA export and splicing, is a substrate for PRMT1 [50]. CARM1 catalyzes the methylation of three splicing factors: SmB, U1-C and SF3B4/SAP49 [51]. PRMT9 forms a complex with splicing factors SF3B2/SAP145 and SF3B4/SAP49 and methylates SF3B2/SAP145 [52]. The methylation marks catalyzed by distinct PRMTs affect the subcellular localization of both serine/arginine-rich splicing factors 1 and 2 (SRSF1 and SRSF2, respectively) and occur between the two RRM domains in SRSF1 and in the RRM domain of SRSF2 [32,53]. They also impact the RNA binding functions of SRSF2.

Using a proteomic approach in HeLa cells, Vhuiyan and coworkers found associations between PRMT2 via the SH3 domain and different splicing-related proteins, some of which are also methylated by other PRMTs [54]. The list includes the Sm core snRNP protein SmB/B'; snRNP components; splicing regulators, such as hnRNPs; and other proteins involved in splicing, such as the heterogeneous nuclear ribonucleoprotein U-like 1 (HNRNPUL1), an hnRNP which represses basic transcription driven by several virus and cellular promoters, initially identified as an interactant by Kzhyshkowska et al [8]. The most characterized implication of hPRMT2 in splicing is the interaction with SAM68 (Src-associated in mitosis 68 kDa protein), a PRMT1 substrate that mediates the alternative splicing of the apoptosis regulator Bcl-X [54]. hPRMT2 promotes an increase in the BCL-X(L)/BCL-X(s) ratio in TNF-σ and LPS stimulated cells. This suggests an involvement of PRMT2 in regulating BCL-X alternative splicing in cells under inflammatory conditions and is consistent with the effect on the NF-κB pathway as previously described [35].

A few years ago, while purifying mouse PRMT2 from the insect expression host *Spodoptera frugiperda*, we found a 16 kDa contaminant co-eluting with mPRMT2 [10]. This polypeptide has been identified by mass spectrometry as repressor splicing factor 1 (RSF1). This insect-specific splicing repressor antagonizes serine and arginine-rich (SR) protein function [55] or coregulates alternative splicing with the other SR proteins in drosophila [56]. It contains an N-terminal domain folded into an RNA recognition motif (RRM) and a disordered arginine/glycine-rich C-terminal part. RSF1 is related to the serine/arginine-rich (SR) family of splicing regulators, in particular with the RRM domain of serine/arginine-rich splicing factors 7 and 3 (SRSF7 and SRSF3, respectively), which are involved in pre-mRNA splicing and mRNA export. Furthermore, six arginines methylated by PRMT2 were identified on RSF1, making it a usable substrate to detect PRMT2 enzymatic activity, as we showed with PRMT2 from mouse and *Danio rerio* [10]. However, it is still unclear whether PRMT2 releases the methylated RSF1 after the enzymatic reaction. In addition, the deletion of the SH3 domain leads to a sevenfold decrease in RSF1 methylation compared with the full-length enzyme, indicating that the SH3 domain could stabilize the interaction with RSF1. Thus, although RSF1 cannot be a natural substrate for PRMT2 due to the absence of this enzyme in insect cells, it could nevertheless shed light on the interaction between PRMT2 and a potential splicing regulator.

5. Diseases

5.1. Breast Cancer

PRMT2 has been identified as a coactivator of several nuclear receptors, such as ERα and androgen receptors, which are involved in the development of hormone-dependent cancers [34]. The implication of PRMT2 in breast carcinogenesis has been described in several studies and remains complex.

In addition to full-length human PRMT2, four alternatively spliced PRMT2 enzymatically inactive isoforms (PRMT2L2, PRMT2α, β and γ) have been identified [13,14] (Figure 3). All of the PRMT2 isoforms showed increased expression in breast tumor compared to normal tissues and are all able to enhance ERα-mediated transactivation activity in the presence of estradiol. PRMT2L2 is predominantly localized in the cytoplasm, and PRMT2β exhibits an even distribution between the nucleus, including the nucleoli, and the cytoplasm, while full-length PRMT2, PRMT2α and γ are mainly present in the nucleus. This suggests that the alternatively spliced C-terminus would influence PRMT2 localization, while N-terminus extremity could control the transcriptional regulatory activity of PRMT2 isoforms. PRMT2 and PRMT2β expression suppresses the cell proliferation and colony formation of MCF7 cells, providing these isoforms with a tumor-suppressive role [57,58].

The loss of PRMT2 nuclear expression in breast cancer cells is linked to increased cyclin D1 expression via indirectly binding to the AP-1 site on the cyclin D1 promoter, thus promoting breast tumor cell proliferation. Inconsistently with these results, Ho et al. correlate PRMT2 depletion with decreased cyclin D1 expression [59].

The increased expression of the total amount of PRMT2 reported in breast cancer tissue could be explained by the high level of PRMT2 in the cytoplasm, since PRMT2 is clearly decreased in cell nuclei compared with normal breast tissue [58]. Thus, PRMT2 mRNA alternative splicing could be at least partially responsible for breast tumor development.

PRMT2 was able to reverse tamoxifen resistance in breast cancer cells generated by ER-α36, an estrogen receptor isoform lacking transcription activation functions AF-1 and AF-2 but still containing the DNA-binding domain and most of the hormone-binding domain [60]. This study revealed the interaction between PRMT2 and ER-α36 to suppress its non-genomic signaling pathways, PI3K/Akt and MAPK/ERK. Despite the confirmation of a direct association between PRMT2 and ER-α36, the PRMT2-mediated ER-α36 inhibition mechanism remains unknown.

While these studies all highlighted a critical role of PRMT2 expression in breast cancer, the mechanism remains widely unknown.

5.2. Other Pathologies

PRMT2 expression is upregulated in glioblastoma multiforme (GBM) [25] and in hepatocellular carcinoma (HCC) tissues and cells [26]. In both cases, PRMT2, through its catalytic product, H3R8me2a, is implicated in tumorigenesis. Hu et al. showed that PRMT2 is recruited to the Bcl-2 promoter and generates H3R8 dimethylation, which maintains Bcl-2 gene expression by inducing STAT3 accessibility, thereby promoting cell proliferation in HCC.

Very recently, a decrease in PRMT2 expression in cardia gastric cancer tumors has been observed, which suggests a potential antitumor activity played by PRMT2 [61].

Zeng and coworkers revealed that PRMT2 provides protection against the proliferation of vascular smooth muscle cells and reduces the production of proinflammatory cytokines induced with angiotensin II [62]. These results show the ability of PRMT2 to reduce inflammation mediated by angiotensin II and suggest that it is as a potential target for cardiovascular diseases associated with vascular smooth muscle cell proliferation and inflammation.

6. Conclusions

PRMT2 is one of the least studied PRMTs, essentially because its methyl transferase activity is difficult to detect in vivo and no efficient substrate is available to determine enzymatic constants in vitro. RSF1 is, to date, the only interactant that can be used to reconstitute a complex with PRMT2 and that can be methylated in vitro. However, it is still unclear whether PRMT2 releases the methylated RSF1 after the enzymatic reaction, limiting its use in enzymology studies. It is therefore necessary to carry on investigations in order to identify an authentic substrate of PRMT2. On this point, techniques developed to analyze PRMT interactomes and methylomes succeeded in identifying interactants and substrates for different PRMTs and would certainly help in the discovery of substrates for PRMT2. This protein is known to interact with a multitude of splicing factors and splicing-related proteins, but there is no evidence of methylation by PRMT2, indicating possible functions that are independent of its catalytic activity. The role of the SH3 domain should also be clarified. This PRMT2-specific domain seems dispensable for PRMT2 coactivator function, but it has been demonstrated to be important for interactions with partner proteins. On this issue, isolation and structure determination of complexes would make a real breakthrough in the understanding of the SH3 domain's function in PRMT2. Additionally, as a transcriptional coactivator of genes involved in oncogenesis, PRMT2 has been implicated in cancer pathogenesis and is, therefore, a potential target for cancer therapy. Thus, a better characterization of its physiological role in nuclear receptor signaling could encourage the development of therapeutic strategies.

Author Contributions: Writing—original draft preparation, V.C. and J.C. Writing—review and editing, V.C. and J.C. Visualization, V.C. and J.C. All authors have read and agreed to the published version of the manuscript.

Funding: This research was funded by grants from CNRS, Université de Strasbourg, INSERM, Instruct-ERIC, part of the European Strategy Forum on Research Infrastructures (ESFRI) supported by national member subscriptions as well as the French Infrastructure for Integrated Structural Biology (FRISBI) (ANR-10-INSB-005, grant ANR-10-LABX-0030-INRT); a French State fund managed by the Agence Nationale de la Recherche under the frame program Investissements d'Avenir labelled ANR-19-CE11-0010-01 JC and IGBMC; grants from Association pour la Recherche contre le Cancer (ARC) (ARC 2016, no. PJA 20161204817); and grants from "Ligue d'Alsace contre le Cancer".

Institutional Review Board Statement: Not applicable.

Informed Consent Statement: Not applicable.

Acknowledgments: We thank our colleagues Luc Bonnefond, Nils Marechal and Nathalie Troffer-Charlier for comments and discussions.

Conflicts of Interest: The authors declare no conflict of interest.

References

1. Gayatri, S.; Bedford, M.T. Readers of histone methylarginine marks. *Biochim. Biophys. Acta* **2014**, *1839*, 702–710. [CrossRef] [PubMed]
2. Lorton, B.M.; Shechter, D. Cellular consequences of arginine methylation. *Cell Mol. Life Sci.* **2019**, *76*, 2933–2956. [CrossRef] [PubMed]
3. Price, O.M.; Hevel, J.M. Toward Understanding Molecular Recognition between PRMTs and their Substrates. *Curr. Protein Pept. Sci.* **2020**, *21*, 713–724. [CrossRef] [PubMed]
4. Wei, H.-H.; Fan, X.-J.; Hu, Y.; Tian, X.-X.; Guo, M.; Mao, M.-W.; Fang, Z.-Y.; Wu, P.; Gao, S.-X.; Peng, C.; et al. A systematic survey of PRMT interactomes reveals the key roles of arginine methylation in the global control of RNA splicing and translation. *Sci. Bull.* **2021**, *66*, 1342–1357. [CrossRef]
5. Katsanis, N.; Yaspo, M.L.; Fisher, E.M. Identification and mapping of a novel human gene, HRMT1L1, homologous to the rat protein arginine N-methyltransferase 1 (PRMT1) gene. *Mamm. Genome* **1997**, *8*, 526–529. [CrossRef] [PubMed]
6. Wang, Y.C.; Wang, J.D.; Chen, C.H.; Chen, Y.W.; Li, C. A novel BLAST-Based Relative Distance (BBRD) method can effectively group members of protein arginine methyltransferases and suggest their evolutionary relationship. *Mol. Phylogenet. Evol.* **2015**, *84*, 101–111. [CrossRef]
7. Wang, Y.C.; Li, C. Evolutionarily conserved protein arginine methyltransferases in non-mammalian animal systems. *FEBS J.* **2012**, *279*, 932–945. [CrossRef]
8. Kzhyshkowska, J.; Schütt, H.; Liss, M.; Kremmer, E.; Stauber, R.; Wolf, H.; Dobner, T. Heterogeneous nuclear ribonucleoprotein E1B-AP5 is methylated in its Arg-Gly-Gly (RGG) box and interacts with human arginine methyltransferase HRMT1L1. *Biochem. J.* **2001**, *358*, 305–314. [CrossRef]
9. Meyer, R.; Wolf, S.S.; Obendorf, M. PRMT2, a member of the protein arginine methyltransferase family, is a coactivator of the androgen receptor. *J. Steroid Biochem. Mol. Biol.* **2007**, *107*, 1–14. [CrossRef]
10. Cura, V.; Marechal, N.; Troffer-Charlier, N.; Strub, J.M.; van Haren, M.J.; Martin, N.I.; Cianferani, S.; Bonnefond, L.; Cavarelli, J. Structural studies of protein arginine methyltransferase 2 reveal its interactions with potential substrates and inhibitors. *FEBS J.* **2017**, *284*, 77–96. [CrossRef]
11. Beitz, E. TEXshade: Shading and labeling of multiple sequence alignments using LATEX2 epsilon. *Bioinformatics* **2000**, *16*, 135–139. [CrossRef]
12. Senior, A.W.; Evans, R.; Jumper, J.; Kirkpatrick, J.; Sifre, L.; Green, T.; Qin, C.; Zidek, A.; Nelson, A.W.R.; Bridgland, A.; et al. Improved protein structure prediction using potentials from deep learning. *Nature* **2020**, *577*, 706–710. [CrossRef]
13. Zhong, J.; Cao, R.X.; Hong, T.; Yang, J.; Zu, X.Y.; Xiao, X.H.; Liu, J.H.; Wen, G.B. Identification and expression analysis of a novel transcript of the human PRMT2 gene resulted from alternative polyadenylation in breast cancer. *Gene* **2011**, *487*, 1–9. [CrossRef]
14. Zhong, J.; Cao, R.X.; Zu, X.Y.; Hong, T.; Yang, J.; Liu, L.; Xiao, X.H.; Ding, W.J.; Zhao, Q.; Liu, J.H.; et al. Identification and characterization of novel spliced variants of PRMT2 in breast carcinoma. *FEBS J.* **2012**, *279*, 316–335. [CrossRef] [PubMed]
15. Troffer-Charlier, N.; Cura, V.; Hassenboehler, P.; Moras, D.; Cavarelli, J. Functional insights from structures of coactivator-associated arginine methyltransferase 1 domains. *EMBO J.* **2007**, *26*, 4391–4401. [CrossRef]
16. Carducci, M.; Perfetto, L.; Briganti, L.; Paoluzi, S.; Costa, S.; Zerweck, J.; Schutkowski, M.; Castagnoli, L.; Cesareni, G. The protein interaction network mediated by human SH3 domains. *Biotechnol. Adv.* **2012**, *30*, 4–15. [CrossRef] [PubMed]
17. Meirson, T.; Bomze, D.; Markel, G.; Samson, A.O. kappa-helix and the helical lock and key model: A pivotal way of looking at polyproline II. *Bioinformatics* **2020**, *36*, 3726–3732. [CrossRef]
18. Zhang, X.; Zhou, L.; Cheng, X. Crystal structure of the conserved core of protein arginine methyltransferase PRMT3. *EMBO J.* **2000**, *19*, 3509–3519. [CrossRef]
19. Rust, H.L.; Zurita-Lopez, C.I.; Clarke, S.; Thompson, P.R. Mechanistic Studies on Transcriptional Coactivator Protein Arginine Methyltransferase 1. *Biochemistry* **2011**, *50*, 3332–3345. [CrossRef] [PubMed]
20. Scott, H.S.; Antonarakis, S.E.; Lalioti, M.D.; Rossier, C.; Silver, P.A.; Henry, M.F. Identification and Characterization of Two Putative Human Arginine Methyltransferases (HRMT1L1 and HRMT1L2). *Genomics* **1998**, *48*, 330–340. [CrossRef]
21. Frankel, A.; Brown, J.I. Evaluation of kinetic data: What the numbers tell us about PRMTs. *Biochim. Biophys. Acta Proteins Proteom.* **2019**, *1867*, 306–316. [CrossRef]
22. Lakowski, T.M.; Frankel, A. Kinetic analysis of human protein arginine N-methyltransferase 2: Formation of monomethyl- and asymmetric dimethyl-arginine residues on histone H4. *Biochem. J.* **2009**, *421*, 253–261. [CrossRef]
23. Blythe, S.A.; Cha, S.W.; Tadjuidje, E.; Heasman, J.; Klein, P.S. Beta-Catenin primes organizer gene expression by recruiting a histone H3 arginine 8 methyltransferase, Prmt2. *Dev. Cell* **2010**, *19*, 220–231. [CrossRef]
24. Su, X.; Zhu, G.; Ding, X.; Lee, S.Y.; Dou, Y.; Zhu, B.; Wu, W.; Li, H. Molecular basis underlying histone H3 lysine-arginine methylation pattern readout by Spin/Ssty repeats of Spindlin1. *Genes Dev.* **2014**, *28*, 622–636. [CrossRef] [PubMed]
25. Dong, F.; Li, Q.; Yang, C.; Huo, D.; Wang, X.; Ai, C.; Kong, Y.; Sun, X.; Wang, W.; Zhou, Y.; et al. PRMT2 links histone H3R8 asymmetric dimethylation to oncogenic activation and tumorigenesis of glioblastoma. *Nat. Commun.* **2018**, *9*, 4552. [CrossRef] [PubMed]
26. Hu, G.; Yan, C.; Xie, P.; Cao, Y.; Shao, J.; Ge, J. PRMT2 accelerates tumorigenesis of hepatocellular carcinoma by activating Bcl2 via histone H3R8 methylation. *Exp. Cell Res.* **2020**, *394*, 112152. [CrossRef] [PubMed]

27. Bonnefond, L.; Stojko, J.; Mailliot, J.; Troffer-Charlier, N.; Cura, V.; Wurtz, J.M.; Cianferani, S.; Cavarelli, J. Functional insights from high resolution structures of mouse protein arginine methyltransferase 6. *J. Struct. Biol.* **2015**, *191*, 175–183. [CrossRef]
28. Frankel, A.; Yadav, N.; Lee, J.; Branscombe, T.L.; Clarke, S.; Bedford, M.T. The novel human protein arginine N-methyltransferase PRMT6 is a nuclear enzyme displaying unique substrate specificity. *J. Biol. Chem.* **2002**, *277*, 3537–3543. [CrossRef]
29. Dillon, M.B.; Rust, H.L.; Thompson, P.R.; Mowen, K.A. Automethylation of protein arginine methyltransferase 8 (PRMT8) regulates activity by impeding S-adenosylmethionine sensitivity. *J. Biol. Chem.* **2013**, *288*, 27872–27880. [CrossRef]
30. Kuhn, P.; Chumanov, R.; Wang, Y.; Ge, Y.; Burgess, R.; Xu, W. Automethylation of CARM1 allows coupling of transcription and mRNA splicing. *Nucleic Acids Res.* **2011**, *39*, 2717–2726. [CrossRef]
31. Guo, A.; Gu, H.; Zhou, J.; Mulhern, D.; Wang, Y.; Lee, K.A.; Yang, V.; Aguiar, M.; Kornhauser, J.; Jia, X.; et al. Immunoaffinity enrichment and mass spectrometry analysis of protein methylation. *Mol. Cell Proteom.* **2014**, *13*, 372–387. [CrossRef]
32. Larsen, S.C.; Sylvestersen, K.B.; Mund, A.; Lyon, D.; Mullari, M.; Madsen, M.V.; Daniel, J.A.; Jensen, L.J.; Nielsen, M.L. Proteome-wide analysis of arginine monomethylation reveals widespread occurrence in human cells. *Sci. Signal.* **2016**, *9*, rs9. [CrossRef]
33. van Haren, M.; van Ufford, L.Q.; Moret, E.E.; Martin, N.I. Synthesis and evaluation of protein arginine N-methyltransferase inhibitors designed to simultaneously occupy both substrate binding sites. *Org. Biomol. Chem.* **2015**, *13*, 549–560. [CrossRef]
34. Qi, C.; Chang, J.; Zhu, Y.; Yeldandi, A.V.; Rao, S.M.; Zhu, Y.-J. Identification of Protein Arginine Methyltransferase 2 as a Coactivator for Estrogen Receptor α. *J. Biol. Chem.* **2002**, *277*, 28624–28630. [CrossRef] [PubMed]
35. Ganesh, L.; Yoshimoto, T.; Moorthy, N.C.; Akahata, W.; Boehm, M.; Nabel, E.G.; Nabel, G.J. Protein methyltransferase 2 inhibits NF-kappaB function and promotes apoptosis. *Mol. Cell. Biol.* **2006**, *26*, 3864–3874. [CrossRef]
36. Dalloneau, E.; Pereira, P.L.; Brault, V.; Nabel, E.G.; Herault, Y. Prmt2 regulates the lipopolysaccharide-induced responses in lungs and macrophages. *J. Immunol.* **2011**, *187*, 4826–4834. [CrossRef] [PubMed]
37. Iwasaki, H.; Kovacic, J.C.; Olive, M.; Beers, J.K.; Yoshimoto, T.; Crook, M.F.; Tonelli, L.H.; Nabel, E.G. Disruption of protein arginine N-methyltransferase 2 regulates leptin signaling and produces leanness in vivo through loss of STAT3 methylation. *Circ. Res.* **2010**, *107*, 992–1001. [CrossRef] [PubMed]
38. Hussein, M.A.; Shrestha, E.; Ouimet, M.; Barrett, T.J.; Leone, S.; Moore, K.J.; Herault, Y.; Fisher, E.A.; Garabedian, M.J. LXR-Mediated ABCA1 Expression and Function Are Modulated by High Glucose and PRMT2. *PLoS ONE* **2015**, *10*, e0135218. [CrossRef]
39. Li, Y.Y.; Zhou, S.H.; Chen, S.S.; Zhong, J.; Wen, G.B. PRMT2 inhibits the formation of foam cell induced by ox-LDL in RAW 264.7 macrophage involving ABCA1 mediated cholesterol efflux. *Biochem. Biophys. Res. Commun.* **2020**, *524*, 77–82. [CrossRef]
40. Pak, M.L.; Lakowski, T.M.; Thomas, D.; Vhuiyan, M.I.; Husecken, K.; Frankel, A. A protein arginine N-methyltransferase 1 (PRMT1) and 2 heteromeric interaction increases PRMT1 enzymatic activity. *Biochemistry* **2011**, *50*, 8226–8240. [CrossRef]
41. Yoshimoto, T.; Boehm, M.; Olive, M.; Crook, M.F.; San, H.; Langenickel, T.; Nabel, E.G. The arginine methyltransferase PRMT2 binds RB and regulates E2F function. *Exp. Cell Res.* **2006**, *312*, 2040–2053. [CrossRef]
42. Hou, W.; Nemitz, S.; Schopper, S.; Nielsen, M.L.; Kessels, M.M.; Qualmann, B. Arginine Methylation by PRMT2 Controls the Functions of the Actin Nucleator Cobl. *Dev. Cell* **2018**, *45*, 262–275. [CrossRef]
43. Ahuja, R.; Pinyol, R.; Reichenbach, N.; Custer, L.; Klingensmith, J.; Kessels, M.M.; Qualmann, B. Cordon-bleu is an actin nucleation factor and controls neuronal morphology. *Cell* **2007**, *131*, 337–350. [CrossRef]
44. Yildirim, A.O.; Bulau, P.; Zakrzewicz, D.; Kitowska, K.E.; Weissmann, N.; Grimminger, F.; Morty, R.E.; Eickelberg, O. Increased protein arginine methylation in chronic hypoxia: Role of protein arginine methyltransferases. *Am. J. Respir. Cell Mol. Biol.* **2006**, *35*, 436–443. [CrossRef] [PubMed]
45. Bedford, M.; Richard, S. Arginine methylation an emerging regulator of protein function. *Mol. Cell* **2005**, *18*, 263–272. [CrossRef] [PubMed]
46. Yu, M.C. The Role of Protein Arginine Methylation in mRNP Dynamics. *Mol. Biol. Int.* **2011**, *2011*, 163827. [CrossRef] [PubMed]
47. Blackwell, E.; Ceman, S. Arginine methylation of RNA-binding proteins regulates cell function and differentiation. *Mol. Reprod. Dev.* **2012**, *79*, 163–175. [CrossRef]
48. Friesen, W.J.; Paushkin, S.; Wyce, A.; Massenet, S.; Pesiridis, G.S.; Duyne, G.V.; Rappsilber, J.; Mann, M.; Dreyfuss, G. The methylosome, a 20S complex containing JBP1 and pICln, produces dimethylarginine-modified Sm proteins. *Mol. Cell. Biol.* **2001**, *21*, 8289–8300. [CrossRef]
49. Brahms, H.; Meheus, L.; de Brabandere, V.; Fischer, U.; Luhrmann, R. Symmetrical dimethylation of arginine residues in spliceosomal Sm protein B/B′ and the Sm-like protein LSm4, and their interaction with the SMN protein. *RNA* **2001**, *7*, 1531–1542. [CrossRef]
50. Zhang, L.; Tran, N.T.; Su, H.; Wang, R.; Lu, Y.; Tang, H.; Aoyagi, S.; Guo, A.; Khodadadi-Jamayran, A.; Zhou, D.; et al. Cross-talk between PRMT1-mediated methylation and ubiquitylation on RBM15 controls RNA splicing. *eLife* **2015**, *4*. [CrossRef]
51. Cheng, D.; Cote, J.; Shaaban, S.; Bedford, M.T. The arginine methyltransferase CARM1 regulates the coupling of transcription and mRNA processing. *Mol. Cell* **2007**, *25*, 71–83. [CrossRef]
52. Hadjikyriacou, A.; Yang, Y.; Espejo, A.; Bedford, M.T.; Clarke, S.G. Unique Features of Human Protein Arginine Methyltransferase 9 (PRMT9) and Its Substrate RNA Splicing Factor SF3B2. *J. Biol. Chem.* **2015**, *290*, 16723–16743. [CrossRef]
53. Sinha, R.; Allemand, E.; Zhang, Z.; Karni, R.; Myers, M.P.; Krainer, A.R. Arginine Methylation Controls the Subcellular Localization and Functions of the Oncoprotein Splicing Factor SF2/ASF. *Mol. Cell. Biol.* **2010**, *30*, 2762–2774. [CrossRef] [PubMed]

54. Vhuiyan, M.I.; Pak, M.L.; Park, M.A.; Thomas, D.; Lakowski, T.M.; Chalfant, C.E.; Frankel, A. PRMT2 interacts with splicing factors and regulates the alternative splicing of BCL-X. *J. Biochem.* **2017**, *162*, 17–25. [CrossRef]
55. Labourier, E.; Bourbon, H.-M.; Gallouzi, I.-E.; Fostier, M.; Allemand, E.; Tazi, J. Antagonism between RSF1 and SR proteins for both splice-site recognition in vitro and Drosophila development. *Genes Dev.* **1999**, *13*, 740–753. [CrossRef] [PubMed]
56. Bradley, T.; Cook, M.E.; Blanchette, M. SR proteins control a complex network of RNA-processing events. *RNA* **2015**, *21*, 75–92. [CrossRef]
57. Zhong, J.; Cao, R.X.; Liu, J.H.; Liu, Y.B.; Wang, J.; Liu, L.P.; Chen, Y.J.; Yang, J.; Zhang, Q.H.; Wu, Y.; et al. Nuclear loss of protein arginine N-methyltransferase 2 in breast carcinoma is associated with tumor grade and overexpression of cyclin D1 protein. *Oncogene* **2014**, *33*, 5546–5558. [CrossRef]
58. Zhong, J.; Chen, Y.J.; Chen, L.; Shen, Y.Y.; Zhang, Q.H.; Yang, J.; Cao, R.X.; Zu, X.Y.; Wen, G.B. PRMT2beta, a C-terminal splice variant of PRMT2, inhibits the growth of breast cancer cells. *Oncol. Rep.* **2017**, *38*, 1303–1311. [CrossRef]
59. Ho, M.C.; Wilczek, C.; Bonanno, J.B.; Xing, L.; Seznec, J.; Matsui, T.; Carter, L.G.; Onikubo, T.; Kumar, P.R.; Chan, M.K.; et al. Structure of the arginine methyltransferase PRMT5-MEP50 reveals a mechanism for substrate specificity. *PLoS ONE* **2013**, *8*, e57008. [CrossRef]
60. Shen, Y.; Zhong, J.; Liu, J.; Liu, K.; Zhao, J.; Xu, T.; Zeng, T.; Li, Z.; Chen, Y.; Ding, W.; et al. Protein arginine N-methyltransferase 2 reverses tamoxifen resistance in breast cancer cells through suppression of ER-alpha36. *Oncol. Rep.* **2018**, *39*, 2604–2612. [CrossRef]
61. Bednarz-Misa, I.; Fleszar, M.G.; Fortuna, P.; Lewandowski, Ł.; Mierzchala-Pasierb, M.; Diakowska, D.; Krzystek-Korpacka, M. Altered L-Arginine Metabolic Pathways in Gastric Cancer: Potential Therapeutic Targets and Biomarkers. *Biomolecules* **2021**, *11*, 1086. [CrossRef]
62. Zeng, S.Y.; Luo, J.F.; Quan, H.Y.; Xiao, Y.B.; Liu, Y.H.; Lu, H.Q.; Qin, X.P. Protein Arginine Methyltransferase 2 Inhibits Angiotensin II-Induced Proliferation and Inflammation in Vascular Smooth Muscle Cells. *BioMed Res. Int.* **2018**, *2018*, 1547452. [CrossRef]

Review

The Structure and Functions of PRMT5 in Human Diseases

Aishat Motolani [1], Matthew Martin [1], Mengyao Sun [1] and Tao Lu [1,2,3,4,*]

[1] Department of Pharmacology & Toxicology, Indiana University School of Medicine, Indianapolis, IN 46202, USA; amotolan@iu.edu (A.M.); mm217@iupui.edu (M.M.); sun19@iu.edu (M.S.)
[2] Department of Biochemistry & Molecular Biology, Indiana University School of Medicine, Indianapolis, IN 46202, USA
[3] Department of Medical & Molecular Genetics, Indiana University School of Medicine, Indianapolis, IN 46202, USA
[4] Indiana University Melvin and Bren Simon Comprehensive Cancer Center, Indiana University School of Medicine, Indianapolis, IN 46202, USA
* Correspondence: lut@iu.edu; Tel.: +1-317-278-0520

Abstract: Since the discovery of protein arginine methyltransferase 5 (PRMT5) and the resolution of its structure, an increasing number of papers have investigated and delineated the structural and functional role of PRMT5 in diseased conditions. PRMT5 is a type II arginine methyltransferase that catalyzes symmetric dimethylation marks on histones and non-histone proteins. From gene regulation to human development, PRMT5 is involved in many vital biological functions in humans. The role of PRMT5 in various cancers is particularly well-documented, and investigations into the development of better PRMT5 inhibitors to promote tumor regression are ongoing. Notably, emerging studies have demonstrated the pathological contribution of PRMT5 in the progression of inflammatory diseases, such as diabetes, cardiovascular diseases, and neurodegenerative disorders. However, more research in this direction is needed. Herein, we critically review the position of PRMT5 in current literature, including its structure, mechanism of action, regulation, physiological and pathological relevance, and therapeutic strategies.

Keywords: PRMT5; cancer; cardiovascular disease; neurodegenerative diseases; diabetes; inflammation

1. Introduction

Arginine methylation is a ubiquitous post-translational modification (PTM) that occurs on both nuclear and cytoplasmic proteins [1]. It is catalyzed by a family of enzymes named protein arginine methyltransferases (PRMTs) [1,2]. So far, nine PRMTs have been identified in human cells: PRMT1, 2, 3, 4 (also known as co-activator-associated arginine methyltransferase 1, CARM1), 5, 6, 7, 8, and 9 (also known as F-box only protein 11, FBXO11) [3].

Depending on the type of methylarginine that is introduced, PRMTs are classified into four types: type I, II, III, and IV. Type I, II, and III PRMTs transfer methyl groups from S-adenosylmethionine (SAM) to a terminal ω-guanidine nitrogen of protein arginine residue, generating methylarginine and another, product S-adenosylhomocysteine (SAH) [4]. Whereas a type IV PRMT methylates the internal (or δ) guanidino nitrogen, generating monomethylarginine, which has been only described in yeast [3,5]. In a stepwise manner, type I and type II PRMTs first methylate arginine residues to result in ω-NG-mono- methylarginine (MMA), which acts as an intermediate. Subsequently, type I PRMTs (PRMT1, 2, 3, 4, 6, and 8) catalyze the formation of asymmetric ω-NG, NG-dimethylarginine (ADMA), while type II PRMTs (PRMT5, PRMT7, and PRMT9) catalyze the formation of symmetric ω-NG, N'G-dimethylarginine (SDMA) [6]. PRMT7 also exhibits type III PRMT activity by catalyzing the monomethylation of certain substrates without further formation of SDMA [3,7,8] (Table 1). Notably, with an assigned designation from the HUGO Gene Nomenclature Committee, PRMT10 is now also referred to as PRMT9.

Table 1. Human PRMTs superfamily.

Member	Type	Methylation Pattern	References
PRMT1	I	Monomethylation and asymmetric dimethylation	[3]
PRMT2	I	Monomethylation and asymmetric dimethylation	[3]
PRMT3	I	Monomethylation and asymmetric dimethylation	[3]
PRMT4	I	Monomethylation and asymmetric dimethylation	[3]
PRMT5	II	Monomethylation and symmetric dimethylation	[3]
PRMT6	I	Monomethylation and asymmetric dimethylation	[3]
PRMT7	II and III	Monomethylation OR symmetric dimethylation	[3]
PRMT8	I	Monomethylation and asymmetric dimethylation	[3]
PRMT9/FBXO11/PRMT10	II	Monomethylation and symmetric dimethylation	[3]

PRMTs are generally expressed in tissues and can methylate both histone and non-histone proteins. Methylation of arginine residues is critical to various biological processes, including cellular signaling transduction, mRNA splicing, transcription, DNA damage repair, cell proliferation and differentiation, and protein-protein interactions [9–12]. Notably, the deregulation of PRMT enzymes is implicated in the pathogenesis of different diseases, including cancer, metabolic diseases, neurodegenerative diseases, cardiovascular diseases, aging, and so on [13]. Particularly, the role of arginine methylation has been extensively researched in various human diseases, such as cancer, diabetes, Alzheimer's disease, and cardiovascular disease [14]. Further insights into the distinct roles of the different PRMTs are anticipated to provide new approaches to disease prevention, diagnosis, and treatment [15]. Among the PRMT family members, PRMT5 has become increasingly attractive as a therapeutic target for small molecule inhibition [16,17]. This review discusses PRMT5's structure and function, including its role in human diseases and promising therapeutic treatments.

2. Structure and General Function of PRMT5
2.1. Human PRMT5 Structure and Mechanism of Action

Approximately 20 years ago, Pollack and colleagues studied the function of Janus kinase 2 (JAK2) in cell signaling, wherein they aimed to identify proteins that interact with JAK2. In this study, PRMT5, then known as Jak-binding protein 1 (JBP), was discovered and shown to possess methyltransferase activity [18]. Ten years later, the first crystal structure of human PRMT5 was resolved and described by Antonysamy et al. [19]. The human PRMT5, which is 637 amino acids long, commonly associates with methylosome protein 50 (MEP50) in a 435 kDa heterooctameric complex. MEP50 is a tryptophan-aspartic acid (WD) repeat-containing protein and acts to stabilize the PRMT5 complex and potentiate its methyltransferase activity. Thus, PRMT5 oligomerizes to form an inner core tetramer, with four surrounding MEP50 molecules bound to the N-terminal of PRMT5 (Figure 1). This unique PRMT5:MEP50 complex further interacts with several partner proteins—such as B lymphocyte-induced maturation protein (Blimp1), RIO kinase 1 (RioK1), menin, pICLn, methyl-CpG-binding domain/nucleosome remodeling deacetylase (MBD/NuRD), and coordinator of PRMT5 (COPR5)—in a context-dependent manner to enable a diverse range of substrate specificities and biological functions [19]. As shown in Figure 1, the structural composition of PRMT5 includes an N-terminal TIM barrel domain, which aids the assembly of the PRMT5:MEP50 complex. This is followed by a middle Rossman fold domain that is responsible for binding SAM and catalysis. The Rossman fold domain also has a general methyltransferase structure that facilitates the catalytic activity of PRMT5 [19]. Then, a C-terminal β-barrel domain aids dimerization of PRMT5 to form the inner core tetramer in the PRMT5:MEP50 complex [20].

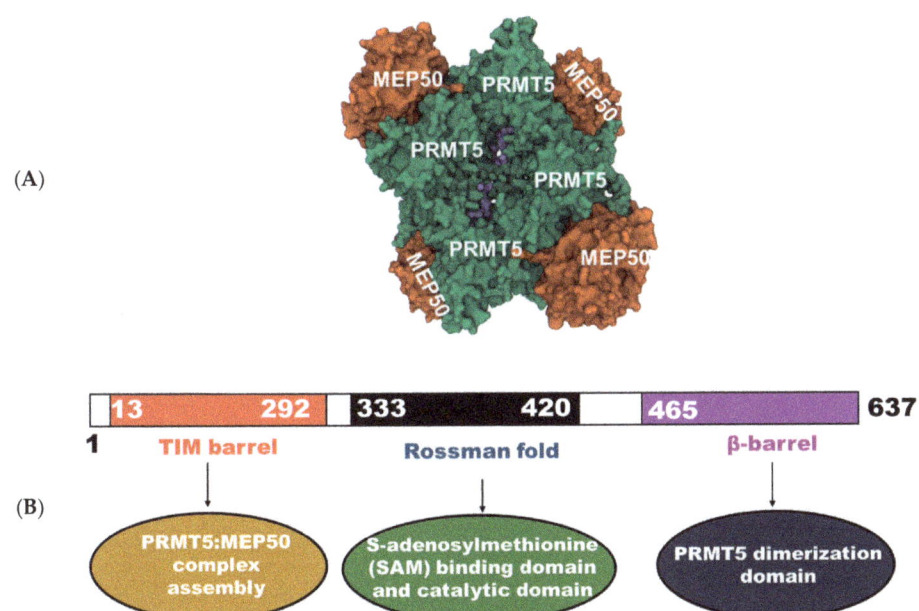

Figure 1. (**A**) Crystal structure of human PRMT5 in a heterooctameric complex with MEP50 (PDB ID: 4GQB). (**B**) Structural and functional domains of PRMT5.

As already mentioned, PRMT5 is a type II PRMT that symmetrically mono- or dimethylates its substrates by catalyzing the transfer of methyl groups from SAM to arginine residues on a substrate protein (Figure 2). This results in the formation of mono- or dimethylarginine and one or two molecules of SAH [21]. The methyltransferase function of PRMT5 is facilitated by critical conserved residues within the Rossman fold and the β-barrel domain. For example, during the methylation of a substrate protein such as histone 4 (H4), the arginine residue of H4 binds into a tunnel-like region composed of tryptophan (W) 579, phenylalanine (F) 327, and leucine (L) 312. These residues form favorable interactions with the substrate protein, thus enabling the arginine residue access to PRMT5's active site. The active site of PRMT5 contains two conserved glutamate (E) residues, E435 and E444, which form two hydrogen bond with the terminal guanidino nitrogen atom of arginine [9]. Then, F327 further orients the substrate's arginine for efficient transfer of the methyl group. F327 has been shown to reinforce the specificity of PRMT5 for the generation of symmetric dimethylation products. The product specificity role of F327 is evident from the formation of asymmetric dimethylated H4R3 when F327 is mutated to methionine [20]. Collectively, these critical residues of PRMT5 position a substrate protein in the best conformation for efficient catalysis, leading to the modulation of a protein's function.

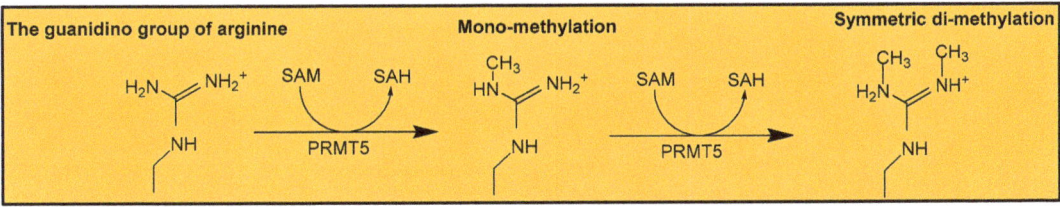

Figure 2. Scheme of the generation of dimethylarginine by PRMT5.

Similarly, with regard to sequence specificity, PRMT5 has an increased preference to methylate arginine residues that are sandwiched between two glycine residues (GRG motif) on RNA processing proteins [22]. A separate study with *C. elegans* PRMT5 (cPRMT5), which has 34% sequence similarity with human PRMT5, showed that the presence of positively charged residues downstream of arginine is essential for high-affinity binding to a substrate [23]. These studies indicate that PRMT5 targets specific arginine residues based on the surrounding sequences of a substrate.

2.2. Cellular Function of PRMT5

The activity of PRMT5 is essential to a wide range of biological processes, including cellular growth and development, differentiation, chromatin regulation, splicing, translation, DNA damage response, protein trafficking, and cell signaling [24]. PRMT5 is responsible for the methylation of various proteins like histones H2A, H3, and H4, transcription factors, cell receptors, etc., to regulate their physiological functions [24]. It has been well-established that the symmetric dimethylation of arginine 3 of H2A (H2AR3me2s) and H4 (H4R3me2s) suppresses gene transcription [25]. These repressive methylation marks exert their biological effects on different systems and cellular processes. For example, in conjunction with MEP50, PRMT5 maintains embryonic stem cell (ESC) pluripotency via dimethylation of H2AR3, leading to the downregulation of several differentiation genes such as *GATA4, 6,* and *HOXD9*. The knock-out of the *PRMT5* gene in mice also causes early embryonic lethality as it prevents the pluripotency of blastocysts [25]. Because of its extensive role in embryonic development and differentiation, PRMT5 has been termed the guardian of the germline [26]. In addition, PRMT5 plays a role in immune function. A recent study demonstrated that PRMT5 is essential for T-cell activation and proliferation and necessary for activated B-cell survival, maturation, proliferation, and antibody production [27,28]. PRMT5 is also expressed in the nervous system tissues, and it promotes the development and differentiation of oligodendrocyte progenitor cells—a major cell type responsible for myelin production, via deposition of methylation marks on H4R3 and inhibition of lysine acetylation on H4 [29].

In addition to histone methylation and its role in cellular development, PRMT5 coordinates other cellular processes through the methylation of diverse proteins. For example, PRMT5 activity is critical to hematopoiesis. It helps to maintain the viability and functions of hematopoietic stem cells (HSCs) through the repression of tumor protein p53 (p53), mechanistic target of rapamycin kinase (mTOR) signaling, and via regulation of the splicing of DNA repair genes [30]. Similarly, PRMT5 methylates several transcription factors, such as p53, E2F transcription factor 1 (E2F-1), and nuclear factor kappa B (NF-κB), to regulate cellular apoptosis, cell cycle progression, and inflammation [24,31]. The growing evidence of PRMT5 function on different physiological processes demonstrates its potential as a viable drug target in diseases.

3. Regulation of PRMT5

Since PRMT5 is highly involved in orchestrating several critical biological processes, its cellular function needs to be tightly regulated. Multiple sources of evidence in the literature have shown that the PRMT5's activity is positively or negatively regulated via different mechanisms, including the alteration of its subcellular localization, turnover rate, and substrate specificity [32]. These distinct mechanisms can be mediated by PTMs, microRNAs (miRNA), and/or interaction with partner proteins (Table 2).

3.1. Regulation by PTMs

PTM involves the addition or removal of distinct chemical groups on proteins to broaden proteins' functions [14,33]. To date, two main PTMs have been reported to modulate PRMT5 function: methylation and phosphorylation. For instance, the asymmetric methylation of PRMT5 by coactivator-associated arginine methyltransferase 1 (CARM1) at the evolutionary conserved arginine(R) 505 is essential for PRMT5 homodimerization, and the abrogation of this R505 methylation results in impaired methyltransferase activity [34]. Similarly, there are several lines of evidence documenting the regulatory role of phosphorylation on PRMT5's cellular activity. For example, Espejo and colleagues reported that the phosphorylation of PRMT5 by Akt and serum- and glucocorticoid-inducible kinases (SGK) at threonine (T)634 regulates PRMT5 subcellular localization and promotes its interaction with proteins containing 14-3-3 motifs rather than those containing PDZ motifs [35]. This switch in PRMT5 protein interaction may result in the sequestration of PRMT5 or aid the methylation of a specific set of substrates. Also, two tyrosine(Y) residues on PRMT5, Y304 and 307, known to be phosphorylated by JAK2, facilitate the substrate protein-binding for efficient catalysis [19]. Notably, our laboratory recently uncovered another key PRMT5 residue modified by phosphorylation. In this study, we discovered that PRMT5 is phosphorylated on serine (S)15 by protein kinase C iota (PKCι). This S15 phosphorylation was shown to be critical for NF-κB activation and to regulate the expression of a subgroup of NF-κB target genes in HEK 293 cells [36]. Additionally, in breast cancer cells, PRMT5 is phosphorylated at T139 and 144 by LKB1, and the mutation of these sites leads to a significant decrease in PRMT5 catalytic activity and reduces the interaction with its cofactors—MEP50, pICln, and RioK1 [37]. Although less characterized, the ubiquitination of PRMT5 by carboxyl terminus of heat shock cognate 70-interacting protein (CHIP) at multiple lysine residues has been reported to cause PRMT5 proteasomal degradation in prostate cancer cells [38]. Taken together, these studies document the diverse regulatory impact of PTMs on the functions of PRMT5 in cells.

3.2. Regulation by miRNA

PRMT5 modulates the expression of several miRNA via its methyltransferase activity on histones at miRNA promoters [39]. Conversely, miRNA can also regulate PRMT5 expression through distinct mechanisms. In a previous study, Pal and colleagues showed that low PRMT5 levels in normal B cells were maintained by significantly higher expression levels of miR-92b and miR-96, both of which are low in transformed B-cells. miR-92b and miR-96 inhibit PRMT5 translation by binding to the 3'UTR of PRMT5 mRNA, and this inhibition, in turn, suppresses the expression of the PRMT5 target gene—the suppressor of tumorigenicity 7 (ST7) [40]. The same group also reported that miR-19a, miR-25, and miR-32 inhibit PRMT5 expression in transformed B cells compared to normal B cells by a similar mechanism [41]. These studies underscore the importance of miRNA in maintaining the normal function and turnover of PRMT5, and how miRNA dysregulation can lead to aberrant PRMT5 expression.

3.3. Regulation by Interacting Proteins

The substrate specificity and cellular function of PRMT5 are often directed by its associating, binding partner proteins [32]. For example, PRMT5, when bound to the cooperator of PRMT5 (COPR5), preferentially methylates H4R3 rather than H3R8. Thus, COPR5 serves as an important adaptor for the recruitment of PRMT5 to the chromatin [42]. Also, PRMT5 interacts with the human SWItch/Sucrose Non-Fermentable (hSWI/SNF) chromatin, remodeling enzymes to methylate H3R8 and H4R3 at the promoters of ST7 and the nonmetastatic 23 (NM23) gene, to regulate cell growth [43]. Another chromatin remodeling complex, MBD2/NuRD, associates with the PRMT5:MEP50 complex and recruits the complex to the CpG islands of p14^{ARF} and p16^{INK4a}, thereby suggesting a role of MBD2/NuRD in regulating PRMT5 repressive activity on the endogenous inhibitors of the cell cycle [44]. Similarly, a scaffold protein, menin, is known to bind to PRMT5 and recruit it to the promoter of growth arrest specific 1 (Gas1) gene as a corepressor, to antagonize Sonic Hedgehog signaling in pancreatic islets [45]. Also, Blimp1 binds to PRMT5 in primordial germ cells to promote the symmetric dimethylation of H2AR3 and H4R3, and Blimp1 directs PRMT5-mediated repression of a subset of genes involved in the cell cycle, cell signaling, metabolism, and transcription [46]. In HeLa cells, Ski proteins were found to associate with PRMT5, alongside histone deacetylase 3 (HDAC3) and mothers against decapentaplegic homolog 2/3/4 (Smad2/3/4) proteins, to maintain the transcriptionally repressive state of Smad7 in the absence of TGF-β [47]. On the other hand, PRMT5 functions as a co-activator in its cooperation with pICln via the symmetric dimethylation of H4R3 at the promoter of genes involved in DNA double-stranded break repair [48]. Also, the association of PRMT5 with pICln and WDR77 promotes the recruitment of spliceosomal proteins, such as SmD1 and SmD3, to PRMT5 for methylation. This event is integral to the assembly of pre-mRNA splicing machinery [49]. Notably, RioK1 and pICln bind competitively to the PRMT5:MEP50 complex, with RioK1 interaction, promoting the recruitment of nucleolin to the complex for methylation [50]. Nucleolin is essential to ribosomal maturation and synthesis [50]. Thus, the aforementioned studies suggest that PRMT5's role in transcription, RNA processing, or translation can be modulated by specific proteins interacting with the PRMT5:MEP50 complex (Table 2).

Table 2. Summary of the distinct mechanisms of PRMT5 regulation.

Regulation	Regulators	Mechanism	Effect	Reference
PTMs	Coactivator-associated arginine methyltransferase 1 (CARM1)	Methylation of R505	Promotes PRMT5 homodimerization	[34]
	Protein kinase B/Akt	Phosphorylation of T634	Aids interaction with 14-3-3-proteins	[35]
	Janus kinase 2 (JAK2)	Phosphorylation of Y304 and Y307	Increases substrate binding	
	Protein Kinase C iota (PKCι)	Phosphorylation of S15	Promotes NF-κB activation	[36]
	Liver kinase B1 (LKB1)	Phosphorylation of T139 and T144	Increases methyltransferase activity and interaction with co-factors	[37]
miRNAs	miR-19a, miR-25, miR-32 miR-92b, and miR-96	Binds to 3′UTR of PRMT5 mRNA	Reduces PRMT5 levels	[40]
Protein Interactions	Coordinator Of PRMT5 (COPR5)	Serves as an adaptor for PRMT5 recruitment to chromatin	Causes PRMT5 preferential methylation of H4R3	[42]
	Human SWItch/Sucrose Non-Fermentable (hSWI/SNF)	Recruits PRMT5 to H3R8 and H4R3 at ST7 and NM23 promoters	Reduces expression of ST7 and NM23	[43]
	Methyl-CpG-binding domain protein 2/nucleosome remodeling and deacetylase (MBD2/NuRD)	Recruits PRMT5 to CpG islands of p14^{ARF} and p16^{INK4a}	Reduces expression of p14^{ARF} and p16^{INK4a}	[44]
	Menin	Recruits PRMT5 to Gas1 promoter	Reduces Gas1 gene expression and enhances Sonic Hedgehog signaling	[45]
	B-Lymphocyte induced maturation protein-1 (Blimp1)	Recruits PRMT5 to H2AR3 and H4R3	Repression of genes in cell cycle, cell signaling, metabolism and transcription	[46]
	pICln	Recruits PRMT5 to H4R3 and spliceosomal proteins	Repression of genes in DNA double-stranded break; Assembly of pre-mRNA splicing machinery	[48,49]
	RIO kinase 1 (RioK1)	Recruits PRMT5 to nucleolin for methylation	Ribosomal synthesis and maturation	[50]

4. Role of PRMT5 in Human Diseases

4.1. PRMT5 in Cancer

A growing number of studies have established the role of PRMT5 as a tumor promoting factor in several types of cancers. Owing to its methyltransferase activity on the histones and oncoproteins, PRMT5's role in cancers is entrenched in distinct cellular processes like cell signaling, DNA damage response, gene regulation, and splicing, among others [51]. Particularly, dysregulation of PRMT5 is critical for the progression of hematologic malignancies. For example, in lymphoma cell lines, PRMT5 is upregulated and increases the expression of pro-survival proteins like cyclin D1, c-myc, and survivin. This occurs via the deposition of repressive methylation marks on H3R8 in the promoter region of *AXIN2* and *WIF1*, both of which are negative regulators of wnt/β-catenin signaling [52]. Similarly, in vivo studies showed that the tumorigenesis in lymphocytes driven by oncogenes such as cyclin D1 requires high PRMT5 expression, and that this increased PRMT5 expression further antagonizes the apoptotic function of p53 via arginine methylation [53]. In mantle cell lymphoma (MCL), low levels of miR-92b and miR-96 drive increased PRMT5 expression, promoting cell proliferation [40]. Another study reported that PRMT5 interaction with tripartite motif-containing protein 21 (TRIM21), an IKKβ ubiquitin ligase, inhibits IKKβ degradation in multiple myeloma, thereby inducing NF-κB signaling and cell growth of multiple myeloma cells [54]. Taken together, convincing literature documents the pertinent role of aberrant PRMT5 expression in promoting major cancer hallmarks in hematologic cancers.

In addition, PRMT5 promotes oncogenicity in various solid cancers, including colon, breast, prostate, lung, liver, bone, skin, ovarian, gastric, brain, and pancreatic cancers, among others [55]. For instance, PRMT5 aids cellular proliferation, migration, and invasion in breast cancer cells by inhibiting the expression of Dickkopf WNT signaling pathway inhibitor 1 (DKK1) and DKK3, known antagonists of the wnt/β-catenin pathway [56]. In hepatocellular carcinoma, PRMT5-catalyzed repressive dimethylation on H4R3 at the B-cell translocation gene 2 (BTG2) promoter increases cell proliferation through the ERK signaling pathway [57]. Similarly, in lung cancer, PRMT5 induces the downregulation of tumor suppressor genes, such as GLI pathogenesis related 1 (GLIPR1), leprecan-like 1 (Leprel1), and BTG2, and the upregulation of growth factors such as fibroblast growth factor receptor substrate 1/2/3/4 (FGFR1/2/3/4) and human epidermal growth factor receptor 2/3 (HER2/3), thereby enhancing cell growth [58]. Particularly, PRMT5-mediated increased FGFR3 signaling is caused by silencing of the miR-99 family, which negatively regulates the expression of FGFR3 in lung cancer [59]. Interestingly, our laboratory reported that PRMT5-catalyzed dimethylation on R30 of the NF-κB p65 subunit and R205 of YBX1 promotes cell proliferation, migration, and anchorage-independent growth in CRC, suggesting the versatility of a PRMT5-regulated substrate in CRC tumors [31,60]. We further showed that PRMT5 inhibition significantly decreases the survival of colorectal and pancreatic cancer cell lines [17]. From a clinical perspective, high PRMT5 has been associated with a poor prognosis in patients with breast cancer, hepatocellular carcinoma, lung cancer, ovarian, and gastric cancer [61–63]. Notably, PRMT5's high nuclear expression has been suggested as a potential biomarker for assessing submucosal invasion of tumors resected in the early stage of CRC. A similar prognostic potential of high PRMT5 expression in the nucleus and/or the cytoplasm has been reported in brain, lung, ovarian, skin, and prostate cancers [64]. Collectively, these lines of evidence demonstrate the extensive oncogenic role of PRMT5 in cancers and its potential value as a clinical biomarker to improve patients' treatment modalities.

4.2. PRMT5 in Diabetes

Beyond cancer, evidence suggests that PRMT5 plays an important role in diabetes. Type 2 diabetes mellitus (T2DM), which accounts for over 90% of diabetic patients, is mainly characterized by the dysfunction of pancreatic β-cells, resulting in defective insulin release and insulin resistance [65]. Other associating pathophysiologies of T2DM include hyperglycemia, hyperlipidemia, mitochondrial dysfunction, inflammation, and increased reactive oxygen species (ROS) levels [66]. Interestingly, PRMT5 has been reported to play a role in metabolic pathways that perpetuate the pathologies of T2DM. For instance, in white adipose tissue, PRMT5 methylates sterol regulatory element-binding transcription factor 1a (SREBP1a) to enhance triacylglycerol formation [67]. PRMT5 also methylates the transcription elongation factor SPT5 to promote lipid droplet biogenesis [67]. A separate study also reported that PRMT5 serves as a coactivator for the expression of adipogenic genes, such as peroxisome proliferator-activated receptor γ2 (PPARγ2), adipocyte protein 2 (aP2), adiponectin, leptin, and resistin [68]. This demonstrates the regulatory role of PRMT5 in the metabolism of fatty acids and thus insulin sensitivity. Similarly, in association with the menin scaffold protein, PRMT5 reduces the expression of glucagon-like peptide-1 (GLP1) and dimethylates cAMP responsive element binding protein (CREB) and forkhead box O1 (FOXO1) to block protein kinase A (PKA)-mediated phosphorylation [69]. Another group showed that, in response to glucagon, PRMT5 promotes phosphorylation of CREB via PRMT5 binding to the CREB regulated transcription coactivator 2 (CRTC2) promoter, leading to increased expression of gluconeogenic genes. Notably, the increased activity of CREB and CRTC2 is also observed in diabetes and contributes to hyperglycemia [70]. In summary, these events orchestrated by PRMT5 play a key role in suppressing pancreatic β-cell function and in regulating glucose homeostasis. On the contrary, a study conducted by Ma and colleagues reported that the conditional knockout of PRMT5 in islet cells of the pancreas caused defects in glucose tolerance and glucose-stimulated insulin release in β-cells [71]. The proposed mechanism suggests that PRMT5 dimethylates H3R8 to increase the binding of the brahma-related gene-1 (BRG1) chromatin remodeling enzyme to the insulin promoter, to increase insulin production. This unusual observation was attributed to a compensatory mechanism of elevated β-cell proliferation induced by impaired insulin production on the PRMT5 knockout mice [71]. Thus, further studies are required to understand the nuanced role of PRMT5 in T2DM models, to aid better exploration of PRMT5 as a viable therapeutic target in diabetes.

4.3. PRMT5 in Cardiovascular Diseases

Cardiovascular disease is a category of disease that occurs in the heart or blood vessels [72]. The number of studies investigating the role of PRMT5 in cardiovascular diseases is limited. However, reports published recently suggest that differential PRMT5 levels may serve as a risk indicator for developing certain cardiovascular diseases or may reduce/promote its related morbidities. In cellular models of cardiomyocyte hypertrophy, a form of heart enlargement that causes heart failure, overexpression of PRMT5 results in the reduction of isoprenaline-induced hypertrophy through repressive methylation of *HOXA9*, a gene that plays a critical role in the development of several cardiovascular diseases [73]. This finding corroborates a previous report that demonstrates the role of PRMT5 in suppressing the expression of hypertrophic genes in cardiomyocytes via methylation of *GATA4*, a transcription factor that regulates cardiac remodeling [74]. Notably, in the peripheral blood obtained from 178 patients with acute myocardial infarction (AMI) and stable coronary artery disease (CAD), PRMT5 was significantly lower in AMI patients compared to stable CAD patients [75]. Thus, this study suggests that low PRMT5 expression in the blood may serve as a biomarker for individuals with increased risk for AMI development. In contrast, increased PRMT5 expression may enhance the pathological progression of inflammatory-driven cardiovascular diseases. For example, the high expression of C-X-C motif chemokine ligand 10 (CXCL10), a chemokine that extensively contributes to atherosclerosis and coronary artery disease in endothelial cells, is driven in part by PRMT5 methylation of NF-κB at R30 and R35. This event was observed in response to tumor necrosis factor-alpha (TNF-α), a potent activator of NF-κB signaling [76]. A follow-up study by the same group reported that, in response to TNF-α and interferon gamma (IFN-γ), PRMT5-induced methylation of p65 at R174 increases the expression of CXCL11, another chemokine that worsens the atherosclerosis pathology [77]. Collectively, these discussed studies suggest that the role of PRMT5 is context-dependent, and therefore, the elucidation of the PRMT5 molecular activity in contribution to different cardiovascular diseases is an area worthy of further exploration.

4.4. PRMT5 in Neurodegenerative Diseases

The role of PRMT5 in neurodegenerative diseases is not well studied. Given that a high expression of PRMT5 has been identified in the human brain, a few studies delineated the role of PRMT5 in neurodegenerative disorders [78]. For instance, Alzheimer's disease (AD) is one of the most common neurodegenerative diseases. It is often characterized by the accumulation of amyloid-β production, which induces neuronal death [79]. Interestingly, in a human AD cell model, the depletion of PRMT5 was reported to induce cell death and trigger apoptosis in neurons when indued by Aβ [80]. It is worthwhile to note that in microglial cells, activation of NF-κB signaling exacerbates the AD pathology via upregulation of cytokines that aid neuroinflammation and the formation of Aβ plaques [81,82]. However, the link between PRMT5 and NF-κB is yet to be examined in glial cells. In a human neuroblastoma cell line with overexpression of Swedish mutant of human amyloid-β precursor protein, PRMT5 overexpression resulted in reduced expression of E2F-1, p53, and Bax, and increased levels of glycogen synthase kinase 3β (GSK-3β), all of which prevent apoptosis [80].

In addition, the importance of PRMT5 in preventing Huntington's disease (HD) has been suggested. HD is an autosomal dominant neurogenerative disease, the cause of which is often attributed to the presence of mutant polyglutamine sequence in the huntingtin (Htt) protein [83]. According to a study by Ratovitski and colleagues, the methyltransferase activity of PRMT5 on histones is severely impaired by the mutant Htt protein, and the ectopic expression of PRMT5 enhances the survival of neuronal cells expressing mutant Htt [84]. In conclusion, in several studies, PRMT5 has been suggested to play a protective role in terms of neuronal cells. However, studies to examine its role in other brain cell types, such as microglia and astrocytes, within the context of neurodegeneration, are missing. This research gap certainly warrants future clarification of the overall role of PRMT5 in human neurodegenerative diseases. As the brain has several important cell types, the overall effect of PRMT5 in the human brain is determined by its integrated role in all the cell types, instead of in just a couple of them.

5. Targeting PRMT5 in Human Diseases

Because PRMT5 is a critical regulator of several systems, it is unsurprising that the overexpression or dysregulation of PRMT5 is associated with several disease states, including several hematologic and solid-state cancers [61,85,86]. Furthermore, the dysregulation and upregulation of PRMT5 in several cardiovascular disorders [75] and neurological disorders [80,84] raise the question of how PRMT5 may be therapeutically targeted for disease treatment. The breakdown of potential therapeutics for PRMT5 can be broadly classified into targeted and non-specific. These therapeutics are shown in Table 3. Several small molecule inhibitors of PRMT5 have been developed and are currently going into or through clinical trials. Many of these small molecules act by blocking the binding of SAM to PRMT5, either through competitive or non-competitive binding. In some cases, the mechanism of inhibition is unknown. However, most of the inhibitors are known to work through direct binding to the catalytic region of PRMT5 (direct inhibition). The GSK inhibitor (EPZ015938/GSK3326595) was the first PRMT5 inhibitor to be studied in a clinical setting, but there were mixed results of the Meteor-1 Phase I trial. Adenoid cystic carcinoma (ACC) was the primary indication for use of this compound. High dosages were required (400 mg, QD) to see any benefit, and adverse events were observed at multiple dosing levels, including the 400 mg dose (Table 3). Currently, this GSK inhibitor, EPZ015938/GSK3326595, is in phase I/II trials for leukemias as well as solid state cancers [87]. Furthermore, EPZ015938/GSK3326595 is on course for planned clinical trials relating to breast cancer and solid-state cancers (NCT02783300, NCT04676516). Another example is GSK3186000A, which was used in leukemia cells to reduce PRMT5 activity and may represent a potentially useful future therapy [88]. JNJ-64619178 is a small molecule inhibitor used in phase I clinical trials in brain cancers and advanced solid-state tumors [89]. Clinical trials are also underway for a new PRMT5 small molecule inhibitor PF-06939999 developed to treat esophageal cancers as well as small lung cell carcinoma [90]. Other therapeutics have been developed to directly target the SAM binding pocket by larger pharmaceutical companies but have not yet gone into clinical trials. These include LLY-283 by Eli Lilly and company, which has been shown to reduce tumor growth in skin cancer [91]. CMP-5 is a compound developed by Merck et al. that has been used to block SAM binding in glioblastoma [92,93]. MRTX9768, a compound developed by Miratis Therapeutics, Inc., is focused on inhibiting the methylthioadenosine phosphorylase (MTA)-PRMT5 complex in MTAP^{del} cancer cells, resulting in their targeting for destruction [94]. PRT543 and PRT811, developed by Prelude Therapeutics, are further inhibitors that directly bind to the SAM binding pocket; these are slated for clinical trials soon (NCT03886831 and NCT04089449) [95]. PR5-LL-CM01, developed by our lab and further licensed to EQon Pharmaceuticals, shows great efficacy in tumor inhibition in pancreatic cancer, colon cancer, and breast cancer [17]. Furthermore, the use of most PRMT5 inhibitors is almost exclusively focused on the treatment of cancers, and in almost all cases, the mode of inhibition is direct. This opens up a wide range of therapeutic applications for PRMT5 inhibitors in other diseases such as cardiovascular disease, and perhaps neurodegenerative disorders, etc.

Table 3. Novel PRMT5 therapeutics and their current stages of development.

Company	Compound Name	Mode of Inhibition	Clinical Trial Stage	Trial Duration	Country	Types of Indications	References
Epizyme (sponsored by GSK)	EPZ015938/GSK3326595 (Pemrametostat)	Direct	Phase I/II	Phase I: 08/30/2016–04/29/2025 Phase II (03/21/2021–12/31/2022)	USA	Phase I: Solid tumors and non-Hodgkin's lymphoma (with drug: pembrolizumab) Phase II: Early-stage breast cancer	(NCT02783300/ NCT04676516) [87]
Johnson & Johnson	JNJ-64619178 (Onametostat)	Direct	Phase I	07/13/2018–12/30/2022	USA	Solid tumor, adult; non-Hodgkin lymphoma; myelodysplastic syndromes	NCT03573310 [89]
Pfizer	PF-06939999	Direct	Phase I	03/14/2019–04/14/2026	USA	Advanced or metastatic non-small cell lung cancer, head and neck squamous cell carcinoma, esophageal cancer, endometrial cancer, cervical cancer, bladder cancer (monotherapy, in combination with docetaxel)	NCT03854227 [90]
Prelude	PRT811	Direct	Phase I	11/06/2019–10/2022	USA	Advanced solid tumors, CNS lymphoma, and recurrent high-grade gliomas	NCT04089449
Prelude	PRT543	Direct	Phase I	02/11/2019–08/11/2022	USA	Relapsed/refractory advanced solid tumors; relapsed/refractory diffuse large B-cell lymphoma; relapsed/refractory myelodysplasia; relapsed/refractory myelofibrosis; adenoid cystic carcinoma; relapsed/refractory mantle cell lymphoma; relapsed/refractory acute myeloid leukemia; refractory chronic myelomonocytic leukemia	NCT03886831 [95]
EQon Pharmaceuticals	PR5-LL-CM01	Direct	Preclinical		USA	Pancreatic cancer/colorectal cancer (breast cancer)	[17]
GSK (license with Epizyme)	GSK3186000A	Direct	Preclinical		USA	Leukemia	[89]
Eli Lilly	LLY-283	Direct	Preclinical		USA	Skin cancer	[91]
Merck	CMP-5	Direct	Preclinical		USA	Glioblastoma	[93]
Miratis	MRTX9768	Targets PRMT5-MTA complex	Preclinical		USA	MTAPdel cancer cells	[94]

6. Perspective and Conclusions

The recent explosion of PRMT5 research in various disease model systems is indicative of its enormous potential as a drug target to inhibit the progression of human diseases. Undoubtedly, PRMT5 is an important type II arginine methyltransferase with diverse substrates and functions in humans. PRMT5's unique heterooctameric structure facilitates its interaction with co-factors, partner proteins, and substrates, thus helping to maintain the genomic integrity, signal transduction, and development of cells. The clinical importance of PRMT5 is evident from its extensive role in driving or reducing inflammatory diseases, such as cancer, diabetes, neurodegeneration, and cardiovascular disease. Thus, the current knowledge on PRMT5 in the literature can be translated to improve the outcomes of patients with related diseases. Notably, our group was one of the first to establish the link of PRMT5 to inflammation through its methyltransferase activity on the NF-κB signaling pathway. Together with other groups, we have demonstrated the pathological relevance of the PRMT5/NF-κB axis in pancreatic cancer, colon cancer, and heart disease [31,76]. However, more studies are needed to buttress the role of the PRMT5/NF-κB signaling axis in AD, considering that the activation of NF-κB in surrounding glia cells contributes to neuroinflammation [81]. Hence, future work studying PRMT5 should investigate how the differential regulation of PRMT5 leads to its tissue-dependent function, and vice versa, thus defining under what conditions PRMT5 plays a role in diabetes and cardiovascular disease, as well as examining strategies to overcome the existential barriers to effective PRMT5 targeted therapies in human diseases.

Author Contributions: Conceptualization, A.M. and T.L.; writing, A.M., M.M., M.S. and T.L.; project administration, T.L.; funding acquisition, T.L. All authors have read and agreed to the published version of the manuscript.

Funding: This review article is funded by NIH-NIGMS grant no. 1R01GM120156-01A1 (to T.L.).

Institutional Review Board Statement: Not applicable.

Informed Consent Statement: Not applicable.

Data Availability Statement: Not applicable.

Conflicts of Interest: T.L. is the founder of EQon Pharmaceuticals, LLC.

References

1. Yang, Y.; Bedford, M.T. Protein arginine methyltransferases and cancer. *Nat. Rev. Cancer* **2012**, *13*, 37–50. [CrossRef]
2. Morettin, A.; Baldwin, R.M.; Côté, J. Arginine methyltransferases as novel therapeutic targets for breast cancer. *Mutagenesis* **2015**, *30*, 177–189. [CrossRef] [PubMed]
3. Bedford, M.T. Arginine methylation at a glance. *J. Cell Sci.* **2007**, *120*, 4243–4246. [CrossRef] [PubMed]
4. Stopa, N.; Krebs, J.E.; Shechter, D. The PRMT5 arginine methyltransferase: Many roles in development, cancer and beyond. *Cell. Mol. Life Sci.* **2015**, *72*, 2041–2059. [CrossRef] [PubMed]
5. Niewmierzycka, A.; Clarke, S. S-adenosylmethionine-dependent methylation in *Saccharomyces cerevisiae*. *J. Biol. Chem.* **1999**, *274*, 814–824. [CrossRef] [PubMed]
6. Motolani, A.; Martin, M.; Sun, M.; Lu, T. Phosphorylation of the regulators, a complex facet of NF-κB signaling in cancer. *Biomolecules* **2020**, *11*, 15. [CrossRef] [PubMed]
7. Lee, J.-H.; Cook, J.R.; Yang, Z.-H.; Mirochnitchenko, O.; Gunderson, S.I.; Felix, A.M.; Herth, N.; Hoffmann, R.; Pestka, S. PRMT7, a new protein arginine methyltransferase that synthesizes symmetric dimethylarginine. *J. Biol. Chem.* **2005**, *280*, 3656–3664. [CrossRef] [PubMed]
8. Zurita-Lopez, C.I.; Sandberg, T.; Kelly, R.; Clarke, S.G. Human protein arginine methyltransferase 7 (PRMT7) is a type III enzyme forming omega-NG-monomethylated arginine residues. *J. Biol. Chem.* **2012**, *287*, 7859–7870. [CrossRef] [PubMed]
9. Cheung, N.; Chan, L.C.; Thompson, A.; Cleary, M.L.; So, C.W.E. Protein arginine-methyltransferase-dependent oncogenesis. *Nat. Cell Biol.* **2007**, *9*, 1208–1215. [CrossRef]
10. Kim, J.H.; Yoo, B.C.; Yang, W.S.; Kim, E.; Hong, S.; Cho, J.Y. The role of protein arginine methyltransferases in inflammatory responses. *Mediat. Inflamm.* **2016**, *2016*, 4028353. [CrossRef] [PubMed]
11. Le Romancer, M.; Treilleux, I.; Bouchekioua-Bouzaghou, K.; Sentis, S.; Corbo, L. Methylation, a key step for nongenomic estrogen signaling in breast tumors. *Steroids* **2010**, *75*, 560–564. [CrossRef]

12. Mitchell, T.R.H.; Glenfield, K.; Jeyanthan, K.; Zhu, X.-D. Arginine methylation regulates telomere length and stability. *Mol. Cell. Biol.* **2009**, *29*, 4918–4934. [CrossRef]
13. Blanc, R.S.; Richard, S. Arginine methylation: The coming of age. *Mol. Cell* **2017**, *65*, 8–24. [CrossRef] [PubMed]
14. Wei, H.; Mundade, R.; Lange, K.C.; Lu, T. Protein arginine methylation of non-histone proteins and its role in diseases. *Cell Cycle* **2014**, *13*, 32–41. [CrossRef] [PubMed]
15. Cheng, D.; Yadav, N.; King, R.W.; Swanson, M.S.; Weinstein, E.J.; Bedford, M.T. Small molecule regulators of protein arginine methyltransferases. *J. Biol. Chem.* **2004**, *279*, 23892–23899. [CrossRef]
16. Wang, Y.; Lianjun, Z.; Liu, C.; Han, F.; Chen, M.; Zhang, L.; Cui, X.; Qin, Y.; Bao, S.; Gao, F. Protein arginine methyltransferase 5 (Prmt5) is required for germ cell survival during mouse embryonic development. *Biol. Reprod.* **2015**, *92*, 104. [CrossRef]
17. Prabhu, L.; Wei, H.; Chen, L.; Demir, Ö.; Sandusky, G.; Sun, E.; Wang, J.; Mo, J.; Zeng, L.; Fishel, M.; et al. Adapting AlphaLISA high throughput screen to discover a novel small-molecule inhibitor targeting protein arginine methyltransferase 5 in pancreatic and colorectal cancers. *Oncotarget* **2017**, *8*, 39963–39977. [CrossRef]
18. Pollack, B.P.; Kotenko, S.V.; He, W.; Izotova, L.S.; Barnoski, B.L.; Pestka, S. The human homologue of the yeast proteins Skb1 and Hsl7p interacts with Jak kinases and contains protein methyltransferase activity. *J. Biol. Chem.* **1999**, *274*, 31531–31542. [CrossRef]
19. Antonysamy, S.; Bonday, Z.; Campbell, R.M.; Doyle, B.; Druzina, Z.; Gheyi, T.; Han, B.; Jungheim, L.N.; Qian, Y.; Rauch, C.; et al. Crystal structure of the human PRMT5: MEP50 complex. *Proc. Natl. Acad. Sci. USA* **2012**, *109*, 17960–17965. [CrossRef]
20. Sun, L.; Wang, M.; Lv, Z.; Yang, N.; Liu, Y.; Bao, S.; Gong, W.; Xu, R.-M. Structural insights into protein arginine symmetric dimethylation by PRMT5. *Proc. Natl. Acad. Sci. USA* **2011**, *108*, 20538–20543. [CrossRef]
21. Eddershaw, A.R.; Stubbs, C.J.; Edwardes, L.V.; Underwood, E.; Hamm, G.R.; Davey, P.R.J.; Clarkson, P.N.; Syson, K. Characterization of the kinetic mechanism of human protein arginine methyltransferase 5. *Biochemistry* **2020**, *59*, 4775–4786. [CrossRef]
22. Musiani, D.; Bok, J.; Massignani, E.; Wu, L.; Tabaglio, T.; Ippolito, M.R.; Cuomo, A.; Ozbek, U.; Zorgati, H.; Ghoshdastider, U.; et al. Proteomics profiling of arginine methylation defines PRMT5 substrate specificity. *Sci. Signal.* **2019**, *12*, eaat8388. [CrossRef]
23. Wang, M.; Xu, R.-M.; Thompson, P.R. Substrate specificity, processivity, and kinetic mechanism of protein arginine methyltransferase 5. *Biochemistry* **2013**, *52*, 5430–5440. [CrossRef]
24. Koh, C.M.; Bezzi, M.; Guccione, E. The where and the how of PRMT5. *Curr. Mol. Biol. Rep.* **2015**, *1*, 19–28. [CrossRef]
25. Tee, W.-W.; Pardo, M.; Theunissen, T.; Yu, L.; Choudhary, J.; Hajkova, P.; Surani, M.A. Prmt5 is essential for early mouse development and acts in the cytoplasm to maintain ES cell pluripotency. *Genes Dev.* **2010**, *24*, 2772–2777. [CrossRef] [PubMed]
26. Berrens, R.V.; Reik, W. Prmt5: A guardian of the germline protects future generations. *EMBO J.* **2015**, *34*, 689–690. [CrossRef] [PubMed]
27. Litzler, L.; Zahn, A.; Meli, A.; Hébert, S.; Patenaude, A.-M.; Methot, S.P.; Sprumont, A.; Bois, T.; Kitamura, D.; Costantino, S.; et al. PRMT5 is essential for B cell development and germinal center dynamics. *Nat. Commun.* **2019**, *10*, 22. [CrossRef] [PubMed]
28. Tanaka, H.; Fujita, N.; Tsuruo, T. 3-phosphoinositide-dependent protein kinase-1-mediated IκB kinase β (IKKβ) phosphorylation activates NF-κB signaling. *J. Biol. Chem.* **2005**, *280*, 40965–40973. [CrossRef]
29. Scaglione, A.; Patzig, J.; Liang, J.; Frawley, R.; Bok, J.; Mela, A.; Yattah, C.; Zhang, J.; Teo, S.X.; Zhou, T.; et al. PRMT5-mediated regulation of developmental myelination. *Nat. Commun.* **2018**, *9*, 2840. [CrossRef] [PubMed]
30. Tan, D.Q.; Li, Y.; Yang, C.; Li, J.; Tan, S.H.; Chin, D.W.; Nakamura-Ishizu, A.; Yang, H.; Suda, T. Abstract 971: PRMT5 modulates splicing for genome integrity and preserves proteostasis of hematopoietic stem cells. *Tumor Biol.* **2019**, *26*, 2316–2328. [CrossRef]
31. Wei, H.; Wang, B.; Miyagi, M.; She, Y.; Gopalan, B.; Huang, D.-B.; Ghosh, G.; Stark, G.R.; Lu, T. PRMT5 dimethylates R30 of the p65 subunit to activate NF-κB. *Proc. Natl. Acad. Sci. USA* **2013**, *110*, 13516–13521. [CrossRef] [PubMed]
32. Hartley, A.-V.; Lu, T. Modulating the modulators: Regulation of protein arginine methyltransferases by post-translational modifications. *Drug Discov. Today* **2020**, *25*, 1735–1743. [CrossRef]
33. Motolani, A.A.; Sun, M.; Martin, M.; Sun, S.; Lu, T. Discovery of small molecule inhibitors for histone methyltransferases in cancer. In *Translational Research in Cancer*; BoD—Books on Demand: Norderstedt, Germany, 2021. [CrossRef]
34. Nie, M.; Wang, Y.; Guo, C.; Li, X.; Wang, Y.; Deng, Y.; Yao, B.; Gui, T.; Ma, C.; Liu, M.; et al. CARM1-mediated methylation of protein arginine methyltransferase 5 represses human γ-globin gene expression in erythroleukemia cells. *J. Biol. Chem.* **2018**, *293*, 17454–17463. [CrossRef]
35. Espejo, A.B.; Gao, G.; Black, K.; Gayatri, S.; Veland, N.; Kim, J.; Chen, T.; Sudol, M.; Walker, C.; Bedford, M.T. PRMT5 C-terminal phosphorylation modulates a 14-3-3/PDZ interaction switch. *J. Biol. Chem.* **2017**, *292*, 2255–2265. [CrossRef] [PubMed]
36. Hartley, A.-V.; Wang, B.; Jiang, G.; Wei, H.; Sun, M.; Prabhu, L.; Martin, M.; Safa, A.; Sun, S.; Liu, Y.; et al. Regulation of a PRMT5/NF-κB axis by phosphorylation of PRMT5 at serine 15 in colorectal cancer. *Int. J. Mol. Sci.* **2020**, *21*, 3684. [CrossRef]
37. Lattouf, H.; Poulard, C.; le Romancer, M. PRMT5 prognostic value in cancer. *Oncotarget* **2019**, *10*, 3151–3153. [CrossRef]
38. Zhang, H.-T.; Zeng, L.-F.; He, Q.-Y.; Tao, W.A.; Zha, Z.-G.; Hu, C.-D. The E3 ubiquitin ligase CHIP mediates ubiquitination and proteasomal degradation of PRMT5. *Biochim. Biophys. Acta BBA Bioenerg.* **2016**, *1863*, 335–346. [CrossRef]
39. Jin, J.; Martin, M.; Hartley, A.-V.; Lu, T. PRMTs and miRNAs: Functional cooperation in cancer and beyond. *Cell Cycle* **2019**, *18*, 1676–1686. [CrossRef] [PubMed]
40. Pal, S.; Baiocchi, R.A.; Byrd, J.C.; Grever, M.R.; Jacob, S.T.; Sif, S. Low levels of miR-92b/96 induce PRMT5 translation and H3R8/H4R3 methylation in mantle cell lymphoma. *EMBO J.* **2007**, *26*, 3558–3569. [CrossRef]
41. Wang, L.; Pal, S.; Sif, S. Protein arginine methyltransferase 5 suppresses the transcription of the RB family of tumor suppressors in leukemia and lymphoma cells. *Mol. Cell. Biol.* **2008**, *28*, 6262–6277. [CrossRef]

42. Lacroix, M.; El Messaoudi, S.; Rodier, G.; le Cam, A.; Sardet, C.; Fabbrizio, E. The histone-binding protein COPR5 is required for nuclear functions of the protein arginine methyltransferase PRMT5. *EMBO Rep.* **2008**, *9*, 452–458. [CrossRef]
43. Pal, S.; Vishwanath, S.N.; Erdjument-Bromage, H.; Tempst, P.; Sif, S. Human SWI/SNF-associated PRMT5 methylates histone H3 arginine 8 and negatively regulates expression of ST7 and NM23 tumor suppressor genes. *Mol. Cell. Biol.* **2004**, *24*, 9630–9645. [CrossRef]
44. Le Guezennec, X.; Vermeulen, M.; Brinkman, A.B.; Hoeijmakers, W.A.M.; Cohen, A.; Lasonder, E.; Stunnenberg, H.G. MBD2/NuRD and MBD3/NuRD, two distinct complexes with different biochemical and functional properties. *Mol. Cell. Biol.* **2006**, *26*, 843–851. [CrossRef]
45. Matkar, S.; Thiel, A.; Hua, X. Menin: A scaffold protein that controls gene expression and cell signaling. *Trends Biochem. Sci.* **2013**, *38*, 394–402. [CrossRef]
46. Ancelin, K.; Lange, U.C.; Hajkova, P.; Schneider, R.J.; Bannister, A.; Kouzarides, T.; Surani, A. Blimp1 associates with Prmt5 and directs histone arginine methylation in mouse germ cells. *Nat. Cell Biol.* **2006**, *8*, 623–630. [CrossRef] [PubMed]
47. Tabata, T.; Kokura, K.; Dijke, P.T.; Ishii, S. Ski co-repressor complexes maintain the basal repressed state of the TGF-β target gene, SMAD7, via HDAC3 and PRMT5. *Genes Cells* **2008**, *14*, 17–28. [CrossRef] [PubMed]
48. Owens, J.L.; Beketova, E.; Liu, S.; Tinsley, S.L.; Asberry, A.M.; Deng, X.; Huang, J.; Li, C.; Wan, J.; Hu, C.-D. PRMT5 cooperates with pICln to function as a master epigenetic activator of DNA double-strand break repair genes. *iScience* **2020**, *23*, 100750. [CrossRef] [PubMed]
49. Friesen, W.J.; Paushkin, S.; Wyce, A.; Massenet, S.; Pesiridis, G.S.; van Duyne, G.; Rappsilber, J.; Mann, M.; Dreyfuss, G. The methylosome, a 20S complex containing JBP1 and pICln, produces dimethylarginine-modified Sm proteins. *Mol. Cell. Biol.* **2001**, *21*, 8289–8300. [CrossRef]
50. Guderian, G.; Peter, C.; Wiesner, J.; Sickmann, A.; Schulze-Osthoff, K.; Fischer, U.; Grimmler, M. RioK1, a new interactor of protein arginine methyltransferase 5 (PRMT5), competes with pICln for binding and modulates PRMT5 complex composition and substrate specificity. *J. Biol. Chem.* **2011**, *286*, 1976–1986. [CrossRef] [PubMed]
51. Kim, H.; Ronai, Z.A. PRMT5 function and targeting in cancer. *Cell Stress* **2020**, *4*, 199–215. [CrossRef]
52. Chung, J.; Karkhanis, V.; Baiocchi, R.A.; Sif, S. Protein arginine methyltransferase 5 (PRMT5) promotes survival of lymphoma cells via activation of WNT/β-catenin and AKT/GSK3β proliferative signaling. *J. Biol. Chem.* **2019**, *294*, 7692–7710. [CrossRef] [PubMed]
53. Li, Y.; Chitnis, N.; Nakagawa, H.; Kita, Y.; Natsugoe, S.; Yang, Y.; Li, Z.; Wasik, M.; Klein-Szanto, A.J.P.; Rustgi, A.K.; et al. PRMT5 is required for lymphomagenesis triggered by multiple oncogenic drivers. *Cancer Discov.* **2015**, *5*, 288–303. [CrossRef] [PubMed]
54. Gullà, A.; Hideshima, T.; Bianchi, G.; Fulciniti, M.; Samur, M.K.; Qi, J.; Tai, Y.-T.; Harada, T.; Morelli, E.; Amodio, N.; et al. Protein arginine methyltransferase 5 has prognostic relevance and is a druggable target in multiple myeloma. *Leukemia* **2018**, *32*, 996–1002. [CrossRef] [PubMed]
55. Shailesh, H.; Zakaria, Z.Z.; Baiocchi, R.; Sif, S. Protein arginine methyltransferase 5 (PRMT5) dysregulation in cancer. *Oncotarget* **2018**, *9*, 36705–36718. [CrossRef]
56. Shailesh, H.; Siveen, K.S.; Sif, S. Protein arginine methyltransferase 5 (PRMT5) activates WNT/β-catenin signalling in breast cancer cells via epigenetic silencing of DKK1 and DKK3. *J. Cell. Mol. Med.* **2021**, *25*, 1583–1600. [CrossRef] [PubMed]
57. Jiang, H.; Zhu, Y.; Zhou, Z.; Xu, J.; Jin, S.; Xu, K.; Zhang, H.; Sun, Q.; Wang, J.; Xu, J. PRMT5 promotes cell proliferation by inhibiting BTG2 expression via the ERK signaling pathway in hepatocellular carcinoma. *Cancer Med.* **2018**, *7*, 869–882. [CrossRef]
58. Sheng, X.; Wang, Z. Protein arginine methyltransferase 5 regulates multiple signaling pathways to promote lung cancer cell proliferation. *BMC Cancer* **2016**, *16*, 567. [CrossRef]
59. Jing, P.; Zhao, N.; Ye, M.; Zhang, Y.; Zhang, Z.; Sun, J.; Wang, Z.; Zhang, J.; Gu, Z. Protein arginine methyltransferase 5 promotes lung cancer metastasis via the epigenetic regulation of miR-99 family/FGFR3 signaling. *Cancer Lett.* **2018**, *427*, 38–48. [CrossRef]
60. Hartley, A.-V.; Wang, B.; Mundade, R.; Jiang, G.; Sun, M.; Wei, H.; Sun, S.; Liu, Y.; Lu, T. PRMT5-mediated methylation of YBX1 regulates NF-κB activity in colorectal cancer. *Sci. Rep.* **2020**, *10*, 15934. [CrossRef]
61. Bao, X.; Zhao, S.; Liu, T.; Liu, Y.; Liu, Y.; Yang, X. Overexpression of PRMT5 promotes tumor cell growth and is associated with poor disease prognosis in epithelial ovarian cancer. *J. Histochem. Cytochem.* **2013**, *61*, 206–217. [CrossRef]
62. Lattouf, H.; Kassem, L.; Jacquemetton, J.; Choucair, A.; Poulard, C.; Tredan, O.; Corbo, L.; Diab-Assaf, M.; Hussein, N.; Treilleux, I.; et al. LKB1 regulates PRMT5 activity in breast cancer. *Int. J. Cancer* **2019**, *144*, 595–606. [CrossRef]
63. Shimizu, D.; Kanda, M.; Sugimoto, H.; Shibata, M.; Tanaka, H.; Takami, H.; Iwata, N.; Hayashi, M.; Tanaka, C.; Kobayashi, D.; et al. The protein arginine methyltransferase 5 promotes malignant phenotype of hepatocellular carcinoma cells and is associated with adverse patient outcomes after curative hepatectomy. *Int. J. Oncol.* **2017**, *50*, 381–386. [CrossRef] [PubMed]
64. Pak, M.G.; Lee, H.W.; Roh, M.S. High nuclear expression of protein arginine methyltransferase-5 is a potentially useful marker to estimate submucosal invasion in endoscopically resected early colorectal carcinoma. *Pathol. Int.* **2015**, *65*, 541–548. [CrossRef]
65. Chatterjee, S.; Khunti, K.; Davies, M.J. Type 2 diabetes. *Lancet* **2017**, *389*, 2239–2251. [CrossRef]
66. Galicia-Garcia, U.; Benito-Vicente, A.; Jebari, S.; Larrea-Sebal, A.; Siddiqi, H.; Uribe, K.B.; Ostolaza, H.; Martín, C. Pathophysiology of type 2 diabetes mellitus. *Int. J. Mol. Sci.* **2020**, *21*, 6275. [CrossRef]
67. Jia, Z.; Yue, F.; Chen, X.; Narayanan, N.; Qiu, J.; Syed, S.A.; Imbalzano, A.N.; Deng, M.; Yu, P.; Hu, C.; et al. Protein arginine methyltransferase PRMT5 regulates fatty acid metabolism and lipid droplet biogenesis in white adipose tissues. *Adv. Sci.* **2020**, *7*, 2002602. [CrossRef]

68. Leblanc, S.E.; Konda, S.; Wu, Q.; Hu, Y.-J.; Oslowski, C.M.; Sif, S.; Imbalzano, A.N. Protein arginine methyltransferase 5 (Prmt5) promotes gene expression of peroxisome proliferator-activated receptor γ2 (PPARγ2) and its target genes during adipogenesis. *Mol. Endocrinol.* **2012**, *26*, 583–597. [CrossRef] [PubMed]
69. Muhammad, A.B.; Xing, B.; Liu, C.; Naji, A.; Ma, X.; Simmons, R.; Hua, X. Menin and PRMT5 suppress GLP1 receptor transcript and PKA-mediated phosphorylation of FOXO1 and CREB. *Am. J. Physiol. Metab.* **2017**, *313*, E148–E166. [CrossRef] [PubMed]
70. Tsai, W.-W.; Niessen, S.; Goebel, N.; Yates, J.R.; Guccione, E.; Montminy, M. PRMT5 modulates the metabolic response to fasting signals. *Proc. Natl. Acad. Sci. USA* **2013**, *110*, 8870–8875. [CrossRef]
71. Ma, J.; He, X.; Cao, Y.; O'Dwyer, K.; Szigety, K.M.; Wu, Y.; Gurung, B.; Feng, Z.; Katona, B.; Hua, X.; et al. Islet-specific Prmt5 excision leads to reduced insulin expression and glucose intolerance in mice. *J. Endocrinol.* **2020**, *244*, 41–52. [CrossRef]
72. Nabel, E.G. Cardiovascular disease. *N. Engl. J. Med.* **2003**, *349*, 60–72. [CrossRef]
73. Cai, S.; Liu, R.; Wang, P.; Li, J.; Xie, T.; Wang, M.; Cao, Y.; Li, Z.; Liu, P. PRMT5 prevents cardiomyocyte hypertrophy via symmetric dimethylating HoxA9 and repressing HoxA9 expression. *Front. Pharmacol.* **2020**, *11*, 2140. [CrossRef] [PubMed]
74. Chen, M.; Yi, B.; Sun, J. Inhibition of cardiomyocyte hypertrophy by protein arginine methyltransferase 5. *J. Biol. Chem.* **2014**, *289*, 24325–24335. [CrossRef]
75. Tan, B.; Liu, Q.; Yang, L.; Yang, Y.; Liu, D.; Liu, L.; Meng, F. Low expression of PRMT5 in peripheral blood may serve as a potential independent risk factor in assessments of the risk of stable CAD and AMI. *BMC Cardiovasc. Disord.* **2019**, *19*, 31. [CrossRef] [PubMed]
76. Harris, D.P.; Bandyopadhyay, S.; Maxwell, T.J.; Willard, B.; DiCorleto, P.E. Tumor necrosis factor (TNF)-α induction of CXCL10 in endothelial cells requires protein arginine methyltransferase 5 (PRMT5)-mediated nuclear factor (NF)-κB p65 methylation. *J. Biol. Chem.* **2014**, *289*, 15328–15339. [CrossRef] [PubMed]
77. Harris, D.P.; Chandrasekharan, U.M.; Bandyopadhyay, S.; Willard, B.; DiCorleto, P.E. PRMT5-mediated methylation of NF-κB p65 at Arg174 is required for endothelial CXCL11 gene induction in response to TNF-α and IFN-γ costimulation. *PLoS ONE* **2016**, *11*, e0148905. [CrossRef]
78. Han, X.; Li, R.; Zhang, W.; Yang, X.; Wheeler, C.G.; Friedman, G.K.; Province, P.; Ding, Q.; You, Z.; Fathallah-Shaykh, H.M.; et al. Expression of PRMT5 correlates with malignant grade in gliomas and plays a pivotal role in tumor growth in vitro. *J. Neuro-Oncol.* **2014**, *118*, 61–72. [CrossRef]
79. Haass, C.; Selkoe, D.J. Soluble protein oligomers in neurodegeneration: Lessons from the Alzheimer's amyloid β-peptide. *Nat. Rev. Mol. Cell Biol.* **2007**, *8*, 101–112. [CrossRef]
80. Quan, X.; Yue, W.; Luo, Y.; Cao, J.; Wang, H.; Wang, Y.; Lu, Z. The protein arginine methyltransferase PRMT5 regulates Aβ-induced toxicity in human cells and *Caenorhabditis elegans* models of Alzheimer's disease. *J. Neurochem.* **2015**, *134*, 969–977. [CrossRef] [PubMed]
81. Lukiw, W.J. *Bacteroides fragilis* lipopolysaccharide and inflammatory signaling in Alzheimer's disease. *Front. Microbiol.* **2016**, *7*, 1544. [CrossRef]
82. Zhan, X.; Stamova, B.; Sharp, F.R. Lipopolysaccharide associates with amyloid plaques, neurons and oligodendrocytes in Alzheimer's disease brain: A review. *Front. Aging Neurosci.* **2018**, *10*, 42. [CrossRef]
83. Ross, C.A.; Aylward, E.H.; Wild, E.; Langbehn, D.; Long, J.; Warner, J.H.; Scahill, R.; Leavitt, B.R.; Stout, J.; Paulsen, J.; et al. Huntington disease: Natural history, biomarkers and prospects for therapeutics. *Nat. Rev. Neurol.* **2014**, *10*, 204–216. [CrossRef]
84. Ratovitski, T.; Arbez, N.; Stewart, J.C.; Chighladze, E.; Ross, C.A. PRMT5- mediated symmetric arginine dimethylation is attenuated by mutant huntingtin and is impaired in Huntington's disease (HD). *Cell Cycle* **2015**, *14*, 1716–1729. [CrossRef] [PubMed]
85. Zhu, F.; Rui, L. PRMT5 in gene regulation and hematologic malignancies. *Genes Dis.* **2019**, *6*, 247–257. [CrossRef] [PubMed]
86. Demetriadou, C.; Pavlou, D.; Mpekris, F.; Achilleos, C.; Stylianopoulos, T.; Zaravinos, A.; Papageorgis, P.; Kirmizis, A. NAA40 contributes to colorectal cancer growth by controlling PRMT5 expression. *Cell Death Dis.* **2019**, *10*, 236. [CrossRef]
87. Watts, J.M.; Bradley, T.J.; Thomassen, A.; Brunner, A.M.; Minden, M.D.; Papadantonakis, M.N.; Abedin, S.; Baines, A.J.; Barbash, O.; Gorman, S.; et al. A phase I/II study to investigate the safety and clinical activity of the protein arginine methyltransferase 5 inhibitor GSK3326595 in subjects with myelodysplastic syndrome and acute myeloid leukemia. *Blood* **2019**, *134*, 2656. [CrossRef]
88. Hamard, P.-J.; Santiago, G.E.; Liu, F.; Karl, D.L.; Martinez, C.; Man, N.; Mookhtiar, A.K.; Duffort, S.; Greenblatt, S.; Verdun, R.E.; et al. PRMT5 regulates DNA repair by controlling the alternative splicing of histone-modifying enzymes. *Cell Rep.* **2018**, *24*, 2643–2657. [CrossRef]
89. Millar, H.J.; Brehmer, D.; Verhulst, T.; Haddish-Berhane, N.; Greway, T.; Gaffney, D.; Packman, K. In vivo efficacy and pharmacodynamic modulation of JNJ-64619178, a selective PRMT5 inhibitor, in human lung and hematologic preclinical models. *Cancer Res.* **2019**, *79*, 950.
90. Ahnert, J.R.; Perez, C.A.; Wong, K.M.; Maitland, M.L.; Tsai, F.; Berlin, J.; Liao, K.H.; Wang, I.-M.; Markovtsova, L.; Jacobs, I.A.; et al. PF-06939999, a potent and selective PRMT5 inhibitor, in patients with advanced or metastatic solid tumors: A phase 1 dose escalation study. *J. Clin. Oncol.* **2021**, *39*, 3019. [CrossRef]
91. Pastore, F.; Bhagwat, N.; Pastore, A.; Radzisheuskaya, A.; Karzai, A.; Krishnan, A.; Li, B.; Bowman, R.L.; Xiao, W.; Viny, A.D.; et al. PRMT5 inhibition modulates E2F1 methylation and gene-regulatory networks leading to therapeutic efficacy in JAK2V617F-mutant MPN. *Cancer Discov.* **2020**, *10*, 1742–1757. [CrossRef]

92. Bonday, Z.Q.; Cortez, G.S.; Grogan, M.J.; Antonysamy, S.; Weichert, K.; Bocchinfuso, W.P.; Li, F.; Kennedy, S.; Li, B.; Mader, M.M.; et al. LLY-283, a potent and selective inhibitor of arginine methyltransferase 5, PRMT5, with antitumor activity. *ACS Med. Chem. Lett.* **2018**, *9*, 612–617. [CrossRef] [PubMed]
93. Banasavadi-Siddegowda, Y.K.; Welker, A.; An, M.; Yang, X.; Zhou, W.; Shi, G.; Imitola, J.; Li, C.; Hsu, S.; Wang, J.; et al. PRMT5 as a druggable target for glioblastoma therapy. *Neuro-Oncology* **2017**, *20*, 753–763. [CrossRef] [PubMed]
94. Smith, C.R.; Kulyk, S.; Lawson, J.D.; Engstrom, L.D.; Aranda, R.; Briere, D.M.; Gunn, R.; Moya, K.; Rahbaek, L.; Waters, L.; et al. Abstract LB003: Fragment based discovery of MRTX9768, a synthetic lethal-based inhibitor designed to bind the PRMT5-MTA complex and selectively target MTAP/CDKN2A-deleted tumors. *Cancer Chem.* **2021**, *81*, LB003. [CrossRef]
95. Carter, J.; Ito, K.; Thodima, V.; Bhagwat, N.; Rager, J.; Burr, N.S.; Kaufman, J.; Ruggeri, B.; Scherle, P.; Vaddi, K. PRMT5 inhibition downregulates MYB and NOTCH1 signaling, key molecular drivers of adenoid cystic carcinoma. *Cancer Res.* **2021**, *81* (Suppl. S13), 1138. [CrossRef]

Review

Structure, Activity and Function of the Protein Arginine Methyltransferase 6

Somlee Gupta [1], Rajashekar Varma Kadumuri [2], Anjali Kumari Singh [2], Sreenivas Chavali [2] and Arunkumar Dhayalan [1,*]

[1] Department of Biotechnology, Pondicherry University, Puducherry 605014, India; somleegupta@gmail.com
[2] Department of Biology, Indian Institute of Science Education and Research (IISER) Tirupati, Tirupati 517507, India; rajashekar@labs.iisertirupati.ac.in (R.V.K.); anjaliksingh@students.iisertirupati.ac.in (A.K.S.); schavali@iisertirupati.ac.in (S.C.)
* Correspondence: arun.dbt@pondiuni.edu.in; Tel.: +91-413-2654789

Abstract: Members of the protein arginine methyltransferase (PRMT) family methylate the arginine residue(s) of several proteins and regulate a broad spectrum of cellular functions. Protein arginine methyltransferase 6 (PRMT6) is a type I PRMT that asymmetrically dimethylates the arginine residues of numerous substrate proteins. PRMT6 introduces asymmetric dimethylation modification in the histone 3 at arginine 2 (H3R2me2a) and facilitates epigenetic regulation of global gene expression. In addition to histones, PRMT6 methylates a wide range of cellular proteins and regulates their functions. Here, we discuss (i) the biochemical aspects of enzyme kinetics, (ii) the structural features of PRMT6 and (iii) the diverse functional outcomes of PRMT6 mediated arginine methylation. Finally, we highlight how dysregulation of PRMT6 is implicated in various types of cancers and response to viral infections.

Keywords: protein arginine methylation; PRMT6; post-translational modification; H3R2me2a; epigenetics; cancer

1. Introduction

Protein arginine methyltransferases (PRMTs) are the family of enzymes which methylate the arginine residue(s) of substrate proteins and regulate a wide range of cellular events including transcription, splicing, translation, DNA damage responses and phase separation [1]. PRMTs are classified into three types depending on the nature of the methylarginine that they form. Type I PRMTs (PRMT1, PRMT2, PRMT3, PRMT4, PRMT6 and PRMT8) and Type II PRMTs (PRMT5 and PRMT9) generate asymmetric dimethylarginine and symmetric dimethylarginine, respectively, in addition to the monomethylarginine. Type III enzyme (PRMT7) exclusively generates monomethylarginines in proteins [2,3].

Protein arginine methyltransferase 6 (PRMT6) is a type I PRMT which is involved in epigenetic regulation of gene expression [4–6], alternative splicing [7,8], development and differentiation [9–13], DNA repair [14], cell proliferation and senescence [15–20], DNA methylation [21], mitosis [22,23], inflammation [24–26], innate antiviral immunity [27], spermatogenesis [28], transactivation of nuclear receptors [7,29,30] and cell signaling [31,32]. A high throughput yeast two-hybrid screening of PRMT6 identified 36 proteins as potential interaction partners of PRMT6 suggesting the role of PRMT6 in various additional, so far unidentified cellular functions [33]. In addition to these physiological functions, the dysregulation of PRMT6 is implicated in viral diseases [34–41], cancers [42] and cardiac hypertrophy [43].

The human PRMT6 gene, located on Chromosome 1, encodes for the 41.9 kDa PRMT6 enzyme. PRMT6 is predominantly localized to the nucleus, in stark contrast to PRMT3 and PRMT5 which are preponderantly cytosolic, while other PRMTs are found in both nucleus and cytosol [44]. PRMT6 is expressed in a wide range of tissues with high expression in

kidney and testes [44]. The F-box proteins, FBXO24 and FBXW17, regulate the protein levels of PRMT6 by facilitating its proteasomal degradation [45,46]. PRMT6 generates asymmetric dimethylation modifications in histone 3 at arginine 2, arginine 17 and arginine 42 (H3R2me2a, H3R17me2a and H3R42me2a) [4–6,47,48] and in histone H2A at arginine 26 (H2AR26me2a) [49] and participates in the epigenetic regulation of gene expression. In addition to the histones, PRMT6 methylates a wide range of non-histone proteins and regulates various biological functions. In the following sections, we discuss the biochemical features, structural aspects, epigenetic functions of PRMT6, functional outcomes of PRMT6-mediated methylation of non-histone substrates and viral proteins and the role of PRMT6 in various types of cancers.

2. Biochemical Features

PRMT6 was identified as a protein arginine methyltransferase based on the presence of conserved catalytic core of PRMT family members. In vitro studies using the GST-GAR substrate, showed that the recombinant PRMT6 enzyme generates monomethylarginines and asymmetric dimethylarginines, establishing PRMT6 as a bona fide type I PRMT enzyme [44]. GST-GAR, an in vitro substrate used to investigate the enzymatic activity of PRMTs, is a GST fusion of N-terminal region of the human rRNA 2′-O-methyltransferase fibrillarin enzyme (amino acids 1 to 148). GST-GAR contains nine arginine residues in the sequence context of "RGG" and six arginine residues in the sequence context "RG" [50–52].

Kinetic studies of PRMT6 mediated methylation of GST-GAR substrate revealed that it generates asymmetric dimethylarginines in a processive manner [44]. Furthermore, investigation of kinetics of recombinant PRMT6 with substrate peptides containing a single arginine or monomethylarginine residue [53] revealed that PRMT6 follows an ordered sequential mechanism of catalysis in which the S-adenosyl-L-methionine (SAM) binds to the enzyme first, followed by the binding of the substrate peptide resulting in the formation of a ternary complex (Figure 1A) [53,54]. Consequent to the methyl group transfer, the methylated substrate peptide dissociates first from the enzyme followed by the S-adenosyl-L-homocysteine (SAH) [53,54]. The catalytic efficiency of PRMT6 on the monomethylarginine containing substrate peptides is much higher than that of the substrate peptides with unmethylated arginine [53]. Notably, peptide products with asymmetric dimethylarginine could not be detected in the enzymatic reaction of PRMT6 with unmethylated arginine containing peptides. Collectively these findings suggest that the PRMT6 enzyme generates asymmetric dimethylarginine in a distributive manner rather than introducing two methyl groups processively in a single arginine residue in a single binding event [53].

However, this ordered sequential kinetic mechanism of PRMT6 enzyme was contradicted by a subsequent study which showed that the pre-incubation of SAM with the PRMT6 enzyme did not alter the IC_{50} of C21 peptide on the activity of PRMT6 [55]. C21 is a modified H4 tail peptide which contains chloroacetamidine modified ornithine at arginine 3 position and this C21 peptide inhibits the enzymatic activity of PRMTs irreversibly with high selectivity for PRMT1 and modest selectivity for PRMT6 [55,56]. This suggests that the binding of the SAM with the enzyme is not a pre-requisite step for the substrate peptide binding, challenging the ordered sequential kinetic mechanism of PRMT6 [55]. This observation prompted further detailed investigations of the kinetic mechanism of PRMT6 catalysis. Product inhibition studies and usage of dead-end analogs revealed that PRMT6 follows a rapid equilibrium and random kinetic mechanism with the generation of dead-end complexes (Figure 1B) [55]. These studies led to the proposal that PRMT6 binds to the substrate peptides and SAM in a random fashion to produce a ternary complex and products dissociate randomly after the methyl group transfer to generate the free enzyme [55]. This necessitates detailed structural studies of PRMT6 in complex with the proper substrate peptide and thorough kinetic studies to resolve this conflict and delineate the actual kinetic mechanism of PRMT6 catalysis.

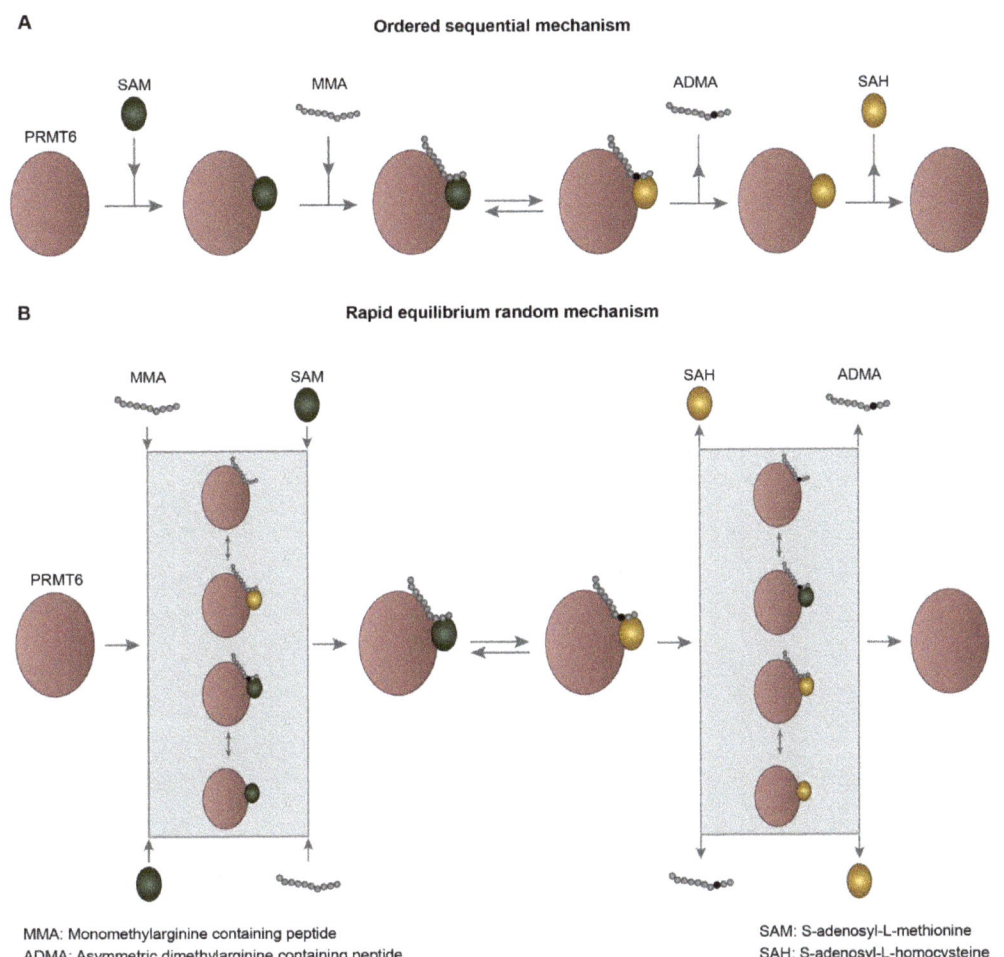

Figure 1. The proposed kinetic mechanisms of PRMT6 catalysis. Schematic representation of (**A**) Sequential ordered kinetic mechanism and (**B**) Rapid equilibrium random mechanism with dead-end complexes.

In addition to methylating its substrates, PRMT6 also undergoes automethylation at Arg35. This automethylation of PRMT6 regulates its stability and anti-HIV1 function without affecting its catalytic activity [44,57]. High throughput mass spectrometry analysis of PRMT6 activity on the H3 peptides carrying different amino acid exchanges at different positions showed that PRMT6 is relatively a non-specific enzyme in vitro as it tolerated most of the amino acid substitutions, with preference for positively charged amino acids and bulkier amino acids around the target arginine. PRMT6 exhibits a higher preference to methylate RG or RGK motif containing peptides more efficiently than the RGG motif containing peptides in vitro [58]. Contrarily, there is a higher preference of PRMT1 to methylate RGG motif containing substrates, while PRMT4 shows a preference to methylate the substrates in which the target arginine is flanked by the proline residues [58–61].

Given that the PRMT6 exhibits a relaxed substrate specificity, it is highly likely that it might methylate several hitherto unidentified substrate proteins, including the substrates of other PRMTs. Indeed, an in vitro methylation assay on rat cell extracts revealed that PRMT1, PRMT4 and PRMT6 methylate some common substrate proteins, alongside protein substrates that are unique to each of these PRMTs [44]. Both PRMT4 and PRMT6 are involved in the generation of asymmetric dimethylation at Arg17 of histone 3 (H3R17me2a) and asymmetric dimethylation at Arg42 of histone 3 (H3R42me2a) in vivo [47,48]. Examples of common substrates include (i) the type1A topoisomerase enzyme, TOP3B, which is methylated by PRMT1, PRMT3 and PRMT6 enzymes [62] and (ii) the cell cycle inhibitor, p16 is methylated PRMT1, PRMT4 and PRMT6 enzymes [15]. All these observations suggest that PRMT6 can also methylate the substrates of other PRMTs in addition to its unique substrates. Nevertheless, the biochemical and structural mechanisms underlying how overlapping substrates are methylated by different PRMTs remains elusive.

3. Structural Attributes

PRMT6 of *Trypanosoma brucei*, mouse and human in complex with sinefungin (SNF) or SAH and/or short substrate peptide have been structurally characterized [63–65]. Mouse PRMT6 shares 91.5% sequence homology with human PRMT6 and 29% sequence homology with *T. brucei* PRMT6 [64]. All members of the PRMT family share a conserved catalytic PRMT core region with variable N-terminal regions. The overall structure of human PRMT6 is similar to that of other PRMT family members. PRMT6 consists of three structural components viz. (i) the N-terminal Rossmann fold, which contains the SAM binding pocket, (ii) the C-terminal β-barrel domain and (iii) a dimerization helix, which are located between the β6 strand and β7 strand of the C-terminal β-barrel domain (Figure 2A). A conserved proline (Pro186) in cis-conformation connects the N-terminal Rossmann fold and the C-terminal β-barrel domain [64,65]. The invariant residues in the SAM binding pocket of Rossmann fold interacts with the homocysteine carboxylate, adenine ring and ribose of SAH through hydrogen bonds and salt bridges [65] (Figure 2B).

Structural comparison of PRMT6 with other type I PRMT structures revealed two interesting features of PRMT6: (i) The conserved aromatic motif Y(F/Y)xxY at the N-terminal region covers the adenosine moiety of SAH as a lid in case of PRMT3 and PRMT4 structures [66–68], while in PRMT6 and PRMT1 this motif is positioned outwards from the SAH binding pocket [65,69,70] (Figure 2B). Since this aromatic motif is important for the SAH binding and catalysis of PRMT1 [70], it might adopt different conformations dynamically to facilitate the release of SAH during catalysis of PRMT1 and PRMT6. (ii) Similar to other type I PRMT members, PRMT6 also forms a dimer through the interaction of dimerization arm helices of one monomer with the N-terminal helices and the helices of Rossmann fold of the other monomer [65]. The dimerization arm of PRMT6 exhibits a different conformation and forms a flat ring dimer structure with a central cavity in contrast to the concave surface cavity formed by the other type I PRMT dimers [65,70,71]. This unique arrangement of the central cavity of PRMT6 dimer might influence the substrate selectivity of PRMT6.

Superimposition of the *T. brucei* PRMT7 (TbPRMT7) structure (PDBID: 4M38) [72] and modelling of the arginine residue into the active site of the human PRMT6 structure (PDBID: 5HZM) [65] revealed that the conserved glutamate Glu155 in the active site forms a hydrogen bond with the substrate arginine residue (Figure 2C). However, the other conserved glutamate, Glu164, which is part of the double E loop is directed outward from arginine in the active site. An outward direction of Glu164 was also observed upon the superimposition of the PRMT1-arginine-SAH complex structure [69] on the PRMT6 structure [65]. The corresponding glutamate Glu181 of TbPRMT7 and Glu444 of PRMT5 forms a hydrogen bond with the target arginine in TbPRMT7-H4R3 peptide-SAH and PRMT5-H4R3 peptide-SAH complexes respectively (Figure 2C) [72,73]. Moreover, the Glu164 residue points towards the active site in PRMT6 in complex with SAH and the inhibitor, EPZ0204111 [65]. Collectively, these observations suggest that the conserved Glu164 residue of the double E loop points towards the target arginine and forms a hydrogen bond or it points away from the active site, and this flexible nature of this residue might be important for the catalysis.

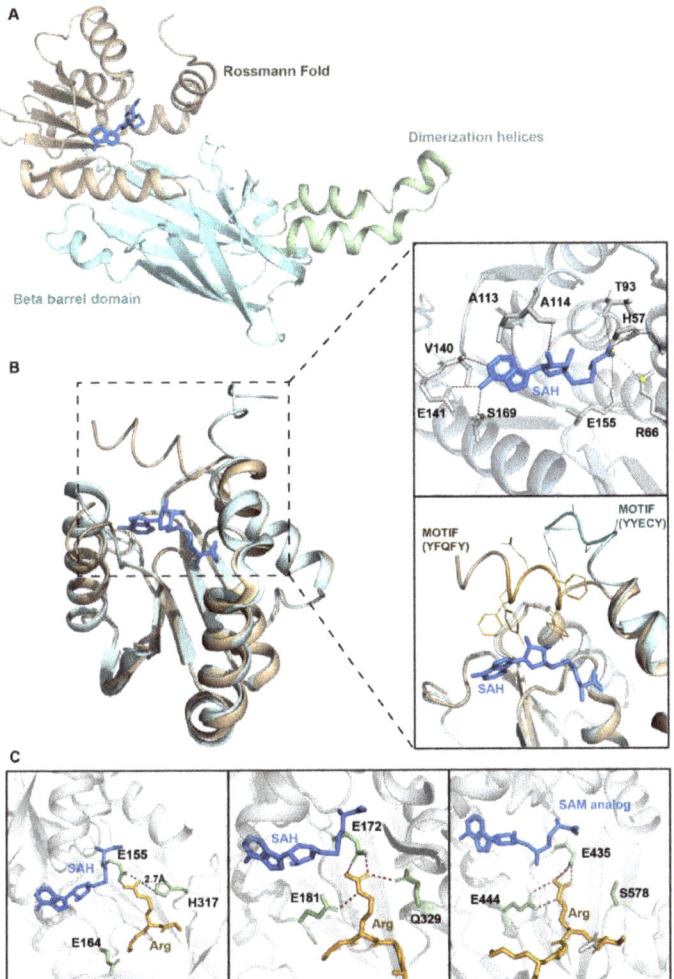

Figure 2. Structural attributes of PRMT6. (**A**) Crystal structure of the human PRMT6 (PDBID: 6W6D) in complex with SAH (marine blue). (**B**) Structural superimposition of the Rossmann fold region of PRMT6 (PDBID: 6W6D, cyan), PRMT4 (PDBID: 6IZQ, brown). Top inset shows the detailed interactions between PRMT6 and SAH (Top; PDBID: 6W6D). Polar contacts (red) and salt bridges (yellow) are displayed as dashed lines. PRMT6 is represented in cartoon and side chains of the residues (white) interacting with S-adenosyl-L-homocysteine (SAH; marine blue) are represented as sticks. Bottom inset shows the structural superimposition of the conserved motif from PRMT6 (Motif: YYECY) and PRMT4 (Motif: YFQFY). The sidechains of the motif residues are highlighted as wire-representations. (**C**) Structural comparison of the human PRMT6 (Left panel; PDBID: 5HZM) with *T. brucei* PRMT7 (Middle panel; PDBID: 4M38) and human PRMT5 (Right panel; PDBID: 4GQB) highlighting the active site in complex with the arginine (orange). SAH in the first two panels and S-adenosyl-L-methionine (SAM) analog in the third panel have been highlighted in marine blue. For the PRMT6 (left panel), arginine is modeled into the active site of human PRMT6 (PDBID: 5HZM) by superimposing the *T. brucei* PRMT7 structure (PDBID: 4M38). Distance between His317 and arginine Nη2 atom (left panel) is shown as black colored dashed line. Polar contacts are displayed as red colored dashed lines. Interacting residues side chains are represented as sticks (green).

The THW motif, present in the loop that connects the β11 and β12 of the β-barrel domain, is located in the active site of PRMT6 and is conserved among the type I PRMTs. His317 of the THW motif is likely to form a hydrogen bond with the Nη2 atom of the target arginine as it is within the distance of 2.7 Å (Figure 2C). This might impede the swapping of Nη2 and methyl Nη1 positions and thus preclude symmetric dimethylation. The corresponding residue in TbPRMT7 complex, Gln329, also forms a hydrogen bond with the Nη2 atom of the target arginine in a similar manner and hence prevents the symmetric dimethylation (Figure 2C) [65,72]. However, in case of the type II PRMT5 enzyme complex, the corresponding residue of His317 is Ser578, which does not form a hydrogen bond with Nη2 atom of the target arginine as it points away from the target arginine (Figure 2C). Such an arrangement allows the swapping of Nη2 and methyl Nη1 positions and facilitates the generation of symmetric dimethyl arginine [65,73]. Taken together, these findings and comparisons provide the plausible structural basis for the mechanism underlying PRMT6 generation of asymmetric dimethyl arginine.

4. Biological Roles of PRMT6

4.1. Epigenetic Functions of PRMT6

Post-translational modifications of the histone tails regulate the chromatin structure and control various chromatin-dependent processes including gene expression. PRMT6 is the major enzyme that generates asymmetric dimethylation at Arg2 of histone 3 (H3R2me2a) in vivo and regulates the global levels of H3R2me2a [4–6]. PRMT6-mediated H3R2me2a modifications are enriched in the gene body and promoter regions of inactive genes and inversely correlate with active H3K4me3 (trimethylation of histone 3 at Lys4) modifications in the promoter regions. The H3R2me2a modification is a repressive mark as its presence in the promoters negatively correlates with the transcript levels of the associated genes [4]. Not surprisingly, presence of the active histone mark H3K4me3 inhibits the PRMT6 activity on the H3 peptides [4,5]. The SET1/MLL family of enzymes catalyze the formation of the active H3K4me3 modification and these enzymes function as a complex with other protein factors [74–77]. The MLL complex does not show activity on the H3 peptides with H3R2me2a modification [4,6]. These findings explain the observed counter-correlation between the H3K4me3 and H3R2me2a modifications and the corresponding roles of methyltransferases at the promoter regions.

The majority of the reader proteins which recognize the H3K4me3 modification exhibit a reduced or non-detectable binding with the H3 peptides carrying the dual H3K4me3 and H3R2me2a modifications. This indicates that the presence of H3R2me2a modification inhibits the recognition of H3K4me3 modification by the reader proteins [5]. MLL complex activates the expression of HoxA genes and Myc targets genes by generating H3K4me3 modifications at their promoters [76,78–80] and the PRMT6 down regulates the expression of these genes [6]. The over-expression of PRMT6 represses the expression of HoxA2 by increasing the level of H3R2me2 marks and reducing H3K4me3 modifications near the transcription start site of the HoxA2 gene [6] (Figure 2A).

PRMT6 mediated H3R2me2a modifications tend to co-occur with H3K27me3 in a subset of silent gene promoters [12,80,81]. PRMT6 interacts with the polycomb repressive complexes (PRC) and silences the transcription of rostral *HOXA* genes by generating H3R2me2a and H3K27me3 modifications in their promoters [82]. Thrombospondin-1 (TSP-1) is a secretory protein which inhibits angiogenesis strongly and negatively affects cell migration [83–85]. In the U2OS osteosarcoma cells, PRMT6 has been shown to down regulate the expression of TSP-1 by introducing H3R2me2a modifications and reducing the active H3K4me3 modification in the TSP-1 promoter regions (Figure 3A). This negative regulation of TSP-1 by PRMT6 increases the migration and invasive properties of the U2OS cells [85]. However, this effect appears to be cell/cancer-type specific, as the over-expression of PRMT6 in human estrogen-sensitive breast cancer cells (MCF7) and the human prostate cancer cells (PC3) increases the TSP-1 expression and hence inhibits the movement and invasion of cancer cells [86]. Additional studies are required to resolve this conundrum as to why enhanced levels of PRMT6 have opposing effects in different cancers.

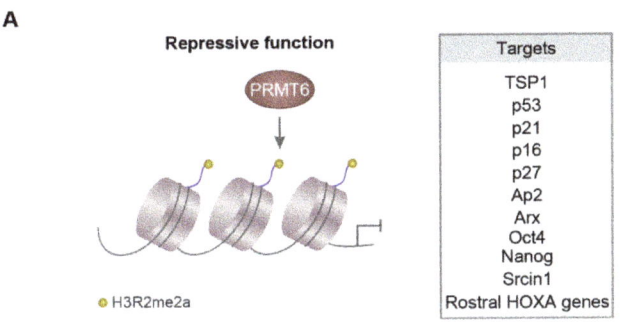

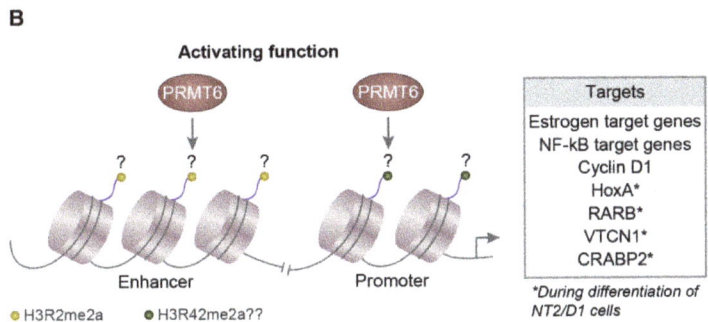

Figure 3. PRMT6 mediated epigenetic regulation of gene expression. (**A**) Schema highlighting the repressive role of PRMT6. PRMT6 generates H3R2me2a modifications at the promoters of the target genes and suppresses their expression. (**B**) Schematic representation of the transcriptional activation function of PRMT6. PRMT6 generates H3R2me2a modifications at the enhancers and probably H3R42me2a modifications at the promoters of the target genes and activates their expression. '?' indicates that the highlighted histone modifications are putative and need further characterization.

From a molecular point of view, PRMT6 promotes cell proliferation and contributes to the tumorigenic properties and prevents premature cellular senescence by downregulating the expression of tumor suppressor genes, p53, p21, p16 and p27. PRMT6 downregulates these tumor suppressors by generating the repressive H3R2me2a modifications in their promoters and other regulatory regions [16,17,20,87]. PRMT6 regulates adipocyte differentiation by interacting with PPARγ and inhibiting its functions. PRMT6 suppresses the expression of the PPARγ target gene, adipocyte protein 2 (Ap2) by binding to the PPAR responsive regulatory element in the promoter of the Ap2 along with PPARγ and by adding repressive H3R2me2a modifications [88]. Aristaless Related Homeobox (Arx) is a lineage determining gene which is expressed exclusively in the pancreatic α cells. DNA methylation and PRMT6-mediated H3R2me2a modifications in the upstream regulatory regions of Arx gene suppresses its expression in pancreatic β cells [89]. Repeated cocaine exposure decreases the PRMT6 in dopamine D2 expressing medium spiny neurons (D2-MSNs) present in the nucleus accumbens of the basal forebrain. This leads to the decrease of H3R2me2a modifications and an increase in H3K4me3 modifications in the promoter region of Src kinase signaling inhibitor 1 (Srcin1) which in turn upregulates the Srcin 1 expression. The elevated levels of Srcin 1 inhibits the Src signaling, thereby reducing cocaine reward and the intent for the self-administration of cocaine [90] (Figure 3A).

PRMT6 and the associated H3R2me2a modifications increase globally during the differentiation of the mouse embryonic stem cells. Specifically, PRMT6 regulates the expression of the pluripotency genes, Oct4 and Nanog by modulating the levels of H3R2me2a and H3K4me3 modifications at their promoter regions [10]. During the megakaryocytic/erythroid lineage bifurcation of common hematopoietic progenitor cells, PRMT6 facilitates megakaryocytic differentiation by inhibiting the expression of erythroid genes. In the CD34 positive hematopoietic progenitor cells, the transcription factor RUNX1 interacts with PRMT6 and establishes the H3R2me2a modifications at the promoters of megakaryocytic genes [12,13]. During the megakaryocytic differentiation of the progenitor cells, PRMT6 dissociates from the RUNX1 co-repressor complex and facilitates the expression of megakaryocytic genes [12,13]. In addition, PRMT6 inhibits the expression of erythroid genes by generating the repressive H3R2me2a modifications in their promoter regions during megakaryocytic differentiation of the progenitor cells [91] (Figure 3A).

UHRF1 is the multi-domain protein factor which facilitates the recruitment of DNMT1 to the hemi-methylated CpG sites and the maintenance of DNA methylation pattern [92,93]. PRMT6 mediated H3R2me2a modifications inhibit the binding of UHRF1 to the chromatin which in turn negatively affects the DNA methylation by DNMT1 [21,94–97]. The high levels of PRMT6 in cancer cells lead to the global DNA hypomethylation and contribute to the carcinogenesis, possibly through the passive DNA demethylation [21]. The H3R2me2a modifications generated by PRMT6 play an important role in chromosome condensation during mitosis, because H3R2me2a recruits chromosomal passenger complex (CPC) to the chromosome upon mitotic entry and augments the H3S10 phosphorylation by Aurora B kinase which in turn facilitates the chromosome condensation [22].

In addition to the repressive epigenetic roles, PRMT6 also functions as a co-activator for steroid hormone receptors and this co-activator function requires the methyltransferase activity of PRMT6. PRMT6 promotes the expression of estrogen target genes and cell proliferation in an estrogen-dependent manner in MCF7 cells [7]. PRMT6 interacts with the transcription factor NF-κB and serves as co-activator to facilitate the expression of NF-κB target genes [24]. The transcription factor LEF1 activates the expression of cyclin D1 by recruiting PRMT6 to the promoter of cyclin D1 [98]. This co-activator function of PRMT6 might be due to the ability of PRMT6 to generate the active H3R42me2a modifications as well in the chromatin (Figure 3B).

A recent thorough genome wide study of PRMT6 mediated H3R2me2a modification in human embryonal carcinoma NT2/D1 cells strikingly revealed that this modification is associated with promoters, transcriptional start site and enhancer elements of active genes rather than repressed genes [99]. The promoter and TSS site associated H3R2me2a modifications suppress the transcription of the associated active genes by preventing the generation of active H3K4me3 modifications at these locations. However, the enhancer associated H3R2me2a modifications activate the transcription of the associated genes by facilitating the deposition of H3K4me1 and H3K27ac modifications [99]. Interestingly, the PRMT6 generated H3R2me2a modifications tend not to co-occur at the promoter and enhancer regions of the same genes [99]. Thus, the transcriptional outcome of H3R2me2a modifications is dependent on the genomic location of these modifications (Figure 3B).

In addition to H3R2 methylation, PRMT6 also methylates H4 at Arg3 residue (H4R3me2a) in vitro [6], H3 at Arg42 residue (H3R42me2a) [47] and H2A at Arg29 residue (H2AR29me2a) both in vitro and in vivo [49]. The PRMT6 mediated H2AR29me2a modifications are enriched in the promoter regions of the specific genes which are down regulated [49]. PRMT6 also methylates H3 at Arg17 residue (H3R17me2a) in vitro alongside the PRMT4 enzyme. H3R17me2a levels are increased during the mitosis [100] and this increase of H3R17me2a requires both PRMT4 and PRMT6 enzymes. Moreover, over-expression of PRMT6 increases the H3R17me2a modifications globally. All these findings suggest that PRMT6 is also involved in the deposition of H3R17me2a alongside PRMT4 [48]. Thus, PRMT6 mediated histone modifications affect diverse biological processes by regulating the gene expression, deposition of other histone modifications and DNA methylation.

4.2. Functional Outcomes of PRMT6 Mediated Methylation of Its Substrates

Below, we will discuss the different non-histone substrates of PRMT6, identified through targeted molecular and/or biochemical studies and their functional consequences, as applicable. Firstly, we will discuss the human substrates and then shed light on the current understanding of the viral proteins that are methylated by PRMT6 (Table 1).

Table 1. Non-histone substrate proteins of PRMT6 and the functional outcomes of their methylation by PRMT6.

S. No.	Substrate Proteins	Functional Outcome(s) of PRMT6 Mediated Methylation	Reference (PMIDs)
1.	HMGA1a	Might regulate the binding of HMGA1a with DNA [101–103].	16157300, 16293633, 17550233
2.	DNA Polymerase β	Increases the polymerase activity and facilitates the base excision repair [14].	16600869
3.	P16	Promotes the cell cycle by inhibiting P16 interaction with CDK4 [15,18].	23032699, 26622834
4.	P21	Promotes the phosphorylation of P21 and accumulation of P21 in cytoplasm [19].	26436589
5.	CRTC2	Promotes the CRTC2-CREG interaction and enhances the expression of gluconeogenic enzymes in hepatocytes [104].	24570487
6.	ERα	Might promote estrogen dependent functions of ERα [7,30].	24742914, 20047962
7.	AR	Inhibits the phosphorylation of AR and promotes the hormone dependent transactivation of AR [29].	25569348
8.	GPS2	Enhances the protein stability of GPS2 [105].	26070566
9.	TOP3B	Enhances the topoisomerase activity and facilitates TOP3B localization in stress granules [62].	29471495
10.	SIRT7	Inhibits SIRT7 deacetylase activity, thereby promoting mitochondrial biogenesis [106].	30420520
11.	FOXO3	Enhances FOXO3 activity and contributes to the muscle atrophy [107].	30653406
12.	PTEN	Inhibits Akt signaling and modulates global alternative splicing [31].	30886105
13.	HTT	Facilitates the axonal transport of organelles by HTT and enhances the neuronal viability [108].	33852844
14.	CRAF	Inhibits CRAF-RAS interaction and suppresses the MEK/ERK signaling in Hepatocellular Carcinoma (HCC) [32].	30332648
15.	BAG5	Promotes the degradation of the BAG5 interaction partner, HSC70 which in turn inhibits autophagy in HCC [109].	33186656
16.	RCC1	Facilitates association of RCC1 with chromatin and promotes mitosis in Glioblastoma [23].	33539787
17.	HIV1-TAT	(i) Increases TAT1 stability, (ii) excludes the Tat from the nucleolus and (iii) decreases its transactivation function [34,35,37,38].	19726520, 15596808, 26611710, 17267505
18.	HIV1-REV	Inhibits nuclear export function of REV [39].	17176473
19.	HIV1-Nucleocapsid protein (NC)	Decreases the capacity of NC to anneal the tRNA Lys to the primer site of viral RNA [40].	17415034
20.	pUL69 of human cytomegalovirus	The functional consequence of this methylation is unknown [41].	26178996

4.2.1. Substrates in Human Cells

(i). *DNA repair protein-DNA Polymerase β*

DNA Polymerase β plays an important role in the base excision repair [110–112]. PRMT6 interacts with and methylates DNA Polymerase β at Arg83 and Arg152 residues. Methylation of DNA Polymerase β by PRMT6 enhances its polymerase activity by increasing its processivity. The residues which are methylated by PRMT6 are important for efficient repair of DNA damage introduced by the alkylating agents [14].

(ii). *Chromatin modifiers-HMGA1a, SIRT7*

HMGA1 proteins are nuclear non-histone proteins which regulate the chromatin structure and gene expression [113–115]. PRMT6 has been shown to methylate HMGA1a protein at Arg57 and Arg59 residues, both in vitro and in vivo [101–103]. These PRMT6 target residues are located in the second AT hook region of HMGA1a which is important for the binding of HMGA1a with DNA, suggesting that PRMT6 mediated methylation of HMGA1a might affect its binding with the DNA [102,103].

SIRT7, a deacetylase, catalyzes the removal of the acetylation modification of histone 3 at Lys18 (H3K18ac) and regulates many cellular processes including mitochondrial biogenesis [116–121]. PRMT6 interacts with SIRT7 and methylates it at Arg388. This methylation inhibits the deacetylase activity of SIRT7 and positively regulates mitochondrial biogenesis. The glucose availability regulates the PRMT6-mediated methylation of SIRT7 in an AMPK dependent manner [106]. Thus, the methylation of SIRT7 by PRMT6 serves as a link that connects mitochondrial biogenesis with glucose levels [106].

(iii). *Transcription regulators-CRTC2, FOXO3, GPS2 and TOP3B*

The transcription factor CREB (cAMP response element binding protein) and the CREB-regulated transcriptional coactivator 2 (CRTC2) activate the expression of gluconeogenic enzymes in the liver during fasting [122–125]. PRMT6 interacts with CRTC2 and generates asymmetric dimethylation modifications at several arginine residues of CRTC2. These PRMT6-mediated methylations of CRTC2 aid its association with CREB at the promoter regions and enhance the expression of gluconeogenic enzymes in hepatocytes [104,126].

FOXO3 is a multifunctional transcription factor and is implicated in various biological processes including gluconeogenesis, DNA repair, cell cycle autophagy, redox balance and proteostasis [127,128]. It is also involved in muscle atrophy by inducing protein degradation through the expression of muscle specific ubiquitin ligases [129–131]. The muscle specific knock out of PRMT1 increases the expression of autophagic markers and muscle specific ubiquitin ligases and promotes the muscle atrophy. Depletion of PRMT1 upregulates the expression of PRMT6 in muscle cells which in turn methylates FOXO3 at Arg118, Arg218 and Arg249. The PRMT6 mediated methylation of FOXO3 enhances its activity and contributes to the muscle atrophy [107].

G protein pathway suppressor 2 (GPS2) is a transcriptional regulator and is implicated in several cellular processes including cell cycle, apoptosis, bile acid synthesis and inflammation [132–138]. Methylation of GPS2 at Arg312 and Arg323 residues by PRMT6 promotes the association of GPS2 with TBL1, which inhibits the proteasomal degradation of GPS2, thereby leading to its enhanced stability, with implications across diverse cellular functions [105].

TOP3B is a type1A topoisomerase enzyme which resolves the topological strains of both DNA and RNA [139,140]. The histone arginine methylation reader protein, TDRD3 interacts with TOP3B and recruits it to the target chromatin regions wherein TOP3B executes its functions [141,142]. PRMT1, PRMT3 and PRMT6 methylates the Arg833 and Arg835 in the C-terminal region of TOP3B. The methylation of TOP3B by PRMT6 is required for (i) the efficient relaxation of supercoiled DNA and preventing the R-loop formation during transcription and (ii) localization of TOP3B to the stress granules [62]. The Tudor domain of TDRD3 recognizes the methylarginines of TOP3B which enhances TOP3B-TDRD3 interaction and facilitates the localization of TOP3B to stress granules [62].

(iv). *Cell cycle inhibitors and tumor suppressor-P16, P21 and PTEN*

In addition to suppressing the expression of the cell cycle inhibitors p16 and p21 through epigenetic modifications [16,17], PRMT6 also methylates p16 and p21 and negatively regulates their functions. PRMT6 methylates p16 at Arg22, Arg131 and Arg138 residues and promotes the cell cycle by inhibiting the interaction of p16 with CDK4 [15,18]. PRMT6 has been shown to methylate p21 at Arg156 residue both in vitro and in vivo. This methylation promotes the phosphorylation of p21 resulting in the accumulation of the protein in cytoplasm [19].

The phosphatase PTEN is a tumor suppressor gene which is often mutated in cancers [143,144]. PTEN negatively regulates AKT signaling by dephosphorylating the phosphatidylinositol-3,4,5-trisphosphate (PIP3) [145,146]. PRMT6 methylates PTEN at Arg135, which is frequently mutated in cancer. The methylation of PTEN by PRMT6 is required for the efficient suppression of AKT signaling by PTEN and modulates the global alternative splicing of pre-mRNAs [31].

(v). *Hormonal receptors-ERα and AR*

PRMT6 interacts and methylates the nuclear receptor ERα. PRMT6 promotes the estrogen-dependent and estrogen-independent activities of ERα in a methyltransferase activity dependent and independent manner, respectively. The interaction of PRMT6 with ERα inhibits the ERα-HSP90 interaction and promotes ligand independent functions of ERα [30]. Though the methyltransferase activity of PRMT6 is required for the enhancement of estrogen dependent activities of ERα [7,30], the precise functional outcome(s) of the PRMT6 mediated methylation of ERα is unknown.

The expansion of a polyglutamine tract in the androgen receptor (AR) causes the spinobulbar muscular atrophy (SBMA) disease [147–150]. PRMT6 promotes the hormone dependent transactivation function of normal AR as well as the polyglutamine expanded AR (mutant AR) in a methyltransferase activity dependent manner. The enhancement of transactivation function by PRMT6 is more pronounced for mutant AR compared to the normal AR [29]. PRMT6 forms a complex with AR and methylates AR at the arginine residues in the Akt consensus motif and inhibits the phosphorylation of AR by the Akt. PRMT6 contributes to the toxicity of polyglutamine expanded AR through its enhanced transactivation and interaction with the mutant AR [29].

(vi). *Scaffold protein-HTT*

The polyglutamine expansion in the huntingtin (HTT) protein causes the neurodegenerative disorder Huntington's disease (HD). The scaffold protein, HTT facilitates the transport of organelles in the axons and dendrites of the neurons [108,151]. The polyglutamine expansion in the mutant HTT affects its axonal transport [152–155]. PRMT6 interacts with HTT protein and deposits asymmetric dimethylation at Arg118. PRMT6 mediated methylation of HTT is required for the efficient axonal transport of vesicles and the viability of neurons [108]. Overexpression of PRMT6 in HD cells rescued the axonal trafficking and the neuronal viability [108].

Besides the aforementioned substrates, PRMT6 also methylates snRNPB, MIF, TUBB2A and HSJ-2 in vitro. However, the functional consequences of these methylations are unknown and remain to be investigated [33].

4.2.2. Viral Substrates

Several proteins of human immunodeficiency virus type I (HIV-1) are methylated by PRMT6, as part of the host response to suppress viral infection/propagation. For instance, PRMT6 interacts with the Tat protein of HIV-1, a transcriptional activator that stimulates the transcription of the viral genes and facilitates the viral replication [156], and methylates it at Arg52 and Arg53 positions [34–36]. This PRMT6 mediated methylation of Tat (i) increases its stability, (ii) excludes the Tat from the nucleolus and (iii) decreases its transactivation function resulting in the reduced production of viral particles [34,35,37,38]. Mechanistically, PRMT6 mediated methylation of Tat inhibits its interaction with Tat transactivation region (TAR) of HIV-1 RNA and inhibits formation of Tat-TAR-cyclin T1 ternary complex, resulting in the compromised transactivation function of Tat [35]. However, ectopic expression of PRMT6 in A549 cells and HeLa cells did not show inhibition of transactivation function of Tat [36] (Table 1).

Besides Tat protein, PRMT6 also methylates other HIV-1 viral proteins [39,40], including the Rev protein at its N-terminal arginine rich motif [39]. The Rev protein of HIV1 facilitates the nuclear export of intron containing viral RNAs [157–159]. PRMT6-mediated methylation of Rev decreases its binding with the Rev response element (RRE) of the viral RNA and inhibits the nuclear export function of the Rev [39]. The nucleocapsid protein (NC) of HIV-1 is methylated at Arg10 and Arg32 positions by PRMT6. The NC protein plays an important role in the packing of the HIV1 RNAs and in the annealing of tRNA Lys to the primer binding site of viral

RNA [160–166]. PRMT6-mediated methylation of NC protein hampers its ability to anneal the tRNA Lys to the primer site of viral RNA [40] (Table 1).

The pUL69 protein of the human cytomegalovirus facilitates the export of the unspliced mRNAs from nucleus to the cytoplasm by interacting with host mRNA export factor UAP56 or URH49 [167,168]. PRMT6 interacts with pUL69 at its N-terminal region which is important for the pUL69-UAP56 or URH49 interactions [41]. It is possible that the pUL69-PRMT6 interaction might affect the mRNA export function of pUL69. PRMT6 also methylates the N-terminal region of pUL69 [41] but the functional consequence of this methylation is hitherto not known (Table 1).

Taken together, these findings demonstrate that PRMT6 restricts HIV1 replication and propagation of viral particles by methylating the viral proteins and inhibiting their functions. Hence, any intervention strategy which maintains the optimal level of PRMT6 in cells such as preventing the proteasomal degradation of PRMT6 or promoting the expression of PRMT6 might serve well to restrict HIV1 infection and hence disease progression.

5. Role of PRMT6 in Cancers

PRMT6 levels are elevated in several types of cancers and the depletion of PRMT6 inhibits the proliferation of lung and bladder cancer cells [42]. In the following sections, we will discuss the role of PRMT6 in different cancer types (Figure 4).

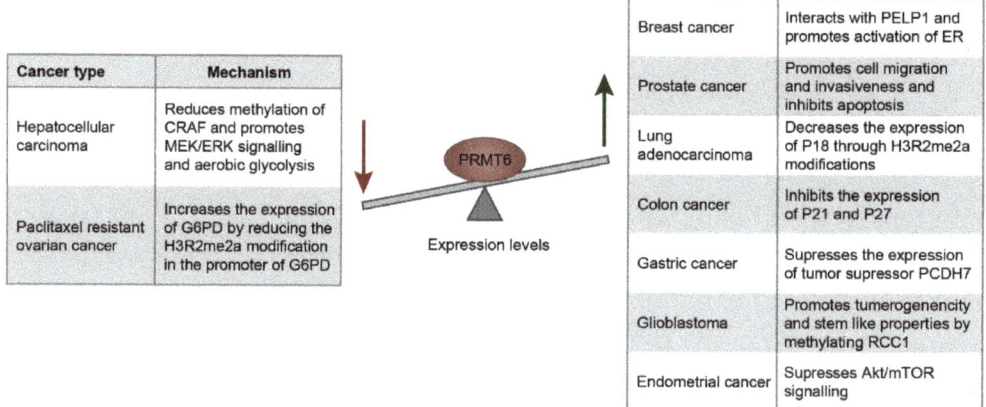

Figure 4. The role of PRMT6 in cancers. Schema representing the role of PRMT6 dysregulation in different cancers. The green arrow represents the upregulation of PRMT6 levels, while red arrow indicates the down regulation of PRMT6 levels in the corresponding cancer types, provided alongside in the table.

5.1. Breast, Prostate, Endometrial and Ovarian Cancers

PRMT6-mediated gene expression and alternative splicing changes are implicated in the pathophysiology of breast cancer [8]. The protooncogene PELP1 interacts with PRMT6 and promotes the activation of estrogen receptor, cell proliferation and clonogenic capacity of the breast cancer cells [169]. Overexpression of PRMT6 in the mammary glands of the mouse models promoted the tumorigenesis of mammary glands and deregulated Akt signaling in mammary epithelial cells [170].

PRMT6 levels are up-regulated in prostate cancer [171,172] and in endometrial cancer [173]. Depletion of PRMT6 in prostate cancer cells promotes apoptosis and decreases the cell migration and invasiveness properties [172]. Depletion of PRMT6 inhibits the endometrial cancer cell proliferation and migration by negatively regulating AKT/mTOR signaling [173].

Glucose-6-phosphate dehydrogenase (G6PD) levels are higher in paclitaxel resistant ovarian cancer cells compared to that of paclitaxel sensitive cancer cells. Depletion of G6PD in the paclitaxel resistant ovarian cancer cells increases its sensitivity to the paclitaxel treatment. PRMT6 is less abundant in paclitaxel resistant ovarian cancer cells compared to the sensitive cells. These low levels of PRMT6 lead to reduction of the repressive H3R2me2a modifications in the promoter region of G6PD resulting in the enhanced expression of G6PD which contributes to the paclitaxel resistance of the ovarian cancer cells [174].

5.2. Lung Cancer

PRMT6 levels are elevated in lung cancer [42,175]. Over-expression of PRMT6 in the lungs of mouse model enhanced the cell proliferation, promoting the lung tumor growth upon induction with a chemical carcinogen. PRMT6 interacts with interleukin enhancer binding protein 2 (ILF2) and activates the tumor associated macrophages [175]. PRMT6 levels are higher in the lung tissues of the patients with lung adenocarcinoma, which is associated with poor clinical outcomes [176]. Increased levels of PRMT6 reduces the expression of the cell cycle regulator p18 by increasing the amount of H3R2me2a and reducing the levels of H3K4me3 modifications. Depletion of PRMT6 activates the expression of p18 and inhibits the proliferation of lung adenocarcinoma cells [176].

5.3. Colon and Gastric Cancers

PRMT6 levels are elevated in colon cancer possibly due to the hypomethylation of PRMT6 promoter regions [177,178]. An abundance of PRMT6 is correlated with the shorter disease-free survival of colon cancer patients [178]. PPARα regulates the expression of DNMT1 and PRMT6 in the intestinal cells and protects against colon cancer. Loss or depletion of PPARα increases the levels of DNMT1 and PRMT6 which suppresses the expression of the cell cycle inhibitors p21 and p27 respectively and contributes to the colon carcinogenesis [179].

PRMT6 and H3R2me2a levels are elevated in gastric cancer (GC) and are correlated with poor prognosis. PRMT6 promotes the tumorigenicity and invasiveness of GC cells by silencing the expression of the tumor suppressor PCDH7 by depositing the repressive H3R2me2a marks at its promoter regions [180].

5.4. Glioblastoma

Regulator of chromosome condensin1 (RCC1) is a guanine nucleotide exchange factor for RAN-GTPase [181], which plays an important role in mitosis [182]. PRMT6 promotes the tumorigenicity and stem-like properties of the Glioblastoma stem cells (GSCs) by interacting with RCC1 and methylating it at Arg214. PRMT6-mediated methylation of RCC1 facilitates its association with chromatin and promotes mitosis through the generation of RAN-GTP [23,183].

5.5. Hepatocellular Carcinoma

RAF kinases activate the MEK/ERK signaling pathway and its dysregulation is implicated in cancers [184]. PRMT6 is downregulated in Hepatocellular carcinoma (HCC) and this downregulation is correlated with aggressive features of HCC. PRMT6 negatively regulates tumorigenic and stem-like properties of HCC. PRMT6 interacts and methylates CRAF at Arg100 which negatively regulates MEK/ERK signaling by inhibiting its interaction with RAS [32]. This negative regulation of MEK/ERK signaling also inhibits the aerobic glycolysis by preventing the ERK dependent nuclear localization of pyruvate kinase M2 (PKM2) [185]. In addition, the downregulation of PRMT6 promotes the autophagy and contributes to the tumorigenicity and cell survival in the tumor microenvironment. PRMT6 interacts with and methylates the co-chaperone Bcl-2, which is associated with athanogene 5 (BAG5) [186] at Arg15 and Arg24 and inhibits autophagy [109]. BAG5 forms a complex with the autophagic inducer, HSC70 [187,188] and the PRMT6-mediated methylation of BAG5 decreases the stability of HSC70 which in turn negatively regulates autophagy [109]. Thus, the downregulation of PRMT6 contributes to the tumorigenic and stem-like properties of HCC.

Taken together, the elevated levels of PRMT6 are linked to carcinogenesis of many different types of cancers mediated through methylation of histone or non-histone substrates and/or protein interactions. Not surprisingly, downregulation of PRMT6 across these cancer types leads to a reduction in cell proliferation and invasiveness through different mechanisms (Figure 4). Depletion or knock-out of PRMT6 reduced (i) the cell proliferation and clonogenic capacity of the breast cancer cells [169], (ii) the cell migration and invasiveness of prostate cancer cells [172], (iii) endometrial cancer cell proliferation and migration [173], (iv) proliferation of lung adenocarcinoma cells [176] and (v) tumorigenic properties of gastric cancer cells [180]. It was reported that the inhibition of PRMT6 activity by the PRMT6 specific inhibitor EPZ020411 reduces the tumorigenicity of glioblastoma and improves its response to radiotherapy [23]. All these findings establish that PRMT6 is an important potential therapeutic target for various cancers. This necessitates studies that investigate the efficacy of PRMT6 specific inhibitors for cancer therapy. Contrary to these observations, downregulation of PRMT6 was observed in melanoma [189] and HCC [32].

6. Future Perspectives

PRMT6 downregulates the expression of the target genes by generating H3R2me2a modifications at their promoters. In addition to the repressive functions, PRMT6 also activates the expression of certain target genes. While the repressive functions of PRMT6 are well studied, we are beginning to understand the transcriptional activation functions of PRMT6. A combination of biochemical, structural and molecular studies are required to obtain a thorough understanding of the spatio-temporal context of the transcriptional activation functions of PRMT6. In addition to H3R2me2a modifications, PRMT6 also generates H2AR29me2a and H3R42me2a in vivo. We envision that future studies will delineate (i) the biological functions of these modifications, (ii) their regulatory roles on various cellular processes and (iii) their connections to the diseases. Since PRMT6 exhibits a relaxed substrate specificity in high throughput studies with peptides, it is very likely that PRMT6 has many more cellular protein substrates. This necessitates extensive efforts to identify the hitherto unidentified substrate proteins of PRMT6 and characterize the functional consequences of their PRMT6 mediated methylation. Structural studies of PRMT6 in complex with the substrate peptides and thorough kinetic studies are needed to delineate the actual kinetic mechanism of PRMT6 catalysis. The fact that PRMT6 methylates proteins of the RNA virus HIV-1 and modulates their functions necessitates a systematic investigation on the regulatory role(s) of PRMT6 on other RNA viral infections, especially SARS-CoV2. PRMT6 is elevated in several types of cancers and contributes to the tumorigenesis through various mechanisms. Hence, investigations based on the existing as well as novel potent and cell active PRMT6 inhibitors [190–194] for their therapeutic activities against various cancers especially in combination with the standard anti-cancer drugs could pave a way for targeted cancer interventions.

Author Contributions: S.G. and A.D. pursued literature search and prepared the first draft of the review. All authors were involved in the preparation of the illustrations. All authors edited the manuscript and agreed to the published version of the manuscript.

Funding: We gratefully acknowledge funding support from Board of Research in Nuclear Sciences (Grant No 37(1)/14/17/2017-BRNS/37019 to AD), Indian Council of Medical Research (Senior Research Fellowship to SG), Prime Minister's Research Fellowship (to AKS), Science & Engineering Research Board (National Post-Doctoral Fellowship to RVK, SRG/2019/001785 to SC), and Department of Biotechnology (Ramalingaswami Re-entry Fellowship, BT/RLF/Re-entry/05/2018 to SC), Government of India.

Institutional Review Board Statement: Not applicable.

Informed Consent Statement: Not applicable.

Data Availability Statement: Not applicable.

Conflicts of Interest: The authors declare no conflict of interest.

References

1. Guccione, E.; Richard, S. The regulation, functions and clinical relevance of arginine methylation. *Nat. Rev. Mol. Cell Biol.* **2019**, *20*, 642–657. [CrossRef] [PubMed]
2. Fulton, M.D.; Brown, T.; Zheng, Y.G. The Biological Axis of Protein Arginine Methylation and Asymmetric Dimethylarginine. *Int. J. Mol. Sci.* **2019**, *20*, 3322. [CrossRef] [PubMed]
3. Bedford, M.T.; Clarke, S.G. Protein Arginine Methylation in Mammals: Who, What, and Why. *Mol. Cell* **2009**, *33*, 1–13. [CrossRef] [PubMed]
4. Guccione, E.; Bassi, C.; Casadio, F.; Martinato, F.; Cesaroni, M.; Schuchlautz, H.; Lüscher, B.; Amati, B. Methylation of histone H3R2 by PRMT6 and H3K4 by an MLL complex are mutually exclusive. *Nature* **2007**, *449*, 933–937. [CrossRef] [PubMed]
5. Iberg, A.N.; Espejo, A.; Cheng, D.; Kim, D.; Michaud-Levesque, J.; Richard, S.; Bedford, M.T. Arginine methylation of the histone H3 tail impedes effector binding. *J. Biol. Chem.* **2008**, *283*, 3006–3010. [CrossRef] [PubMed]
6. Hyllus, D.; Stein, C.; Schnabel, K.; Schiltz, E.; Imhof, A.; Dou, Y.; Hsieh, J.; Bauer, U.M. PRMT6-mediated methylation of R2 in histone H3 antagonizes H3 K4 trimethylation. *Genes Dev.* **2007**, *21*, 3369–3380. [CrossRef]
7. Harrison, M.J.; Tang, Y.H.; Dowhan, D.H. Protein arginine methyltransferase 6 regulates multiple aspects of gene expression. *Nucleic Acids Res.* **2010**, *38*, 2201–2216. [CrossRef]
8. Dowhan, D.H.; Harrison, M.J.; Eriksson, N.A.; Bailey, P.; Pearen, M.A.; Fuller, P.J.; Funder, J.W.; Simpson, E.R.; Leedman, P.J.; Tilley, W.D.; et al. Protein arginine methyltransferase 6-dependent gene expression and splicing: Association with breast cancer outcomes. *Endocr. Relat. Cancer* **2012**, *19*, 509–526. [CrossRef]
9. Zhao, X.X.; Zhang, Y.B.; Ni, P.L.; Wu, Z.L.; Yan, Y.C.; Li, Y.P. Protein arginine methyltransferase 6 (Prmt6) is essential for early zebrafish development through the direct suppression of gadd45αa stress sensor gene. *J. Biol. Chem.* **2016**, *291*, 402–412. [CrossRef]
10. Lee, Y.H.; Ma, H.; Tan, T.Z.; Ng, S.S.; Soong, R.; Mori, S.; Fu, X.Y.; Zernicka-Goetz, M.; Wu, Q. Protein arginine methyltransferase 6 regulates embryonic stem cell identity. *Stem Cells Dev.* **2012**, *21*, 2613–2622. [CrossRef] [PubMed]
11. Li, Z.; Wang, P.; Li, J.; Xie, Z.; Cen, S.; Li, M.; Liu, W.; Ye, G.; Zheng, G.; Ma, M.; et al. The N 6-methyladenosine demethylase ALKBH5 negatively regulates the osteogenic differentiation of mesenchymal stem cells through PRMT6. *Cell Death Dis.* **2021**, *12*, 578. [CrossRef]
12. Herglotz, J.; Kuvardina, O.N.; Kolodziej, S.; Kumar, A.; Hussong, H.; Grez, M.; Lausen, J. Histone arginine methylation keeps RUNX1 target genes in an intermediate state. *Oncogene* **2013**, *32*, 2565–2575. [CrossRef]
13. Lausen, J. Contributions of the histone arginine methyltransferase PRMT6 to the epigenetic function of RUNX1. *Crit. Rev. Eukaryot. Gene Expr.* **2013**, *23*, 265–274. [CrossRef] [PubMed]
14. El-Andaloussi, N.; Valovka, T.; Toueille, M.; Steinacher, R.; Focke, F.; Gehrig, P.; Covic, M.; Hassa, P.O.; Schär, P.; Hübscher, U.; et al. Arginine Methylation Regulates DNA Polymerase β. *Mol. Cell* **2006**, *22*, 51–62. [CrossRef] [PubMed]
15. Ma, W.L.; Wang, L.; Liu, L.X.; Wang, X.L. Effect of phosphorylation and methylation on the function of the p16INK4a protein in non-small cell lung cancer A549 cells. *Oncol. Lett.* **2015**, *10*, 2277–2282. [CrossRef] [PubMed]
16. Stein, C.; Riedl, S.; Rüthnick, D.; Nötzold, R.R.; Bauer, U.M. The arginine methyltransferase PRMT6 regulates cell proliferation and senescence through transcriptional repression of tumor suppressor genes. *Nucleic Acids Res.* **2012**, *40*, 9522–9533. [CrossRef]
17. Kleinschmidt, M.A.; de Graaf, P.; van Teeffelen, H.A.A.M.; Timmers, H.T.M. Cell cycle regulation by the PRMT6 arginine methyltransferase through repression of cyclin-dependent kinase inhibitors. *PLoS ONE* **2012**, *7*, e0041446. [CrossRef]
18. Wang, X.; Huang, Y.; Zhao, J.; Zhang, Y.; Lu, J.; Huang, B. Suppression of PRMT6-mediated arginine methylation of p16 protein potentiates its ability to arrest A549 cell proliferation. *Int. J. Biochem. Cell Biol.* **2012**, *44*, 2333–2341. [CrossRef]
19. Nakakido, M.; Deng, Z.; Suzuki, T.; Dohmae, N.; Nakamura, Y.; Hamamoto, R. PRMT6 increases cytoplasmic localization of p21CDKN1A in cancer cells through arginine methylation and makes more resistant to cytotoxic agents. *Oncotarget* **2015**, *6*, 30957–30967. [CrossRef]
20. Neault, M.; Mallette, F.A.; Vogel, G.; Michaud-Levesque, J.; Richard, S. Ablation of PRMT6 reveals a role as a negative transcriptional regulator of the p53 tumor suppressor. *Nucleic Acids Res.* **2012**, *40*, 9513–9521. [CrossRef]
21. Veland, N.; Hardikar, S.; Zhong, Y.; Gayatri, S.; Dan, J.; Strahl, B.D.; Rothbart, S.B.; Bedford, M.T.; Chen, T. The Arginine Methyltransferase PRMT6 Regulates DNA Methylation and Contributes to Global DNA Hypomethylation in Cancer. *Cell Rep.* **2017**, *21*, 3390–3397. [CrossRef] [PubMed]
22. Kim, S.; Kim, N.H.; Park, J.E.; Hwang, J.W.; Myung, N.; Hwang, K.T.; Kim, Y.A.Y.K.; Jang, C.Y.; Kim, Y.A.Y.K. PRMT6-mediated H3R2me2a guides Aurora B to chromosome arms for proper chromosome segregation. *Nat. Commun.* **2020**, *11*, 612. [CrossRef] [PubMed]
23. Huang, T.; Yang, Y.; Song, X.; Wan, X.; Wu, B.; Sastry, N.; Horbinski, C.M.; Zeng, C.; Tiek, D.; Goenka, A.; et al. PRMT6 methylation of RCC1 regulates mitosis, tumorigenicity, and radiation response of glioblastoma stem cells. *Mol. Cell* **2021**, *81*, 1276–1291.e9. [CrossRef]
24. Di Lorenzo, A.; Yang, Y.; Macaluso, M.; Bedford, M.T. A gain-of-function mouse model identifies PRMT6 as a NF-κB coactivator. *Nucleic Acids Res.* **2014**, *42*, 8297–8309. [CrossRef]

25. Tsai, K.D.; Lee, W.X.; Chen, W.; Chen, B.Y.; Chen, K.L.; Hsiao, T.C.; Wang, S.H.; Lee, Y.J.; Liang, S.Y.; Shieh, J.C.; et al. Upregulation of PRMT6 by LPS suppresses Klotho expression through interaction with NF-κB in glomerular mesangial cells. *J. Cell. Biochem.* **2018**, *119*, 3404–3416. [CrossRef] [PubMed]
26. Stavride, P.; Arampatzi, P.; Papamatheakis, J. Differential regulation of MHCII genes by PRMT6, via an AT-hook motif of RFX5. *Mol. Immunol.* **2013**, *56*, 390–398. [CrossRef]
27. Zhang, H.; Han, C.; Li, T.; Li, N.; Cao, X. The methyltransferase PRMT6 attenuates antiviral innate immunity by blocking TBK1–IRF3 signaling. *Cell. Mol. Immunol.* **2019**, *16*, 800–809. [CrossRef] [PubMed]
28. Luo, M.; Li, Y.; Guo, H.; Lin, S.; Chen, J.; Ma, Q.; Gu, Y.; Jiang, Z.; Gui, Y. Protein arginine methyltransferase 6 involved in germ cell viability during spermatogenesis and down-regulated by the androgen receptor. *Int. J. Mol. Sci.* **2015**, *16*, 29467–29481. [CrossRef]
29. Scaramuzzino, C.; Casci, I.; Parodi, S.; Lievens, P.M.J.; Polanco, M.J.; Milioto, C.; Chivet, M.; Monaghan, J.; Mishra, A.; Badders, N.; et al. Protein Arginine Methyltransferase 6 Enhances Polyglutamine-Expanded Androgen Receptor Function and Toxicity in Spinal and Bulbar Muscular Atrophy. *Neuron* **2015**, *85*, 88–100. [CrossRef]
30. Sun, Y.; Chung, H.H.; Woo, A.R.E.; Lin, V.C.L. Protein arginine methyltransferase 6 enhances ligand-dependent and -independent activity of estrogen receptor α via distinct mechanisms. *Biochim. Biophys. Acta-Mol. Cell Res.* **2014**, *1843*, 2067–2078. [CrossRef]
31. Feng, J.; Dang, Y.; Zhang, W.; Zhao, X.; Zhang, C.; Hou, Z.; Jin, Y.; McNutt, M.A.; Marks, A.R.; Yin, Y. PTEN arginine methylation by PRMT6 suppresses PI3K–AKT signaling and modulates pre-mRNA splicing. *Proc. Natl. Acad. Sci. USA* **2019**, *116*, 6868–6877. [CrossRef] [PubMed]
32. Chan, L.H.; Zhou, L.; Ng, K.Y.; Wong, T.L.; Lee, T.K.; Sharma, R.; Loong, J.H.; Ching, Y.P.; Yuan, Y.F.; Xie, D.; et al. PRMT6 Regulates RAS/RAF Binding and MEK/ERK-Mediated Cancer Stemness Activities in Hepatocellular Carcinoma through CRAF Methylation. *Cell Rep.* **2018**, *25*, 690–701.e8. [CrossRef] [PubMed]
33. Lo Sardo, A.; Altamura, S.; Pegoraro, S.; Maurizio, E.; Sgarra, R.; Manfioletti, G. Identification and Characterization of New Molecular Partners for the Protein Arginine Methyltransferase 6 (PRMT6). *PLoS ONE* **2013**, *8*, e0053750. [CrossRef]
34. Boulanger, M.-C.; Liang, C.; Russell, R.S.; Lin, R.; Bedford, M.T.; Wainberg, M.A.; Richard, S. Methylation of Tat by PRMT6 Regulates Human Immunodeficiency Virus Type 1 Gene Expression. *J. Virol.* **2005**, *79*, 124–131. [CrossRef] [PubMed]
35. Xie, B.; Invernizzi, C.F.; Richard, S.; Wainberg, M.A. Arginine Methylation of the Human Immunodeficiency Virus Type 1 Tat Protein by PRMT6 Negatively Affects Tat Interactions with both Cyclin T1 and the Tat Transactivation Region. *J. Virol.* **2007**, *81*, 4226–4234. [CrossRef] [PubMed]
36. Sivakumaran, H.; Lin, M.H.; Apolloni, A.; Cutillas, V.; Jin, H.; Li, D.; Wei, T.; Harrich, D. Overexpression of PRMT6 does not suppress HIV-1 Tat transactivation in cells naturally lacking PRMT6. *Virol. J.* **2013**, *10*, 207. [CrossRef] [PubMed]
37. Sivakumaran, H.; van der Horst, A.; Fulcher, A.J.; Apolloni, A.; Lin, M.-H.; Jans, D.A.; Harrich, D. Arginine Methylation Increases the Stability of Human Immunodeficiency Virus Type 1 Tat. *J. Virol.* **2009**, *83*, 11694–11703. [CrossRef]
38. Fulcher, A.J.; Sivakumaran, H.; Jin, H.; Rawle, D.J.; Harrich, D.; Jans, D.A. The protein arginine methyltransferase PRMT6 inhibits HIV-1 Tat nucleolar retention. *Biochim. Biophys. Acta-Mol. Cell Res.* **2016**, *1863*, 254–262. [CrossRef]
39. Invernizzi, C.F.; Xie, B.; Richard, S.; Wainberg, M.A. PRMT6 diminishes HIV-1 Rev binding to and export of viral RNA. *Retrovirology* **2006**, *3*, 93. [CrossRef]
40. Invernizzi, C.F.; Xie, B.; Frankel, F.A.; Feldhammer, M.; Roy, B.B.; Richard, S.; Wainberg, M.A. Arginine methylation of the HIV-1 nucleocapsid protein results in its diminished function. *AIDS* **2007**, *21*, 795–805. [CrossRef]
41. Thomas, M.; Sonntag, E.; Müller, R.; Schmidt, S.; Zielke, B.; Fossen, T.; Stamminger, T. pUL69 of Human Cytomegalovirus Recruits the Cellular Protein Arginine Methyltransferase 6 via a Domain That Is Crucial for mRNA Export and Efficient Viral Replication. *J. Virol.* **2015**, *89*, 9601–9615. [CrossRef]
42. Yoshimatsu, M.; Toyokawa, G.; Hayami, S.; Unoki, M.; Tsunoda, T.; Field, H.I.; Kelly, J.D.; Neal, D.E.; Maehara, Y.; Ponder, B.A.J.; et al. Dysregulation of PRMT1 and PRMT6, Type I arginine methyltransferases, is involved in various types of human cancers. *Int. J. Cancer* **2011**, *128*, 562–573. [CrossRef]
43. Raveendran, V.V.; Al-Haffar, K.; Kunhi, M.; Belhaj, K.; Al-Habeeb, W.; Al-Buraiki, J.; Eyjolsson, A.; Poizat, C. Protein arginine methyltransferase 6 mediates cardiac hypertrophy by differential regulation of histone H3 arginine methylation. *Heliyon* **2020**, *6*, e03864. [CrossRef]
44. Frankel, A.; Yadav, N.; Lee, J.; Branscombe, T.L.; Clarke, S.; Bedford, M.T. The novel human protein arginine N-methyltransferase PRMT6 is a nuclear enzyme displaying unique substrate specificity. *J. Biol. Chem.* **2002**, *277*, 3537–3543. [CrossRef] [PubMed]
45. Chen, W.; Gao, D.; Xie, L.; Wang, A.; Zhao, H.; Guo, C.; Sun, Y.; Nie, Y.; Hong, A.; Xiong, S. SCF-FBXO24 regulates cell proliferation by mediating ubiquitination and degradation of PRMT6. *Biochem. Biophys. Res. Commun.* **2020**, *530*, 75–81. [CrossRef] [PubMed]
46. Li, T.; He, X.; Luo, L.; Zeng, H.; Ren, S.; Chen, Y. F-Box Protein FBXW17-Mediated Proteasomal Degradation of Protein Methyltransferase PRMT6 Exaggerates CSE-Induced Lung Epithelial Inflammation and Apoptosis. *Front. Cell Dev. Biol.* **2021**, *9*, 599020. [CrossRef]
47. Casadio, F.; Lu, X.; Pollock, S.B.; LeRoy, G.; Garcia, B.A.; Muir, T.W.; Roeder, R.G.; Allis, C.D. H3R42me2a is a histone modification with positive transcriptional effects. *Proc. Natl. Acad. Sci. USA* **2013**, *110*, 14894–14899. [CrossRef] [PubMed]
48. Cheng, D.; Gao, G.; Di Lorenzo, A.; Jayne, S.; Hottiger, M.O.; Richard, S.; Bedford, M.T. Genetic evidence for partial redundancy between the arginine methyltransferases CARM1 and PRMT6. *J. Biol. Chem.* **2020**, *295*, 17060–17070. [CrossRef] [PubMed]

49. Waldmann, T.; Izzo, A.; Kamieniarz, K.; Richter, F.; Vogler, C.; Sarg, B.; Lindner, H.; Young, N.L.; Mittler, G.; Garcia, B.A.; et al. Methylation of H2AR29 is a novel repressive PRMT6 target. *Epigenetics Chromatin* **2011**, *4*, 11. [CrossRef]
50. Lischwe, M.A.; Ahn, Y.S.; Yeoman, L.C.; Busch, H.; Cook, R.G. Clustering of Glycine and NG,NG-Dimethylarginine in Nucleolar Protein C23. *Biochemistry* **1985**, *24*, 6025–6028. [CrossRef]
51. Tang, J.; Gary, J.D.; Clarke, S.; Herschman, H.R. PRMT 3, a type I protein arginine N-methyltransferase that differs from PRMT1 in its oligomerization, subcellular localization, substrate specificity, and regulation. *J. Biol. Chem.* **1998**, *273*, 16935–16945. [CrossRef] [PubMed]
52. Zhang, X.; Zhou, L.; Cheng, X. Crystal structure of the conserved core of protein arginine methyltransferase PRMT3. *EMBO J.* **2000**, *19*, 3509–3519. [CrossRef] [PubMed]
53. Lakowski, T.M.; Frankel, A. A kinetic study of human protein arginine N-methyltransferase 6 reveals a distributive mechanism. *J. Biol. Chem.* **2008**, *283*, 10015–10025. [CrossRef] [PubMed]
54. Frankel, A. Inconvenient truths for PRMT6 kinetic studies. *J. Biol. Chem.* **2012**, *287*, 9512. [CrossRef]
55. Obianyo, O.; Thompson, P.R. Kinetic mechanism of protein arginine methyltransferase 6 (PRMT6). *J. Biol. Chem.* **2012**, *287*, 6062–6071. [CrossRef] [PubMed]
56. Obianyo, O.; Causey, C.P.; Osborne, T.C.; Jones, J.E.; Lee, Y.H.; Stallcup, M.R.; Thompson, P.R. A chloroacetamidine-based inactivator of protein arginine methyltransferase 1: Design, synthesis, and in vitro and in vivo evaluation. *ChemBioChem* **2010**, *11*, 1219–1223. [CrossRef]
57. Singhroy, D.N.; Mesplède, T.; Sabbah, A.; Quashie, P.K.; Falgueyret, J.P.; Wainberg, M.A. Automethylation of protein arginine methyltransferase 6 (PRMT6) regulates its stability and its anti-HIV-1 activity. *Retrovirology* **2013**, *10*, 27880. [CrossRef]
58. Hamey, J.J.; Rakow, S.; Bouchard, C.; Senst, J.M.; Kolb, P.; Bauer, U.M.; Wilkins, M.R.; Hart-Smith, G. Systematic investigation of PRMT6 substrate recognition reveals broad specificity with a preference for an RG motif or basic and bulky residues. *FEBS J.* **2021**. [CrossRef]
59. Wooderchak, W.L.; Zang, T.; Zhou, Z.S.; Acuña, M.; Tahara, S.M.; Hevel, J.M. Substrate profiling of PRMT1 reveals amino acid sequences that extend beyond the "RGG" paradigm. *Biochemistry* **2008**, *47*, 9456–9466. [CrossRef]
60. Hamey, J.J.; Separovich, R.J.; Wilkins, M.R. MT-MAMS: Protein Methyltransferase Motif Analysis by Mass Spectrometry. *J. Proteome Res.* **2018**, *17*, 3485–3491. [CrossRef]
61. Shishkova, E.; Zeng, H.; Liu, F.; Kwiecien, N.W.; Hebert, A.S.; Coon, J.J.; Xu, W. Global mapping of CARM1 substrates defines enzyme specificity and substrate recognition. *Nat. Commun.* **2017**, *8*, 15571. [CrossRef]
62. Huang, L.; Wang, Z.; Narayanan, N.; Yang, Y. Arginine methylation of the C-terminus RGG motif promotes TOP3B topoisomerase activity and stress granule localization. *Nucleic Acids Res.* **2018**, *46*, 3061–3074. [CrossRef]
63. Wang, C.; Zhu, Y.; Chen, J.; Li, X.; Peng, J.; Chen, J.; Zou, Y.; Zhang, Z.; Jin, H.; Yang, P.; et al. Crystal structure of arginine methyltransferase 6 from Trypanosoma brucei. *PLoS ONE* **2014**, *9*, e0087267. [CrossRef]
64. Bonnefond, L.; Stojko, J.; Mailliot, J.; Troffer-Charlier, N.; Cura, V.; Wurtz, J.M.; Cianférani, S.; Cavarelli, J. Functional insights from high resolution structures of mouse protein arginine methyltransferase 6. *J. Struct. Biol.* **2015**, *191*, 175–183. [CrossRef]
65. Wu, H.; Zheng, W.; Eram, M.S.; Vhuiyan, M.; Dong, A.; Zeng, H.; He, H.; Brown, F.; Frankel, A.; Vedadi, M.; et al. Structural basis of arginine asymmetrical dimethylation by PRMT6. *Biochem. J.* **2016**, *473*, 3049–3063. [CrossRef] [PubMed]
66. Goulet, I.; Gauvin, G.; Boisvenue, S.; Côté, J. Alternative splicing yields protein arginine methyltransferase 1 isoforms with distinct activity, substrate specificity, and subcellular localization. *J. Biol. Chem.* **2007**, *282*, 33009–33021. [CrossRef] [PubMed]
67. Siarheyeva, A.; Senisterra, G.; Allali-Hassani, A.; Dong, A.; Dobrovetsky, E.; Wasney, G.A.; Chau, I.; Marcellus, R.; Hajian, T.; Liu, F.; et al. An allosteric inhibitor of protein arginine methyltransferase 3. *Structure* **2012**, *20*, 1425–1435. [CrossRef]
68. Yue, W.W.; Hassler, M.; Roe, S.M.; Thompson-Vale, V.; Pearl, L.H. Insights into histone code syntax from structural and biochemical studies of CARM1 methyltransferase. *EMBO J.* **2007**, *26*, 4402–4412. [CrossRef] [PubMed]
69. Weiss, V.H.; McBride, A.E.; Soriano, M.A.; Filman, D.J.; Silver, P.A.; Hogle, J.M. The structure and oligomerization of the yeast arginine methyltransferase, Hmt1. *Nat. Struct. Biol.* **2000**, *7*, 1165–1171. [CrossRef] [PubMed]
70. Zhang, X.; Cheng, X. Structure of the predominant protein arginine methyltransferase PRMT1 and analysis of its binding to substrate peptides. *Structure* **2003**, *11*, 509–520. [CrossRef]
71. Troffer-Charlier, N.; Cura, V.; Hassenboehler, P.; Moras, D.; Cavarelli, J. Functional insights from structures of coactivator-associated arginine methyltransferase 1 domains. *EMBO J.* **2007**, *26*, 4391–4401. [CrossRef]
72. Wang, C.; Zhu, Y.; Caceres, T.B.; Liu, L.; Peng, J.; Wang, J.; Chen, J.; Chen, X.; Zhang, Z.; Zuo, X.; et al. Structural determinants for the strict monomethylation activity by trypanosoma brucei protein arginine methyltransferase 7. *Structure* **2014**, *22*, 756–768. [CrossRef] [PubMed]
73. Antonysamy, S.; Bonday, Z.; Campbell, R.M.; Doyle, B.; Druzina, Z.; Gheyi, T.; Han, B.; Jungheim, L.N.; Qian, Y.; Rauch, C.; et al. Crystal structure of the human PRMT5:MEP50 complex. *Proc. Natl. Acad. Sci. USA* **2012**, *109*, 17960–17965. [CrossRef] [PubMed]
74. Tenney, K.; Shilatifard, A. A COMPASS in the voyage of defining the role of trithorax/MLL-containing complexes: Linking leukemogensis to covalent modifications of chromatin. *J. Cell. Biochem.* **2005**, *95*, 429–436. [CrossRef]
75. Wysocka, J.; Myers, M.P.; Laherty, C.D.; Eisenman, R.N.; Herr, W. Human Sin3 deacetylase and trithorax-related Set1/Ash2 histone H3-K4 methyltransferase are tethered together selectively by the cell-proliferation factor HCF-1. *Genes Dev.* **2003**, *17*, 896–911. [CrossRef]

76. Hughes, C.M.; Rozenblatt-Rosen, O.; Milne, T.A.; Copeland, T.D.; Levine, S.S.; Lee, J.C.; Hayes, D.N.; Shanmugam, K.S.; Bhattacharjee, A.; Biondi, C.A.; et al. Menin associates with a trithorax family histone methyltransferase complex and with the Hoxc8 locus. *Mol. Cell* **2004**, *13*, 587–597. [CrossRef]
77. Steward, M.M.; Lee, J.S.; O'Donovan, A.; Wyatt, M.; Bernstein, B.E.; Shilatifard, A. Molecular regulation of H3K4 trimethylation by ASH2L, a shared subunit of MLL complexes. *Nat. Struct. Mol. Biol.* **2006**, *13*, 852–854. [CrossRef]
78. Milne, T.A.; Briggs, S.D.; Brock, H.W.; Martin, M.E.; Gibbs, D.; Allis, C.D.; Hess, J.L. MLL targets SET domain methyltransferase activity to Hox gene promoters. *Mol. Cell* **2002**, *10*, 1107–1117. [CrossRef]
79. Nakamura, T.; Mori, T.; Tada, S.; Krajewski, W.; Rozovskaia, T.; Wassell, R.; Dubois, G.; Mazo, A.; Croce, C.M.; Canaani, E. ALL-1 is a histone methyltransferase that assembles a supercomplex of proteins involved in transcriptional regulation. *Mol. Cell* **2002**, *10*, 1119–1128. [CrossRef]
80. Guccione, E.; Martinato, F.; Finocchiaro, G.; Luzi, L.; Tizzoni, L.; Dall' Olio, V.; Zardo, G.; Nervi, C.; Bernard, L.; Amati, B. Myc-binding-site recognition in the human genome is determined by chromatin context. *Nat. Cell Biol.* **2006**, *8*, 764–770. [CrossRef]
81. Kuvardina, O.N.; Herglotz, J.; Kolodziej, S.; Kohrs, N.; Herkt, S.; Wojcik, B.; Oellerich, T.; Corso, J.; Behrens, K.; Kumar, A.; et al. RUNX1 represses the erythroid gene expression program during megakaryocytic differentiation. *Blood* **2015**, *125*, 3570–3579. [CrossRef]
82. Stein, C.; Nötzold, R.R.; Riedl, S.; Bouchard, C.; Bauer, U.M. The arginine methyltransferase PRMT6 cooperates with polycomb proteins in regulating HOXA gene expression. *PLoS ONE* **2016**, *11*, e0148892. [CrossRef]
83. Dawson, D.W.; Pearce, S.F.A.; Zhong, R.; Silverstein, R.L.; Frazier, W.A.; Bouck, N.P. CD36 mediates the in vitro inhibitory effects of thrombospondin-1 on endothelial cells. *J. Cell Biol.* **1997**, *138*, 707–717. [CrossRef] [PubMed]
84. Dawson, D.W.; Volpert, O.V.; Frieda, S.; Schneider, A.J.; Silverstein, R.L.; Henkin, J.; Bouck, N.P. Three distinct D-amino acid substitutions confer potent antiangiogenic activity on an inactive peptide derived from a thrombospondin-1 type 1 repeat. *Mol. Pharmacol.* **1999**, *55*, 332–338. [CrossRef]
85. Michaud-Levesque, J.; Richard, S. Thrombospondin-1 is a transcriptional repression target of PRMT6. *J. Biol. Chem.* **2009**, *284*, 21338–21346. [CrossRef] [PubMed]
86. Kim, N.H.; Kim, S.N.; Seo, D.W.; Han, J.W.; Kim, Y.K. PRMT6 overexpression upregulates TSP-1 and downregulates MMPs: Its implication in motility and invasion. *Biochem. Biophys. Res. Commun.* **2013**, *432*, 60–65. [CrossRef]
87. Phalke, S.; Mzoughi, S.; Bezzi, M.; Jennifer, N.; Mok, W.C.; Low, D.H.P.; Thike, A.A.; Kuznetsov, V.A.; Tan, P.H.; Voorhoeve, P.M.; et al. P53-Independent regulation of p21Waf1/Cip1 expression and senescence by PRMT6. *Nucleic Acids Res.* **2012**, *40*, 9534–9542. [CrossRef] [PubMed]
88. Hwang, J.W.; So, Y.S.; Bae, G.U.; Kim, S.N.; Kim, Y.K. Protein arginine methyltransferase 6 suppresses adipogenic differentiation by repressing peroxisome proliferator-activated receptor γ activity. *Int. J. Mol. Med.* **2019**, *43*, 2462–2470. [CrossRef]
89. Dhawan, S.; Georgia, S.; Tschen, S.; Fan, G.; Bhushan, A. Pancreatic β Cell Identity Is Maintained by DNA Methylation-Mediated Repression of Arx. *Dev. Cell* **2011**, *20*, 419–429. [CrossRef] [PubMed]
90. Damez-Werno, D.M.; Sun, H.S.; Scobie, K.N.; Shao, N.; Rabkin, J.; Dias, C.; Calipari, E.S.; Maze, I.; Pena, C.J.; Walker, D.M.; et al. Histone arginine methylation in cocaine action in the nucleus accumbens. *Proc. Natl. Acad. Sci. USA* **2016**, *113*, 9623–9628. [CrossRef]
91. Herkt, S.C.; Kuvardina, O.N.; Herglotz, J.; Schneider, L.; Meyer, A.; Pommerenke, C.; Salinas-Riester, G.; Seifried, E.; Bonig, H.; Lausen, J. Protein arginine methyltransferase 6 controls erythroid gene expression and differentiation of human CD34+ progenitor cells. *Haematologica* **2018**, *103*, 18–29. [CrossRef]
92. Bostick, M.; Jong, K.K.; Estève, P.O.; Clark, A.; Pradhan, S.; Jacobsen, S.E. UHRF1 plays a role in maintaining DNA methylation in mammalian cells. *Science* **2007**, *317*, 1760–1764. [CrossRef]
93. Sharif, J.; Muto, M.; Takebayashi, S.I.; Suetake, I.; Iwamatsu, A.; Endo, T.A.; Shinga, J.; Mizutani-Koseki, Y.; Toyoda, T.; Okamura, K.; et al. The SRA protein Np95 mediates epigenetic inheritance by recruiting Dnmt1 to methylated DNA. *Nature* **2007**, *450*, 908–912. [CrossRef] [PubMed]
94. Lallous, N.; Legrand, P.; McEwen, A.G.; Ramón-Maiques, S.; Samama, J.P.; Birck, C. The PHD finger of human UHRF1 reveals a new subgroup of unmethylated histone H3 tail readers. *PLoS ONE* **2011**, *6*, e0027579. [CrossRef] [PubMed]
95. Wang, C.; Shen, J.; Yang, Z.; Chen, P.; Zhao, B.; Hu, W.; Lan, W.; Tong, X.; Wu, H.; Li, G.; et al. Structural basis for site-specific reading of unmodified R2 of histone H3 tail by UHRF1 PHD finger. *Cell Res.* **2011**, *21*, 1379–1382. [CrossRef]
96. Rajakumara, E.; Wang, Z.; Ma, H.; Hu, L.; Chen, H.; Lin, Y.; Guo, R.; Wu, F.; Li, H.; Lan, F.; et al. PHD Finger Recognition of Unmodified Histone H3R2 Links UHRF1 to Regulation of Euchromatic Gene Expression. *Mol. Cell* **2011**, *43*, 275–284. [CrossRef] [PubMed]
97. Hu, L.; Li, Z.; Wang, P.; Lin, Y.; Xu, Y. Crystal structure of PHD domain of UHRF1 and insights into recognition of unmodified histone H3 arginine residue 2. *Cell Res.* **2011**, *21*, 1374–1378. [CrossRef]
98. Schneider, L.; Herkt, S.; Wang, L.; Feld, C.; Wesely, J.; Kuvardina, O.N.; Meyer, A.; Oellerich, T.; Häupl, B.; Seifried, E.; et al. PRMT6 activates cyclin D1 expression in conjunction with the transcription factor LEF1. *Oncogenesis* **2021**, *10*, 42. [CrossRef]
99. Bouchard, C.; Sahu, P.; Meixner, M.; Nötzold, R.R.; Rust, M.B.; Kremmer, E.; Feederle, R.; Hart-Smith, G.; Finkernagel, F.; Bartkuhn, M.; et al. Genomic Location of PRMT6-Dependent H3R2 Methylation Is Linked to the Transcriptional Outcome of Associated Genes. *Cell Rep.* **2018**, *24*, 3339–3352. [CrossRef] [PubMed]

100. Sakabe, K.; Hart, G.W. O-GlcNAc transferase regulates mitotic chromatin dynamics. *J. Biol. Chem.* **2010**, *285*, 34460–34468. [CrossRef]
101. Miranda, T.B.; Webb, K.J.; Edberg, D.D.; Reeves, R.; Clarke, S. Protein arginine methyltransferase 6 specifically methylates the nonhistone chromatin protein HMGA1a. *Biochem. Biophys. Res. Commun.* **2005**, *336*, 831–835. [CrossRef]
102. Sgarra, R.; Lee, J.; Tessari, M.A.; Altamura, S.; Spolaore, B.; Giancotti, V.; Bedford, M.T.; Manfioletti, G. The AT-hook of the chromatin architectural transcription factor high mobility group A1a is arginine-methylated by protein arginine methyltransferase 6. *J. Biol. Chem.* **2006**, *281*, 3764–3772. [CrossRef]
103. Zou, Y.; Webb, K.; Perna, A.D.; Zhang, Q.; Clarke, S.; Wang, Y. A mass spectrometric study on the in vitro methylation of HMGA1a and HMGA1b proteins by PRMTs: Methylation specificity, the effect of binding to AT-rich duplex DNA, and the effect of C-terminal phosphorylation. *Biochemistry* **2007**, *46*, 7896–7906. [CrossRef] [PubMed]
104. Han, H.S.; Jung, C.Y.; Yoon, Y.S.; Choi, S.; Choi, D.; Kang, G.; Park, K.G.; Kim, S.T.; Koo, S.H. Arginine methylation of CRTC2 is critical in the transcriptional control of hepatic glucose metabolism. *Sci. Signal.* **2014**, *7*, ra19. [CrossRef] [PubMed]
105. Huang, J.; Cardamone, M.D.; Johnson, H.E.; Neault, M.; Chan, M.; Floyd, Z.E.; Mallette, F.A.; Perissi, V. Exchange factor TBL1 and arginine methyltransferase PRMT6 cooperate in protecting g protein pathway suppressor 2 (GPS2) from proteasomal degradation. *J. Biol. Chem.* **2015**, *290*, 19044–19054. [CrossRef] [PubMed]
106. Yan, W.; Liang, Y.; Zhang, Q.; Wang, D.; Lei, M.; Qu, J.; He, X.; Lei, Q.; Wang, Y. Arginine methylation of SIRT 7 couples glucose sensing with mitochondria biogenesis. *EMBO Rep.* **2018**, *19*, e46377. [CrossRef] [PubMed]
107. Choi, S.; Jeong, H.J.; Kim, H.; Choi, D.; Cho, S.C.; Seong, J.K.; Koo, S.H.; Kang, J.S. Skeletal muscle-specific Prmt1 deletion causes muscle atrophy via deregulation of the PRMT6-FOXO3 axis. *Autophagy* **2019**, *15*, 1069–1081. [CrossRef]
108. Migazzi, A.; Scaramuzzino, C.; Anderson, E.N.; Tripathy, D.; Hernández, I.H.; Grant, R.A.; Roccuzzo, M.; Tosatto, L.; Virlogeux, A.; Zuccato, C.; et al. Huntingtin-mediated axonal transport requires arginine methylation by PRMT6. *Cell Rep.* **2021**, *35*, 108980. [CrossRef]
109. Che, N.; Ng, K.Y.; Wong, T.L.; Tong, M.; Kau, P.W.; Chan, L.H.; Lee, T.K.; Huen, M.S.; Yun, J.P.; Ma, S. PRMT6 deficiency induces autophagy in hostile microenvironments of hepatocellular carcinoma tumors by regulating BAG5-associated HSC70 stability. *Cancer Lett.* **2021**, *501*, 247–262. [CrossRef] [PubMed]
110. Hübscher, U.; Maga, G.; Spadari, S. Eukaryotic DNA polymerases. *Annu. Rev. Biochem.* **2002**, *71*, 133–163. [CrossRef]
111. Idriss, H.T.; Al-Assar, O.; Wilson, S.H. DNA polymerase β. *Int. J. Biochem. Cell Biol.* **2002**, *34*, 321–324. [CrossRef]
112. Sobol, R.W.; Horton, J.K.; Kühn, R.; Gu, H.; Singhal, R.K.; Prasad, R.; Rajewsky, K.; Wilson, S.H. Requirement of mammalian DNA polymerase-β in base-excision repair. *Nature* **1996**, *379*, 183–186. [CrossRef]
113. Bustin, M. Regulation of DNA-Dependent Activities by the Functional Motifs of the High-Mobility-Group Chromosomal Proteins. *Mol. Cell. Biol.* **1999**, *19*, 5237–5246. [CrossRef]
114. Thanos, D.; Maniatis, T. Virus induction of human IFNβ gene expression requires the assembly of an enhanceosome. *Cell* **1995**, *83*, 1091–1100. [CrossRef]
115. Reeves, R. Molecular biology of HMGA proteins: Hubs of nuclear function. *Gene* **2001**, *277*, 63–81. [CrossRef]
116. Yoshizawa, T.; Karim, M.F.; Sato, Y.; Senokuchi, T.; Miyata, K.; Fukuda, T.; Go, C.; Tasaki, M.; Uchimura, K.; Kadomatsu, T.; et al. SIRT7 controls hepatic lipid metabolism by regulating the ubiquitin-proteasome pathway. *Cell Metab.* **2014**, *19*, 712–721. [CrossRef] [PubMed]
117. Shin, J.; He, M.; Liu, Y.; Paredes, S.; Villanova, L.; Brown, K.; Qiu, X.; Nabavi, N.; Mohrin, M.; Wojnoonski, K.; et al. SIRT7 represses myc activity to suppress er stress and prevent fatty liver disease. *Cell Rep.* **2013**, *5*, 654–665. [CrossRef] [PubMed]
118. Mohrin, M.; Shin, J.; Liu, Y.; Brown, K.; Luo, H.; Xi, Y.; Haynes, C.M.; Chen, D. A mitochondrial UPR-mediated metabolic checkpoint regulates hematopoietic stem cell aging. *Science* **2015**, *347*, 1374–1377. [CrossRef] [PubMed]
119. Vazquez, B.N.; Thackray, J.K.; Simonet, N.G.; Kane-Goldsmith, N.; Martinez-Redondo, P.; Nguyen, T.; Bunting, S.; Vaquero, A.; Tischfield, J.A.; Serrano, L. SIRT 7 promotes genome integrity and modulates non-homologous end joining DNA repair. *EMBO J.* **2016**, *35*, 1488–1503. [CrossRef]
120. Barber, M.F.; Michishita-Kioi, E.; Xi, Y.; Tasselli, L.; Kioi, M.; Moqtaderi, Z.; Tennen, R.I.; Paredes, S.; Young, N.L.; Chen, K.; et al. SIRT7 links H3K18 deacetylation to maintenance of oncogenic transformation. *Nature* **2012**, *487*, 114–118. [CrossRef]
121. Chen, S.; Seiler, J.; Santiago-Reichelt, M.; Felbel, K.; Grummt, I.; Voit, R. Repression of RNA Polymerase I upon Stress Is Caused by Inhibition of RNA-Dependent Deacetylation of PAF53 by SIRT7. *Mol. Cell* **2013**, *52*, 303–313. [CrossRef]
122. Quinn, P.G.; Granner, D.K. Cyclic AMP-dependent protein kinase regulates transcription of the phosphoenolpyruvate carboxykinase gene but not binding of nuclear factors to the cyclic AMP regulatory element. *Mol. Cell. Biol.* **1990**, *10*, 3357–3364. [CrossRef]
123. Montminy, M.; Koo, S.H.; Zhang, X. The CREB Family: Key regulators of hepatic metabolism. *Ann. Endocrinol. (Paris)* **2004**, *65*, 73–75. [CrossRef]
124. Koo, S.H.; Flechner, L.; Qi, L.; Zhang, X.; Screaton, R.A.; Jeffries, S.; Hedrick, S.; Xu, W.; Boussouar, F.; Brindle, P.; et al. The CREB coactivator TORC2 is a key regulator of fasting glucose metabolism. *Nature* **2005**, *437*, 1109–1114. [CrossRef]
125. Herzig, S.; Long, F.; Jhala, U.S.; Hedrick, S.; Quinn, R.; Bauer, A.; Rudolph, D.; Schutz, G.; Yoon, C.; Puigserver, P.; et al. CREB regulates hepatic gluconeogenesis through the coactivator PGC-1. *Nature* **2001**, *413*, 179–183. [CrossRef] [PubMed]
126. Han, H.S.; Choi, D.; Choi, S.; Koo, S.H. Roles of protein arginine methyltransferases in the control of glucose metabolism. *Endocrinol. Metab.* **2014**, *29*, 435–440. [CrossRef] [PubMed]

127. Morris, B.J.; Willcox, D.C.; Donlon, T.A.; Willcox, B.J. FOXO3: A Major Gene for Human Longevity-A Mini-Review. *Gerontology* **2015**, *61*, 515–525. [CrossRef] [PubMed]
128. Moon, K.M.; Lee, M.-K.; Hwang, T.; Choi, C.W.; Kim, M.S.; Kim, H.-R.; Lee, B. The multi-functional roles of forkhead box protein O in skin aging and diseases. *Redox Biol.* **2021**, *46*, 102101. [CrossRef]
129. Sandri, M.; Sandri, C.; Gilbert, A.; Skurk, C.; Calabria, E.; Picard, A.; Walsh, K.; Schiaffino, S.; Lecker, S.H.; Goldberg, A.L. Foxo transcription factors induce the atrophy-related ubiquitin ligase atrogin-1 and cause skeletal muscle atrophy. *Cell* **2004**, *117*, 399–412. [CrossRef]
130. Lecker, S.H.; Jagoe, R.T.; Gilbert, A.; Gomes, M.; Baracos, V.; Bailey, J.; Price, S.R.; Mitch, W.E.; Goldberg, A.L. Multiple types of skeletal muscle atrophy involve a common program of changes in gene expression. *FASEB J.* **2004**, *18*, 39–51. [CrossRef]
131. Senf, S.M.; Dodd, S.L.; Judge, A.R. FOXO signaling is required for disuse muscle atrophy and is directly regulated by Hsp70. *Am. J. Physiol.-Cell Physiol.* **2010**, *298*, C45. [CrossRef]
132. Peng, Y.-C.; Kuo, F.; Breiding, D.E.; Wang, Y.-F.; Mansur, C.P.; Androphy, E.J. AMF1 (GPS2) Modulates p53 Transactivation. *Mol. Cell. Biol.* **2001**, *21*, 5913–5924. [CrossRef]
133. Peng, Y.-C.; Breiding, D.E.; Sverdrup, F.; Richard, J.; Androphy, E.J. AMF-1/Gps2 Binds p300 and Enhances Its Interaction with Papillomavirus E2 Proteins. *J. Virol.* **2000**, *74*, 5872–5879. [CrossRef] [PubMed]
134. Sanyal, S.; Båvner, A.; Haroniti, A.; Nilsson, L.M.; Lundåsen, T.; Rehnmark, S.; Witt, M.R.; Einarsson, C.; Talianidis, I.; Gustafsson, J.Å.; et al. Involvement of corepressor complex subunit GPS2 in transcriptional pathways governing human bile acid biosynthesis. *Proc. Natl. Acad. Sci. USA* **2007**, *104*, 15665–15670. [CrossRef] [PubMed]
135. Zhang, J.; Kalkum, M.; Chait, B.T.; Roeder, R.G. The N-CoR-HDAC3 nuclear receptor corepressor complex inhibits the JNK pathway through the integral subunit GPS2. *Mol. Cell* **2002**, *9*, 611–623. [CrossRef]
136. Jakobsson, T.; Venteclef, N.; Toresson, G.; Damdimopoulos, A.E.; Ehrlund, A.; Lou, X.; Sanyal, S.; Steffensen, K.R.; Gustafsson, J.Å.; Treuter, E. GPS2 Is Required for Cholesterol Efflux by Triggering Histone Demethylation, LXR Recruitment, and Coregulator Assembly at the ABCG1 Locus. *Mol. Cell* **2009**, *34*, 510–518. [CrossRef] [PubMed]
137. Cardamone, D.M.; Krones, A.; Tanasa, B.; Taylor, H.; Ricci, L.; Ohgi, K.A.; Glass, C.K.; Rosenfeld, M.G.; Perissi, V. A Protective Strategy against Hyperinflammatory Responses Requiring the Nontranscriptional Actions of GPS2. *Mol. Cell* **2012**, *46*, 91–104. [CrossRef]
138. Cardamone, M.D.; Tanasa, B.; Chan, M.; Cederquist, C.T.; Andricovich, J.; Rosenfeld, M.G.; Perissi, V. GPS2/KDM4A pioneering activity regulates promoter-specific recruitment of PPARγ. *Cell Rep.* **2014**, *8*, 163–176. [CrossRef]
139. Pommier, Y.; Sun, Y.; Huang, S.Y.N.; Nitiss, J.L. Roles of eukaryotic topoisomerases in transcription, replication and genomic stability. *Nat. Rev. Mol. Cell Biol.* **2016**, *17*, 703–721. [CrossRef]
140. Baker, N.M.; Rajan, R.; Mondragńn, A. Structural studies of type I topoisomerases. *Nucleic Acids Res.* **2009**, *37*, 693–701. [CrossRef]
141. Yang, Y.; McBride, K.M.; Hensley, S.; Lu, Y.; Chedin, F.; Bedford, M.T. Arginine Methylation Facilitates the Recruitment of TOP3B to Chromatin to Prevent R Loop Accumulation. *Mol. Cell* **2014**, *53*, 484–497. [CrossRef] [PubMed]
142. Yang, Y.; Lu, Y.; Espejo, A.; Wu, J.; Xu, W.; Liang, S.; Bedford, M.T. TDRD3 Is an Effector Molecule for Arginine-Methylated Histone Marks. *Mol. Cell* **2010**, *40*, 1016–1023. [CrossRef]
143. Li, J.; Yen, C.; Liaw, D.; Podsypanina, K.; Bose, S.; Wang, S.I.; Puc, J.; Miliaresis, C.; Rodgers, L.; McCombie, R.; et al. PTEN, a putative protein tyrosine phosphatase gene mutated in human brain, breast, and prostate cancer. *Science* **1997**, *275*, 1943–1947. [CrossRef] [PubMed]
144. Steck, P.A.; Pershouse, M.A.; Jasser, S.A.; Yung, W.K.A.; Lin, H.; Ligon, A.H.; Langford, L.A.; Baumgard, M.L.; Hattier, T.; Davis, T.; et al. Identification of a candidate tumour suppressor gene, MMAC1, at chromosome 10q23.3 that is mutated in multiple advanced cancers. *Nat. Genet.* **1997**, *15*, 356–362. [CrossRef] [PubMed]
145. Sulis, M.L.; Parsons, R. PTEN: From pathology to biology. *Trends Cell Biol.* **2003**, *13*, 478–483. [CrossRef]
146. Hopkins, B.D.; Hodakoski, C.; Barrows, D.; Mense, S.M.; Parsons, R.E. PTEN function: The long and the short of it. *Trends Biochem. Sci.* **2014**, *39*, 183–190. [CrossRef]
147. Orr, H.T.; Zoghbi, H.Y. Trinucleotide repeat disorders. *Annu. Rev. Neurosci.* **2007**, *30*, 575–621. [CrossRef]
148. Katsuno, M.; Adachi, H.; Kume, A.; Li, M.; Nakagomi, Y.; Niwa, H.; Sang, C.; Kobayashi, Y.; Doyu, M.; Sobue, G. Testosterone reduction prevents phenotypic expression in a transgenic mouse model of spinal and bulbar muscular atrophy. *Neuron* **2002**, *35*, 843–854. [CrossRef]
149. Kennedy, W.R.; Alter, M.; Sung, J.H. Progressive proximal spinal and bulbar muscular atrophy of late onset: A sex-linked recessive trait. *Neurology* **1968**, *18*, 671–680. [CrossRef]
150. Spada, A.R.L.; Wilson, E.M.; Lubahn, D.B.; Harding, A.E.; Fischbeck, K.H. Androgen receptor gene mutations in X-linked spinal and bulbar muscular atrophy. *Nature* **1991**, *352*, 77–79. [CrossRef]
151. Saudou, F.; Humbert, S. The Biology of Huntingtin. *Neuron* **2016**, *89*, 910–926. [CrossRef] [PubMed]
152. Gauthier, L.R.; Charrin, B.C.; Borrell-Pagès, M.; Dompierre, J.P.; Rangone, H.; Cordelières, F.P.; De Mey, J.; MacDonald, M.E.; Leßmann, V.; Humbert, S.; et al. Huntingtin controls neurotrophic support and survival of neurons by enhancing BDNF vesicular transport along microtubules. *Cell* **2004**, *118*, 127–138. [CrossRef]
153. Gunawardena, S.; Goldstein, L.S.B. Disruption of axonal transport and neuronal viability by amyloid precursor protein mutations in Drosophila. *Neuron* **2001**, *32*, 389–401. [CrossRef]

154. Gunawardena, S.; Her, L.S.; Brusch, R.G.; Laymon, R.A.; Niesman, I.R.; Gordesky-Gold, B.; Sintasath, L.; Bonini, N.M.; Goldstein, L.S.B. Disruption of axonal transport by loss of huntingtin or expression of pathogenic polyQ proteins in Drosophila. *Neuron* **2003**, *40*, 25–40. [CrossRef]
155. Trushina, E.; Dyer, R.B.; Badger, J.D.; Ure, D.; Eide, L.; Tran, D.D.; Vrieze, B.T.; Legendre-Guillemin, V.; McPherson, P.S.; Mandavilli, B.S.; et al. Mutant Huntingtin Impairs Axonal Trafficking in Mammalian Neurons In Vivo and In Vitro. *Mol. Cell. Biol.* **2004**, *24*, 8195–8209. [CrossRef]
156. Liang, C.; Wainberg, M.A. The role of Tat in HIV-1 replication: An activator and/or a suppressor? *AIDS Rev.* **2002**, *4*, 41–49.
157. Kalland, K.H.; Szilvay, A.M.; Langhoff, E.; Haukenes, G. Subcellular distribution of human immunodeficiency virus type 1 Rev and colocalization of Rev with RNA splicing factors in a speckled pattern in the nucleoplasm. *J. Virol.* **1994**, *68*, 1475–1485. [CrossRef]
158. Meggio, F.; D'Agostino, D.M.; Ciminale, V.; Chieco-Bianchi, L.; Pinna, L.A. Phosphorylation of HIV-1 Rev protein: Implication of protein kinase CK2 and pro-directed kinases. *Biochem. Biophys. Res. Commun.* **1996**, *226*, 547–554. [CrossRef]
159. Fernandes, J.D.; Jayaraman, B.; Frankel, A.D. The HIV-1 Rev response element: An RNA scaffold that directs the cooperative assembly of a homo-oligomeric ribonucleoprotein complex. *RNA Biol.* **2012**, *9*, 6–11. [CrossRef] [PubMed]
160. Shubsda, M.F.; Paoletti, A.C.; Hudson, B.S.; Borer, P.N. Affinities of packaging domain loops in HIV-1 RNA for the nucleocapsid protein. *Biochemistry* **2002**, *41*, 5276–5282. [CrossRef] [PubMed]
161. Roldan, A.; Warren, O.U.; Russell, R.S.; Liang, C.; Wainberg, M.A. A HIV-1 minimal gag protein is superior to nucleocapsid at in vitro tRNA3Lys annealing and exhibits multimerization-induced inhibition of reverse transcription. *J. Biol. Chem.* **2005**, *280*, 17488–17496. [CrossRef]
162. De Rocquigny, H.; Gabus, C.; Vincent, A.; Fournie-Zaluski, M.C.; Roques, B.; Darlix, J.L. Viral RNA annealing activities of human immunodeficiency virus type 1 nucleocapsid protein require only peptide domains outside the zinc fingers. *Proc. Natl. Acad. Sci. USA* **1992**, *89*, 6472–6476. [CrossRef]
163. Cen, S.; Khorchid, A.; Gabor, J.; Rong, L.; Wainberg, M.A.; Kleiman, L. Roles of Pr55 gag and NCp7 in tRNA 3 Lys Genomic Placement and the Initiation Step of Reverse Transcription in Human Immunodeficiency Virus Type 1. *J. Virol.* **2000**, *74*, 10796–10800. [CrossRef] [PubMed]
164. Sakaguchi, K.; Zambrano, N.; Baldwin, E.T.; Shapiro, B.A.; Erickson, J.W.; Omichinski, J.G.; Clore, G.M.; Gronenborn, A.M.; Appella, E. Identification of a binding site for the human immunodeficiency virus type 1 nucleocapsid protein. *Proc. Natl. Acad. Sci. USA* **1993**, *90*, 5219–5223. [CrossRef]
165. Amarasinghe, G.K.; De Guzman, R.N.; Turner, R.B.; Chancellor, K.J.; Wu, Z.R.; Summers, M.F. NMR structure of the HIV-1 nucleocapsid protein bound to stem-loop SL2 of the Ψ-RNA packaging signal. Implications for genome recognition. *J. Mol. Biol.* **2000**, *301*, 491–511. [CrossRef] [PubMed]
166. Maki, A.H.; Ozarowski, A.; Misra, A.; Urbaneja, M.A.; Casas-Finet, J.R. Phosphorescence and optically detected magnetic resonance of HIV-1 nucleocapsid protein complexes with stem-loop sequences of the genomic ψ-recognition element. *Biochemistry* **2001**, *40*, 1403–1412. [CrossRef] [PubMed]
167. Lischka, P.; Toth, Z.; Thomas, M.; Mueller, R.; Stamminger, T. The UL69 Transactivator Protein of Human Cytomegalovirus Interacts with DEXD/H-Box RNA Helicase UAP56 To Promote Cytoplasmic Accumulation of Unspliced RNA. *Mol. Cell. Biol.* **2006**, *26*, 1631–1643. [CrossRef]
168. Toth, Z.; Lischka, P.; Stamminger, T. RNA-binding of the human cytomegalovirus transactivator protein UL69, mediated by arginine-rich motifs, is not required for nuclear export of unspliced RNA. *Nucleic Acids Res.* **2006**, *34*, 1237–1249. [CrossRef]
169. Mann, M.; Zou, Y.; Chen, Y.; Brann, D.; Vadlamudi, R. PELP1 oncogenic functions involve alternative splicing via PRMT6. *Mol. Oncol.* **2014**, *8*, 389–400. [CrossRef]
170. Bao, J.; Di Lorenzo, A.; Lin, K.; Lu, Y.; Zhong, Y.; Sebastian, M.M.; Muller, W.J.; Yang, Y.; Bedford, M.T. Mouse models of overexpression reveal distinct oncogenic roles for different type I protein arginine methyltransferases. *Cancer Res.* **2019**, *79*, 21–32. [CrossRef]
171. Vieira, F.Q.; Costa-Pinheiro, P.; Ramalho-Carvalho, J.; Pereira, A.; Menezes, F.D.; Antunes, L.; Carneiro, I.; Oliveira, J.; Henrique, R.; Jerónimo, C. Deregulated expression of selected histone methylases and demethylases in prostate carcinoma. *Endocr. Relat. Cancer* **2014**, *21*, 51–61. [CrossRef] [PubMed]
172. Almeida-Rios, D.; Graça, I.; Vieira, F.Q.; Ramalho-Carvalho, J.; Pereira-Silva, E.; Martins, A.T.; Oliveira, J.; Gonçalves, C.S.; Costa, B.M.; Henrique, R.; et al. Histone methyltransferase PRMT6 plays an oncogenic role of in prostate cancer. *Oncotarget* **2016**, *7*, 53018–53028. [CrossRef] [PubMed]
173. Jiang, N.; Li, Q.L.; Pan, W.; Li, J.; Zhang, M.F.; Cao, T.; Su, S.G.; Shen, H. PRMT6 promotes endometrial cancer via AKT/mTOR signaling and indicates poor prognosis. *Int. J. Biochem. Cell Biol.* **2020**, *120*, 105681. [CrossRef]
174. Feng, Q.; Li, X.; Sun, W.; Sun, M.; Li, Z.; Sheng, H.; Xie, F.; Zhang, S.; Shan, C. Targeting G6PD reverses paclitaxel resistance in ovarian cancer by suppressing GSTP1. *Biochem. Pharmacol.* **2020**, *178*, 114092. [CrossRef] [PubMed]
175. Avasarala, S.; Wu, P.Y.; Khan, S.Q.; Yanlin, S.; Van Scoyk, M.; Bao, J.; Di Lorenzo, A.; David, O.; Bedford, M.T.; Gupta, V.; et al. PRMT6 promotes lung tumor progression via the alternate activation of tumor-associated macrophages. *Mol. Cancer Res.* **2020**, *18*, 166–178. [CrossRef] [PubMed]
176. Tang, J.; Meng, Q.; Shi, R.; Xu, Y. PRMT6 serves an oncogenic role in lung adenocarcinoma via regulating p18. *Mol. Med. Rep.* **2020**, *22*, 3161–3172. [CrossRef]

177. Pan, R.; Yu, H.; Dai, J.; Zhou, C.; Ying, X.; Zhong, J.; Zhao, J.; Zhang, Y.; Wu, B.; Mao, Y.; et al. Significant association of PRMT6 hypomethylation with colorectal cancer. *J. Clin. Lab. Anal.* **2018**, *32*, e22590. [CrossRef]
178. Lim, Y.; Yu, S.; Yun, J.A.; Do, I.G.; Cho, L.; Kim, Y.H.; Kim, H.C. The prognostic significance of protein arginine methyltransferase 6 expression in colon cancer. *Oncotarget* **2018**, *9*, 9010–9020. [CrossRef]
179. Luo, Y.; Xie, C.; Brocker, C.N.; Fan, J.; Wu, X.; Feng, L.; Wang, Q.; Zhao, J.; Lu, D.; Tandon, M.; et al. Intestinal PPARα Protects Against Colon Carcinogenesis via Regulation of Methyltransferases DNMT1 and PRMT6. *Gastroenterology* **2019**, *157*, 744–759. [CrossRef]
180. Okuno, K.; Akiyama, Y.; Shimada, S.; Nakagawa, M.; Tanioka, T.; Inokuchi, M.; Yamaoka, S.; Kojima, K.; Tanaka, S. Asymmetric dimethylation at histone H3 arginine 2 by PRMT6 in gastric cancer progression. *Carcinogenesis* **2019**, *40*, 15–26. [CrossRef]
181. Hadjebi, O.; Casas-Terradellas, E.; Garcia-Gonzalo, F.R.; Rosa, J.L. The RCC1 superfamily: From genes, to function, to disease. *Biochim. Biophys. Acta-Mol. Cell Res.* **2008**, *1783*, 1467–1479. [CrossRef] [PubMed]
182. Clarke, P.R.; Zhang, C. Spatial and temporal coordination of mitosis by Ran GTPase. *Nat. Rev. Mol. Cell Biol.* **2008**, *9*, 464–477. [CrossRef]
183. Clarke, P.R. Keep it focused: PRMT6 drives the localization of RCC1 to chromosomes to facilitate mitosis, cell proliferation, and tumorigenesis. *Mol. Cell* **2021**, *81*, 1128–1129. [CrossRef]
184. Lavoie, H.; Therrien, M. Regulation of RAF protein kinases in ERK signalling. *Nat. Rev. Mol. Cell Biol.* **2015**, *16*, 281–298. [CrossRef] [PubMed]
185. Wong, T.L.; Ng, K.Y.; Tan, K.V.; Chan, L.H.; Zhou, L.; Che, N.; Hoo, R.L.C.; Lee, T.K.; Richard, S.; Lo, C.M.; et al. CRAF Methylation by PRMT6 Regulates Aerobic Glycolysis–Driven Hepatocarcinogenesis via ERK-Dependent PKM2 Nuclear Relocalization and Activation. *Hepatology* **2020**, *71*, 1279–1296. [CrossRef] [PubMed]
186. Kabbage, M.; Dickman, M.B. The BAG proteins: A ubiquitous family of chaperone regulators. *Cell. Mol. Life Sci.* **2008**, *65*, 1390–1402. [CrossRef] [PubMed]
187. Kalia, S.K.; Lee, S.; Smith, P.D.; Liu, L.; Crocker, S.J.; Thorarinsdottir, T.E.; Glover, J.R.; Fon, E.A.; Park, D.S.; Lozano, A.M. BAG5 inhibits parkin and enhances dopaminergic neuron degeneration. *Neuron* **2004**, *44*, 931–945. [CrossRef]
188. Arakawa, A.; Handa, N.; Ohsawa, N.; Shida, M.; Kigawa, T.; Hayashi, F.; Shirouzu, M.; Yokoyama, S. The C-Terminal BAG Domain of BAG5 Induces Conformational Changes of the Hsp70 Nucleotide- Binding Domain for ADP-ATP Exchange. *Structure* **2010**, *18*, 309–319. [CrossRef]
189. Limm, K.; Ott, C.; Wallner, S.; Mueller, D.W.; Oefner, P.; Hellerbrand, C.; Bosserhoff, A.K. Deregulation of protein methylation in melanoma. *Eur. J. Cancer* **2013**, *49*, 1305–1313. [CrossRef]
190. Mitchell, L.H.; Drew, A.E.; Ribich, S.A.; Rioux, N.; Swinger, K.K.; Jacques, S.L.; Lingaraj, T.; Boriack-Sjodin, P.A.; Waters, N.J.; Wigle, T.J.; et al. Aryl Pyrazoles as Potent Inhibitors of Arginine Methyltransferases: Identification of the First PRMT6 Tool Compound. *ACS Med. Chem. Lett.* **2015**, *6*, 655–659. [CrossRef]
191. Shen, Y.; Szewczyk, M.M.; Eram, M.S.; Smil, D.; Kaniskan, H.Ü.; Ferreira De Freitas, R.; Senisterra, G.; Li, F.; Schapira, M.; Brown, P.J.; et al. Discovery of a Potent, Selective, and Cell-Active Dual Inhibitor of Protein Arginine Methyltransferase 4 and Protein Arginine Methyltransferase 6. *J. Med. Chem.* **2016**, *59*, 9124–9139. [CrossRef] [PubMed]
192. Shen, Y.; Li, F.; Szewczyk, M.M.; Halabelian, L.; Park, K.S.; Chau, I.; Dong, A.; Zeng, H.; Chen, H.; Meng, F.; et al. Discovery of a First-in-Class Protein Arginine Methyltransferase 6 (PRMT6) Covalent Inhibitor. *J. Med. Chem.* **2020**, *63*, 5477–5487. [CrossRef] [PubMed]
193. Gong, S.; Maegawa, S.; Yang, Y.; Gopalakrishnan, V.; Zheng, G.; Cheng, D. Licochalcone A is a natural selective inhibitor of arginine methyltransferase 6. *Biochem. J.* **2021**, *478*, 389–406. [CrossRef] [PubMed]
194. Shen, Y.; Li, F.; Szewczyk, M.M.; Halabelian, L.; Chau, I.; Eram, M.S.; Dela Seña, C.; Park, K.S.; Meng, F.; Chen, H.; et al. A First-in-Class, Highly Selective and Cell-Active Allosteric Inhibitor of Protein Arginine Methyltransferase 6. *J. Med. Chem.* **2021**, *64*, 3697–3706. [CrossRef]

Review

Structure and Function of Protein Arginine Methyltransferase PRMT7

Levon Halabelian [1] and Dalia Barsyte-Lovejoy [1,2,*]

1. Structural Genomics Consortium, Temerty Faculty of Medicine, University of Toronto, Toronto, ON M5S 1A8, Canada; l.halabelian@utoronto.ca
2. Department of Pharmacology and Toxicology, University of Toronto, Toronto, ON M5S 1A8, Canada
* Correspondence: d.barsyte@utoronto.ca

Abstract: PRMT7 is a member of the protein arginine methyltransferase (PRMT) family, which methylates a diverse set of substrates. Arginine methylation as a posttranslational modification regulates protein–protein and protein–nucleic acid interactions, and as such, has been implicated in various biological functions. PRMT7 is a unique, evolutionarily conserved PRMT family member that catalyzes the mono-methylation of arginine. The structural features, functional aspects, and compounds that inhibit PRMT7 are discussed here. Several studies have identified physiological substrates of PRMT7 and investigated the substrate methylation outcomes which link PRMT7 activity to the stress response and RNA biology. PRMT7-driven substrate methylation further leads to the biological outcomes of gene expression regulation, cell stemness, stress response, and cancer-associated phenotypes such as cell migration. Furthermore, organismal level phenotypes of PRMT7 deficiency have uncovered roles in muscle cell physiology, B cell biology, immunity, and brain function. This rapidly growing information on PRMT7 function indicates the critical nature of context-dependent functions of PRMT7 and necessitates further investigation of the PRMT7 interaction partners and factors that control PRMT7 expression and levels. Thus, PRMT7 is an important cellular regulator of arginine methylation in health and disease.

Keywords: protein arginine methylation; PRMT7; epigenetics; cancer; immunity; pluripotency

Citation: Halabelian, L.; Barsyte-Lovejoy, D. Structure and Function of Protein Arginine Methyltransferase PRMT7. *Life* **2021**, *11*, 768. https://doi.org/10.3390/life11080768

Academic Editors: Albert Jeltsch and Arunkumar Dhayalan

Received: 6 July 2021
Accepted: 28 July 2021
Published: 30 July 2021

Publisher's Note: MDPI stays neutral with regard to jurisdictional claims in published maps and institutional affiliations.

Copyright: © 2021 by the authors. Licensee MDPI, Basel, Switzerland. This article is an open access article distributed under the terms and conditions of the Creative Commons Attribution (CC BY) license (https://creativecommons.org/licenses/by/4.0/).

1. Introduction

Arginine methylation of proteins is a post-translational modification that is introduced by protein arginine methyltransferases (PRMTs). By altering hydrogen bonding, introducing bulk, and some hydrophobicity, arginine methylation can influence protein–protein and protein–nucleic acid interactions, thus playing a role in chromatin, RNA biology, and other phenomena such as phase separation [1,2]. PRMTs regulate normal physiological processes such as myogenesis, embryonic development, and immune system function and play roles in pathologies such as cancer, neurodegeneration, and inflammation [1–5]. Recent knowledge on shared and unique arginine-methylated substrates of PRMTs has shed light on the individual members of the PRMT family. The nine members of the PRMT family are divided into type I represented by PRMT1-4, 6, and 8 that asymmetrically di-methylate the guanidino group of arginine, while type II PRMT5 and 9 engage in symmetric dimethylation of arginines. The sole representative of the type III group is the PRMT7 enzyme that only monomethylates arginine [3,4,6–9]. The unique structure of this enzyme and the substrate repertoire reflects its function in cells and organisms. This review aims to discuss the recent findings on PRMT7 structure and function, as well as the progress in understanding the roles this enzyme plays in cell biology, disease, and physiological processes.

2. Structural Features of PRMT7

2.1. Domain Architecture and Evolution

Most PRMTs contain one catalytic seven β strand Rossman fold domain, but require homodimerization to form an active enzyme [10]. However, two family members, PRMT7 and PRMT9, underwent gene duplication in metazoans and thus, contain two tandem domains that fold together, forming a homodimer-like structure. While the N-terminal domain of PRMT7 is catalytically active, the C-terminal domain is considered inactive (see structure discussion below) [11,12] (Figure 1).

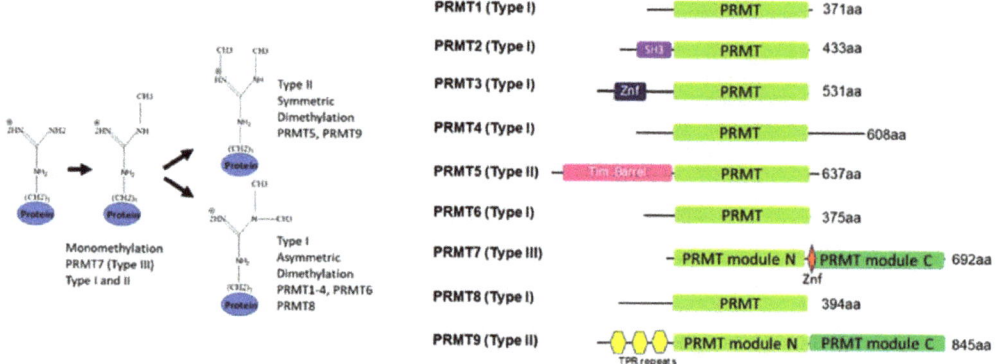

Figure 1. General overview of protein arginine methylation modes (**left**) and PRMT family domain structure (**right**).

Interestingly, outside of metazoans, PRMT7 has been identified in protozoan Kinetoplastida *Trypanosoma* sp. (Figure 2) where this double-domain structure is not conserved, and only one catalytically active domain is present. In the representative species from the animal, fungi, and plant kingdoms, the PRMT7 gene duplication exemplifies the classical double-domain PRMT7 structure. Although PRMT7 has not been described in yeasts, such as *S. cerevisiae*, it is present in several fungi, particularly mold species (Figure 2). Likewise, PRMT7 seems to be absent from the non-vascular plants or even vascular non-seed-bearing plants. Thus, the evolutionary origin and the duplication of PRMT7 warrants further investigation.

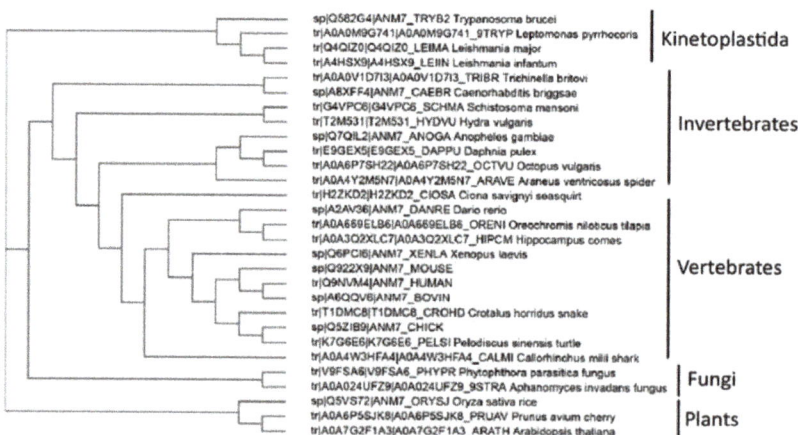

Figure 2. Evolutionary analysis of known PRMT7 proteins. Known PRMT7 sequences from the UniProt database were aligned, and cladogram-rendered using ClustalW2 and simple phylogeny software (EBI).

2.2. Structure

PRMT7 contains two tandem PRMT modules (N and C) that are connected by a 19-residue linker. Each PRMT module in PRMT7 consists of an N-terminal Rossmann fold that is responsible for the cofactor S-adenosylmethionine (SAM)-binding, a C-terminal β-barrel domain for substrate recognition and binding, and a dimerization arm (Figure 3A,B). PRMT7 also contains an additional zinc-finger motif at the junction between the two PRMT modules (Figure 3A), which was shown to lock the module-C in an inactive conformation compared to module-N in the crystal structure of MmPRMT7 in complex with S-adenosylhomocysteine (SAH), the demethylated product of SAM (PDB ID: 4C4A) [11].

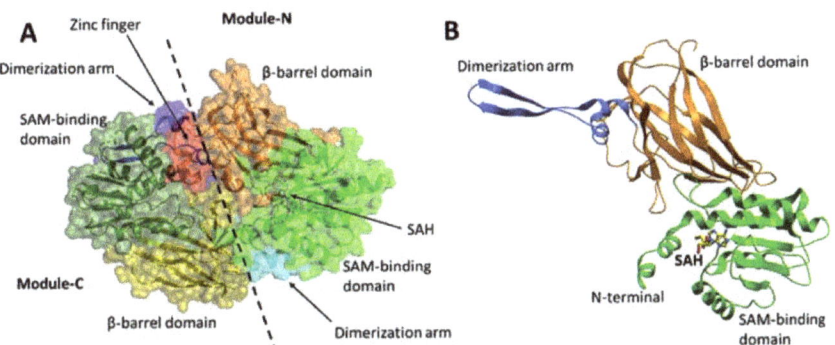

Figure 3. Crystal structure of MmPRMT7 in the complex with SAH (PDB ID: 4C4A). (**A**) Overall structure of MmPRMT7 is shown in the surface representation color-coded according to its domain boundaries. The two catalytic modules (N and C) in MmPRMT7 are divided by a dashed line. (**B**) Module-N of MmPRMT7 is shown as cartoon representation in green (SAM-binding domain), orange (β-barrel domain) and blue (dimerization arm), and SAH in yellow sticks.

Furthermore, several key PRMT signature motifs crucial for SAM/SAH binding are not conserved in module-C [11]. Accordingly, only one SAH molecule was reported to bind to the module-N SAM-binding pocket of the MmPRMT7-SAH complex (PDB ID: 4C4A). Overlay of the SAM-binding domains of modules N and C revealed several residues in module-C, such as D410, P459, F481, and F481 directly overlap with SAM/SAH in the SAM-binding pocket (Figure 4A), indicating that module-C in MmPRMT7 is unable to bind SAM and hence is catalytically inactive.

Recently SGC3027, a highly potent, selective, and cell-active chemical probe for PRMT7 was reported [13]. It represents a cell-permeable prodrug that converts into SGC8158 within the cells. In the crystal structure of MmPRMT7 in complex with SGC8158 (PDB ID: 6OGN), the adenosyl moiety of SGC8158 binds to the SAM-binding pocket of the catalytically active module-N by directly competing with SAM, thus explaining its activity as a SAM-competitive inhibitor (Figure 4B). Moreover, its biphenylmethylamine moiety inserts into an adjacent hydrophobic pocket in the conserved THW motif region, known for substrate arginine coordination in other PRMTs [14,15]. Structural comparison of SGC8158-bound MmPRMT7 with that of TbPRMT7 in complex with H4 peptide (PDB ID: 4M38) shows that only the flexible linker region of SGC8158 overlaps with Arginine sidechain of histone H4 peptide, which may or may not be sufficient to compete with SGC8158 (Figure 4B). Thus, despite the presence of the biphenylmethylamine moiety in the above-mentioned hydrophobic pocket, SGC8158 did not act as a peptide competitive inhibitor.

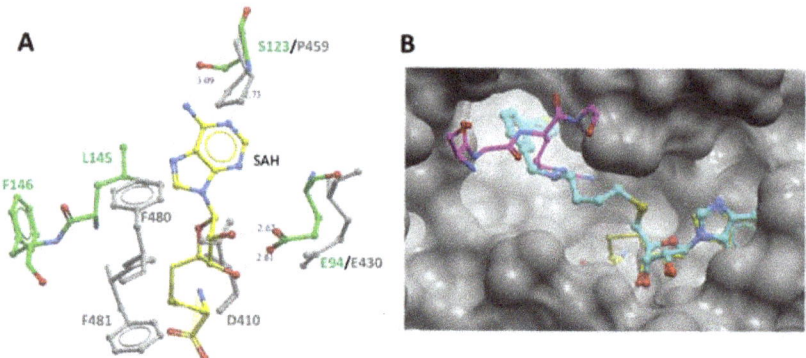

Figure 4. PRMT7 SAM-binding pocket comparison and occupancy by the inhibitory compound SGC 8158. (**A**) Comparison of SAM-binding domains of modules N and C. Overlay of module-N (in green) and module-C (in grey) SAM-binding domains of MmPRMT7 in complex with SAH (yellow). (**B**) Close-up view of module-N SAM-binding domain in complex with SGC8158 chemical probe (cyan) (PDB ID: 6OGN). The SAM-binding domains for both MmPRMT7-SGC8158 and TbPRMT7-SAH in complex with H4 peptide (PDB ID: 4M38) were superimposed to show the SGC8158 binding mode relative to SAH (yellow) and H4 peptide (magenta).

3. Enzyme Function of PRMT7

3.1. Regulation, Enzymatic Properties, and Crosstalk with Other PRMTs

Automethylation of PRMT7 R531 was reported to play a role in breast cancer cell migration [16]. Remarkably, the Phosphosite database indicates that human PRMT7 is also monomethylated at R7 and R32, and several ubiquitylation and phosphorylation sites are present. However, the enzymes responsible for these PTMs or their functional outcomes are now known.

Extensive investigations of recombinant PRMT7 in methylation assays indicated that both bacteria and insect cell-expressed recombinant PRMT7 is highly active with a preference for basic arginine-rich substrates such as histones H4 or H2B (KKDGKKRKRSRKESYK peptide) [8,17]. Overall, the PRMT7 enzymatic activity parameters indicate micromolar affinity to SAM and H2B substrate and relatively slow reaction turnover (see references [8,17] for excellent discussion).

One of the more remarkable, recently discovered features of the PRMT7 enzyme is the crosstalk with PRMT5. PRMT5 symmetrically dimethylates H4R3, H2AR3, and H3R8 which are associated with transcriptional repression [18]. Interestingly, PRMT7-directed H3R17 monomethylation drastically increased PRMT5-mediated H4R3 symmetric dimethylation through an allosteric mechanism [19]. Other PRMTs may be similarly affected by PRMT7 monomethylation of neighboring histone arginine residues. Further studies using more complex reaction conditions, substrates, and including PRMT7 binding partners may be able to further address the reaction kinetics of PRMT7, intriguing preference for low reaction temperature, and non-physiological salt concentration preference (discussed in detail in an excellent recent review [20]).

3.2. PRMT7 Substrates

The highest enzymatic activity of PRMT7 is found with histone peptides as in vitro substrates. *T. brucei* PRMT7 has also been co-crystalized with the SAM cofactor and H4 peptide (PDB:4M38). Early studies have reported PRMT7-mediated dimethylation of arginines [21,22]; however, subsequent evidence on enzymatic activity, structure, and mutagenesis has unequivocally shown monomethylation activity PRMT7 [6–9,20]. While histone–peptide substrates have been extensively reported as PRMT7 substrates in vitro, the evidence of PRMT7 dependent histone methylation in cells relies on antibody-based detection in chromatin immunoprecipitation (ChIP) experiments. Several studies noted that

PRMT7 modulates H4R3me2s levels at specific loci [23–26]. In addition, PRMT7 dependent regulation of H2AR3me2s at DNA damage response associated loci was observed [26]. An early study described PRMT5 and PRMT7 regulation of H3R2me2s in association with transcriptional activation Mixed Lineage Leukemia (MLL) complex [27]. In light of the above discussed PRMT7 and PRMT5 crosstalk, it is possible that the observed regulation of H4R3me2s by PRMT7 was an outcome of PRMT7 activating the PRMT5 dimethyltransferase function. Another early study has reported PRMT7-mediated methylation of small nuclear ribonucleoproteins (snRNPs) [21] that was subsequently confirmed in the large-scale mass spectrometry study [28].

Proteomics and antibody-based approaches of methyl-arginine identification have expanded the number of PRMT7 substrates beyond histone proteins (Table 1). However, it should be noted that histone methylation detection by mass spectrometry is challenging as arginine-rich histones are poorly suited for traditional trypsin digestion, and overall differences in modifications at specific genomic loci often fall below detection limits. The most commonly used proteomic approach combining antibody-based enrichment of methyl arginine-containing peptides followed by the mass spectrometry analysis may also introduce an antibody bias. Recently reported antibody-independent methods of methyl arginine detection employing NMR may overcome these limitations [29].

Table 1. Selected PRMT7 substrates in cells.

Substrate	R Methylation Sites	Function, Disease Relevance, Reference
DVL3	271, 342, 614	DVL3 localization, wnt signaling, cancer [30]
EIF2S1 (EIF2 alpha)	55	Translation arrest, stress granule regulation, [31]
G3BP2	432, 438, 452, 468	Wnt signaling, cancer, [32]
GLI2	225/227	Cell senescence, [33]
Histone H4, H2A	H4R3, H2AR3	Gene expression, [23–27]
HNRNPA1	194, 206, 218, 225	Splicing, [28]
HSP70	469	Stress response, [13]
NALCN	1653	Neuronal excitability, [34]
MAVS	52	Viral infection, [35]
MRPS23	21	Oxidative phosphorylation, cell invasion, cancer [36]
P38MAPK	70	Myoblast differentiation, [37]
PRMT7	531	Cell migration, cancer [16]

Nevertheless, three recent studies on PRMT7 dependent methylome in mammalian cell lines and *Leishmania* sp. parasite highlight the broad diversity of PRMT7 substrates [13,28,38]. The largest category of proteins enriched in the PRMT7 methylated hits was RNA binding and metabolism-associated proteins, a finding that was consistent in all three studies. Interestingly, these studies also enabled elucidation of the preferred methylation motif of PRMT7. Numerous in vitro experiments have attributed the RXR motif as highly methylated by PRMT7 [8,39]. However, the proteomic studies indicate an overall preference for methyl arginine to reside in glycine-rich regions [28,38], whilst in mammalian cells, there is a slight preference for proline to precede the methyl arginine [28].

3.3. Inhibitor Compounds for PRMT7

The discovery of potent and selective inhibitors for PRMT enzymes has enabled experimental approaches to specifically address the catalytic functions of these enzymes and facilitated therapeutic development [17,40]. One of the first compounds described as a dual inhibitor for PRMT5 and PRMT7, DS-437 was based on the SAM cofactor design. Although relatively potent in vitro (6 μM), the compound required high cellular concentrations to inhibit PRMT5 activity in cells, and specific PRMT7 activity was not addressed [41]. Further exploration and optimization of compounds occupying the SAM binding pocket of PRMT7 yielded extremely potent (2.5 nM) in vitro compounds selective for PRMT7 over other PRMT family members and other methyltransferases (Figure 4B). Due to the poor cellular permeability of the chemical SAM scaffold, a prodrug strategy was employed to generate cell-active inhibitors. SGC3027 is a prodrug compound that, upon reduction by abundant cellular reductases, releases the active component of SGC8158. Importantly the negative control compound is also available to ensure meaningful experimental data [13]. SGC3027 compound has been demonstrated to inhibit PRMT7-dependent methylation of HSP70 and other protein substrates in cells [13,28].

4. Cellular Roles of PRMT7

4.1. The Role of PRMT7 in Gene Expression and Genome Maintenance

PRMT7 methylates histones and results in gene transcription regulation. Repressive H4R3me1 and H4R3me2s marks were associated with PRMT7 activity on the BCL6 promoter, although the latter could be due to the allosteric activation of PRMT5. PRMT7 regulates B cell development, and overexpression in the B cell lineage cell lines resulted in lower BCL6 levels and higher H4R3me2s at the promoter of *Bcl6* [26]. Another study found that PRMT7 dimethylated H2AR3, and H4R3 were enriched on DNA repair genes. Knockdown of PRMT7 upregulated the expression of multiple transcripts involved in DNA repair, including ALKBH5, APEX2, POLD1, and POLD2 [25]. Expression of these genes, especially DNA polymerase (POLD1), could mediate the sensitivity to DNA damage conferred by PRMT7 [25]. PRMT7 can also antagonize the action of the MLL methyltransferase complex. MLL4 is the H3K4 methyltransferase that plays a role in cellular differentiation. Knockdown of PRMT7 enhanced the levels of H3K4me3, decreased H4R3me1, and increased the expression of MLL4 target genes, promoting neuronal differentiation [24]. The factors that recruit PRMT7 to these complexes and gene loci are not always clear, as is the context of other histone modifications that may influence PRMT7-driven arginine methylation.

A study investigating the function of PRMT7 in muscle physiology determined that PRMT7 knockdown in C2C12 cells resulted in a decrease of H4R3me2s on the promoters of several genes, including *Dnmt3b* and *Cdkn1a*. However, in PRMT7-deficient cells, the activating mark H3K4me3 was decreased at the *Dnmt3b* promoter, while it was increased at the *Cdnk1a* promoter, thus, correlating with reduced expression of *Dnmt3b* and increased Cdkn1a mRNA levels that resulted in premature senescence [23]. Another study examining epigenetic regulation of imprinted genes identified CTCFL/BORIS, a paralog of CTCF, as PRMT7 binding partners. PRMT7 was recruited to imprinting control regions during embryonic male germ cell development. This resulted in the symmetric dimethylation of H4R3 at nearby nucleosomes, thereby facilitating the recruitment of the de novo DNA methyltransferases 3 (DNMT3a/b). DNA methylation of the imprinting control region determines the parental specific expression of *Igf2* in male germ cells [42]. Such tissue-specific roles of PRMT7 through the interaction with tissue-restricted binding partners may prove to be more widespread than previously thought, as PRMT7 knockout studies indicate distinct functional outcomes in distinct cell types (see below).

4.2. Regulation of Pluripotency, Cell Differentiation, and Senescence

The balance between cellular states is controlled by the intricate orchestration of cellular signaling and transcription factors. PRMT7 is highly expressed in pluripotent cells [43]. Examination of candidate reprogramming factors in mammalian oocytes determined that PRMT7 protein levels were the highest in the pluripotent oocyte and changed substantially during mouse embryogenesis. In addition, the authors found that PRMT7 replaced SOX2 as one of the Yamanaka factors in generating induced pluripotent stem cells (iPSC) from mouse embryonic fibroblasts [44].

By modifying H4R3me2s, PRMT7 repressed the *miR-24-2* gene cluster that downregulates the expression of *Oct4, Nanog, Klf4,* and *c-Myc*. These miRNAs also targeted the 3′UTR of their repressor gene *Prmt7* thus forming a double-negative feedback loop where miR-24-3p/miR24-2-5p downregulates PRMT7 and vice versa to control Oct4, Nanog, Klf4, and c-Myc in pluripotency [45]. PRMT7-mediated repression of another miRNA cluster, miR-221-3p and miR-221-5p, also plays a critical role in pluripotency factor Oct4,

Nanog, and Sox2 levels and mouse embryonic stem cell stemness [46]. Remarkably PRMT7, together with PRMT5, regulates mouse embryonic development from the 2-cell to 4-cell stages and plays a role in early human embryonic developmental arrest [47,48].

Several reports highlight the role of PRMT7 in normal tissue homeostasis. PRMT7 is preferentially expressed in injury-activated muscle satellite cells and is required for muscle regeneration. As mentioned above, PRMT7 regulates histone methylation and thus p21CIP and DNMT3b expression, leading to cell-cycle arrest and premature cellular senescence, consequently resulting in a deficiency of regenerating myofibers, a reduced pool of PAX7-positive cells, and a failure of satellite cells to self-renew [23]. Interestingly another mouse knockout study demonstrated that PRMT7-deficient muscle exhibit decreased oxidative metabolism, which is associated with reduced expression of PGC-1α, a critical regulator of the mitochondria. Changes in muscle structure and fiber type were attributed to PGC-1α that in turn was regulated by p38MAPK. The authors provided a link between PRMT7 methylation of p38 mitogen-activated protein kinase (p38MAPK), which activates Activating Transcription Factor 2 (ATF2), an upstream transcriptional regulator for PGC-1α [37]. The cellular senescence phenotype observed in PRMT7 deficiency was also found in mouse embryonic fibroblasts where premature senescence coincided with reduced levels of sonic hedgehog (SHH) pathway regulator GLI2. The authors have shown that PRMT7 promotes SHH signaling via GLI2 methylation regulating the localization of GLI2 [33]. This study is especially important in light of the complex regulation of GLI1 and GLI2 by PRMT1, PRMT5, and PRMT7 that controls cell senescence, self-renewal with potentially far-reaching implications in pluripotency and cancer-initiating cell biology [49].

Conditional knockout of PRMT7 in the B-cell lineage resulted in impaired B-cell differentiation and hyperplasia of the germinal center. In contrast, over-expression of PRMT7 triggered an increase in BCL6 in germinal center-derived B-cell lines. Thus, PRMT7 overexpression impairs lymphoid differentiation, and normal PRMT7 function is needed for B cell development [26]. Interestingly, PRMT7 also plays a role in adipogenesis by controlling C/EBP-β activity or PPAR-γ2 expression [50]. Overall, PRMT7 plays an essential role in regulating the cellular states of pluripotency, differentiation, senescence, and the epithelial–mesenchymal transition discussed below.

4.3. PRMT7 and Stress Response

One of the earliest functional descriptions of PRMT7 came from studies in Chinese hamster cell line DC-3F, linking low levels of PRMT7 to resistance to the topoisomerase II inhibitors 9-OH-ellipticine, etoposide, and cisplatin [51,52]. Genomic linkage studies indicated that PRMT7 resides in the susceptibility to etoposide-induced cytotoxicity loci [53]. In contrast, another study showed that downregulation of PRMT7 isoforms in DC-3F hamster cells was associated with increased sensitivity to the topoisomerase inhibitor camptothecin [54]. Subsequent work by Karkhanis et al. demonstrated that PRMT7 regulates DNA damage response genes and thus the sensitivity to DNA damage [25]. In addition to the above-mentioned regulation of POLD1, PRMT7 interacts with BRG1 and BAF, SWI/SNF chromatin remodeling subunits to regulate methylation H2AR3 and H4R3 and suppress DNA repair gene expression, subsequently resulting in the sensitization to the DNA damage stress [25].

Several other regulators involved in cellular stress response have been associated with PRMT7 function. PRMT7 interacts with and can methylate eukaryotic translation initiation factor 2 alpha (EIF2S1) at R55 and in neighboring arginine [31]. Various stresses can result in EIF2S1 phosphorylation, translational shutdown, and unfolded protein response [55]. Haghandish et al. showed a regulatory interplay between EIF2S1 arginine methylation by PRMT7 and the S51 phosphorylation status of eIF2α. Upon translational stress, EIF2S1 is phosphorylated, and PRMT7 is required for EIF2S1-dependent stress granule formation that sequesters transcripts, translational machinery, and initiates a protective response program [31]. Interestingly the eukaryotic elongation factor 2 (EEF2) and stress granule protein G3BP2 were also reported methylated by PRMT7 [32,56].

PRMT7 methylates HSP70 protein family members HSPA1A/B, 6, 8 [13,57]. HSP70 protein chaperones are critical for folding new proteins, maintaining protein homeostasis, and are highly upregulated in response to various stressors [58]. Methylation of HSP70 R469 by PRMT7 facilitated the correct substrate refolding after heat shock and regulated the magnitude of the stress granule response after proteostasis perturbations due to proteasome inhibition. These protective features of methylated HSP70 were associated with higher resistance of wild-type cells to proteasome inhibition when compared to PRMT7 knockout cells [13].

It is possible that other PRMT7 substrates are also involved in stress response, and incidentally, the largest category of PRMT7 methylated proteins are RNA binding proteins [13,28,38] that do play a variety of roles in the stress response [59]. Interestingly the prominent role of PRMT7 in muscle cell physiology (see above) may be linked to extensive regulation of proteostasis in this tissue. Evolutionary adaptation to stress may underlie such phenomena as the noted preference of PRMT7 enzymatic activity to lower temperatures than 37 °C [8,60]. The role of PRMT7 in stress and adaptation would be consistent with highly context-dependent PRMT7 phenotypes in cells and organisms.

5. Connection to Disease and Organismal Phenotypes

5.1. Knockout Phenotypes

Several studies have addressed the organismal role of PRMT7. PRMT7-knockout mice generated by a gene-trap approach displayed significantly reduced body size, reduced weight, and shortened fifth metatarsals. These mice were subviable with surviving adult mice exhibiting increased fat mass, limb and bone anomalies, such as reduced bone mineral content [61]. The subviable nature of PRMT7 knockouts was also noted in another study that subsequently derived B cell-specific PRMT7 knockouts [26]. In this context, PRMT7 loss did not result in changes in frequency and number of early B cell subpopulations but led to decreased mature marginal zone B cells, increased follicular B cells, and promoted germinal center formation. In addition to the aforementioned repressive histone methylation on the *Bcl6* promoter by PRMT7, the authors provided clear links to downstream gene expression programs that involved integrin-mediated cell adhesion. Another exciting avenue discussed in this study was PRMT7 association with the misregulation of DNA damage response that may play an essential role in resting B cells [26].

The role of PRMT7 in muscle physiology has been investigated by several groups. Here, whole-body knockouts of PRMT7 were found viable, possibly indicating that the mouse strain context is important. The mutant animals had decreased muscle regeneration after injury due to the loss of a stem cell population of satellite cells, see above for mechanistic discussion [23]. Jeong et al. noted age-associated obesity in PRMT7-deficient mice and changes in overall muscle structure with the shift from fast-twitch glycolytic fibers to slow-twitch oxidative phosphorylation dependent fibers [37]. These studies indicate the important role PRMT7 plays in normal adult muscle function.

PRMT7 knockout mouse brain dentate granule cells displayed increased firing frequency attributed to enhanced NALCN, the resting membrane potential regulator, and overall hyperexcitability in the knockout brain granule cells. PRMT7 methylates a highly conserved Arg1653 of the NALCN, leading to NALCN Ser1652 phosphorylation, NALCN inhibition, and reduced neuronal excitability [34]. It will be interesting to see how this PRMT7 control of intrinsic excitability in hippocampal neurons translates to organismal phenotype.

Interestingly PRMT7 knockout zebrafish were more resistant to spring viremia of carp virus (SVCV) and grass carp reovirus (GCRV) infections and exhibited enhanced expression of critical antiviral genes. Thus, PRMT7 negatively regulates antiviral responses in zebrafish that involve retinoic acid-inducible gene 1 (RIG1) [62]. A recent study in mice also indicates increased resistance to viral infection-induced lethality in PRMT7 knockout animals [35]. These experiments indicate an important organismal role of PRMT7 in innate immunity regulation that may extend to other organisms (also discussed below).

5.2. Human Syndromes

A study on human genetic developmental disorders identified PRMT7 mutations as responsible for a phenotype that phenocopies pseudohypoparathyroidism with mild intellectual disability, obesity, and symmetrical shortening of the digits, posterior metacarpals, and metatarsals. Sexually dimorphic features, including changes in bone mineral content, bone mineral density, and fifth metacarpal length in females were also noted [61]. Overall, several studies established the links between PRMT7 mutations and OMIM phenotype 617157 of Short Stature, Brachydactyly, Intellectual Developmental Disability and Seizures (SBIDDS) [61,63–66]. Further research is needed to draw parallels between this phenotype and the mouse studies to better understand PRMT7 function and evolution.

5.3. Cancer

Several studies report overexpression of PRMT7 in cancer. Although the Cancer Dependency Map [67] does not indicate that PRMT7 is essential for cell survival under normal growth conditions, it is possible that in specific contexts, PRMT7 plays a role in tumorigenesis. One of these contexts is the epithelial–mesenchymal transition (EMT) which occurs during embryonic development, tissue regeneration, organ fibrosis, and cancer metastasis and survival. In EMT, epithelial cells lose polarity, cell-cell junctions, epithelial markers, and gain cell motility, a spindle-cell shape, and mesenchymal markers [68]. In breast carcinoma cells, increased PRMT7-mediated EMT and metastasis by losing E-cadherin expression due to altered histone methylation, specifically elevated H4R3me2s levels [69]. Baldwin et al. highlighted that PRMT7 is overexpressed in basal breast cancer cells, and the knockdown of PRMT7 reduces cell motility and invasion. Conversely, overexpression of PRMT7 in epithelial breast cancer cells promotes cell invasion by upregulating the MMP9 matrix metalloproteinase that is responsible for the degradation of the extracellular matrix enabling cancer cells to invade tissues [70]. PRMT7 automethylation itself also seems to play a role in breast cancer metastasis [16]. The allosteric regulation between PRMT7 and PRMT5 should be considered, especially in breast cancer, where the role of PRMT5 in cancer-initiating cells and disease progression has been well established [71,72].

PRMT7-dependent methylation of R21 in mitochondrial ribosomal protein S23 MRPS23 accelerated the polyubiquitin-dependent degradation of MRPS23. MRPS23 degradation inhibited oxidative phosphorylation and increased mitochondrial reactive oxygen species (ROS) levels, consequently increasing breast cancer cell invasion and metastasis. As low levels of MRPS23 result in breast cancer cell survival through regulating oxidative phosphorylation, PRMT7 overexpression inhibited oxidative phosphorylation and increased breast cancer cell invasion [36]. PRMT7 overexpression also promoted the invasion and colony formation in lung cancer, and the authors of this study found that PRMT7 interacted with mitochondria localized HSP70 family member HSPA5 and elongation factor 2 EEF2 [73]. Thus, overall, the evidence indicates that PRMT7 plays an important role in tumorigenesis by regulating the cellular differentiation states.

5.4. Immunity and Infection

Arginine methylation regulates immune cell function and antiviral response [1,40]. The above-mentioned study on zebrafish PRMT7 downregulating the viral response genes and conferring susceptibility to infection indicated that PRMT7 plays an important role in the immune response [62].

Recent work demonstrated that PRMT7 methylates mitochondrial antiviral-signaling protein (MAVS) on R52 and attenuates MAVS binding to its partner proteins TRIM31 and RIG1 that is key to the downstream antiviral signaling [51]. Viral component binding to RIG-I and melanoma differentiation-associated gene 5 (MDA5) induces their interaction with MAVS via N-terminal caspase recruitment domains (CARDs). The activated MAVS CARD rapidly forms aggregates converting other MAVS on the mitochondrial outer membrane into prion-like aggregates [74]. PRMT7-mediated methylation affects the aggregation of MAVS that is essential for the biological functions of MAVS. Consequently, the PRMT7 inhibitor SGC3027 enhanced interferon signaling, *Ifnb1*, *Isg56*, and *Cxcl10* gene expression downstream of MAVS, at the same time reducing the viral titers from infected cells. Interestingly, the authors show that the loss of one PRMT7 allele protects mice from viral infections [35].

Thus, both studies in mice and zebrafish above indicate that PRMT7 confers susceptibility to viral infections and enhances the immune response. The evolutionary origins of this phenomenon and its biological significance warrant further investigation. Given the complexity of the immune and antiviral responses, as well as links to B cell biology, it will be interesting to see if PRMT7 plays a role in autoimmunity and viral mimicry in cancer.

6. Conclusions and Outlook

A wide variety of biological processes associated with PRMT7 function are consistent with the broad range of proteins methylated by PRMT7 in cells. This repertoire will undoubtedly be further expanded in the course of future research. In light of the emerging pattern of context-dependent function of PRMT7, it will be interesting to explore tissue, cell state, or stimulus-specific roles of PRMT7. So far, the organismal knockouts of PRMT7 have provided a wealth of phenotypic information of which only some are mechanistically accounted for, such as the role of PRMT7 in muscle physiology or B cell biology. However, further studies on the processes underlying neuronal or bone development as well as metabolic phenotypes of PRMT7 knockouts are needed. Concurrently understanding PRMT7 expression patterns and the transcription factors that determine cell-specific or stimulus-driven PRMT7 regulation will enable deeper mechanistic knowledge. And as indicated above, the context-specific function of PRMT7 may stem from its interacting proteins. Several studies have begun to address binding partner proteins of PRMT7 through co-immunoprecipitation or proximity biotinylation techniques, where the former has further delved into the biological outcome of such interactions identifying EIF2S1 methylation and stress response regulation [40]. Such mechanistic research will be necessary to elucidate the significance of PRMT7-driven methylation. Thus, addressing PRMT7 expression pattern, protein stability, substrates, and interaction partners and interplay with other PRMT family members will enable a deeper understanding of PRMT7 function.

Author Contributions: Writing—original draft preparation, L.H. and D.B.-L.; writing—review and editing, L.H. and D.B.-L.; visualization, L.H. and D.B.-L. All authors have read and agreed to the published version of the manuscript.

Funding: This research was funded by grants from Natural Sciences and Engineering Research Council to D.B.-L. The Structural Genomics Consortium is a registered charity (no: 1097737) that receives funds from Bayer AG, Boehringer Ingelheim, Bristol Myers Squibb, Genentech, Genome Canada through Ontario Genomics Institute [OGI-196], EU/EFPIA/OICR/McGill/KTH/Diamond Innovative Medicines Initiative 2 Joint Undertaking [EUbOPEN grant 875510], Janssen, Merck KGaA (aka EMD in Canada and US), Pfizer and Takeda.

Institutional Review Board Statement: Not applicable.

Informed Consent Statement: Not applicable.

Data Availability Statement: Not applicable.

Acknowledgments: We thank our colleagues for their helpful comments and suggestions.

Conflicts of Interest: The authors declare no conflict of interest.

References

1. Guccione, E.; Richard, S. The regulation, functions and clinical relevance of arginine methylation. *Nat. Rev. Mol. Cell Biol.* **2019**, *20*, 642–657. [CrossRef] [PubMed]
2. Lorton, B.M.; Shechter, D. Cellular consequences of arginine methylation. *Cell. Mol. Life Sci.* **2019**, *76*, 2933–2956. [CrossRef] [PubMed]
3. Bedford, M.T.; Clarke, S.G. Protein arginine methylation in mammals: Who, what, and why. *Mol. Cell* **2009**, *33*, 1–13. [CrossRef] [PubMed]

4. Bedford, M.T.; Richard, S. Arginine methylation an emerging regulator of protein function. *Mol. Cell* **2005**, *18*, 263–272. [CrossRef] [PubMed]
5. Hwang, J.W.; Cho, Y.; Bae, G.U.; Kim, S.N.; Kim, Y.K. Protein arginine methyltransferases: Promising targets for cancer therapy. *Exp. Mol. Med.* **2021**, *53*, 788–808. [CrossRef]
6. Debler, E.W.; Jain, K.; Warmack, R.A.; Feng, Y.; Clarke, S.G.; Blobel, G.; Stavropoulos, P. A glutamate/aspartate switch controls product specificity in a protein arginine methyltransferase. *Proc. Natl. Acad. Sci. USA* **2016**, *113*, 2068–2073. [CrossRef]
7. Zurita-Lopez, C.I.; Sandberg, T.; Kelly, R.; Clarke, S.G. Human protein arginine methyltransferase 7 (PRMT7) is a type III enzyme forming omega-NG-monomethylated arginine residues. *J. Biol. Chem.* **2012**, *287*, 7859–7870. [CrossRef] [PubMed]
8. Feng, Y.; Hadjikyriacou, A.; Clarke, S.G. Substrate specificity of human protein arginine methyltransferase 7 (PRMT7): The importance of acidic residues in the double E loop. *J. Biol. Chem.* **2014**, *289*, 32604–32616. [CrossRef] [PubMed]
9. Fisk, J.C.; Sayegh, J.; Zurita-Lopez, C.; Menon, S.; Presnyak, V.; Clarke, S.G.; Read, L.K. A type III protein arginine methyltransferase from the protozoan parasite Trypanosoma brucei. *J. Biol. Chem.* **2009**, *284*, 11590–11600. [CrossRef]
10. Tewary, S.K.; Zheng, Y.G.; Ho, M.C. Protein arginine methyltransferases: Insights into the enzyme structure and mechanism at the atomic level. *Cell. Mol. Life Sci.* **2019**, *76*, 2917–2932. [CrossRef]
11. Cura, V.; Troffer-Charlier, N.; Wurtz, J.M.; Bonnefond, L.; Cavarelli, J. Structural insight into arginine methylation by the mouse protein arginine methyltransferase 7: A zinc finger freezes the mimic of the dimeric state into a single active site. *Acta Cryst. D Biol. Cryst.* **2014**, *70*, 2401–2412. [CrossRef]
12. Hasegawa, M.; Toma-Fukai, S.; Kim, J.D.; Fukamizu, A.; Shimizu, T. Protein arginine methyltransferase 7 has a novel homodimer-like structure formed by tandem repeats. *FEBS Lett.* **2014**, *588*, 1942–1948. [CrossRef]
13. Szewczyk, M.M.; Ishikawa, Y.; Organ, S.; Sakai, N.; Li, F.; Halabelian, L.; Ackloo, S.; Couzens, A.L.; Eram, M.; Dilworth, D.; et al. Pharmacological inhibition of PRMT7 links arginine monomethylation to the cellular stress response. *Nat. Commun.* **2020**, *11*, 2396. [CrossRef] [PubMed]
14. Boriack-Sjodin, P.A.; Jin, L.; Jacques, S.L.; Drew, A.; Sneeringer, C.; Scott, M.P.; Moyer, M.P.; Ribich, S.; Moradei, O.; Copeland, R.A. Structural Insights into Ternary Complex Formation of Human CARM1 with Various Substrates. *ACS Chem. Biol.* **2016**, *11*, 763–771. [CrossRef]
15. Wang, C.; Zhu, Y.; Caceres, T.B.; Liu, L.; Peng, J.; Wang, J.; Chen, J.; Chen, X.; Zhang, Z.; Zuo, X.; et al. Structural determinants for the strict monomethylation activity by trypanosoma brucei protein arginine methyltransferase 7. *Structure* **2014**, *22*, 756–768. [CrossRef]
16. Geng, P.; Zhang, Y.; Liu, X.; Zhang, N.; Liu, Y.; Liu, X.; Lin, C.; Yan, X.; Li, Z.; Wang, G.; et al. Automethylation of protein arginine methyltransferase 7 and its impact on breast cancer progression. *FASEB J.* **2017**, *31*, 2287–2300. [CrossRef] [PubMed]
17. Li, A.S.M.; Li, F.; Eram, M.S.; Bolotokova, A.; Dela Sena, C.C.; Vedadi, M. Chemical probes for protein arginine methyltransferases. *Methods* **2020**, *175*, 30–43. [CrossRef] [PubMed]
18. Tarighat, S.S.; Santhanam, R.; Frankhouser, D.; Radomska, H.S.; Lai, H.; Anghelina, M.; Wang, H.; Huang, X.; Alinari, L.; Walker, A.; et al. The dual epigenetic role of PRMT5 in acute myeloid leukemia: Gene activation and repression via histone arginine methylation. *Leukemia* **2016**, *30*, 789–799. [CrossRef]
19. Jain, K.; Jin, C.Y.; Clarke, S.G. Epigenetic control via allosteric regulation of mammalian protein arginine methyltransferases. *Proc. Natl. Acad. Sci. USA* **2017**, *114*, 10101–10106. [CrossRef]
20. Jain, K.; Clarke, S.G. PRMT7 as a unique member of the protein arginine methyltransferase family: A review. *Arch. Biochem. Biophys.* **2019**, *665*, 36–45. [CrossRef] [PubMed]
21. Gonsalvez, G.B.; Tian, L.; Ospina, J.K.; Boisvert, F.M.; Lamond, A.I.; Matera, A.G. Two distinct arginine methyltransferases are required for biogenesis of Sm-class ribonucleoproteins. *J. Cell Biol.* **2007**, *178*, 733–740. [CrossRef]
22. Lee, J.H.; Cook, J.R.; Yang, Z.H.; Mirochnitchenko, O.; Gunderson, S.I.; Felix, A.M.; Herth, N.; Hoffmann, R.; Pestka, S. PRMT7, a new protein arginine methyltransferase that synthesizes symmetric dimethylarginine. *J. Biol. Chem.* **2005**, *280*, 3656–3664. [CrossRef]
23. Blanc, R.S.; Vogel, G.; Chen, T.; Crist, C.; Richard, S. PRMT7 Preserves Satellite Cell Regenerative Capacity. *Cell Rep.* **2016**, *14*, 1528–1539. [CrossRef] [PubMed]
24. Dhar, S.S.; Lee, S.H.; Kan, P.Y.; Voigt, P.; Ma, L.; Shi, X.; Reinberg, D.; Lee, M.G. Trans-tail regulation of MLL4-catalyzed H3K4 methylation by H4R3 symmetric dimethylation is mediated by a tandem PHD of MLL4. *Genes Dev.* **2012**, *26*, 2749–2762. [CrossRef] [PubMed]
25. Karkhanis, V.; Wang, L.; Tae, S.; Hu, Y.J.; Imbalzano, A.N.; Sif, S. Protein arginine methyltransferase 7 regulates cellular response to DNA damage by methylating promoter histones H2A and H4 of the polymerase delta catalytic subunit gene, POLD1. *J. Biol. Chem.* **2012**, *287*, 29801–29814. [CrossRef]
26. Ying, Z.; Mei, M.; Zhang, P.; Liu, C.; He, H.; Gao, F.; Bao, S. Histone Arginine Methylation by PRMT7 Controls Germinal Center Formation via Regulating Bcl6 Transcription. *J. Immunol.* **2015**, *195*, 1538–1547. [CrossRef]
27. Migliori, V.; Muller, J.; Phalke, S.; Low, D.; Bezzi, M.; Mok, W.C.; Sahu, S.K.; Gunaratne, J.; Capasso, P.; Bassi, C.; et al. Symmetric dimethylation of H3R2 is a newly identified histone mark that supports euchromatin maintenance. *Nat. Struct. Mol. Biol.* **2012**, *19*, 136–144. [CrossRef]

28. Li, W.J.; He, Y.H.; Yang, J.J.; Hu, G.S.; Lin, Y.A.; Ran, T.; Peng, B.L.; Xie, B.L.; Huang, M.F.; Gao, X.; et al. Profiling PRMT methylome reveals roles of hnRNPA1 arginine methylation in RNA splicing and cell growth. *Nat. Commun.* **2021**, *12*, 1946. [CrossRef]
29. Zhang, F.; Kerbl-Knapp, J.; Rodriguez Colman, M.; Macher, T.; Vujić, N.; Fasching, S.; Jany-Luig, E.; Korbelius, M.; Kuentzel, K.; Mack, M.; et al. Global analysis of protein arginine methylation. *Cell Rep. Methods* **2021**, *1*, 100016. [CrossRef]
30. Bikkavilli, R.K.; Avasarala, S.; Vanscoyk, M.; Sechler, M.; Kelley, N.; Malbon, C.C.; Winn, R.A. Dishevelled3 is a novel arginine methyl transferase substrate. *Sci. Rep.* **2012**, *2*, 805. [CrossRef] [PubMed]
31. Haghandish, N.; Baldwin, R.M.; Morettin, A.; Dawit, H.T.; Adhikary, H.; Masson, J.Y.; Mazroui, R.; Trinkle-Mulcahy, L.; Cote, J. PRMT7 methylates eukaryotic translation initiation factor 2alpha and regulates its role in stress granule formation. *Mol. Biol. Cell* **2019**, *30*, 778–793. [CrossRef]
32. Bikkavilli, R.K.; Malbon, C.C. Wnt3a-stimulated LRP6 phosphorylation is dependent upon arginine methylation of G3BP2. *J. Cell Sci.* **2012**, *125*, 2446–2456.
33. Vuong, T.A.; Jeong, H.J.; Lee, H.J.; Kim, B.G.; Leem, Y.E.; Cho, H.; Kang, J.S. PRMT7 methylates and suppresses GLI2 binding to SUFU thereby promoting its activation. *Cell Death Differ.* **2020**, *27*, 15–28. [CrossRef]
34. Lee, S.Y.; Vuong, T.A.; Wen, X.; Jeong, H.J.; So, H.K.; Kwon, I.; Kang, J.S.; Cho, H. Methylation determines the extracellular calcium sensitivity of the leak channel NALCN in hippocampal dentate granule cells. *Exp. Mol. Med.* **2019**, *51*, 1–14. [CrossRef]
35. Zhu, J.; Li, X.; Cai, X.; Zha, H.; Zhou, Z.; Sun, X.; Rong, F.; Tang, J.; Zhu, C.; Liu, X.; et al. Arginine monomethylation by PRMT7 controls MAVS-mediated antiviral innate immunity. *Mol. Cell* **2021**, Online ahead of print. [CrossRef]
36. Liu, L.; Zhang, X.; Ding, H.; Liu, X.; Cao, D.; Liu, Y.; Liu, J.; Lin, C.; Zhang, N.; Wang, G.; et al. Arginine and lysine methylation of MRPS23 promotes breast cancer metastasis through regulating OXPHOS. *Oncogene* **2021**, *40*, 3548–3563. [CrossRef] [PubMed]
37. Jeong, H.J.; Lee, H.J.; Vuong, T.A.; Choi, K.S.; Choi, D.; Koo, S.H.; Cho, S.C.; Cho, H.; Kang, J.S. Prmt7 Deficiency Causes Reduced Skeletal Muscle Oxidative Metabolism and Age-Related Obesity. *Diabetes* **2016**, *65*, 1868–1882. [CrossRef]
38. Ferreira, T.R.; Dowle, A.A.; Parry, E.; Alves-Ferreira, E.V.C.; Hogg, K.; Kolokousi, F.; Larson, T.R.; Plevin, M.J.; Cruz, A.K.; Walrad, P.B. PRMT7 regulates RNA-binding capacity and protein stability in Leishmania parasites. *Nucleic Acids Res.* **2020**, *48*, 5511–5526. [CrossRef] [PubMed]
39. Feng, Y.; Maity, R.; Whitelegge, J.P.; Hadjikyriacou, A.; Li, Z.; Zurita-Lopez, C.; Al-Hadid, Q.; Clark, A.T.; Bedford, M.T.; Masson, J.Y.; et al. Mammalian protein arginine methyltransferase 7 (PRMT7) specifically targets RXR sites in lysine- and arginine-rich regions. *J. Biol. Chem.* **2013**, *288*, 37010–37025. [CrossRef]
40. Wu, Q.; Schapira, M.; Arrowsmith, C.H.; Barsyte-Lovejoy, D. Protein arginine methylation: From enigmatic functions to therapeutic targeting. *Nat. Rev. Drug Discov.* **2021**, *7*, 509–530. [CrossRef] [PubMed]
41. Smil, D.; Eram, M.S.; Li, F.; Kennedy, S.; Szewczyk, M.M.; Brown, P.J.; Barsyte-Lovejoy, D.; Arrowsmith, C.H.; Vedadi, M.; Schapira, M. Discovery of a Dual PRMT5-PRMT7 Inhibitor. *ACS Med. Chem. Lett.* **2015**, *6*, 408–412. [CrossRef]
42. Jelinic, P.; Stehle, J.C.; Shaw, P. The testis-specific factor CTCFL cooperates with the protein methyltransferase PRMT7 in H19 imprinting control region methylation. *PLoS Biol.* **2006**, *4*, e355. [CrossRef]
43. Buhr, N.; Carapito, C.; Schaeffer, C.; Kieffer, E.; Van Dorsselaer, A.; Viville, S. Nuclear proteome analysis of undifferentiated mouse embryonic stem and germ cells. *Electrophoresis* **2008**, *29*, 2381–2390. [CrossRef] [PubMed]
44. Wang, B.; Pfeiffer, M.J.; Drexler, H.C.; Fuellen, G.; Boiani, M. Proteomic Analysis of Mouse Oocytes Identifies PRMT7 as a Reprogramming Factor that Replaces SOX2 in the Induction of Pluripotent Stem Cells. *J. Proteome Res.* **2016**, *15*, 2407–2421. [CrossRef] [PubMed]
45. Lee, S.H.; Chen, T.Y.; Dhar, S.S.; Gu, B.; Chen, K.; Kim, Y.Z.; Li, W.; Lee, M.G. A feedback loop comprising PRMT7 and miR-24-2 interplays with Oct4, Nanog, Klf4 and c-Myc to regulate stemness. *Nucleic Acids Res.* **2016**, *44*, 10603–10618. [CrossRef] [PubMed]
46. Chen, T.Y.; Lee, S.H.; Dhar, S.S.; Lee, M.G. Protein arginine methyltransferase 7-mediated microRNA-221 repression maintains Oct4, Nanog, and Sox2 levels in mouse embryonic stem cells. *J. Biol. Chem.* **2018**, *293*, 3925–3936. [CrossRef]
47. Morita, K.; Hatanaka, Y.; Ihashi, S.; Asano, M.; Miyamoto, K.; Matsumoto, K. Symmetrically dimethylated histone H3R2 promotes global transcription during minor zygotic genome activation in mouse pronuclei. *Sci. Rep.* **2021**, *11*, 10146. [CrossRef] [PubMed]
48. Zhang, W.; Li, S.; Li, K.; Li, L.; Yin, P.; Tong, G. The role of protein arginine methyltransferase 7 in human developmentally arrested embryos cultured in vitro. *Acta Biochim. Biophys. Sin.* **2021**, *53*, 925–932. [CrossRef]
49. Abe, Y.; Tanaka, N. Fine-Tuning of GLI Activity through Arginine Methylation: Its Mechanisms and Function. *Cells* **2020**, *9*, 1973. [CrossRef]
50. Leem, Y.E.; Bae, J.H.; Jeong, H.J.; Kang, J.S. PRMT7 deficiency enhances adipogenesis through modulation of C/EBP-beta. *Biochem. Biophys. Res. Commun.* **2019**, *517*, 484–490. [CrossRef]
51. Gros, L.; Delaporte, C.; Frey, S.; Decesse, J.; de Saint-Vincent, B.R.; Cavarec, L.; Dubart, A.; Gudkov, A.V.; Jacquemin-Sablon, A. Identification of new drug sensitivity genes using genetic suppressor elements: Protein arginine N-methyltransferase mediates cell sensitivity to DNA-damaging agents. *Cancer Res.* **2003**, *63*, 164–171.
52. Gros, L.; Renodon-Corniere, A.; de Saint Vincent, B.R.; Feder, M.; Bujnicki, J.M.; Jacquemin-Sablon, A. Characterization of prmt7alpha and beta isozymes from Chinese hamster cells sensitive and resistant to topoisomerase II inhibitors. *Biochim. Biophys. Acta* **2006**, *1760*, 1646–1656. [CrossRef]
53. Bleibel, W.K.; Duan, S.; Huang, R.S.; Kistner, E.O.; Shukla, S.J.; Wu, X.; Badner, J.A.; Dolan, M.E. Identification of genomic regions contributing to etoposide-induced cytotoxicity. *Hum. Genet.* **2009**, *125*, 173–180. [CrossRef]

54. Verbiest, V.; Montaudon, D.; Tautu, M.T.; Moukarzel, J.; Portail, J.P.; Markovits, J.; Robert, J.; Ichas, F.; Pourquier, P. Protein arginine (N)-methyl transferase 7 (PRMT7) as a potential target for the sensitization of tumor cells to camptothecins. *FEBS Lett.* **2008**, *582*, 1483–1489. [CrossRef] [PubMed]
55. Wek, R.C.; Cavener, D.R. Translational control and the unfolded protein response. *Antioxid. Redox Signal.* **2007**, *9*, 2357–2371. [CrossRef]
56. Jung, G.A.; Shin, B.S.; Jang, Y.S.; Sohn, J.B.; Woo, S.R.; Kim, J.E.; Choi, G.; Lee, K.M.; Min, B.H.; Lee, K.H.; et al. Methylation of eukaryotic elongation factor 2 induced by basic fibroblast growth factor via mitogen-activated protein kinase. *Exp. Mol. Med.* **2011**, *43*, 550–560. [CrossRef] [PubMed]
57. Gao, W.W.; Xiao, R.Q.; Peng, B.L.; Xu, H.T.; Shen, H.F.; Huang, M.F.; Shi, T.T.; Yi, J.; Zhang, W.J.; Wu, X.N.; et al. Arginine methylation of HSP70 regulates retinoid acid-mediated RARbeta2 gene activation. *Proc. Natl. Acad. Sci. USA* **2015**, *112*, E3327–E3336. [CrossRef]
58. Mayer, M.P.; Bukau, B. Hsp70 chaperones: Cellular functions and molecular mechanism. *Cell. Mol. Life Sci.* **2005**, *62*, 670–684. [CrossRef] [PubMed]
59. Gerstberger, S.; Hafner, M.; Tuschl, T. A census of human RNA-binding proteins. *Nat. Rev. Genet.* **2014**, *15*, 829–845. [CrossRef] [PubMed]
60. Hadjikyriacou, A.; Clarke, S.G. Caenorhabditis elegans PRMT-7 and PRMT-9 Are Evolutionarily Conserved Protein Arginine Methyltransferases with Distinct Substrate Specificities. *Biochemistry* **2017**, *56*, 2612–2626. [CrossRef] [PubMed]
61. Akawi, N.; McRae, J.; Ansari, M.; Balasubramanian, M.; Blyth, M.; Brady, A.F.; Clayton, S.; Cole, T.; Deshpande, C.; Fitzgerald, T.W.; et al. Discovery of four recessive developmental disorders using probabilistic genotype and phenotype matching among 4125 families. *Nat. Genet.* **2015**, *47*, 1363–1369. [CrossRef] [PubMed]
62. Zhu, J.; Liu, X.; Cai, X.; Ouyang, G.; Fan, S.; Wang, J.; Xiao, W. Zebrafish prmt7 negatively regulates antiviral responses by suppressing the retinoic acid-inducible gene-I-like receptor signaling. *FASEB J.* **2020**, *34*, 988–1000. [CrossRef]
63. Agolini, E.; Dentici, M.L.; Bellacchio, E.; Alesi, V.; Radio, F.C.; Torella, A.; Musacchia, F.; Tartaglia, M.; Dallapiccola, B.; Nigro, V.; et al. Expanding the clinical and molecular spectrum of PRMT7 mutations: 3 additional patients and review. *Clin. Genet.* **2018**, *93*, 675–681. [CrossRef] [PubMed]
64. Birnbaum, R.; Yosha-Orpaz, N.; Yanoov-Sharav, M.; Kidron, D.; Gur, H.; Yosovich, K.; Lerman-Sagie, T.; Malinger, G.; Lev, D. Prenatal and postnatal presentation of PRMT7 related syndrome: Expanding the phenotypic manifestations. *Am. J. Med. Genet. A* **2019**, *179*, 78–84. [CrossRef] [PubMed]
65. Kernohan, K.D.; McBride, A.; Xi, Y.; Martin, N.; Schwartzentruber, J.; Dyment, D.A.; Majewski, J.; Blaser, S.; Care4Rare Canada, C.; Boycott, K.M.; et al. Loss of the arginine methyltransferase PRMT7 causes syndromic intellectual disability with microcephaly and brachydactyly. *Clin. Genet.* **2017**, *91*, 708–716. [CrossRef]
66. Valenzuela, I.; Segura-Puimedon, M.; Rodriguez-Santiago, B.; Fernandez-Alvarez, P.; Vendrell, T.; Armengol, L.; Tizzano, E. Further delineation of the phenotype caused by loss of function mutations in PRMT7. *Eur. J. Med. Genet.* **2019**, *62*, 182–185. [CrossRef]
67. Tsherniak, A.; Vazquez, F.; Montgomery, P.G.; Weir, B.A.; Kryukov, G.; Cowley, G.S.; Gill, S.; Harrington, W.F.; Pantel, S.; Krill-Burger, J.M.; et al. Defining a Cancer Dependency Map. *Cell* **2017**, *170*, 564–576.e16. [CrossRef]
68. Nieto, M.A.; Huang, R.Y.; Jackson, R.A.; Thiery, J.P. Emt: 2016. *Cell* **2016**, *166*, 21–45. [CrossRef]
69. Yao, R.; Jiang, H.; Ma, Y.; Wang, L.; Wang, L.; Du, J.; Hou, P.; Gao, Y.; Zhao, L.; Wang, G.; et al. PRMT7 induces epithelial-to-mesenchymal transition and promotes metastasis in breast cancer. *Cancer Res.* **2014**, *74*, 5656–5667. [CrossRef]
70. Baldwin, R.M.; Haghandish, N.; Daneshmand, M.; Amin, S.; Paris, G.; Falls, T.J.; Bell, J.C.; Islam, S.; Cote, J. Protein arginine methyltransferase 7 promotes breast cancer cell invasion through the induction of MMP9 expression. *Oncotarget* **2015**, *6*, 3013–3032. [CrossRef] [PubMed]
71. Chiang, K.; Zielinska, A.E.; Shaaban, A.M.; Sanchez-Bailon, M.P.; Jarrold, J.; Clarke, T.L.; Zhang, J.; Francis, A.; Jones, L.J.; Smith, S.; et al. PRMT5 Is a Critical Regulator of Breast Cancer Stem Cell Function via Histone Methylation and FOXP1 Expression. *Cell Rep.* **2017**, *21*, 3498–3513. [CrossRef] [PubMed]
72. Jarrold, J.; Davies, C.C. PRMTs and Arginine Methylation: Cancer's Best-Kept Secret? *Trends Mol. Med.* **2019**, *25*, 993–1009. [CrossRef] [PubMed]
73. Cheng, D.; He, Z.; Zheng, L.; Xie, D.; Dong, S.; Zhang, P. PRMT7 contributes to the metastasis phenotype in human non-small-cell lung cancer cells possibly through the interaction with HSPA5 and EEF2. *Onco Targets Ther.* **2018**, *11*, 4869–4876. [CrossRef] [PubMed]
74. Hou, F.; Sun, L.; Zheng, H.; Skaug, B.; Jiang, Q.X.; Chen, Z.J. MAVS forms functional prion-like aggregates to activate and propagate antiviral innate immune response. *Cell* **2011**, *146*, 448–461. [CrossRef]

Review

Activity and Function of the PRMT8 Protein Arginine Methyltransferase in Neurons

Rui Dong, Xuejun Li and Kwok-On Lai *

Department of Neuroscience, City University of Hong Kong, Hong Kong, China; ruidong@cityu.edu.hk (R.D.); xuejunli2@cityu.edu.hk (X.L.)
* Correspondence: kwokolai@cityu.edu.hk

Abstract: Among the nine mammalian protein arginine methyltransferases (PRMTs), PRMT8 is unusual because it has restricted expression in the nervous system and is the only membrane-bound PRMT. Emerging studies have demonstrated that this enzyme plays multifaceted roles in diverse processes in neurons. Here we will summarize the unique structural features of PRMT8 and describe how it participates in various neuronal functions such as dendritic growth, synapse maturation, and synaptic plasticity. Recent evidence suggesting the potential role of PRMT8 function in neurological diseases will also be discussed.

Keywords: neuron; synapse; dendritic spine; actin cytoskeleton; GTPase; post-translational modification

1. Introduction

Arginine methylation is a prominent protein post-translational modification identified decades ago [1]. Despite being detected in the brain, the function of arginine methylation is not generally well studied in neurons. In recent years, however, an increasing number of studies have unraveled how protein arginine methyltransferases (PRMTs), the enzymes which catalyze arginine, are involved in neuronal function. In particular, our understanding of arginine methylation has increased since the discovery of PRMT8 in 2015. PRMT8 was identified from the expressed sequence tag (EST) databases based on conserved motifs present in PRMTs [2]. It belongs to the type I PRMTs and shares almost 80% amino acid sequence identity with PRMT1. However, unlike PRMT1, which is ubiquitously expressed, the expression of PRMT8 appears to be exclusive in the brain [2]. Besides this unusual tissue distribution, PRMT8 also displays two unique properties: its ability to anchor to the plasma membrane and the presence of not only methyltransferase activity but also the catalytic activity of phospholipase.

2. PRMT8 Characteristics

2.1. Domain and Structure

The PRMT family of enzymes consists of nine members, which are classified into three types (I, II, and III) based on their catalytic activities. Type I PRMTs (PRMT1, 2, 3, 4, 6, and 8) catalyze the formation of both asymmetric dimethylarginine (ADMA) and monomethylated arginine (MMA); type II PRMTs (PRMT5 and 9) catalyze symmetric dimethylarginine (SDMA) and monomethylated arginine (MMA) formation; type III PRMT (PRMT7) only catalyzes monomethylated arginine (MMA) formation [3].

The canonical PRMT core structure adopts a conserved Rossman fold domain followed by a β-barrel domain where the dimerization arm is located [4]. Most PRMTs contain one catalytic Rossman fold domain, but dimerization through the β-barrel is required to compose the active enzyme. While the catalytic core domain of all PRMTs is structurally conserved, the N-terminal non-catalytic domain is very diverse among family members [5]. For example, PRMT3 contains a zinc finger motif which is required for its interaction with the ribosomal protein rpS2 to recognize RNA-associated substrates [6,7]; in contrast,

PRMT5 contains a TIM barrel which is responsible for its interaction with the WD40 repeat protein MEP50 [8,9]. For PRMT8, there are a number of unique structural features in the N-terminal half of the protein (Figure 1): first, there is the presence of a myristoylation site at the N-terminus, which mediates its anchorage to the plasma membrane [2]. Second, phospholipase activity is present within the Rossman fold, making it the only PRMT that contains dual enzyme activities of methyltransferase and phospholipase [10]. These two distinct enzyme activities contribute to different functions of PRMT8 in neurons (see below). Third, its N-terminal region harbours two proline-rich sequences which allow its binding to SH3 domain-containing proteins such as Fyn (a protein tyrosine kinase), p85 (a regulatory subunit of PI3K) and PRMT2 [11]. The functional significance of the interactions between PRMT8 and these SH3 domain-containing proteins remains to be determined.

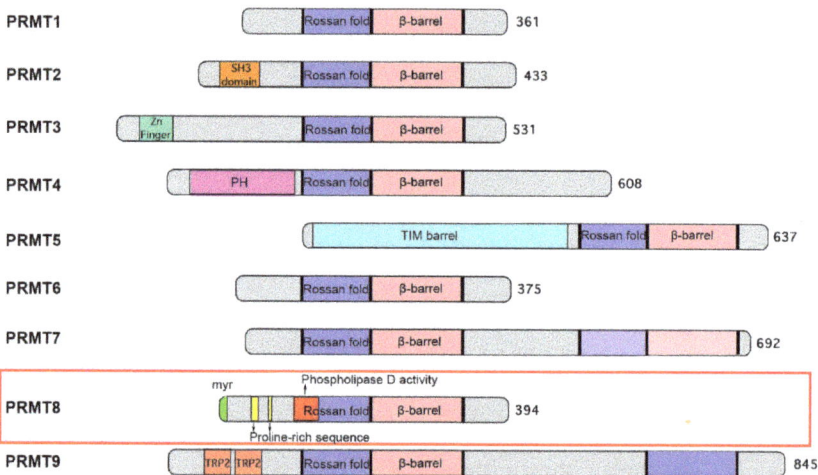

Figure 1. Schematic diagram illustrating the domain architecture of various PRMTs.

2.2. Expression Profile and Subcellular Localization

In contrast to the ubiquitous expression of most PRMTs, PRMT8 has restricted tissue distribution and is specifically expressed in the neurons of the central nervous system (CNS) [10]. PRMT8 is expressed in multiple brain areas such as the cerebral cortex, hippocampus and cerebellum [12]. At the subcellular level, PRMT8 can be localized to the plasma membrane through its unique N-terminal myristoylation motif [2,11]. The N-terminal 20 amino acids of PRMT8 are responsible for plasma membrane targeting by combining its myristoylation with the basic amino acids, and oligomerization/dimerization of PRMT8 enhances the membrane localization [13]. Interestingly, PRMT8 is also present at neuronal synapses [14,15], which are the specialized cellular compartments where neurotransmission between individual neurons takes place. Current evidence suggests that the local methylation of substrates by PRMT8 near the synapse is crucial for the development and plasticity of neuronal connections, which are pivotal to cognitive functions of the brain (see below).

While studies have mostly focused on the neuronal function of PRMT8 in the brain, PRMT8 might be important in glial cells as well, especially in the formation of glioblastoma. Reduced transcript expression PRMT8 is observed in glioblastoma patient tissues, suggesting its downregulation during tumor development in the brain [16]. Moreover, PRMT8 depletion in mouse embryonic stem cells (ESCs) has increased the expression of cellular markers which are associated with gliomagenesis [17]. However, the mechanism underlying PRMT8 in the pathogenesis of glioblastoma is still unclear. Intriguingly, somatic mutations and altered expression of the *PRMT8* gene have been found in cancer cells outside the brain [18–20]. The non-neuronal function of PRMT8, especially in the context of cancer progression, warrants further investigation.

3. Enzymatic Properties of PRMT8

3.1. Regulation and Crosstalk with Other PRMTs

The methyltransferase activity of PRMTs can be regulated in multiple ways. For example, PRMT activity can be regulated by auto-methylation. An auto-inhibitory mechanism for PRMT1, 4, 6, and 8 has been described [11,21]. For PRMT8, its enzymatic activity is negatively regulated by the first 60 amino acid residues at the N terminus [11], which contains two auto-methylation sites. It has been proposed that upon methylation of its substrates, the availability of fewer unmethylated substrates increases PRMT8 auto-methylation. This causes its N-terminus to bind to the enzyme core and prevents the methyl donor AdoMet from entering the catalytic site, thereby reducing PRMT8 activity [22]. The activity of PRMT can also be modulated through interacting with regulatory proteins that can either activate, inhibit, or change the specificity of PRMTs substrates [23,24]. In this regard, it is noteworthy that PRMT8 binds to Fyn and p85 through its proline-rich sequences in the N-terminal region [11]. Given that the N-terminal of PRMT8 is involved in its own auto-inhibition, the binding of these proteins to the N-terminal region may release the auto-inhibition of PRMT8 [11], a possibility that requires further investigation.

Two other PRMTs, namely PRMT1 and PRMT2, have been found to interact with PRMT8. The formation of heterodimers between PRMT8 with PRMT1 may recruit PRMT1 activity to the plasma membrane [2]. PRMT8 can also interact with the SH3 domain of PRMT2 through the two proline-rich sequences in the N-terminal region [11]. However, the precise function of the interaction or cross-talk between PRMT8 and other family members remains unclear.

There are pharmacological inhibitors of the PRMT family of enzymes. These include simple SAM analogs sinefungin, SAH, methylthioadenosine (MTA), and AzaAdoMet, which are often used as tools to change the global methylation levels in cells [25]. However, these are Pan-MTase inhibitors and do not target specific PRMT family members. Up to now, specific inhibitors have been developed for PRMT1, PRMT3, PRMT4, PRMT5, PRMT6, and PRMT7 [3,25]. Although several compounds can inhibit PRMT8, their action also extends to other PRMTs such as PRMT1, PRMT4, PRMT5, and PRMT6 [25]. Therefore, a specific PRMT8 inhibitor has yet to be identified.

3.2. Substrates of PRMT8

As PRMT8 is structurally most similar to PRMT1, they may share similar catalytic activities and substrate specificities. Both enzymes can methylate the same substrates, such as Npl3, GAR, and histone H4, in vitro [2]. Nonetheless, relatively few substrates of PRMT8 have been identified.

The pro-oncoprotein Ewing sarcoma protein (EWS) has been reported to be a PRMT8 substrate. The interaction between EWS and PRMT8 is mediated through the C-terminal RGG3 domain of EWS and the AdoMet binding domain of PRMT8 in the N-terminal region [26–29]. However, the precise role of methylation by PRMT8 towards regulating EWS function remains unknown. The nucleolar protein interacting with the fork head-associated (FHA) domain of Ki-67 (NIFK), a RNA binding protein, has also been demonstrated to act as a PRMT8 substrate by in vitro methylation, and methylation of NIFK is required for large subunit ribosomal RNA maturation [30].

Voltage-gated sodium channel (Nav), which functions to initiate action potentials in neurons, contains methylation sites and is a putative PRMT8 substrate. Although the PRMT that methylates Nav in vivo has not been identified, co-expression of the major brain Nav channel Nav1.2 with PRMT8 causes a striking 3-fold increase in the Nav1.2 current [31], indicating Nav1.2 as a potential candidate of PRMT8 substrate. Notably, increased methylation of Nav1.2 has been detected in mouse brain after acute seizures induced by kainic acid. As a consequence, sodium channel function is altered, which affects neuronal excitability. It would be important to investigate in the future whether alteration of PRMT8 expression or activity and the subsequent changes in arginine methylation of sodium channels may occur in epileptic patients.

The RasGAP SH3 domain-binding protein 1 (G3BP1), an RNA-binding protein crucial for the formation of stress granules that limit protein synthesis during cellular stress, such as oxidative stress, ultraviolet radiation and viral infection, is a substrate of PRMT8 [15,32]. In non-neuronal cells, the arginine methylation of G3BP1 prevents the assembly of stress granules in response to oxidative stress [32]. On the other hand, the effect of PRMT8 in synapse maturation of neurons is mediated, at least in part, through the methylation of G3BP1 and modulation of synaptic actin dynamics [15].

4. Neuronal Functions of PRMT8

4.1. PRMT8 Functions as a Phospholipase to Regulate Purkinje Cell Dendritic Arborization

Among the different PRMTs, PRMT8 is the only family member that possesses both methyltransferase and phospholipase activities. The phospholipase activity of PRMT8 directly catalyzes the hydrolysis of phosphatidylcholine (PC) to generate choline and phosphatidic acid (PA) [10]. PC is a major component of biological membranes and is dynamically regulated through hydrolysis and biosynthesis [33,34]. The metabolism of PC is highly involved in multiple morphogenetic processes during neuron development, such as axonal outgrowth, elongation, and neurite branching [35,36]. PC modulates neuronal functions from two aspects. First, PC-hydrolyzed choline is converted to acetylcholine, a major neurotransmitter that regulates multiple brain functions and animal behaviors [37]. Second, phosphatidic acid modulates morphological changes in neurons through its action on cytoskeleton remodeling and plasma membrane rearrangement [38]. Like other eukaryotic phospholipase D (PLDs), PRMT8 has a typical catalytic HxKxxxxD (HKD) motif which is unique among all PRMTs, suggesting the role of PRMT8 as a phospholipase [39,40]. Indeed, aberrant reduction of acetylcholine and choline levels, as well as increased PC levels, are detected in the cerebellum of $Prmt8^{-/-}$ mice [10]. Using MALDI-QIT-TOF/MS, Kim et al. further identify lysine-107 on PRMT8 as the essential amino acid residue for its phospholipase activity in vitro, and its PC hydrolysis activity promotes neurite branching in PC12 cells upon treatment with nerve growth factor (NGF) [10,41].

Purkinje cells located in the cerebellar cortex have highly elaborate dendritic trees, whose morphological changes are closely related to animal motor performance [42]. Purkinje cells are well known for their critical role in maintaining cerebellar functions including motor coordination and attention [42]. Dendritic arborization of Purkinje cells is regulated by multiple signaling processes including external cues and internal molecular pathways such as PC metabolism [43,44]. In the mouse brain, PRMT8 has predominant expression in the descending axons and dendritic arbors of Purkinje cells along development [12]. In $Prmt8$ homozygous knockout mice, Purkinje cells display stunted dendritic trees and reduced dendritic arborization [10]. Moreover, $Prmt8^{-/-}$ mice exhibit increased spontaneous behavioral hyperactivity and gait abnormalities, supporting an essential role of PRMT8 in cerebellar-related functions [10]. Numerous studies have linked PLD activity to brain development and functions. PLD-deficient mice have delayed brain growth and impaired cognitive functions [45]. Moreover, PLDs are implicated in Alzheimer's disease (AD) with reduced catalytic activities in neurons carrying familial Alzheimer's disease-related PS1 mutation [46]. Given the fact that PRMT8 has dual enzymatic activities of phospholipase and arginine methyltransferase, further studies into the crosstalk between the dual catalytic activities of PRMT8 will provide more information on how PRMT8 regulates neuronal functions and its relationship to neurological diseases [10,47].

4.2. PRMT8 Regulates Synaptic Plasticity and Cognitive Functions

Our brain function relies on communication between neurons through neurotransmission at specific junctions called synapses. Among all the PRMTs, PRMT8 is one of the few that show relatively enriched expression in the mouse synaptoneurosome, a biochemical preparation of synaptic components from the brain [14,15]. Upon N-terminal myristylation, PRMT8 can be membrane-bound [2,4], indicating the possibility that it may be targeted to the synapse. Based on these unique properties, PRMT8 might represent an important regulator in synaptic function. Indeed, Penny et al. have demonstrated that PRMT8 is localized in both pre- and post-synaptic compartments of cultured neurons, whereas relatively little PRMT8 signal is detected in the nuclear fraction of the brain [14]. Similar observations on the synaptic localization and low abundance in the nucleus was confirmed in a later study [15], suggesting the potential role of PRMT8 outside the nucleus, especially at the synapse. In mice with selective deletion of *Prmt8* in nestin-expressing cells, aberrant synaptic plasticity with increased miniature excitatory postsynaptic currents (mEPSC) frequency and amplitude in hippocampal slices has been observed, while no significant difference was detected in the miniature inhibitory postsynaptic currents (mIPSC), suggesting a specific role of PRMT8 in excitatory synapse function. Furthermore, the expression of several synaptic proteins, such as the NMDA receptor subunit GluN2A and eukaryotic translation initiation complex (eIF4E, eIF4G1, and eIF4H), is reduced in *Prmt8* conditional knockout (cKO) brain lysates and synaptosome, while their mRNA levels were not altered [14]. Thus, consistent with its prominent expression outside the neuronal nucleus, PRMT8 can regulate synaptic function through a post-transcriptional mechanism. In the rodent hippocampus, a brain area important for spatial memory, NMDAR is crucial for synaptic plasticity, which is a major cellular mechanism that underlies learning and memory [48,49]. GluN2A is one of the most common NMDAR regulatory subunits and GluN2A-containing NMDARs are enriched in the postsynaptic density (PSD) [50–52]. GluN2A is important in cognitive functions including contextual fear memory formation and spatial working memory [53–55]. Consistent with the altered level of GluN2A, the GluN2A-mediated NMDAR currents are reduced in hippocampal slices from *Prmt8* cKO mice. Consequently, the *Prmt8* cKO mice exhibit impaired contextual fear memory which might be due to the altered synaptic functions [14]. How PRMT8 affects expression of GluN2A and other synaptic proteins at the post-transcriptional level and in turn regulates synaptic function remains to be explored.

4.3. PRMT8 Regulates the Maturation of Synapse and Neural Circuit during Brain Development

In addition to the hippocampus, PRMT8 has been reported to influence the function of interneurons in the developing visual cortex. *Prmt8* ablation disrupts the proteome related to axonal and dendritic development, which might account for the reduced visual acuity and increased parvalbumin neuron complexity [56]. In this study, PRMT8 was found to modulate the transcription of gene encoding Tenascin-R, a component in the peri-neuronal net, thereby influencing dendritic arborization and synaptic functions.

Protein arginine methylation has well-documented roles in the nucleus, including gene transcription, RNA splicing, RNA export, and chromatin remodeling [57–59]. On the other hand, few studies have explored its significance in neurons outside the nucleus. The *Prmt8* mRNA is among the ~2000 transcripts present in the hippocampal neuropil of mouse brain [60], supporting a possible function of locally synthesized PRMT8 in a cellular compartment such as neuronal synapse. Interestingly, PRMT8 expression is substantially upregulated during postnatal days 7 to 21 in the mouse hippocampus [15], which correlates with the maturation of dendritic spines [61]. Dendritic spines are small protrusions on dendrites where most excitatory synapses locate. Dendritic spines are classified based on their morphologies. In general, the elongated filopodia which are prominent in the developing neurons are motile and transient, while the mushroom spines with large heads are more stable and involved in memory consolidation [62]. Filopodia density reaches the peak at an early development stage and then begins to decline, with more mushroom spines being formed for synapse maturation. Loss of PRMT8 after introduction of short-hairpin RNA (shRNA) results in the overproduction of filopodia. Increased filopodia density is also observed in heterozygous or homozygous *Prmt8* knockout neuron [15]. Several studies have reported the association of defective spine maturation with abnormal animal behaviors [63–65]. Indeed, the *Prmt8*-deficient mice display altered anxiety levels in open field test and elevated plus maze test, while sociability is not affected [15]. Thus, PRMT8 is an important regulator of dendritic spine maturation, and PRMT8 deficiency results in selective abnormal animal behaviors.

PRTM8 is present at excitatory synapses and dendritic spines of cultured hippocampal neurons, and it promotes spine maturation through its arginine methyltransferase activity instead of phospholipase activity. Notably, the spine defects caused by the PRMT8 deficiency cannot be rescued by the nuclear-restricted PRMT8, indicating that PRMT8 acts outside the nucleus to promote dendritic spine maturation [15]. Dendritic spines are actin-enriched protrusions and their morphology is tightly controlled by remodeling of the actin cytoskeleton [66,67]. PRMT8 suppresses filopodia formation via the control of Rac1-PAK signaling, which regulates actin dynamics through phosphorylation of the actin-depolymerization factor cofilin. Furthermore, G3BP1, a dendritic RNA-binding protein, has been identified as the downstream substrate of PRMT8 in dendritic spine maturation [15]. G3BP1 is an essential component of stress granules, but little is known about the neuronal function of G3BP1 under normal physiological conditions in the absence of cellular stress [68]. Hippocampal neurons of *G3bp1* knockout mice show exaggerated long-term depression (LTD), indicating the crucial role of G3BP1 in synaptic plasticity [69]. Consistent with this notion, we found that G3BP1 is essential for dendritic spine maturation and actin remodeling, and these functions depend on arginine methylation within the RGG domain of G3BP1 [15]. The precise mechanism by which PRMT8 and G3BP1 regulate Rac1-PAK signaling requires further investigation. However, enhanced eIF4G in the translation initiation complex is observed in the brain upon *Prmt8* deficiency [15]. Since an increase in eIF4E-eIF4G interaction and subsequent elevation of protein synthesis can hyperactivate the Rac1-PAK signaling and impair dendritic spine maturation [70], it is tempting to speculate that the alteration of protein synthesis may contribute to defective actin dynamics in PRMT8- or G3BP1-depleted neurons.

A large number of arginine-methylated proteins have been discovered in the adult mouse brain by mass spectrometry. Notably, among the many putative substrates of PRMTs in the brain are synaptic proteins [71], implying that protein arginine methylation could represent a major post-translational modification in the regulation of synaptic function and PRMT8 may not be the only arginine methyltransferase involved. Indeed, emerging evidence supports the role of other PRMTs in dendritic spine formation and maturation. The coactivator-associated arginine methyltransferase 1 (CARM1/PRMT4) is present in the dendrite and is co-localized with the postsynaptic protein PSD-95. Loss of PRMT4 promotes mushroom spines formation in cultured hippocampal neurons, which is due to the increased number and size of PSD-95 and GluN2B subunits of NMDA receptor at the synapse [72]. On the other hand, PRMT3 modulates dendritic spine maturation in hippocampal neurons through the maintenance of BDNF-dependent local mRNA translation [73]. Furthermore, PRMT2 is crucial for the dendritic arborization of young neurons through the arginine methylation of Cobl, which is an actin nucleation factor [74]. Whether the same PRMT2-Cobl pathway in involved in dendritic spine maturation at later developmental stage remains to be determined. Altogether, these studies provide compelling evidence that multiple PRMTs are critical to neuronal development and synapse maturation that extend beyond their conventional functions as the regulators of transcription and RNA processing within the nucleus (Figure 2).

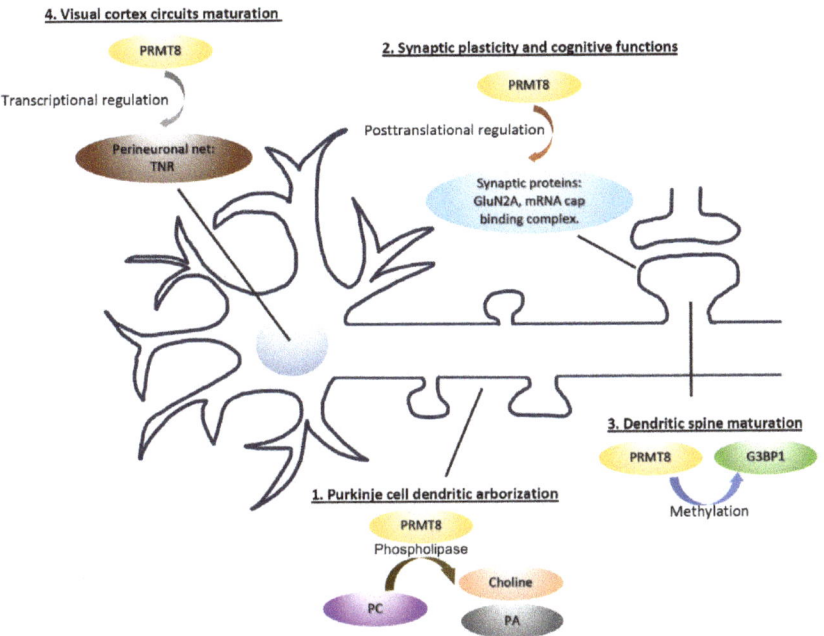

Figure 2. The cellular functions and molecular mechanisms of PRMT8 in neurons. PRMT8 can function both inside and outside the nucleus. The mechanisms of PRMT8 include (1) the production of choline and phosphatidic acid (PA) that promotes dendritic growth; (2) transcription in the nucleus to regulate perineuronal net expression; (3) the regulation of actin dynamics through the RNA-binding protein G3BP1; and (4) the synthesis of synaptic proteins essential for synaptic plasticity and memory formation.

4.4. PRMT8 and Neurological Diseases

Several recent studies have suggested the link between PRMT8 and neurological diseases. One type of motor neuron disease, amyotrophic lateral sclerosis (ALS), has a pathological hallmark of aggregated inclusion bodies in motor neurons. The RNA-binding proteins FUS and TDP-43 are the major components of inclusion bodies in the brain from ALS patients [75]. Various mutations have been identified in TDP-43 and FUS in FUS-TDP-43/ALS-FUS patients [76,77]. Remarkably, PRMT1 and PRMT8 are present in inclusion bodies of cultured COS-1 cells carrying ALS-linked FUS mutations. Furthermore, the FUS-positive inclusion bodies in motor neuron-derived (MN-1) cells and cells from ALS-FUS patient are reduced upon the inhibition of PRMT enzyme activity [78]. In contrast, in a *Drosophila* model carrying ALS-FUS mutations, depletion of endogenous arginine methyltransferase 1 (DART1), which is orthologous to human PRMT1 and PRMT8, enhanced the neurodegenerative phenotype introduced by FUS overexpression in fly eyes [78]. This study thus demonstrated the potential role of PRMT8 in ALS pathogenesis through the regulation of RNA-binding proteins. The importance of PRMT8 in motor neuron function has also been investigated in *Prmt8*-deficient mice. Aged mice (~15 months) lacking PRMT8 exhibit impaired muscle strength with weak limb clasping and muscle atrophy, which are attributed to the dysfunctions of motoneurons in the spinal cord [79]. Increasing DNA double-strand breaks and defective stress tolerance are also found in the motoneurons of *Prmt8*-deficient mice. The dysfunction of motoneurons is tightly related to neurodegenerative disorders [80]. Since PRMT8 influences motor behaviors through its phospholipase activity [10], PRMT8 might represent a potential target for drug discovery in delaying degeneration of motoneurons and treatment for MN-related neurological disorders.

In addition to MN-related degeneration disorders, given that PRMT8 modulates the visual cortical circuit as an epigenetic regulator [56], dendritic spine maturation as regulator of actin dynamics [15], and the control of context-dependent fear learning and anxiety behavior [14,15], it is possible that PRMT8 might also be involved in neurodevelopmental disorders. Indeed, a recent paper has reported that the *PRMT8* gene is located within a common deleted chromosome region from patients with microcephaly [81], suggesting its possible role in human brain development and its dysfunction in neurodevelopmental disorders.

5. Conclusions and Outlook

PRMT8 is unique among all PRMTs due to its brain-restricted expression, the membrane anchorage through myristoylation, and the dual enzymatic activities of methyltransferase and phospholipase. The importance of PRMT8 has now been demonstrated in diverse neuronal functions, including dendritic arborization, dendritic spine maturation, synaptic plasticity, motor performance, and visual acuity. Mass spectrometry has identified numerous cytoplasmic and membrane proteins being arginine-methylated in the brain [71]. These include proteins as diverse as ion channels on the plasma membrane [31] and molecular motors on microtubule [82]. However, relatively few studies have been carried out to elucidate the role of protein arginine methylation outside the nucleus. Therefore, recent studies on the non-nuclear role of PRMT8 in neurons have substantially extended our understanding of the mechanism of protein arginine methylation. Besides synapse development and function, other cellular processes in neurons, including axonal trafficking and dendritic branching, also depend on protein arginine methylation [74,83]. It is worthwhile to further characterize the function of PRMTs in neurons outside the nucleus in the future. In particular, given that PRMT8 can be attached to the plasma membrane, further identification and characterization of membrane proteins as PRMT8 substrates would be pivotal to gain deeper understanding on how this enzyme works in the brain.

Numerous RNA-binding proteins have been identified as arginine-methylated proteins in the cytoplasm [57,84]. Mutations and abnormal expression of RNA-binding proteins have been reported to be engaged in different neurological diseases, including FMRP, TAR DNA binding protein 43 (TDP-43), Hu proteins, and FUS [84–86]. Despite the observations that *Prmt8*-deficient mice exhibited selective abnormal behaviors in motor performance and anxiety test, the in vivo role of PRMT8 in brain function has not yet been well defined, and its precise linkage with diseases is largely unexplored. In addition to G3BP1, searching for other RNA-binding proteins that are present in dendrite and synapses as the downstream targets of PRMT8 would be critical for illuminating the multifaceted functions of PRMT8 in neurodegeneration and neurodevelopmental disorders in the future.

Author Contributions: R.D. and X.L. drafted the manuscript and made the schematic diagrams. K.-O.L. planned the content and organization of the review and revised the manuscript. All authors have read and agreed to the published version of the manuscript.

Funding: This study was supported in part by the Research Grant Council of Hong Kong [General Research Fund (GRF) 17106018 and 17117720]; the Area of Excellence Scheme (Grant AoE/M-604/16) and the Theme-based Research Scheme (Grant T13-605/18-W) of the University Grants Committee of Hong Kong; the Health and Medical Research Fund (06172986) of the Food and Health Bureau of the Hong Kong SAR Government.

Informed Consent Statement: Not applicable.

Conflicts of Interest: The authors declare no conflict of interest.

References

1. Paik, W.K.; Paik, D.C.; Kim, S. Historical review: The field of protein methylation. *Trends Biochem. Sci.* **2007**, *32*, 146–152. [CrossRef] [PubMed]
2. Lee, J.; Sayegh, J.; Daniel, J.; Clarke, S.; Bedford, M.T. PRMT8, a new membrane-bound tissue-specific member of the protein arginine methyltransferase family. *J. Biol. Chem.* **2005**, *280*, 32890–32896. [CrossRef] [PubMed]

3. Hwang, J.W.; Cho, Y.; Bae, G.U.; Kim, S.N.; Kim, Y.K. Protein arginine methyltransferases: Promising targets for cancer therapy. *Exp. Mol. Med.* **2021**, *53*, 788–808. [CrossRef] [PubMed]
4. Bedford, M.T.; Clarke, S.G. Protein arginine methylation in mammals: Who, what, and why. *Mol. Cell* **2009**, *33*, 1–13. [CrossRef] [PubMed]
5. Qualmann, B.; Kessels, M.M. The Role of Protein Arginine Methylation as Post-Translational Modification on Actin Cytoskeletal Components in Neuronal Structure and Function. *Cells* **2021**, *10*, 1079. [CrossRef] [PubMed]
6. Swiercz, R.; Person, M.D.; Bedford, M.T. Ribosomal protein S2 is a substrate for mammalian PRMT3 (protein arginine methyltransferase 3). *Biochem. J.* **2005**, *386*, 85–91. [CrossRef]
7. Frankel, A.; Clarke, S. PRMT3 is a distinct member of the protein arginine N- methyltransferase family. conferral of substrate specificity by a zinc-finger domain. *J. Biol. Chem.* **2000**, *275*, 32974–32982. [CrossRef] [PubMed]
8. Antonysamy, S.; Bonday, Z.; Campbell, R.M.; Doyle, B.; Druzina, Z.; Gheyi, T.; Han, B.; Jungheim, L.N.; Qian, Y.; Rauch, C.; et al. Crystal structure of the human PRMT5:MEP50 complex. *Proc. Natl. Acad. Sci. USA* **2012**, *109*, 17960–17965. [CrossRef]
9. Ho, M.-C.; Wilczek, C.; Bonanno, J.B.; Xing, L.; Seznec, J.; Matsui, T.; Carter, L.G.; Onikubo, T.; Kumar, P.R.; Chan, M.K.; et al. Structure of the arginine methyltransferase PRMT5-MEP50 reveals a mechanism for substrate specificity. *PLoS ONE* **2013**, *8*, 57008–57023. [CrossRef]
10. Kim, J.D.; Park, K.E.; Ishida, J.; Kako, K.; Hamada, J.; Kani, S.; Takeuchi, M.; Namiki, K.; Fukui, H.; Fukuhara, S.; et al. PRMT8 as a phospholipase regulates Purkinje cell dendritic arborization and motor coordination. *Sci. Adv.* **2015**, *1*, e1500615. [CrossRef]
11. Sayegh, J.; Webb, K.; Cheng, D.; Bedford, M.T.; Clarke, S.G. Regulation of protein arginine methyltransferase 8 (PRMT8) activity by its N-terminal domain. *J. Biol. Chem.* **2007**, *282*, 36444–36453. [CrossRef] [PubMed]
12. Kousaka, A.; Mori, Y.; Koyama, Y.; Taneda, T.; Miyata, S.; Tohyama, M. The distribution and characterization of endogenous protein arginine N-methyltransferase 8 in mouse CNS. *Neuroscience* **2009**, *163*, 1146–1157. [CrossRef] [PubMed]
13. Park, S.W.; Jun, Y.W.; Choi, H.E.; Lee, J.A.; Jang, D.J. Deciphering the molecular mechanisms underlying the plasma membrane targeting of PRMT8. *BMB Rep.* **2019**, *52*, 601–606. [CrossRef]
14. Penney, J.; Seo, J.; Kritskiy, O. Loss of Protein Arginine Methyltransferase 8 Alters Synapse Composition and Function, Resulting in Behavioral Defects. *J. Neurosci.* **2017**, *37*, 8655–8666. [CrossRef] [PubMed]
15. Lo, L.H.-Y.; Dong, R.; Lyu, Q.; Lai, K.O. The Protein Arginine Methyltransferase PRMT8 and Substrate G3BP1 Control Rac1-PAK1 Signaling and Actin Cytoskeleton for Dendritic Spine Maturation. *Cell Rep.* **2020**, *31*, 107744. [CrossRef]
16. Samuel, S.F.; Barry, A.; Greenman, J.; Beltran-Alvarez, P. Arginine methylation: The promise of a 'silver bullet' for brain tumours? *Amino Acids* **2021**, *53*, 489–506. [CrossRef]
17. Simandi, Z.; Czipa, E.; Horvath, A.; Koszeghy, A.; Bordas, C.; Póliska, S.; Juhász, I.; Imre, L.; Szabó, G.; Dezso, B.; et al. PRMT1 and PRMT8 regulate retinoic acid-dependent neuronal differentiation with implications to neuropathology. *Stem Cells* **2015**, *33*, 726–741. [CrossRef]
18. Yang, Y.; Bedford, M.T. Protein arginine methyltransferases and cancer. *Nat. Rev. Cancer* **2013**, *13*, 37–50. [CrossRef] [PubMed]
19. Hernandez, S.J.; Dolivo, D.M.; Dominko, T. PRMT8 demonstrates variant-specific expression in cancer cells and correlates with patient survival in breast, ovarian and gastric cancer. *Oncol. Lett.* **2017**, *13*, 1983–1989. [CrossRef]
20. Lin, H.; Wang, B.; Yu, J.; Wang, J.; Li, Q.; Cao, B. Protein arginine methyltransferase 8 gene enhances the colon cancer stem cell (CSC) function by upregulating the pluripotency transcription factor. *J. Cancer* **2018**, *9*, 1394–1402. [CrossRef]
21. Frankel, A.; Yadav, N.; Lee, J.; Branscombe, T.L.; Clarke, S.; Bedford, M.T. The novel human protein arginine N-methyltransferase PRMT6 is a nuclear enzyme displaying unique substrate specificity. *J. Biol. Chem.* **2002**, *277*, 3537–3543. [CrossRef] [PubMed]
22. Dillon, M.B.; Rust, H.L.; Thompson, P.R.; Mowen, K.A. Automethylation of protein arginine methyltransferase 8 (PRMT8) regulates activity by impeding S-adenosylmethionine sensitivity. *J. Biol. Chem.* **2013**, *288*, 27872–27880. [CrossRef] [PubMed]
23. Lin, W.J.; Gary, J.D.; Yang, M.C.; Clarke, S.; Herschman, H.R. The mammalian immediate-early TIS21 protein and the leukemia-associated BTG1 protein interact with a protein-arginine N- methyltransferase. *J. Biol. Chem.* **1996**, *271*, 15034–15044. [CrossRef] [PubMed]
24. Singh, V.; Miranda, T.B.; Jiang, W.; Frankel, A.; Roemer, M.E.; Robb, V.A.; Gutmann, D.H.; Herschman, H.R.; Clarke, S.; Newsham, I.F. DAL-1/4.1B tumor suppressor interacts with protein arginine N-methyltransferase 3 (PRMT3) and inhibits its ability to methylate substrates in vitro and in vivo. *Oncogene* **2004**, *23*, 7761–7771. [CrossRef]
25. Hu, H.; Qian, K.; Ho, M.C.; Zheng, Y.G. Small Molecule Inhibitors of Protein Arginine Methyltransferases. *Expert Opin. Investig. Drugs* **2016**, *25*, 335–358. [CrossRef]
26. Kim, J.D.; Kako, K.; Kakiuchi, M.; Park, G.G.; Fukamizu, A. EWS is a substrate of type I protein arginine methyltransferase, PRMT8. *Int. J. Mol. Med.* **2008**, *22*, 309–315. [CrossRef]
27. Burd, C.G.; Dreyfuss, G. Conserved structures and diversity of functions of RNA-binding proteins. *Science* **1994**, *265*, 615–621. [CrossRef]
28. Najbauer, J.; Johnson, B.A.; Young, A.L.; Aswad, D.W. Peptides with sequences similar to glycine, arginine-rich motifs in proteins interacting with RNA are efficiently recognized by methyltransferase(s) modifying arginine in numerous proteins. *J. Biol. Chem.* **1993**, *268*, 10501–10509. [CrossRef]
29. Gary, J.D.; Clarke, S. RNA and protein interactions modulated by protein arginine methylation. *Prog. Nucl. Acid Res. Mol. Biol.* **1998**, *61*, 65–131.

30. Lee, W.C.; Lin, W.L.; Matsui, T.; Chen, E.S.; Wei, T.Y.; Lin, W.H.; Hu, H.; Zheng, Y.G.; Tsai, M.D.; Ho, M.C. Protein Arginine Methyltransferase 8: Tetrameric Structure and Protein Substrate Specificity. *Biochemistry* **2015**, *54*, 7514–7523. [CrossRef]
31. Baek, J.H.; Rubinstein, M.; Scheuer, T.; Trimmer, J.S. Reciprocal changes in phosphorylation and methylation of mammalian brain sodium channels in response to seizures. *J. Biol. Chem.* **2014**, *289*, 15363–15373. [CrossRef]
32. Tsai, W.C.; Gayatri, S.; Reineke, L.C.; Sbardella, G.; Bedford, M.T.; Lloyd, R.E. Arginine Demethylation of G3BP1 Promotes Stress Granule Assembly. *J. Biol. Chem.* **2016**, *291*, 22671–22685. [CrossRef]
33. Carter, J.M.; Demizieux, L.; Campenot, R.B. Phosphatidylcholine Biosynthesis via CTP: Phosphocholine Cytidylyltransferase β2 Facilitates Neurite Outgrowth and Branching. *J. Biol. Chem.* **2008**, *283*, 202–212. [CrossRef]
34. Gould, R.M.; Connell, F.; Spivack, W. Phospholipid Metabolism in Mouse Sciatic Nerve In Vivo. *J. Neurochem.* **1987**, *48*, 853–859. [CrossRef] [PubMed]
35. Kanaho, Y.; Funakoshi, Y.; Hasegawa, H. Phospholipase D signalling and its involvement in neurite outgrowth. *Biochim. Biophys. Acta BBA Mol. Cell Biol. Lipids* **2009**, *1791*, 898–904. [CrossRef] [PubMed]
36. Vance, J.E.; de Chaves, E.P.; Campenot, R.B.; Vance, D.E. Role of axons in membrane phospholipid synthesis in rat sympathetic neurons. *Neurobiol. Aging* **1995**, *16*, 493–498. [CrossRef]
37. Picciotto, M.R.; Higley, M.J.; Mineur, Y.S. Acetylcholine as a neuromodulator: Cholinergic signaling shapes nervous system function and behavior. *Neuron* **2012**, *76*, 116–129. [CrossRef] [PubMed]
38. Ammar, M.-R.; Kassas, N.; Bader, M.-F.; Vitale, N. Phosphatidic acid in neuronal development: A node for membrane and cytoskeleton rearrangements. *Biochimie* **2014**, *107*, 51–57. [CrossRef]
39. Selvy, P.E.; Lavieri, R.R.; Lindsley, C.W.; Brown, H.A. Phospholipase D—Enzymology, functionality, and chemical modulation. *Chem. Rev.* **2011**, *111*, 6064–6119. [CrossRef] [PubMed]
40. Ponting, C.P.; Kerr, I.D. A novel family of phospholipase D homologues that includes phospholipid synthases and putative endonucleases: Identification of duplicated repeats and potential active site residues. *Protein Sci.* **1996**, *5*, 914–922. [CrossRef]
41. Park, K.-E.; Kim, J.-D.; Nagashima, Y.; Kako, K.; Daitoku, H.; Matsui, M.; Park, G.G.; Fukamizu, A. Detection of choline and phosphatidic acid (PA) catalyzed by phospholipase D (PLD) using MALDI-QIT-TOF/MS with 9-aminoacridine matrix. *Biosci. Biotechnol. Biochem.* **2014**, *78*, 981–988. [CrossRef]
42. Lee, K.J.; Jung, J.G.; Arii, T.; Imoto, K.; Rhyu, I.J. Morphological changes in dendritic spines of Purkinje cells associated with motor learning. *Neurobiol. Learn. Mem.* **2007**, *88*, 445–450. [CrossRef] [PubMed]
43. Cerminara, N.L.; Lang, E.J.; Sillitoe, R.V.; Apps, R. Redefining the cerebellar cortex as an assembly of non-uniform Purkinje cell microcircuits. *Nat. Rev. Neurosci.* **2005**, *16*, 79–93. [CrossRef] [PubMed]
44. Jan, Y.-N.; Jan, L.Y. The Control of Dendrite Development. *Neuron* **2003**, *40*, 229–242. [CrossRef]
45. Burkhardt, U.; Stegner, D.; Hattingen, E.; Beyer, S.; Nieswandt, B.; Klein, J. Impaired brain development and reduced cognitive function in phospholipase D-deficient mice. *Neurosci. Lett.* **2014**, *572*, 48–52. [CrossRef]
46. Cai, D.; Zhong, M.; Wang, R.; Netzer, W.J.; Shields, D.; Zheng, H.; Sisodia, S.S.; Foster, D.A.; Gorelick, F.S.; Xu, H.; et al. Phospholipase D1 corrects impaired βAPP trafficking and neurite outgrowth in familial Alzheimer's disease-linked presenilin-1 mutant neurons. *Proc. Natl. Acad. Sci. USA* **2006**, *103*, 1936–1940. [CrossRef]
47. Lin, Y.; Tsai, Y.-J.; Liu, Y.-F.; Cheng, Y.-C.; Hung, C.-M.; Lee, Y.-J.; Pan, H.; Li, C. The Critical Role of Protein Arginine Methyltransferase prmt8 in Zebrafish Embryonic and Neural Development Is Non-Redundant with Its Paralogue prmt1. *PLoS ONE* **2013**, *8*, e55221. [CrossRef]
48. Morris, R.G.M.; Anderson, E.; Lynch, G.S.; Baudry, M. Selective impairment of learning and blockade of long-term potentiation by an N-methyl-D-aspartate receptor antagonist, AP5. *Nature* **1986**, *319*, 774–776. [CrossRef]
49. Tsien, J.Z.; Huerta, P.T.; Tonegawa, S. The Essential Role of Hippocampal CA1 NMDA Receptor–Dependent Synaptic Plasticity in Spatial Memory. *Cell* **1996**, *87*, 1327–1338. [CrossRef]
50. Akazawa, C.; Shigemoto, R.; Bessho, Y.; Nakanishi, S.; Mizuno, N. Differential expression of five N-methyl-D-aspartate receptor subunit mRNAs in the cerebellum of developing and adult rats. *J. Comp. Neurol.* **1994**, *347*, 150–160. [CrossRef]
51. Monyer, H.; Burnashev, N.; Laurie, D.J.; Sakmann, B.; Seeburg, P.H. Developmental and regional expression in the rat brain and functional properties of four NMDA receptors. *Neuron* **1994**, *12*, 529–540. [CrossRef]
52. Paoletti, P.; Bellone, C.; Zhou, Q. NMDA receptor subunit diversity: Impact on receptor properties, synaptic plasticity and disease. *Nat. Rev. Neurosci.* **2013**, *14*, 383–400. [CrossRef] [PubMed]
53. Baez, M.V.; Oberholzer, M.V.; Cercato, M.C.; Snitcofsky, M.; Aguirre, A.I.; Jerusalinsky, D.A. NMDA receptor subunits in the adult rat hippocampus undergo similar changes after 5 minutes in an open field and after LTP induction. *PLoS ONE* **2013**, *8*, e55244. [CrossRef] [PubMed]
54. Bannerman, D.M.; Niewoehner, B.; Lyon, L.; Romberg, C.; Schmitt, W.B.; Taylor, A.; Sanderson, D.J.; Cottam, J.; Sprengel, R.; Seeburg, P.H.; et al. NMDA receptor subunit NR2A is required for rapidly acquired spatial working memory but not incremental spatial reference memory. *J. Neurosci.* **2008**, *28*, 3623–3630. [CrossRef]
55. Zhang, X.-M.; Yan, X.-Y.; Zhang, B.; Yang, Q.; Ye, M.; Cao, W.; Qiang, W.-B.; Zhu, L.-J.; Du, Y.-L.; Xu, X.-X.; et al. Activity-induced synaptic delivery of the GluN2A-containing NMDA receptor is dependent on endoplasmic reticulum chaperone Bip and involved in fear memory. *Cell Res.* **2015**, *25*, 818–836. [CrossRef]
56. Lee, P.K.M.; Goh, W.W.B.; Sng, J.C.G. Network-based characterization of the synaptic proteome reveals that removal of epigenetic regulator Prmt8 restricts proteins associated with synaptic maturation. *J. Neurochem.* **2017**, *140*, 613–628. [CrossRef]

57. Bedford, M.T.; Richard, S. Arginine methylation an emerging regulator of protein function. *Mol. Cell* **2005**, *18*, 263–272. [CrossRef]
58. Dong, X.; Weng, Z. The correlation between histone modifications and gene expression. *Epigenomics* **2013**, *5*, 113–116. [CrossRef]
59. Poulard, C.; Corbo, L.; Le Romancer, M. Protein arginine methylation/demethylation and cancer. *Oncotarget* **2016**, *7*, 67532–67550. [CrossRef]
60. Cajigas, I.J.; Tushev, G.; Will, T.J.; Dieck, S.T.; Fuerst, N.; Schuman, E.M. The local transcriptome in the synaptic neuropil revealed by deep sequencing and high-resolution imaging. *Neuron* **2012**, *74*, 453–466. [CrossRef]
61. Fiala, J.C.; Feinberg, M.; Popov, V.; Harris, K.M. Synaptogenesis Via Dendritic Filopodia in Developing Hippocampal Area CA1. *J. Neurosci.* **1998**, *18*, 8900–8911. [CrossRef] [PubMed]
62. Bourne, J.N.; Harris, K.M. Balancing structure and function at hippocampal dendritic spines. *Annu. Rev. Neurosci.* **2008**, *31*, 47–67. [CrossRef] [PubMed]
63. Dansie, L.; Phommahaxay, K.; Okusanya, A.; Uwadia, J.; Huang, M.; Rotschafer, S.; Razak, K.; Ethell, D.; Ethell, I. Long-lasting effects of minocycline on behavior in young but not adult Fragile X mice. *Neuroscience* **2013**, *246*, 186–198. [CrossRef] [PubMed]
64. Hutsler, J.J.; Zhang, H. Increased dendritic spine densities on cortical projection neurons in autism spectrum disorders. *Brain Res.* **2010**, *1309*, 83–94. [CrossRef]
65. Wu, H.; Cottingham, C.; Chen, L.; Wang, H.; Che, P.; Liu, K.; Wang, Q. Age-dependent differential regulation of anxiety- and depression-related behaviors by neurabin and spinophilin. *PLoS ONE* **2017**, *12*, e0180638. [CrossRef] [PubMed]
66. Cheadle, L.; Biederer, T. The novel synaptogenic protein Farp1 links postsynaptic cytoskeletal dynamics and transsynaptic organization. *J. Cell Biol.* **2012**, *199*, 985–1001. [CrossRef] [PubMed]
67. Hotulainen, P.; Llano, O.; Smirnov, S.; Tanhuanpää, K.; Faix, J.; Rivera, C.; Lappalainen, P. Defining mechanisms of actin polymerization and depolymerization during dendritic spine morphogenesis. *J Cell Biol.* **2009**, *185*, 323–339. [CrossRef]
68. Matsuki, H.; Takahashi, M.; Higuchi, M.; Makokha, G.N.; Oie, M.; Fujii, M. Both G3BP1 and G3BP2 contribute to stress granule formation. *Genes Cells* **2013**, *18*, 135–146. [CrossRef]
69. Martin, S.; Bellora, N.; González-Vallinas, J.; Irimia, M.; Chebli, K.; de Toledo, M.; Raabe, M.; Eyras, E.; Urlaub, H.; Blencowe, B.J.; et al. Preferential binding of a stable G3BP ribonucleoprotein complex to intron-retaining transcripts in mouse brain and modulation of their expression in the cerebellum. *J. Neurochem.* **2016**, *139*, 349–368. [CrossRef]
70. Santini, E.; Huynh, T.N.; Longo, F.; Koo, S.Y.; Mojica, E.; D'Andrea, L.; Bagni, C.; Klann, E. Reducing eIF4E-eIF4G interactions restores the balance between protein synthesis and actin dynamics in fragile X syndrome model mice. *Sci. Signal.* **2017**, *10*. [CrossRef]
71. Guo, A.; Gu, H.; Zhou, J.; Mulhern, D.; Wang, Y.; Lee, K.A.; Yang, V.; Aguiar, M.; Kornhauser, J.; Jia, X.; et al. Immunoaffinity Enrichment and Mass Spectrometry Analysis of Protein Methylation. *Mol. Cell. Proteom.* **2014**, *13*, 372–387. [CrossRef] [PubMed]
72. Lim, C.S.; Alkon, D.L. Inhibition of coactivator-associated arginine methyltransferase 1 modulates dendritic arborization and spine maturation of cultured hippocampal neurons. *J. Biol. Chem.* **2017**, *292*, 6402–6413. [CrossRef] [PubMed]
73. Miyata, S.; Mori, Y.; Tohyama, M. PRMT3 is essential for dendritic spine maturation in rat hippocampal neurons. *Brain Res.* **2010**, *1352*, 11–20. [CrossRef]
74. Hou, W.; Nemitz, S.; Schopper, S.; Nielsen, M.; Kessels, M.M.; Qualmann, B. Arginine Methylation by PRMT2 Controls the Functions of the Actin Nucleator Cobl. *Dev. Cell* **2018**, *45*, 262–275. [CrossRef]
75. Zhou, H.; Mangelsdorf, M.; Liu, J.; Zhu, L.; Wu, J.Y. RNA-binding proteins in neurological diseases. *Sci. China Life Sci.* **2014**, *57*, 432–444. [CrossRef]
76. Kabashi, E.; Valdmanis, P.; Dion, P.; Spiegelman, D.; McConkey, B.J.; Velde, C.V.; Bouchard, J.-P.; Lacomblez, L.; Pochigaeva, K.; Salachas, F.; et al. TARDBP mutations in individuals with sporadic and familial amyotrophic lateral sclerosis. *Nat. Genet.* **2008**, *40*, 572–574. [CrossRef]
77. Vance, C.; Rogelj, B.; Hortobágyi, T.; De Vos, K.J.; Nishimura, A.L.; Sreedharan, J.; Hu, X.; Smith, B.; Ruddy, D.; Wright, P.; et al. Mutations in FUS, an RNA Processing Protein, Cause Familial Amyotrophic Lateral Sclerosis Type 6. *Science* **2009**, *323*, 1208–1211. [CrossRef] [PubMed]
78. Scaramuzzino, C.; Monaghan, J.; Milioto, C.; Lanson, N.A., Jr.; Maltare, A.; Aggarwal, T.; Casci, I.; Fackelmayer, F.O.; Pennuto, M.; Pandey, U.B. Protein arginine methyltransferase 1 and 8 interact with FUS to modify its sub-cellular distribution and toxicity in vitro and *in vivo*. *PLoS ONE* **2013**, *8*, e61576. [CrossRef]
79. Simandi, Z.; Pajer, K.; Karolyi, K.; Sieler, T.; Jiang, L.-L.; Kolostyak, Z.; Sari, Z.; Fekecs, Z.; Pap, A.; Patsalos, A.; et al. Arginine Methyltransferase PRMT8 Provides Cellular Stress Tolerance in Aging Motoneurons. *J. Neurosci.* **2018**, *38*, 7683–7700. [CrossRef] [PubMed]
80. Shaw, P.J. Molecular and cellular pathways of neurodegeneration in motor neurone disease. *J. Neurol. Neurosurg. Psychiatry* **2005**, *76*, 1046–1057. [CrossRef] [PubMed]
81. Rincic, M.; Rados, M.; Kopic, J.; Krsnik, Z.; Liehr, T. 7p21.3 Together With a 12p13.32 Deletion in a Patient with Microcephaly—Does 12p13.32 Locus Possibly Comprises a Candidate Gene Region for Microcephaly? *Front. Mol. Neurosci.* **2021**, *14*, 2. [CrossRef] [PubMed]
82. Zhao, J.; Fok, A.H.K.; Fan, R.; Kwan, P.Y.; Chan, H.L.; Lo, L.H.; Chan, Y.S.; Yung, W.H.; Huang, J.; Lai, C.S.W.; et al. Specific depletion of the motor protein KIF5B leads to deficits in dendritic transport, synaptic plasticity and memory. *eLife* **2020**, *9*, e53456. [CrossRef] [PubMed]

83. Migazzi, A.; Scaramuzzino, C.; Anderson, E.N. Huntingtin-mediated axonal transport requires arginine methylation by PRMT6. *Cell Rep.* **2021**, *35*, 108980. [CrossRef] [PubMed]
84. Darnell, J.C.; Richter, J.D. Cytoplasmic RNA-Binding Proteins and the Control of Complex Brain Function. *Cold Spring Harb. Perspect. Biol.* **2012**, *4*, a012344. [CrossRef] [PubMed]
85. Dolzhanskaya, N.; Merz, G.; Aletta, J.M.; Denman, R.B. Methylation regulates the intracellular protein-protein and protein-RNA interactions of FMRP. *J. Cell Sci.* **2006**, *119*, 1933–1946. [CrossRef] [PubMed]
86. Mackenzie, I.R.; Rademakers, R.; Neumann, M. TDP-43 and FUS in amyotrophic lateral sclerosis and frontotemporal dementia. *Lancet Neurol.* **2010**, *9*, 995–1007. [CrossRef]

Review

The Structure, Activity, and Function of the SETD3 Protein Histidine Methyltransferase

Apolonia Witecka [1], Sebastian Kwiatkowski [1], Takao Ishikawa [2,*] and Jakub Drozak [1,*]

1 Department of Metabolic Regulation, Institute of Biochemistry, Faculty of Biology, University of Warsaw, Miecznikowa 1, 02-096 Warsaw, Poland; aa.witecka@uw.edu.pl (A.W.); sp.kwiatkowski@uw.edu.pl (S.K.)
2 Department of Molecular Biology, Institute of Biochemistry, Faculty of Biology, University of Warsaw, Miecznikowa 1, 02-096 Warsaw, Poland
* Correspondence: t.ishikawa@uw.edu.pl (T.I.); jdrozak@biol.uw.edu.pl (J.D.); Tel.: +48-22-55-43-111 (T.I.); +48-22-55-43-222 (J.D.)

Abstract: SETD3 has been recently identified as a long sought, actin specific histidine methyltransferase that catalyzes the $N\tau$-methylation reaction of histidine 73 (H73) residue in human actin or its equivalent in other metazoans. Its homologs are widespread among multicellular eukaryotes and expressed in most mammalian tissues. SETD3 consists of a catalytic SET domain responsible for transferring the methyl group from S-adenosyl-L-methionine (AdoMet) to a protein substrate and a RuBisCO LSMT domain that recognizes and binds the methyl-accepting protein(s). The enzyme was initially identified as a methyltransferase that catalyzes the modification of histone H3 at K4 and K36 residues, but later studies revealed that the only bona fide substrate of SETD3 is H73, in the actin protein. The methylation of actin at H73 contributes to maintaining cytoskeleton integrity, which remains the only well characterized biological effect of SETD3. However, the discovery of numerous novel methyltransferase interactors suggests that SETD3 may regulate various biological processes, including cell cycle and apoptosis, carcinogenesis, response to hypoxic conditions, and enterovirus pathogenesis. This review summarizes the current advances in research on the SETD3 protein, its biological importance, and role in various diseases.

Keywords: SETD3; posttranslational modifications; protein histidine methylation; actin; polymerization; cytoskeleton; enteroviruses; oncogenesis

Citation: Witecka, A.; Kwiatkowski, S.; Ishikawa, T.; Drozak, J. The Structure, Activity, and Function of the SETD3 Protein Histidine Methyltransferase. *Life* **2021**, *11*, 1040. https://doi.org/10.3390/life11101040

Academic Editors: Albert Jeltsch and Arunkumar Dhayalan

Received: 13 August 2021
Accepted: 29 September 2021
Published: 2 October 2021

Publisher's Note: MDPI stays neutral with regard to jurisdictional claims in published maps and institutional affiliations.

Copyright: © 2021 by the authors. Licensee MDPI, Basel, Switzerland. This article is an open access article distributed under the terms and conditions of the Creative Commons Attribution (CC BY) license (https://creativecommons.org/licenses/by/4.0/).

1. Introduction

One of the most common posttranslational modifications that modulates the physicochemical properties of proteins and determines their functional diversity, is the transfer of a methyl group from S-adenosyl-L-methionine (AdoMet) to their specific amino acid residues [1]. The primary target sites of methylation are lysine and arginine. However, this process may also occur on other amino acids, namely, cysteine, glutamate, glutamine, and histidine [2]. Decades of research into lysine and arginine methylation on histone tails have led to a fairly good understanding of the importance of such modifications in the epigenetic regulation of gene expression. Furthermore, it has become clear over time that a large number of nonhistone proteins may also be methylated at lysine and arginine residues, which may affect cellular physiology in mammals [2]. On the other hand, our knowledge about the mechanisms and biological significance of methylation on "noncanonical" amino acids has remained surprisingly limited. This seems particularly true for protein histidine. Histidine methylation on the $N\pi$ or $N\tau$ atom of the imidazole ring has been known for many years, but the process has so far been studied in greater detail only for a few proteins, including actin [3], S100A9 [4], myosin [5], MLCK2 [6], and RPL3 [7] (Figure 1). This fact is also indicated by the slow progress of research on actin histidine methylation.

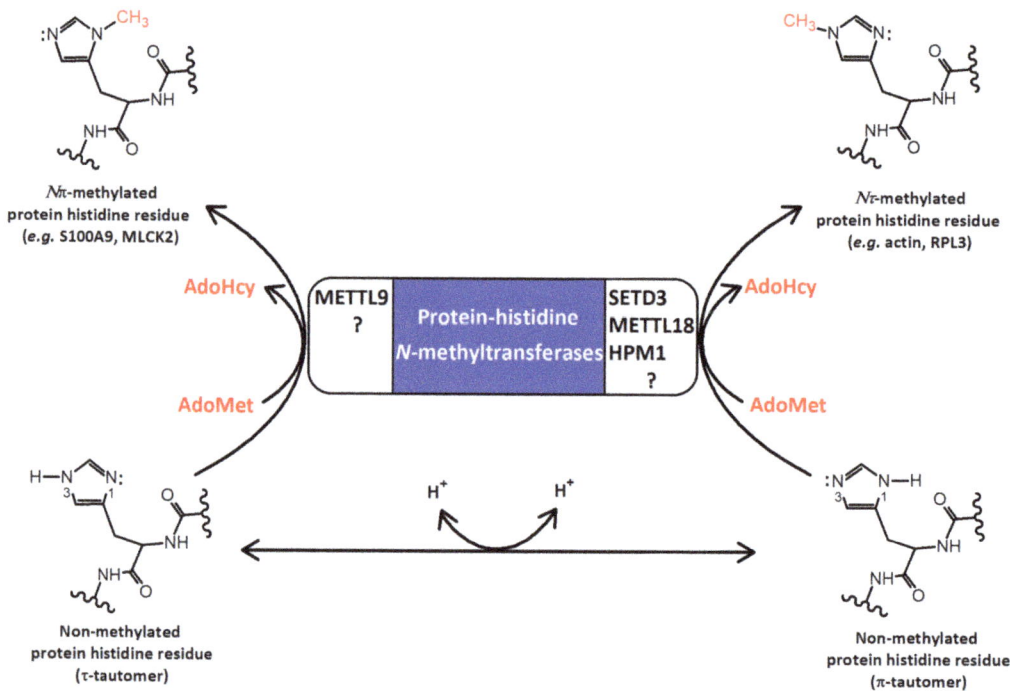

Figure 1. Reactions catalyzed by protein histidine N-methyltransferases. At pH ≈ 7, two neutral tautomers of histidine residues may exist in proteins: the N1-protonated π-tautomer and the N3-protonated τ-tautomer. Data show that different protein histidine methyltransferases catalyze the transfer of a methyl group from S-adenosyl-L-methionine (AdoMet) to specific nitrogen of the imidazole ring. HPM1, SETD3, METTL9, and METTL18 are the only enzymes characterized with this activity so far. AdoHcy—S-adenosyl-L-homocysteine.

The actin cytoskeleton, which is involved in a variety of central cellular processes, such as cell growth, division, and motility, has long been known to undergo different post-translational modifications [8]. In 1967, Johnson and colleagues isolated actin from various vertebrate species, and demonstrated that Nτ-methylhistidine is a natural component of this protein and a product resulting from enzymatic methylation [9]. A similar finding was reported by Asatoor and Armstrong [10]. Later, attempts were made to determine the amino acid sequence around methylhistidine in skeletal muscle actin [11] and establish the biochemical importance of methylation in actin functions [12]. By the late 1970s, it was confirmed that only a single histidine residue in actin is Nτ-methylated, and the residue is located precisely at position 73 of the amino acid sequence [13]. However, it was only in 1987 that the presence of actin histidine methyltransferase in the myofibrillar fraction of rabbit muscle was shown for the first time [14]. The advent of recombinant DNA technology allowed better characterization of a partially purified rabbit enzyme by using nonmethylated recombinant actin which was heterologously expressed in *Escherichia coli* and a synthetic peptide corresponding to residues 69–77 of actin [15]. In addition, it was also proved that rabbit skeletal muscle is a source of two different histidine methyltransferases. The first of these enzymes was specific for actin, while the second one—carnosine N-methyltransferase—converts carnosine (β-alanyl-L-histidine) into anserine (β-alanyl-Nπ-methyl-L-histidine) dipeptides, which are abundantly present in mammalian skeletal muscle. The carnosine-methylating enzyme was later identified as the UPF0596 protein, in eukaryotes [16]. Finally, pioneering studies carried out in 2002, employing actin monomers in methylated or nonmethylated forms, revealed that the methylation of actin at histidine 73 (H73) may facilitate its polymerization [3]. However, since these results were based on a functional comparison of actin monomers isolated from two different species—*Saccharomyces cerevisiae* and cow—their interpretation was difficult and the biological significance of such modification was uncertain.

Only recently, a putative histone lysine methyltransferase, SETD3, has been identified as actin specific histidine *N*-methyltransferase, and shown to regulate cytoskeleton assembly and modulate smooth muscle contractility [17,18] (Figure 1). This finding encouraged the scientific community to conduct more systematic searches for novel protein histidine methyltransferases and their substrates. Indeed, it was recently found that METTL9 methyltransferase acts as a broad specificity enzyme, catalyzing the formation of the majority of $N\pi$-methylhistidine residues in the human proteome, including S100A9 and NDUFB3 proteins [19]. This was also confirmed by Lv and colleagues, who established that METTL9 recognizes an xHxH motif in substrate proteins [20], whereas proteomic studies indicated that the motif is mainly present in human proteins that are methylated at histidine residues [21]. Moreover, the human METTL18 enzyme was shown to $N\tau$-methylate histidine 245 in ribosomal protein RPL3 [22,23], and, thus, resembles its yeast homolog HPM1 protein [7,24]. Histidine methylation has now been found to be prevalent in human cells, involving hundreds of intracellular proteins, which implies that the human proteome may contain several unidentified protein histidine methyltransferases [21].

In this review, we discuss the current advances in research on the SETD3 protein that were stimulated by its identification as the first protein histidine *N*-methyltransferase in metazoans and the renewed interest in histidine methylation as an important mechanism regulating protein functions.

2. The Structural Features of SETD3

SETD3 has a core SET domain (Su(var)3-9, Enhancer-of-zeste (E(z)), and Trithorax (Trx)), which is found in various proteins. In *Drosophila melanogaster*, all these genes code for proteins engaged in posttranslational modifications of histone H3 and transcriptional regulation: (i) Su(var)3-9 encodes [histone H3]-lysine(9) *N*-methyltransferase (EC 2.1.1.355), (ii) E(z) encodes [histone H3]-lysine(27) *N*-trimethyltransferase (EC 2.1.1.356), and (iii) Trx encodes [histone H3]-lysine(4) *N*-methyltransferase (EC 2.1.1.354). The SET domain is typical for enzymes exhibiting methyltransferase activity, and, as indicated by the names of the above mentioned enzymes, the presence of this domain is often associated with methyltransferase activity on lysine residues within the protein substrate. Indeed, SETD3 was initially identified as histone lysine *N*-methyltransferase [25,26], although the enzyme was shown to function as an actin specific histidine *N*-methyltransferase [17,18]. Interestingly, a follow up study by Dai et al. [27] demonstrated that the substitution of histidine by methionine in the actin derived peptide increases its affinity for the SETD3 protein by 76-fold. On the other hand, the substitution of lysine with methionine at K27 and K36 residues was found in histone H3.3 [28,29]. At present, the oncogenic effects of these substitutions are primarily linked with the perturbation of proper lysine methylation [30]. However, the results of Dai et al. [27] suggest that SETD3 in vivo may act as a methionine methyltransferase.

2.1. Domain Architecture

The human SETD3 protein (NCBI Protein: NP_115609.2) consists of 594 amino acid residues and has a molecular weight of 67.26 kDa. In addition to the well characterized isoform 1, there are two isoforms containing 296 and 286 amino acids, respectively. The structural characteristics described hereafter refer to isoform 1.

SETD3 has a 250-residue long SET domain (residues 80–329) which ensures specific recognition of the actin derived peptide, and most probably, the actin molecule itself. This domain is larger than a typical SET domain due to the presence of an inserted region (residues 131–254), designated as iSET. The regions that are responsible for AdoMet binding are located within the SET domain (residues 105–106, 275–279, and 313). Structural studies conducted in recent years have revealed the actual interactions occurring between

SETD3 and *S*-adenosyl-homocysteine (AdoHcy), which is a product of AdoMet demethylation [18,31], or sinefungin (SFG; adenosyl ornithine), which is an AdoMet analog lacking the ability to transfer a methyl group [32,33] and anticipated as a binding site of AdoMet.

The residues 350–475 of SETD3 are folded into a domain that structurally resembles the RuBisCO LSMT (large subunit methyltransferase) substrate binding domain [31]. In LSMT, the substrate binding domain interacts specifically with the RuBisCO large subunit [34,35]. Thus, it seems that the LSMT substrate binding domain present in the SETD3 protein may be involved in the recognition and binding of protein substrates, although experimental data supporting this hypothesis are scarce. The N-terminal and C-terminal regions (residues 1–22 and 549–594, respectively) of the SETD3 protein are considered to be disordered (Figure 2A).

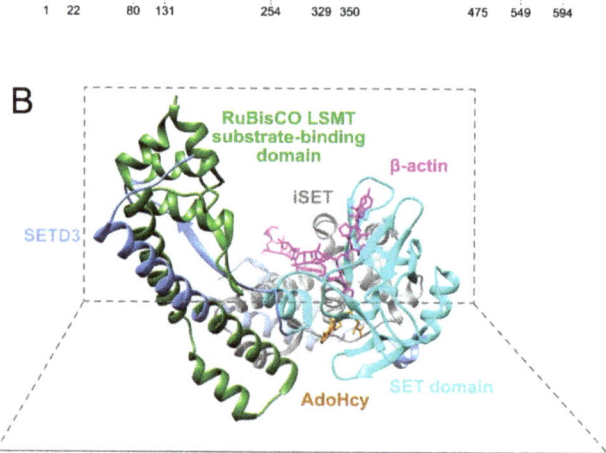

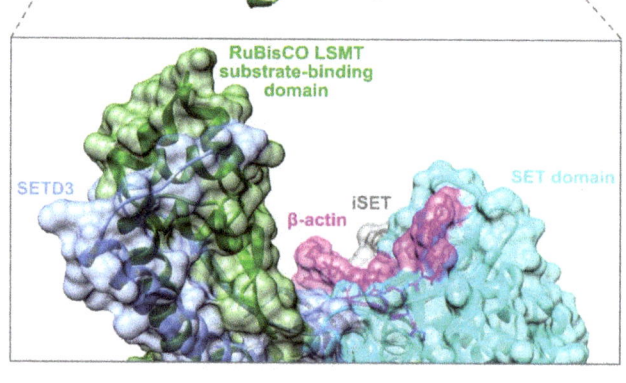

Figure 2. Structure of human SETD3. (**A**) Domain composition of SETD3. Waved lines correspond to the disordered regions at the N-terminal and C-terminal of the protein. Red bars indicate the localization of amino acid residues at which AdoMet binds to SETD3. Data were retrieved from the NCBI Protein database (accession number: XP_011535533.2, accessed on 30 July 2021). (**B**) Conformation of human SETD3 (residues 2–502) in complex with the peptide substrate derived from β-actin (residues 66–88) and close up view of the SETD3 substrate binding cleft with molecular surfaces. The image was created in UCSF Chimera 1.15 software utilizing the coordinates deposited in Protein Data Bank file 6ICV [36]. The color scheme of domains is common in Figures 2 and 3.

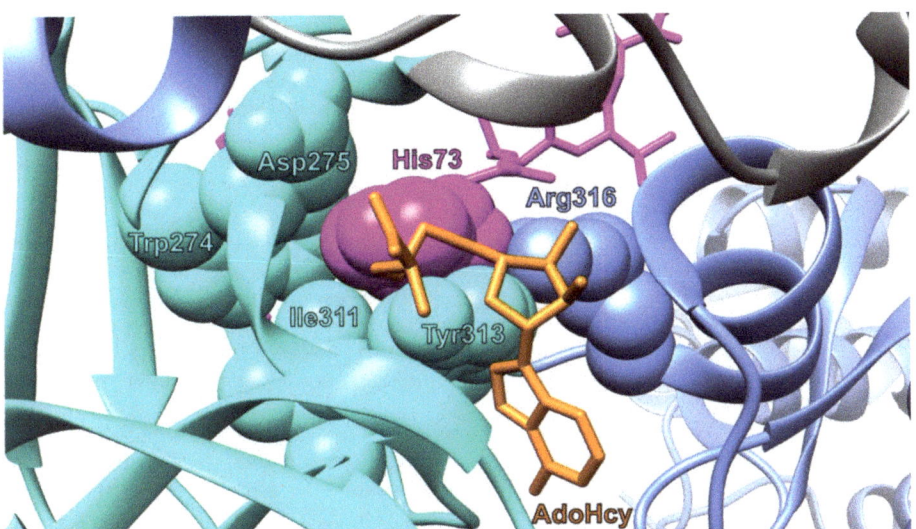

Figure 3. Amino acid residues of SETD3 that are important for proper recognition and alignment of H73 of β-actin to AdoMet prior to methylation. The image was created in UCSF Chimera 1.15 software utilizing the coordinates deposited in Protein Data Bank file 6ICV [36]. The color scheme of domains is common in Figures 2 and 3.

Several single amino acid substitutions can significantly influence the catalytic activity and/or specificity of SETD3. For example, Guo et al. [31] reported that R215A and R316A reduced the affinity of protein histidine N-methyltransferase for the actin derived peptide substrate, and decreased the enzyme activity.

A similar effect of diminished affinity to the actin derived peptide and lower enzyme activity was also found to be triggered by N256A and N256V substitutions [31,32], although the lowest binding affinity was observed with N256D substitution [31]. This finding suggests that the presence of a negative charge at this position may have a detrimental effect on substrate binding. However, the mentioned substitutions allow SETD3 to bind the variants of actin derived peptides with amino acid substitutions within the target sequence, and catalyze the methylation of lysine or methionine, as indicated above [27]. A different substitution at the same amino acid residue (N256F), in combination with W274A substitution, was also shown to trigger protein lysine methyltransferase activity to an actin derived peptide variant containing lysine, instead of histidine, in the target sequence [27].

Wilkinson et al. [18] observed that Y313A substitution affected the activity of SETD3 protein histidine N-methyltransferase, while Y313F substitution, which only removed the hydroxyl group present in the ortho position on the benzene ring, strongly decreased the binding of protein histidine N-methyltransferase to the actin fragment, as well as the enzyme activity [31]. This implies that the hydroxyl group of Y313 is critical for the proper recognition of the substrate by the SETD3 protein, and its catalytic activity.

2.2. Structure

The 3D structures of SETD3 in complex with an unmethylated or methylated actin-derived peptides were successfully determined by applying the X-ray diffraction crystallography technique. Both structures were solved using crystals containing AdoHcy, which was added to the buffer to prevent methylation of a peptide substrate. AdoHcy is one of the products of this reaction and occupies the catalytic pocket of the enzyme, thus preventing the binding of AdoMet [18,31]. Another approach involves the use of SFG, which fits into the catalytic pocket as AdoMet but does not transfer the methyl group [32,33].

Guo et al. [31] reported that the approach of cocrystallizing the SETD3 protein with full length actin was unsuccessful, and the obtained structures contained the core region of

SETD3 (residues 2–502) and the peptide substrate derived from β-actin (residues 66–88). Wilkinson et al. [18] and Dai et al. [32], on the other hand, used full length SETD3 and an actin peptide substrate (residues 66–80). In all these structures, the core region of SETD3 adopted a V shape, resembling the canonical SET domains found in Su(var)3-9, E(z), or Trx [37]. However, the SET domain of the SETD3 protein largely resembles the SET domain of LSMT [34], due to the presence of an inserted α-helical domain bisecting the SET domain, namely, iSET [31,38].

AdoHcy (and also most probably AdoMet) interacts with SETD3 in a cleft formed by the SET domain, which is additionally supported by a fragment of the iSET domain (Figure 2B). Its adenine ring is located between the side chain of E104 and the aromatic ring of F327. The AdoHcy N6 and N7 atoms are supported by hydrogen bonds formed with the main chain carbonyl and amide groups of H279, respectively, while its C8 atom forms a hydrogen bond with the hydroxyl group of Y313 [31]. The mode of interaction of AdoHcy with SETD3 is analogous to that observed in other SET containing enzymes, such as LSMT [34] and SETD6 [39].

The peptide substrate derived from β-actin interacts with SETD3 in a narrow cleft formed by the SET domain including the iSET region—in the same cleft where AdoHcy is located. However, the peptide substrate for histidine methylation is located at the lowest part of a wider cleft on the surface of SETD3. This spacious cleft might serve as an interaction site for larger unidentified protein substrates, together with the RuBisCO LSMT substrate binding domain (Figure 2B).

The methylated H73 residue of β-actin fits into a hydrophobic pocket formed by W274, I311, and Y313 of SETD3 [31] (Figure 3). The imidazole ring of H73 is aligned parallel to the aromatic ring of tyrosine 313. Its orientation is determined by two hydrogen bonds—one formed between the N1 and N3 atoms of the imidazole ring and another between the guanidino group of R316 and the carbonyl group in the main chain of N275 [31]. According to a recent study, the substrate binding pocket of SETD3 is charged in a way that corresponds to the surface charge of the actin fragment fitting to it, which also contributes to the proper alignment of the substrate to the enzyme [33].

Interestingly, the β-actin derived peptide adopts a 3_{10} helix at its C-terminus only when H73 is methylated. However, the overall structure of the complex is very similar to that before methylation, which is confirmed by a root mean square deviation of 0.19 and 0.32 Å over protein and peptide Cα atoms, respectively [31].

SETD3 structural investigations support the notion that the enzyme is primarily a histidine N-methyltransferase [17,18], and not a lysine N-methyltransferase, as it was initially classified [25,26]. The key argument for this is that the substrate-binding site of the SETD3 protein fits very well to the β-actin peptide, but it might be too shallow for the stable binding of the aliphatic side chain of a lysine residue. On the other hand, the wide cleft present above the substrate-binding pocket may allow the interaction of SETD3 with other protein substrates.

It is worth noting, though, that substitutions of N256 in SETD3 to other amino acid residues influence the substrate-binding affinity and/or specificity. Importantly, in the case of structurally similar SET-domain-containing (SETD) enzymes, such as LSMT or SETD6, this position may contain a phenylalanine residue, which is responsible for enzyme interaction with the lysine side chain present in substrate proteins [31]. These findings substantiate the reclassification of SETD3 as a histidine N-methyltransferase.

2.3. Paralogs

The existence of SETD3 paralogs is still unknown. However, based on the amino acid sequence, it can be suggested that the SETD4 protein, with 40% similarity and 24% identity, may be considered as a potential paralog. SETD4 is a histone lysine N-methyltransferase (EC 2.1.1.364), which catalyzes the methylation of histones H3 and H4 at K4 and K30 residues, respectively. It was reported that this enzyme regulates cell proliferation, differentiation, inflammatory response, and heterochromatin formation [40].

The domain structure of SETD4 resembles that of SETD3. Although the amino acid sequence of SETD4 is shorter than that of SETD3 and contains only 440 residues, the SET domain consisting of 226 residues is in the central part of the protein (residues 48–273). The N-terminus of SETD4 is also disordered (residues 1–24), similar to SETD3.

In order to analyze the potential structural and functional convergence of SETD3 and SETD4, we predicted the structure of human SETD4 using the AlphaFold algorithm [41]. Interestingly, three out of the five residues participating in substrate binding in SETD3 (described below) are conserved in SETD4. Moreover, the Y313 residue, which ensures the appropriate alignment of the imidazole ring of histidine substrate in SETD3, is structurally conserved in SETD4 as Y272 (Figure 4). This may signify that SETD4 shows potential SETD3-like protein histidine N-methyltransferase activity, although no experimental evidence is available to confirm this hypothesis.

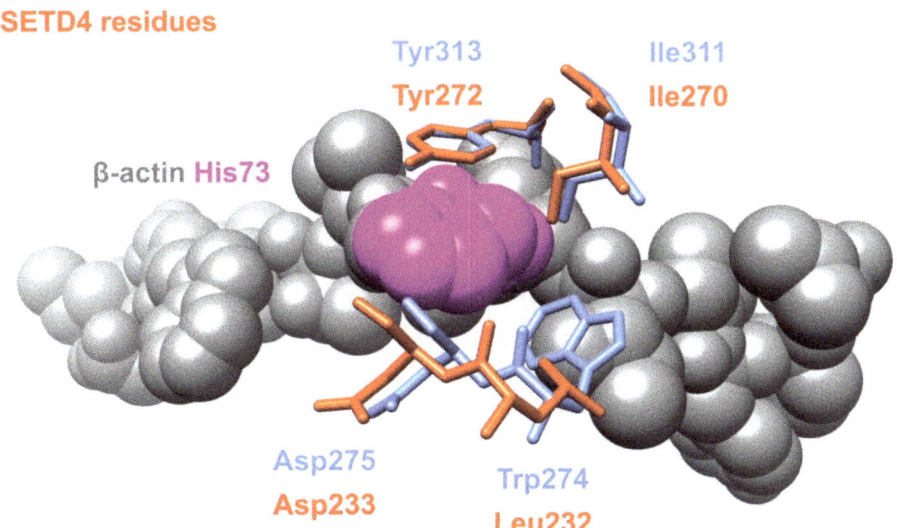

Figure 4. Structural alignment of SETD3 amino acid residues interacting with H73 of β-actin and conserved residues of SETD4. The image was created in UCSF Chimera 1.15 software utilizing the coordinates deposited in Protein Data Bank file 6ICV and the SETD4 structure predicted by AlphaFold [41] using UniProt Q9NVD3 record as an input. Structural alignment was calculated using the MatchMaker tool in UCSF Chimera 1.15 software [36].

Notably, the overall fold of SETD3 is highly similar to that of RuBisCO LSMTs and SETD6, both of which are validated protein lysine methyltransferases. However, SETD3 has low sequence identity with RuBisCO LSMTs and SETD6 (24–25%) [31]. Therefore, these enzymes cannot be listed as closely related paralogs of SETD3, but it can be concluded that the fold of SETD3 is not unique.

3. The Biochemical Features of SETD3

For many years, SETD methyltransferases were exclusively considered as enzymes responsible for the methylation of specific lysine residues at histone proteins and thereby for maintaining and altering the histone code [42]. Nevertheless, this viewpoint gradually changed as more number of nonhistone substrates for SETD methyltransferases were discovered [43]. Not surprisingly, SETD3 was initially thought as an enzyme that catalyzes the modification of histone H3 at K4 and K36 residues and regulates muscle cell differentiation in mice [26]. This was later confirmed by Chen and colleagues [44], who, however, also suggested that SETD3 might act on other nonhistone substrates in the cytoplasm, as the enzyme contains RuBisCO LSMT substrate-binding domain. Once the consensus on its role as a lysine-methylating enzyme began to take shape, SETD3 was identified as a long sought, actin specific histidine N-methyltransferase that catalyzes H73 methylation in the actin protein of metazoans [17,18] (cf. Figures 1 and 5). This discovery was made by two independent research groups with their own dedicated research strategy. Studies performed in our laboratory [17] were based on the extensive purification of the native rat enzyme from leg muscles, using different chromatographic methods, and the subsequent molecular identification of the enzyme by tandem mass spectrometry. After two independent and slightly different rounds of purification, SETD3 methyltransferase was found as the only logical candidate for the enzyme. This discovery was then confirmed by generating recombinant homogenous rat and human SETD3 and determining their actin histidine-methylating activity. Finally, an analysis of SETD3 deficient *D. melanogaster* larvae and the human HAP1 knockout (KO) cell line proved that actin did not undergo histidine methylation in both the examined sources [17]. At the same time, Wilkinson and colleagues [18] analyzed previous evidence supporting the substrate specificity of SETD3 and questioned whether histones were appropriate substrates for this enzyme. To identify the proteins that are methylated by SETD3, recombinant human wild-type and catalytically inactive variants of SETD3 were prepared and incubated with a total cytoplasmic extract of human HT1080 cells in the presence of [^{3}H]AdoMet. Autoradiography analysis revealed that the only detected band corresponded to a protein with a molecular weight of ≈42 kDa. Then, using mass spectrometry, the potential substrates were purified and identified. The most likely candidates were produced in *E. coli* and tested as SETD3 substrates in vitro. It was observed that only actin was methylated by the enzyme. The specific actin residues modified by SETD3 were identified by tandem mass spectrometry. Unexpectedly, no lysine methylation events were detected on the actin protein, and instead, the H73 residue was unambiguously identified as the sole target of SETD3 [18].

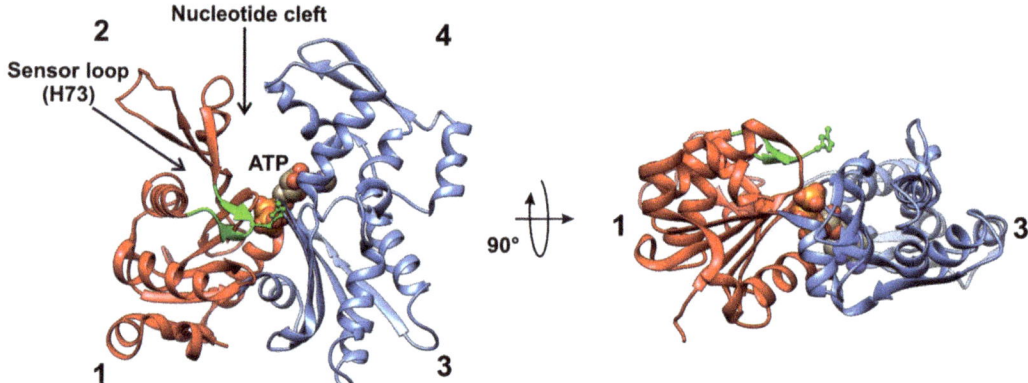

Figure 5. Structures of human β-actin. Ribbon representations of the structures of the actin monomer are shown in different projections. The actin molecule consists of small and large domains (red and blue, respectively), and each one is divided further into two subdomains: 1, 2, and 3, 4, respectively. ATP (or ADP) binds to the cleft between subdomains 2 and 4. The methyl-accepting H73 is located in a sensor loop spanning P70 to N78 (green). This residue is exposed to the surface of the actin monomer and seems to be easily accessible for SETD3. The model was prepared using UCSF Chimera [36] from the Protein Data Bank structures of β-actin (2BTF).

3.1. Substrate Specificity
3.1.1. Actin

In vitro and in vivo experiments have proven that actin is the only known bona fide substrate of SETD3. There are three main isoforms of this protein—α, β, and γ—which differ only by a few amino acids at their N-terminus [45]. Under physiological conditions, actin exists as a 42-kDa monomeric globular protein (G-actin) that binds ATP and spontaneously polymerizes into relatively stable filaments (F-actin). The G-actin molecule consists of small and large domains, which are further subdivided into subdomains 1, 2, and 3, 4, respectively (Figure 5). The cleft between subdomains 2 and 4 is occupied by ATP or ADP. The methyl-accepting H73 residue is located in a sensor loop (P70 to N78), inserted between subdomains 1 and 2. The residue is exposed to the surface of the actin monomer and can thus be easily accessed by SETD3 (Figure 5).

The activity of SETD3 on actin has, so far, been studied using two different substrates: homogenous recombinant human β-actin produced in *E. coli* and an array of synthetic peptides of varying lengths, corresponding to the sensor loop of actin. Of note, full length recombinant actin monomers were purified from bacterial inclusion bodies in denaturing conditions and refolded into a nucleotide free state that represents a quasinative and nonphysiological form of this protein [17]. As actin requires eukaryotic chaperonins for correct folding, it cannot be produced in its native form in bacteria [46].

Radiochemical studies employing quasinative actin and [^{3}H]AdoMet revealed the high affinity of human SETD3 toward both substrates with at least 60- and 300-fold lower K_M values ($\approx$0.8 and $\approx$0.1 μM) than their intracellular concentrations, respectively [17]. The enzyme was also found to exhibit slow activity with a k_{cat} value of about 0.7 min^{-1}, which seems to be typical for methyltransferases acting on protein residues [47]. More interestingly, a comparison of the activity of SETD3 on either recombinant actin produced in *E. coli* or protein produced in *S. cerevisiae*, indicated that the enzyme catalyzed the methylation of only nucleotide free actin from bacteria. Thus, the yeast produced protein, which was nonmethylated due to the lack of SETD3 homolog in *S. cerevisiae* and expected to have a native conformation, could not serve as a substrate for SETD3 unless it was purified in the nucleotide free form [17]. Based on these results, it was interpreted that SETD3 may act on a specific form of actin monomers, plausibly nucleotide free actin, in a complex with one or more actin-binding proteins of unknown identity. This hypothesis is consistent with the current knowledge about SETD methyltransferases. Many of these enzymes form complexes with different proteins, and those interactions are important for their catalytic activity and substrate specificity [42].

Structural and biochemical studies using actin peptides have provided valuable data on the substrate binding and catalytic mechanism of SETD3. It was reported that actin-derived peptides bind in a long groove at the surface of the SET domain of the enzyme, with the H73 residue located within the active site pocket [31,32] (Figure 2B). The affinity of binding was found to increase with increasing peptide length (K_M = 8.7 mM and 21 μM for 9-residue and 15-residue peptide, respectively) [17,32]. However, those peptides containing H73M or H73K mutation were still methylated at position 73 [27,48], which suggests that peptide recognition is mainly sequence specific, rather than targeted residue (histidine)-specific, and, thus, SETD3 can target proteins other than actin, at residues other than histidine. Moreover, the substrate specificity of SETD3 can be altered by engineering critical amino acids in its active site. Only recently, a mutated variant of SETD3 harboring N256F and W274A substitutions was shown to exhibit a 13-fold higher affinity for lysine over histidine [48].

3.1.2. Other Substrates

Studies on SETD3 employing peptide substrates allowed insight into the structural basis of H73 methylation and the catalytic reaction. However, it should be noted that this peptide based approach is a simplification. In fact, such a research model explores only local interactions occurring within the catalytic domain of SETD3, and ignores the entire spectrum of interactions occurring between the enzyme, particularly its RuBisCO LSMT substrate-binding domain, and the protein substrate. Thus, it is not unwise to speculate that RuBisCO LSMT is mainly responsible for controlling the substrate specificity of SETD3, and the enzyme may accept more substrates than only actin. Previous reports based on radiochemical assays have also shown that mammalian core histones, particularly histone

H3, were the substrates for SETD3 [25,26,44]. However, such an activity of the enzyme was not detected in other works [18]. This apparent discrepancy might be explained by different sources of nucleosomes used in enzymatic assays. It seems that SETD3 may act on the isolated native nucleosomes [26,44], but not on recombinant ones [18] or free histone octamers [44]. If true, the targeted amino acid residue(s) must be verified, as data supporting H3 methylation at K4 and K36 sites [25,26] are unconvincing [18]. Finally, Cohn and coworkers [49] have shown that human SETD3 interacts with about 170 different intracellular proteins, including actin, which suggests that there may be many other substrates for this enzyme in mammalian cells.

3.1.3. Inhibitors

Although the M73-containing peptide is a poor substrate for SETD3, it has been found to exhibit strong affinity to the enzyme and inhibit the methylation of the H73 peptide. Based on this observation, actin based peptidomimetics that act as effective substrate competitive inhibitors of human SETD3 were developed [50]. These are 16-residue-long analogs of the actin peptide (66–81), in which the H73 residue is substituted by a simple natural or non-natural amino acid. Among an array of tested peptide analogs, selenomethionine-containing actin peptide was identified as the most potent inhibitor of the human enzyme, with an IC_{50} value of 0.16 µM.

3.2. Reaction Mechanism

The imidazole ring of the histidine residue contains two nitrogen atoms at different positions: 1 (π) and 3 (τ) (Figure 1). These nitrogen atoms can be protonated, resulting in the formation of an imidazolium cation, and each of them can subsequently release a proton to produce a different imidazole tautomer (Figure 1). Both fully protonated and tautomeric forms of the imidazole side chain are believed to be present at physiological pH $\approx$ 7 in proteins [51]. Similar to other AdoMet dependent methyltransferases, SETD3 appears to catalyze a conventional S_N2 methylation reaction, in which the methyl group of AdoMet is transferred to the deprotonated $N\tau$ nitrogen [32] (Figure 6). To facilitate this reaction, the side chain of N256 of the enzyme stabilizes the $N\pi$ nitrogen of the substrate H73 residue in the protonated form, whereas the lone electron pair present at the deprotonated $N\tau$ attacks the methyl group of AdoMet. This model of SETD3 catalysis is consistent with the findings that (i) the enzyme has an optimum pH of 7 and above for H73 methylation (pKa of 6.5 for histidine imidazole) [31], whereas a K73-containing actin peptide is readily methylated only at a pH above 9.5 (pKa of 10.5 for lysine side chain) [52], and (ii) the substitution of N256 by amino acids that cannot form a hydrogen bond with the protonated $N\pi$ nitrogen results in a reduction or complete loss of SETD3 activity toward H73 residue [48].

3.3. Tissue Distribution and Intracellular Localization

The SETD3 protein or its orthologs are present in most of the eukaryotic organisms, including vertebrates (*Homo sapiens*, *Mus musculus*), plants (*Vitis vinifera*), insects (*Onthophagus taurus*, *D. melanogaster*), and fungi (but not in *S. cerevisiae*) [17]. The profile of SETD3 expression in humans shows relatively low tissue specificity (Figure 7).

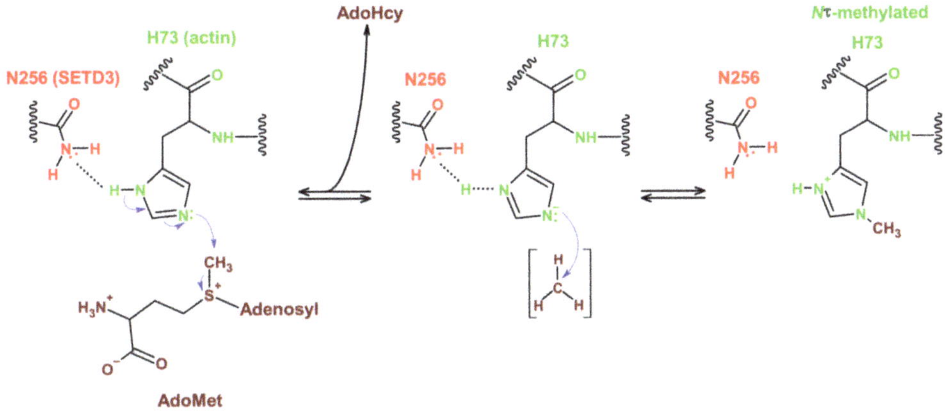

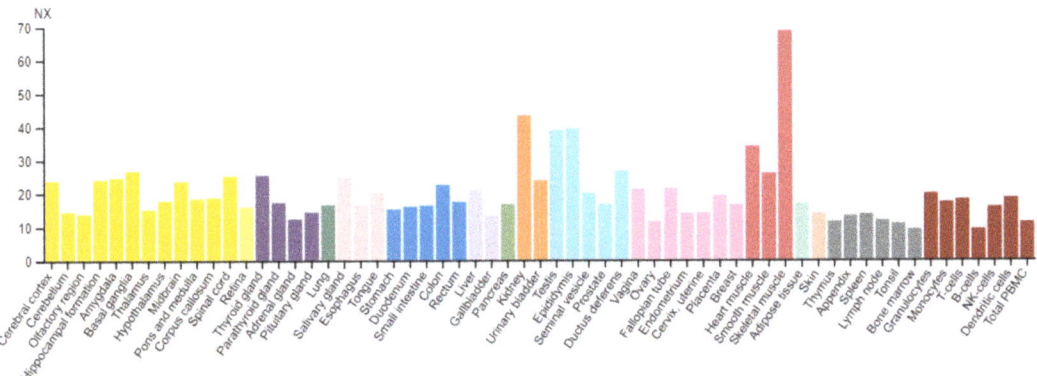

Figure 6. Plausible mechanisms of actin histidine $N\tau$-methylation by SETD3. The enzyme catalyzes the methylation of $N\tau$ nitrogen atom of H73 residue in actin. The methyl group of AdoMet can be transferred only to the deprotonated nitrogen atom. Since each of the two nitrogen atoms of the imidazole ring can hypothetically be protonated, the side chain of N256 residue of SETD3 stabilizes the $N\pi$ nitrogen atom of H73 in the protonated form [32]. Consequently, it enhances the nucleophilicity of the lone pair of electrons present at the deprotonated $N\tau$ nitrogen that may then attack the methyl group of AdoMet.

Figure 7. SETD3 expression in human tissues. RNA data were obtained from the Human Protein Atlas (HPA; https://www.proteinatlas.org, accessed on 29 July 2021) and show consensus normalized expression levels, determined by combining the data from three transcriptomic datasets (HPA, Genotype-Tissue Expression, and FANTOM5) [53]. Color coding is based on tissue groups, each consisting of tissues with common functional features.

The SETD3 mRNA is ubiquitously expressed at a similar basal level in most examined tissues, with the noticeable exception of the skeletal muscle, kidneys, and testes. The widespread expression of the enzyme is consistent with its function as an actin histidine methyltransferase because actin proteins are found in virtually all cells. The expression of STED3 has been shown to be highest in muscles, which is not surprising given the fact that muscle fibers are abundant in actin filaments [54]. This finding is also in good agreement with the enzymatic data, indicating the skeletal muscle as a rich source of actin specific histidine methyltransferase [14]. On the other hand, the augmented expression of SETD3 in kidneys and testes is more puzzling. It could be hypothesized that increased SETD3 expression is related to actin, which is an important protein in these two organs. It is well known that the dynamic remodeling of the actin cytoskeleton is important for efficient mammalian spermatogenesis [55] and for maintaining the functional structure of renal podocytes [56]. However, it cannot be ruled out that higher SETD3 expression in kidneys and testes is due to the role of this enzyme in the methylation of substrates other than actin. The intracellular localization of SETD3 is not well defined yet. Initial studies proposed that the enzyme is localized in the nucleus [26,49]. However, the enzyme was clearly detected in the cytosol [57] and mitochondria of mammalian cells [53].

4. The Cellular Features of SETD3

4.1. Biological Effect of Actin Methylation by SETD3

It is now clear that SETD3 is mainly actin histidine methyltransferase, and actin is its most important physiological substrate. However, the exact role of actin methylation is not clear.

4.1.1. Polymerization of Actin

The presence of actin filaments ensures the stable structure and internal movement of cells [58]. β-Actin is the main cytoskeleton protein [59]. Actin polymerization involves nucleation, elongation, and steady state phases [60], and closely correlates with the concentration of actin monomers. Monomers are stabilized by ATP or ADP binding, but neither dimer nor trimer is stable and are therefore present in an extremely low concentration in the intracellular environment. The oligomer is only partially protected by the addition of four subunits [58]. Actin polymerization is followed by the hydrolysis of ATP to ADP and phosphate [61], which results in the polarity of actin filaments. The pointed end (-) of the actin filament is disassembled more freely, ensuring the presence of subunits that are added at the opposite, barbed end (+). Thus, there exists a balance between filament shortening and elongation [45] (Figure 8). Furthermore, it is well established that the remodeling of filaments requires many different proteins, including myosin, cofilin, profilin, capping proteins, or the Arp2/3 complex. These proteins, for example, promote phosphate dissociation in F-actin or nucleotide exchange in its G form [58]. Methylation of the actin protein at H73 also seems to be implicated in its remodeling, indicating the biological importance of the SETD3 activity.

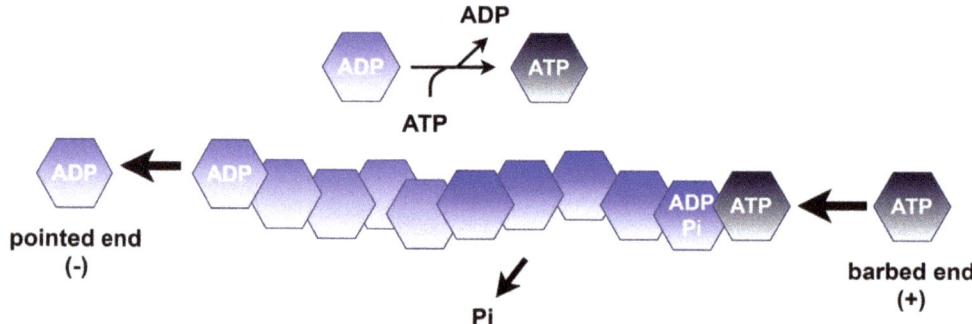

Figure 8. Scheme of nucleotide exchange under steady state during actin polymerization. During the steady state phase of polymerization, ADP–actin complexes dissociate from the pointed end (-) of the filamentous actin. This is followed by nucleotide exchange (from ADP to ATP) and, consequently, ATP–actin associates mainly at the barbed end (+). ATP hydrolysis allows the translocation of subunits between the ends of the filament [45]. SETD3 is found to promote actin polymerization through H73 methylation [18].

4.1.2. Effect of Actin H73 Methylation

Studies performed in the last 50 years attempted to elucidate the importance of H73 methylation in actin. Initially, it was indicated that such methylation is neither obligatory nor necessary for the proper functioning of actin [12,62]. Furthermore, actin with H73 substitutions by arginine or tyrosine residues was shown to polymerize as effectively as the nonmutated protein [62]. By contrast, a recent study revealed that lack of actin methylation affected the stability of actin monomers in SETD3-KO cells. The instability of actin monomers might lead to the accelerated depolymerization of actin fibers, and a loss of cytoskeleton integrity [17]. However, Wilkinson [18] reported that the methylation of actin promotes its polymerization, but without any impact on depolymerization. Thus, further research is needed to better understand the effect of H73 methylation on the stability of actin filaments.

4.2. The Cellular Roles of SETD3 and Association with Signaling Pathways

SETD3 is located mainly in the cytosol, and β-actin is the only cytosolic substrate described for this enzyme so far. However, it seems likely that the enzyme also acts on other substrates. Based on a proteomic approach, it was identified that more than 150 proteins, including cytoskeleton and signal proteins, receptors, hydrolases, and transcription factors, interact with SETD3 [49]. Therefore, it has been postulated that the enzyme may play a role in various biological processes, including myocyte differentiation [26], maintaining cytoskeleton integrity [17], cell cycle regulation and apoptosis [25], response to hypoxic conditions [49], carcinogenesis [44], and enterovirus (EV) pathogenesis [63].

The Functions of Cytosolic SETD3

In addition to its contribution to maintaining cytoskeleton integrity, SETD3 was shown to be involved in the pathogenesis of some EVs [63]. Although several studies have been performed on EVs, the precise mechanisms promoting their replication in target cells are unknown. It was shown that the formation of viral particles was diminished in SETD3-KO cells compared to wild type cells, which indicates that the enzyme supports the replication of viral genomes [63]. More interestingly, the level of replication in cells expressing the catalytically inactive SETD3 mutant was found to be in the control range, suggesting that the methyltransferase activity is not pivotal to viral multiplication. On the other hand, SETD3 was identified to strongly interact with viral protease 2A, and this interaction depends on the presence of both SET and RuBisCO LSMT domains in the enzyme structure [63]. It is well known that viral protease 2A, in combination with protease 3C, is essential for the completion of the EV life cycle. Neither the cleavage of the polyprotein into structural proteins during the replication cycle of EVs, nor the cleavage of the host protein, can occur without the activity of these proteases [64]. Moreover, they are implicated as possibly involved in suppressing stress and antiviral IFN-α/β responses [65]. These findings shed new light on the biological significance of the SETD3 protein, and highlight it as crucial for the successful reproduction of some EVs.

4.3. Other Postulated Functions of the SETD3 Protein

Attempts have been made to explore the potential role of SETD3 in carcinogenesis [44,66–68]. The available information collectively suggests the importance of SETD3 in the development and progression of cancer [44,49], as discussed in the next section.

The other assumed functions of SETD3, including myocyte differentiation, response to hypoxia, and cell cycle regulation, are attributed to the implied histone methylation by this enzyme or its nuclear localization.

As the first proposed activity of SETD3 was H3 methylation, its role in the epigenetic regulation of chromatin was also considered [25,26]. The abundant presence of SETD3 in muscles has been indicated to induce myocyte differentiation. In C2C12 or H9c2 cells, the overexpression of SETD3 activated the transcription of MCK, Myf6, and myogenin genes, which code for proteins involved in myocyte differentiation, whereas SETD3 knockdown was found to inhibit the differentiation of muscle cells. Nevertheless, the transcriptional activation of muscle-related genes by SETD3 needs to be confirmed by further research [26].

It has also been reported that the transcription factor FoxM1 is bound and methylated by SETD3 in vitro [49]. FoxM1 is crucial for the self renewal and proliferation of cells [69]. This is in line with the observation that SETD3 strongly interacted with FoxM1 at chromatin in normoxia, but its association with FoxM1 was weaker under hypoxic conditions. Furthermore, SETD3, along with FoxM1, regulated the expression of VEGF. The dissociation of both SETD3 and FoxM1 from the VEGF promoter was suggested to increase VEGF expression and promote angiogenesis in hypoxic conditions [49].

The functions of SETD3 reported by various studies are summarized in Table 1. Although literature data point out that SETD3 is associated with several signaling pathways, this protein has relatively recently been recognized to act mainly as actin histidine methyltransferase. This implies that its significance in biological processes is largely unexplored and warrants more studies in the future.

Table 1. Summary of the most important reported functions of SETD3 and its association with cell signaling pathways.

Process	SETD3 Activity	Proposed Functions	References
Actin polymerization	Methylation of H73 in β-actin	Stabilization of actin monomers	[17,18]
		Promoting actin polymerization	
Enterovirus pathogenesis	Unknown	Supporting viral genome replication	[63]
		Interaction with viral protease 2A	
Epigenetic regulation of chromatin	Methylation of core histones, plausibly H3	Regulation of gene expression	[25,26,44]
Response to hypoxia conditions	Methylation of FoxM1	Increasing expression of VEGF and promotion of angiogenesis	[49]
Carcinogenesis	Unknown	Regulation of cancer development and progression	[44,49,57,66–68,70–75]

5. The Role of SETD3 in Diseases

The knowledge about the role of SETD3 in the pathogenesis of various diseases remains limited. However, since the discovery and molecular characterization of SETD3 as a histone H3 methyltransferase [25,26,44] and further studies redefining its biological role as an actin H73 methyltransferase [17,18], a growing body of evidence has suggested that the protein may play an ambiguous role in diseases, especially cancer or other abnormalities. Therefore, the following part of the paper summarizes the most current knowledge regarding the potential involvement of the SETD3 protein in pathogenesis, as well as its role as a biomarker in various diseases.

5.1. Cancer

Although the precise role of SETD3 in carcinogenesis is still unclear, available data confirm that the protein might act either as a cancer suppressor or as an oncogenesis-promoting factor. Interestingly, the role of SETD3 varies in different abnormalities and is therefore difficult to comprehend. It was previously shown that an SET-domain-lacking fragment of the SETD3 gene translocated to the immunoglobulin lambda light chain *locus* in B-cell lymphomas [44], which resulted in the disruption of the SETD3 gene and appearance of a shorter form of the SETD3 protein lacking the SET domain. Unexpectedly, this form of the protein accumulated in cancer cells, where the wild type could not. The truncated SETD3 was proposed to act as a dominant negative mutant promoting oncogenesis [44]. Nevertheless, the exact mechanism underlying the oncogenic effect resulting from the overexpression of the short form of SETD3 in lymphoma remains unknown.

The level of the SETD3 protein was observed to fluctuate during the cell cycle [57]. Specifically, it was highest in the S phase, but declined during the progression to the M phase. Such dynamic cell cycle dependent regulation of expression implicates a potential role for SETD3 in carcinogenesis. Indeed, the level of SETD3 was shown to be elevated in hepatocellular carcinoma (HCC) [57]. Two hypothetical mechanisms have been proposed for the decreased degradation of SETD3. The first one involves the mutational burden on the β-isoform of the FBXW7β tumor suppressor protein, which is required for the ubiquitination and proteolysis of SETD3 [57]. On the other hand, a couple of Cdc4 phosphodegrons (CPDs) were identified in the SETD3 sequence, and one of them, CPD1, was shown to be phosphorylated specifically by GSK3β. Not surprisingly, either a decrease in the activity of FBXW7β or GSK3β or mutations within the CPD1 region reduced the extent of degradation of SETD3 [57]. Moreover, it was recently reported that SETD3 is a poor prognostic biomarker in HCC patients [67] and patients with a high level of the protein had lower rates of recurrence free survival and overall survival after surgery. In addition, in vitro and in vivo studies revealed that SETD3 promoted the progression of HCC [57]. The use of SETD3 targeted shRNA resulted in the depletion of the protein and significantly inhibited the variability and colony formation of HCC cells [57]. Similar results were observed with the use of a xenograft tumor model, where the application of shSETD3 resulted in a decreased volume and weight of the abnormal tissues [57]. Surprisingly, the SETD3 protein inhibited metastasis in HCC cells. In vitro studies performed with Hep3B and SK-Hep-1 cell lines showed that SETD3 knockdown led to increased migration and invasion [67]. Furthermore, the SETD3-deficient SK-Hep-1 cells exhibited higher metastatic activity in the mice model than cells containing the functional gene [67]. In addition to promoting metastasis, the SETD3 protein was shown to regulate the expression of serine/threonine-protein kinase DCLK1 by DNA methylation. However, the exact role of SETD3 in DNA methylation remains to be investigated [67], while its DNA-methylating activity has never been described before.

It was recently reported that circRNA transcribed from SETD3 gene exons 2–6 was downregulated in HCC, and the level of the circSETD3 transcript correlated with tumor size and the malignant differentiation of HCC [70]. CircSETD3 is postulated to act as an miRNA sponge that downregulates the level of miR-421, an essential promoter of HCC. Intriguingly, the latest report on the role of circSETD in nasopharyngeal carcinoma revealed the opposite function of circSETD, and indicated that the transcript seems to promote the migration and invasiveness of nasopharyngeal carcinoma [71] by attenuating miR-615-5p and miR-1538. This, in turn, results in the upregulation of MAPRE1 expression and inhibition of α-tubulin acetylation [71]. Thus, the actual role of circSETD3 in carcinogenesis is unclear.

The role of SETD3 in breast cancer is largely determined by the expression of hormone receptors and the mutational status of the p53 protein. In triple negative breast cancer patients with a mutational burden within the p53 protein, the higher level of SETD3 protein was found to correlate with poor prognosis [68]. By contrast, in patients with estrogen receptor positive breast cancer, a higher level of SETD3 correlated with better clinical outcomes [68]. The SETD3 protein has been shown to regulate the expression of various genes associated with cancer progression, including FOXM1, ACTB, ASMA, ACTG, FSCN, and FBXW7. However, the regulation by SETD3 seems to be cell specific [68], and thus, it is difficult to decipher the role and mechanism of this protein.

The SETD3 protein was also implicated in the resistance of cervical cancer (CC) to radiotherapy [72]. With the use of the radioresistant SiHa cell line and a parental cell line lacking radioresistance, it was demonstrated that the level of the SETD3 protein negatively correlated with radioresistance, and its expression was downregulated in radiotherapy-resistant SiHa cells. Analysis of clinical samples from radiotherapy prone and resistant patients revealed comparable results [72]. The finding that SETD3 knockdown decreased the rate of cell death, DNA damage, and apoptosis raised a question regarding the mechanism involved in the protective effect of the SETD3 protein. The elevated level of this protein in CC was associated with decreased expression of KLC4, which was previously shown to participate in cell death by regulating DNA damage response in lung cancer cell lines [73]. However, additional studies are required for further clarification of the function of SETD3 in CC.

The SETD3 protein has been recently proven to act as a regulator of cell apoptosis [74] in colon cancer. Its higher expression was positively correlated with the rate of programmed cell death following doxorubicin treatment. A total of 215 proteins have been identified to interact with the overexpressed SETD3 protein, among which some are linked to RNA metabolism. However, the role of SETD3 in RNA metabolism remains to be investigated [74]. Interestingly, it was also shown that apoptosis was maintained only by the wild type SETD3 protein, while the substitution of tyrosine 313 to alanine (Y313A) attenuated the effect of the protein on the process. This suggests that the methylating activity of SETD3 might be crucial in the regulation of apoptosis [74]. SETD3 was also found to act as a positive regulator of the p53 protein, although it did not directly interact with or methylate the p53 protein [74].

The SETD3 protein may act as a prognostic biomarker in cancer. It was proposed that SETD3, along with the N-lysine methyltransferase SMYD2 and bifunctional lysine specific demethylase and histidyl-hydroxylase NO66, can be helpful in the diagnosis and prognosis of renal cell tumors [66]. Furthermore, clinical data proved that the downregulation of those proteins correlated with shorter disease specific and disease free survival [66]. Similarly, among different methyltransferases, the SETD3 protein was identified to be a key player in the progression of bladder cancer [66]. Nevertheless, the significance of the protein in this particular cancer has not been investigated so far and needs to be studied in the future. The SETD3 protein also seems to have a prognostic value in clear cell ovarian carcinoma [75].

The role of the SETD3 protein in oncogenesis is ambiguous because it may act as an oncoprotein and increase the effectiveness of anticancer therapies (i.e., radiotherapy or doxorubicin treatment). SETD3 might also be helpful to stratify patients according to clinical prognosis. However, additional studies should be performed to obtain more detailed data on the role(s) of SETD3 in the development of various malignancies, their progression, and invasiveness. Several studies published so far have focused on the role of the SETD3 protein in cancer, while only a few have addressed the potential involvement of this protein in other pathologies.

5.2. Other Diseases

As mentioned in Section 4.3, the SETD3 protein has been shown to be involved in the transcriptional regulation of VEGF expression under normoxia and hypoxia [49]. Under hypoxic conditions, the attenuated interaction of the SETD3-FoxM1 complex and promotion of the VEGF expression may result in the onset of hypoxic pulmonary hypertension [76]. On the other hand, overexpression of the SETD3 protein limits VEGF expression and HIF-1 activation and, thus, protects against hypoxic pulmonary hypertension [76].

It was recently shown that the SETD3 protein might be involved in the progression of autoimmune diseases, including systemic lupus erythematosus (SLE) [77]. The disease is associated with an elevated level of CXCR5 in CD4$^+$ follicular helper T cells [77]. CXCR5 promotes the migration and interaction of T cells with B cells which, in turn, results in the formation of plasma cells through the interaction of PD-1 with its ligands (PD-1L and PD-2L) and production of autoantibodies. The SETD3 protein was elevated in the SLE CD4$^+$ cells, and its level correlated with a higher expression of CXCR5 [77].

The SETD3 protein also has a protective effect on ischemia–reperfusion (I/R)-induced brain injury [78]. The level of SETD3 was found to be positively correlated with neuronal survival. The neuroprotective role of the protein was proposed to be related to the actin histidine-methylating activity and regulation of F-actin polymerization [78]. Physiologically, SETD3 expression was downregulated by the activity of PTEN phosphatase as a result of I/R-induced injury. In addition, the downregulation of SETD3 expression results in an increased level of reactive oxygen species, decreased mitochondrial membrane potential, and ATP production [78]. However, further studies are required to understand the mechanism underlying the complex crosstalk between the activity of PTEN phosphatase and the SETD3 protein in neurons.

Recently, it was reported that the actin histidine-methylating activity of the SETD3 protein plays a significant role in dystocia (delayed parturition) [18]. It was reported that the litter sizes of double mutated ($Setd3^{-/-}$) mice were smaller than those of the wild type mice or mice with one functional allele. Nevertheless, this observation was inconsistent with the lack of anatomical abnormalities within the pelvis, and so the association of SETD3 with secondary dystocia was excluded [18]. A relationship between H73 methylation and uterine smooth muscle contraction was also proposed and verified experimentally. It was noted that the depletion of the SETD3 protein and actin H73 methylation resulted in a decreased signal induced contraction of primary human myometrial cells, while the intrinsic contractions were not affected [18]. Moreover, contractions induced by oxytocin and endothelin-1 were restored only by the catalytically active SETD3 protein but not by its mutated inactive form. All these data support the hypothesis that actin H73 methylation influences the signal induced contraction of smooth muscles [18].

The SETD3 protein was also shown to be involved in enteroviral infections [63]. Employing two human EVs—rhinovirus C15 (RV-C15) and EV-D68—SETD3 was selected as a hypothetical host factor essential for the infectiousness of EVs. The potential contribution of SETD3 in the pathogenesis of EVs is described in Section 4. An in vivo study indicated that SETD3 deficient ($Setd3^{-/-}$) mice were viable and showed no symptoms of viral infection [63]. In the context of viral infections, the region encoding the SETD3 protein was recently shown to be an integration site in the precancerous human papillomavirus infections [79]. While only two reports are currently available regarding the importance of the SETD3 protein in viral contagiousness, it is extremely important, taking into account the current pandemic status, to investigate the role of host proteins in the progression of viral infections.

6. Outlook

Although studies have established that SETD3 is the long sought, actin specific histidine N-methyltransferase, the biochemical properties of this protein as well as the cellular processes it regulates are yet to be understood in detail. For instance, the crystal structure of the SETD3–actin complex has not been deciphered and attempts made so far to crystalize the complex were unsuccessful [31]. A possible explanation for this failure could be that the actual physiological form of actin bound and subsequently methylated by SETD3 is not known, and whether the substrate is F-actin, G-actin, or, perhaps, G-actin in a complex with unidentified protein(s) should be verified. However, data collected from experiments involving the purification of native SETD3 showed that the enzyme is tightly bound to myofibrils, suggesting that it forms a relatively stable complex with myofibrillar proteins [14,17].

Further work is needed to explain the functions of SETD3 methyltransferase in the cell nucleus. One may hypothesize that nuclear SETD3 exhibits different substrate specificity and targets histone H3, as has been previously shown for isolated human nucleosomes [44]. Intriguingly, avian histones were reported to undergo $N\tau$–methylation at histidine residues [80], and so it would be interesting to verify whether SETD3 might be responsible for such modification. If true, SETD3 would be recognized as another dual specificity protein methyltransferase whose target activity depends on its interaction with a specific (non)substrate protein(s) [81,82]. Alternatively, the enzyme might work as a scaffold protein, facilitating the formation of a yet unknown protein complex, similar to that observed in the case of enteroviral protease 2A [63].

The regulation of SETD3 activity is another topic that remains to be investigated. All studies to date have focused only on mammalian SETD3. However, the enzyme is prevalent in multicellular eukaryotes. Thus, it would be of considerable interest to analyze the orthologs from more evolutionarily distant species, particularly in the plant kingdom.

It is still unclear whether SETD3 catalyzes the methylation of histidine residues in plant proteins, and if so, what would be the physiological importance of SETD3 in plant species.

In conclusion, at the current research stage, our knowledge of the SETD3 protein seems to be in its infancy. Although a lot is known about the structure of SETD3 and the mechanism of actin H73 methylation, the understanding of the physiological importance of the enzyme is still very limited. Future research will need to address the above questions in more detail in order to gain in depth knowledge about SETD3.

Author Contributions: Writing—original draft preparation, A.W., S.K., T.I. and J.D.; visualization, A.W., T.I. and J.D. All authors have read and agreed to the published version of the manuscript.

Funding: This work was funded by the Opus 14 grant (UMO-2017/27/B/NZ1/00161) from the National Science Centre, Poland. SK was supported by the Foundation for Polish Science (FNP).

Institutional Review Board Statement: Not applicable.

Informed Consent Statement: Not applicable.

Data Availability Statement: Not applicable.

Conflicts of Interest: The authors declare no conflict of interest.

References

1. Walsh, C.T.; Garneau-Tsodikova, S.; Gatto, G.J., Jr. Protein posttranslational modifications: The chemistry of proteome diversifications. *Angew. Chem. Int. Ed. Engl.* **2005**, *44*, 7342–7372. [CrossRef]
2. Clarke, S.G. Protein methylation at the surface and buried deep: Thinking outside the histone box. *Trends Biochem. Sci.* **2013**, *38*, 243–252. [CrossRef]
3. Nyman, T.; Schüler, H.; Korenbaum, E.; Schutt, C.E.; Karlsson, R.; Lindberg, U. The role of MeH73 in actin polymerization and ATP hydrolysis. *J. Mol. Biol.* **2002**, *317*, 577–589. [CrossRef]
4. Raftery, M.J.; Harrison, C.A.; Alewood, P.; Jones, A.; Geczy, C.L. Isolation of the murine S100 protein MRP14 (14 kDa migration-inhibitory-factor-related protein) from activated spleen cells: Characterization of posttranslational modifications and zinc binding. *Biochem. J.* **1996**, *316*, 285–293. [CrossRef] [PubMed]
5. Elzinga, M.; Collins, J.H. Amino acid sequence of a myosin fragment that contains SH-1, SH-2, and Ntau-methylhistidine. *Proc. Natl. Acad. Sci. USA* **1977**, *74*, 4281–4284. [CrossRef]
6. Meyer, H.E.; Mayr, G.W. N pi-methylhistidine in myosin-light-chain kinase. *Biol. Chem. Hoppe-Seyler* **1987**, *368*, 1607–1611. [CrossRef] [PubMed]
7. Webb, K.J.; Zurita-Lopez, C.I.; Al-Hadid, Q.; Laganowsky, A.; Young, B.D.; Lipson, R.S.; Souda, P.; Faull, K.F.; Whitelegge, J.P.; Clarke, S.G. A novel 3-methylhistidine modification of yeast ribosomal protein Rpl3 is dependent upon the YIL110W methyltransferase. *J. Biol. Chem.* **2010**, *285*, 37598–37606. [CrossRef]
8. MacTaggart, B.; Kashina, A. Posttranslational modifications of the cytoskeleton. *Cytoskeleton* **2021**, *78*, 142–173. [CrossRef]
9. Johnson, P.; Harris, C.I.; Perry, S.V. 3-methylhistidine in actin and other muscle proteins. *Biochem. J.* **1967**, *105*, 361–370. [CrossRef] [PubMed]
10. Asatoor, A.M.; Armstrong, M.D. 3-methylhistidine, a component of actin. *Biochem. Biophys. Res. Commun.* **1967**, *26*, 168–174. [CrossRef]
11. Elzinga, M. Amino acid sequence aroung 3-methylhistidine in rabbit skeletal muscle actin. *Biochemistry* **1971**, *10*, 224–229. [CrossRef]
12. Johnson, P.; Perry, S.V. Biological activity and the 3-methylhistidine content of actin and myosin. *Biochem. J.* **1970**, *119*, 293–298. [CrossRef] [PubMed]
13. Elzinga, M.; Collins, J.H.; Kuehl, W.M.; Adelstein, R.S. Complete amino-acid sequence of actin of rabbit skeletal muscle. *Proc. Natl. Acad. Sci. USA* **1973**, *70*, 2687–2691. [CrossRef]
14. Vijayasarathy, C.; Rao, B.S. Partial purification and characterisation of S-adenosylmethionine:protein histidine N-methyltransferase from rabbit skeletal muscle. *Biochim. Biophys. Acta* **1987**, *923*, 156–165. [CrossRef]
15. Raghavan, M.; Lindberg, U.; Schutt, C. The use of alternative substrates in the characterization of actin-methylating and carnosine-methylating enzymes. *Eur. J. Biochem.* **1992**, *210*, 311–318. [CrossRef]
16. Drozak, J.; Piecuch, M.; Poleszak, O.; Kozlowski, P.; Chrobok, L.; Baelde, H.J.; de Heer, E. UPF0586 Protein C9orf41 Homolog Is Anserine-producing Methyltransferase. *J. Biol. Chem.* **2015**, *290*, 17190–17205. [CrossRef]
17. Kwiatkowski, S.; Seliga, A.K.; Vertommen, D.; Terreri, M.; Ishikawa, T.; Grabowska, I.; Tiebe, M.; Teleman, A.A.; Jagielski, A.K.; Veiga-da-Cunha, M.; et al. SETD3 protein is the actin-specific histidine N-methyltransferase. *eLife* **2018**, *7*, e37921. [CrossRef]
18. Wilkinson, A.W.; Diep, J.; Dai, S.; Liu, S.; Ooi, Y.S.; Song, D.; Li, T.M.; Horton, J.R.; Zhang, X.; Liu, C.; et al. SETD3 is an actin histidine methyltransferase that prevents primary dystocia. *Nature* **2019**, *565*, 372–376. [CrossRef] [PubMed]

19. Davydova, E.; Shimazu, T.; Schuhmacher, M.K.; Jakobsson, M.E.; Willemen, H.L.D.M.; Liu, T.; Moen, A.; Ho, A.Y.Y.; Małecki, J.; Schroer, L.; et al. The methyltransferase METTL9 mediates pervasive 1-methylhistidine modification in mammalian proteomes. *Nat. Commun.* **2021**, *12*, 891. [CrossRef]
20. Lv, M.; Cao, D.; Zhang, L.; Hu, C.; Li, S.; Zhang, P.; Zhu, L.; Yi, X.; Li, C.; Yang, A.; et al. METTL9 mediated N1-histidine methylation of zinc transporters is required for tumor growth. *Protein Cell* **2021**, 1–6. [CrossRef]
21. Kapell, S.; Jakobsson, M.E. Large-scale identification of protein histidine methylation in human cells. *NAR Genom. Bioinform.* **2021**, *3*, lqab045. [CrossRef] [PubMed]
22. Małecki, J.M.; Odonohue, M.F.; Kim, Y.; Jakobsson, M.E.; Gessa, L.; Pinto, R.; Wu, J.; Davydova, E.; Moen, A.; Olsen, J.V.; et al. Human METTL18 is a histidine-specific methyltransferase that targets RPL3 and affects ribosome biogenesis and function. *Nucleic Acids Res.* **2021**, *49*, 3185–3203. [CrossRef] [PubMed]
23. Matsuura-Suzuki, E.; Shimazu, T.; Takahashi, M.; Kotoshiba, K.; Suzuki, T.; Kashiwagi, K.; Sohtome, Y.; Akakabe, M.; Sodeoka, M.; Dohmae, N.; et al. METTL18-mediated histidine methylation on RPL3 modulates translation elongation for proteostasis maintenance. *bioRxiv* **2021**, 454307. [CrossRef]
24. Al-Hadid, Q.; Roy, K.; Chanfreau, G.; Clarke, S.G. Methylation of yeast ribosomal protein Rpl3 promotes translational elongation fidelity. *RNA* **2016**, *22*, 489–498. [CrossRef] [PubMed]
25. Kim, D.W.; Kim, K.B.; Kim, J.Y.; Seo, S.B. Characterization of a novel histone H3K36 methyltransferase setd3 in zebrafish. *Biosci. Biotechnol. Biochem.* **2011**, *75*, 289–294. [CrossRef] [PubMed]
26. Eom, G.H.; Kim, K.B.; Kim, J.H.; Kim, J.Y.; Kim, J.R.; Kee, H.J.; Kim, D.W.; Choe, N.; Park, H.J.; Son, H.J.; et al. Histone methyltransferase SETD3 regulates muscle cell differentiation. *J. Biol. Chem.* **2011**, *286*, 34733–34742. [CrossRef]
27. Dai, S.; Holt, M.V.; Horton, J.R.; Woodcock, C.B.; Patel, A.; Zhang, X.; Young, N.L.; Wilkinson, A.W.; Cheng, X. Characterization of SETD3 methyltransferase-mediated protein methionine methylation. *J. Biol. Chem.* **2020**, *295*, 10901–10910. [CrossRef]
28. Schwartzentruber, J.; Korshunov, A.; Liu, X.Y.; Jones, D.T.; Pfaff, E.; Jacob, K.; Sturm, D.; Fontebasso, A.M.; Quang, D.A.; Tonjes, M. Driver mutations in histone H3.3 and chromatin remodeling genes in paediatric glioblastoma. *Nature* **2012**, *482*, 226–231. [CrossRef] [PubMed]
29. Behjati, S.; Tarpey, P.S.; Presneau, N.; Scheipl, S.; Pillay, N.; Van Loo, P.; Wedge, D.C.; Cooke, S.L.; Gundem, G.; Davies, H.; et al. Distinct H3F3A and H3F3B driver mutations define chondroblastoma and giant cell tumor of bone. *Nat. Genet.* **2013**, *45*, 1479–1482. [CrossRef]
30. Lowe, B.R.; Maxham, L.A.; Hamey, J.J.; Wilkins, M.R.; Partridge, J.F. Histone H3 Mutations: An Updated View of Their Role in Chromatin Deregulation and Cancer. *Cancers* **2019**, *11*, 660. [CrossRef]
31. Guo, Q.; Liao, S.; Kwiatkowski, S.; Tomaka, W.; Yu, H.; Wu, G.; Tu, X.; Min, J.; Drozak, J.; Xu, C. Structural insights into SETD3-mediated histidine methylation on β-actin. *eLife* **2019**, *8*, e43676. [CrossRef]
32. Dai, S.; Horton, J.R.; Woodcock, C.B.; Wilkinson, A.W.; Zhang, X.; Gozani, O.; Cheng, X. Structural basis for the target specificity of actin histidine methyltransferase SETD3. *Nat. Commun.* **2019**, *10*, 3541. [CrossRef] [PubMed]
33. Zheng, Y.; Zhang, X.; Li, H. Molecular basis for histidine N3-specific methylation of actin H73 by SETD3. *Cell Discovery* **2020**, *6*, 3. [CrossRef] [PubMed]
34. Trievel, R.C.; Beach, B.M.; Dirk, L.M.; Houtz, R.L.; Hurley, J.H. Structure and catalytic mechanism of a SET domain protein methyltransferase. *Cell* **2002**, *111*, 91–103. [CrossRef]
35. Trievel, R.C.; Flynn, E.M.; Houtz, R.L.; Hurley, J.H. Mechanism of multiple lysine methylation by the SET domain enzyme Rubisco LSMT. *Nat. Struct. Biol.* **2003**, *10*, 545–552. [CrossRef]
36. Pettersen, E.F.; Goddard, T.D.; Huang, C.C.; Couch, G.S.; Greenblatt, D.M.; Meng, E.C.; Ferrin, T.E. UCSF Chimera–a visualization system for exploratory research and analysis. *J. Comput. Chem.* **2004**, *25*, 1605–1612. [CrossRef] [PubMed]
37. Dillon, S.C.; Zhang, X.; Trievel, R.C.; Cheng, X. The SET-domain protein superfamily: Protein lysine methyltransferases. *Genome Biol.* **2005**, *6*, 227. [CrossRef]
38. Raunser, S.; Magnani, R.; Huang, Z.; Houtz, R.L.; Trievel, R.C.; Penczek, P.A.; Walz, T. Rubisco in complex with Rubisco large subunit methyltransferase. *Proc. Nat. Acad. Sci. USA* **2009**, *106*, 3160–3165. [CrossRef]
39. Chang, Y.; Levy, D.; Horton, J.R.; Peng, J.; Zhang, X.; Gozani, O.; Cheng, X. Structural basis of SETD6-mediated regulation of the NF-kB network via methyl-lysine signaling. *Nucleic Acids Res.* **2011**, *39*, 6380–6389. [CrossRef]
40. Ye, S.; Ding, Y.F.; Jia, W.H.; Liu, X.L.; Feng, J.Y.; Zhu, Q.; Cai, S.L.; Yang, Y.S.; Lu, Q.Y.; Huang, X.T.; et al. SET Domain-Containing Protein 4 Epigenetically Controls Breast Cancer Stem Cell Quiescence. *Cancer Res.* **2019**, *79*, 4729–4743. [CrossRef]
41. Jumper, J.; Evans, R.; Pritzel, A.; Green, T.; Figurnov, M.; Ronneberger, O.; Tunyasuvunakool, K.; Bates, R.; Žídek, A.; Potapenko, A.; et al. Highly accurate protein structure prediction with AlphaFold. *Nature* **2021**, *596*, 583–589. [CrossRef]
42. Herz, H.M.; Garruss, A.; Shilatifard, A. SET for life: Biochemical activities and biological functions of SET domain-containing proteins. *Trends Biochem. Sci.* **2013**, *38*, 621–639. [CrossRef]
43. Lukinović, V.; Casanova, A.G.; Roth, G.S.; Chuffart, F.; Reynoird, N. Lysine Methyltransferases Signaling: Histones are Just the Tip of the Iceberg. *Curr. Protein Pept. Sci.* **2020**, *21*, 655–674. [CrossRef] [PubMed]
44. Chen, Z.; Yan, C.T.; Dou, Y.; Viboolsittiseri, S.S.; Wang, J.H. The role of a newly identified SET domain-containing protein, SETD3, in oncogenesis. *Haematologica* **2013**, *98*, 739–743. [CrossRef] [PubMed]
45. Kudryashov, D.S.; Reisler, E. ATP and ADP actin states. *Biopolymers* **2013**, *99*, 245–256. [CrossRef]

46. Stemp, M.J.; Guha, S.; Hartl, F.U.; Barral, J.M. Efficient production of native actin upon translation in a bacterial lysate supplemented with the eukaryotic chaperonin TRiC. *Biol. Chem.* **2005**, *386*, 753–757. [CrossRef]
47. Frankel, A.; Brown, J.I. Evaluation of kinetic data: What the numbers tell us about PRMTs. *Biochim. Biophys. Acta Proteins Proteom.* **2019**, *1867*, 306–316. [CrossRef] [PubMed]
48. Dai, S.; Horton, J.R.; Wilkinson, A.W.; Gozani, O.; Zhang, X.; Cheng, X. An engineered variant of SETD3 methyltransferase alters target specificity from histidine to lysine methylation. *J. Biol. Chem.* **2020**, *295*, 2582–2589. [CrossRef]
49. Cohn, O.; Feldman, M.; Weil, L.; Kublanovsky, M.; Levy, D. Chromatin associated SETD3 negatively regulates VEGF expression. *Sci. Rep.* **2016**, *6*, 37115. [CrossRef]
50. Hintzen, J.C.J.; Moesgaard, L.; Kwiatkowski, S.; Drozak, J.; Kongsted, J.; Mecinović, J. β-Actin Peptide-Based Inhibitors of Histidine Methyltransferase SETD3. *Chem. Med. Chem.* **2021**, *16*, 2695–2702. [CrossRef]
51. Bachovchin, W.W.; Roberts, J.D. Nitrogen-15 nuclear magnetic resonance spectroscopy. The state of histidine in the catalytic triad of .alpha.-lytic protease. Implications for the charge-relay mechanism of peptide-bond cleavage by serine proteases. *J. Am. Chem. Soc.* **1978**, *100*, 8041–8047. [CrossRef]
52. Grimsley, G.R.; Scholtz, J.M.; Pace, C.N. A summary of the measured pK values of the ionizable groups in folded proteins. *Protein Sci.* **2009**, *18*, 247–251. [CrossRef] [PubMed]
53. Human Protein Atlas. Available online: https://www.proteinatlas.org/ENSG00000183576-SETD3/cell (accessed on 29 July 2021).
54. Sanger, J.W.; Wang, J.; Fan, Y.; White, J.; Mi-Mi, L.; Dube, D.K.; Sanger, J.M.; Pruyne, D. Assembly and Maintenance of Myofibrils in Striated Muscle. *Handb. Exp. Pharmacol.* **2017**, *235*, 39–75. [CrossRef]
55. Su, W.; Mruk, D.D.; Cheng, C.Y. Regulation of actin dynamics and protein trafficking during spermatogenesis—Insights into a complex process. *Crit. Rev. Biochem. Mol. Biol.* **2013**, *48*, 153–172. [CrossRef]
56. Sever, S.; Schiffer, M. Actin dynamics at focal adhesions: A common endpoint and putative therapeutic target for proteinuric kidney diseases. *Kidney Int.* **2018**, *93*, 1298–1307. [CrossRef]
57. Cheng, X.; Hao, Y.; Shu, W.; Zhao, M.; Zhao, C.; Wu, Y.; Peng, X.; Yao, P.; Xiao, D.; Qing, G.; et al. Cell cycle-dependent degradation of the methyltransferase SETD3 attenuates cell proliferation and liver tumorigenesis. *J. Biol. Chem.* **2017**, *292*, 9022–9033. [CrossRef] [PubMed]
58. Pollard, T.D. Actin and Actin-Binding Proteins. *Cold Spring Harb. Perspect. Biol.* **2016**, *8*, a018226. [CrossRef]
59. Dominguez, R.; Holmes, K.C. Actin structure and function. *Annu. Rev. Biophys.* **2011**, *40*, 169–186. [CrossRef]
60. DiNubile, M.J. Nucleation and elongation of actin filaments in the presence of high speed supernate from neutrophil lysates: Modulating effects of Ca^{2+} and phosphatidylinositol-4,5-bisphosphate. *Biochim. Biophys. Acta* **1998**, *1405*, 85–98. [CrossRef]
61. Choua, S.Z.; Pollarda, T.D. Mechanism of actin polymerization revealed by cryo-EM structures of actin filaments with three different bound nucleotides. *Proc. Natl. Acad. Sci. USA* **2019**, *116*, 4265–4274. [CrossRef]
62. Solomon, L.R.; Rubenstein, P.A. Studies on the role of actin's N tau-methylhistidine using oligodeoxynucleotide-directed site-specific mutagenesis. *J. Biol. Chem.* **1987**, *262*, 11382–11388. [CrossRef]
63. Diep, J.; Ooi, Y.S.; Wilkinson, A.W.; Peters, C.E.; Foy, E.; Johnson, J.R.; Zengel, J.; Ding, S.; Weng, K.F.; Laufman, O.; et al. Enterovirus pathogenesis requires the host methyltransferase SETD3. *Nat. Microbiol.* **2019**, *4*, 2523–2537. [CrossRef] [PubMed]
64. Laitinen, O.H.; Svedin, E.; Kapell, S.; Nurminen, A.; Hytönen, V.P.; Flodström-Tullberg, M. Enteroviral proteases: Structure, host interactions and pathogenicity. *Rev. Med. Virol.* **2016**, *26*, 251–267. [CrossRef]
65. Visser, L.J.; Langereis, M.A.; Rabouw, H.H.; Wahedi, M.; Muntjewerff, E.M.; de Groot, R.J.; van Kuppeveld, F.J.M. Essential role of enterovirus 2A protease in counteracting stress granule formation and the induction of type I interferon. *J. Virol.* **2019**, *93*, e00222-19. [CrossRef]
66. Pires-Luís, A.S.; Vieira-Coimbra, M.; Vieira, F.Q.; Costa-Pinheiro, P.; Silva-Santos, R.; Dias, P.C.; Antunes, L.; Lobo, F.; Oliveira, J.; Gonçalves, C.S.; et al. Expression of histone methyltransferases as novel biomarkers for renal cell tumor diagnosis and prognostication. *Epigenetics* **2015**, *10*, 1033–1043. [CrossRef]
67. Xu, L.; Wang, P.; Feng, X.; Tang, J.; Li, L.; Zheng, X.; Zhang, Y.; Hu, Y.; Lan, T.; Yuan, K.; et al. SETD3 is regulated by a couple of microRNAs and plays opposing roles in proliferation and metastasis of hepatocellular carcinoma. *Clin. Sci.* **2019**, *133*, 2085–2105. [CrossRef] [PubMed]
68. Hassan, N.; Rutsch, N.; Győrffy, B.; Espinoza-Sánchez, N.A.; Götte, M. SETD3 acts as a prognostic marker in breast cancer patients and modulates the viability and invasion of breast cancer cells. *Sci. Rep.* **2020**, *10*, 2262. [CrossRef]
69. Liao, G.B.; Li, X.Z.; Zeng, S.; Liao, G.B.; Li, X.Z.; Zeng, S.; Liu, C.; Yang, S.M.; Yang, L.; Hu, C.J.; et al. Regulation of the master regulator FOXM1 in cancer. *Cell Commun. Signal.* **2018**, *16*, 57. [CrossRef] [PubMed]
70. Xu, L.; Feng, X.; Hao, X.; Wang, P.; Zhang, Y.; Zheng, X.; Li, L.; Ren, S.; Zhang, M.; Xu, M. CircSETD3 (Hsa_circ_0000567) acts as a sponge for microRNA-421 inhibiting hepatocellular carcinoma growth. *J. Exp. Clin. Cancer Res.* **2019**, *38*, 98. [CrossRef]
71. Tang, L.; Xiong, W.; Zhang, L.; Wang, D.; Wang, Y.; Wu, Y.; Wei, F.; Mo, Y.; Hou, X.; Shi, L.; et al. CircSETD3 regulates MAPRE1 through miR-615-5p and miR-1538 sponges to promote migration and invasion in nasopharyngeal carcinoma. *Oncogene* **2021**, *40*, 307–321. [CrossRef]
72. Li, Q.; Zhang, Y.; Jiang, Q. SETD3 reduces KLC4 expression to improve the sensitization of cervical cancer cell to radiotherapy. *Biochem. Biophys. Res. Commun.* **2019**, *516*, 619–625. [CrossRef]

73. Baek, J.H.; Lee, J.; Yun, H.S.; Lee, C.W.; Song, J.Y.; Um, H.D.; Park, J.K.; Park, I.C.; Kim, J.S.; Kim, E.H.; et al. Kinesin light chain-4 depletion induces apoptosis of radioresistant cancer cells by mitochondrial dysfunction via calcium ion influx. *Cell Death Dis.* **2018**, *9*, 496. [CrossRef]
74. Abaev-Schneiderman, E.; Admoni-Elisha, L.; Levy, D. SETD3 is a positive regulator of DNA-damage-induced apoptosis. *Cell Death Dis.* **2019**, *10*, 74. [CrossRef] [PubMed]
75. Engqvist, H.; Parris, T.Z.; Kovács, A.; Rönnerman, E.W.; Sundfeldt, K.; Karlsson, P.; Helou, K. Validation of Novel Prognostic Biomarkers for Early-Stage Clear-Cell, Endometrioid and Mucinous Ovarian Carcinomas Using Immunohistochemistry. *Front. Oncol.* **2020**, *10*, 162. [CrossRef] [PubMed]
76. Jiang, X.; Li, T.; Sun, J.; Liu, J.; Wu, H. SETD3 negatively regulates VEGF expression during hypoxic pulmonary hypertension in rats. *Hypertens. Res.* **2018**, *41*, 691–698. [CrossRef]
77. Liao, J.; Luo, S.; Yang, M.; Lu, Q. Overexpression of CXCR5 in CD4$^+$ T cells of SLE patients caused by excessive SETD3. *Clin. Immunol.* **2020**, *214*, 108406. [CrossRef]
78. Xu, X.; Cui, Y.; Li, C.; Wang, Y.; Cheng, J.; Chen, S.; Sun, J.; Ren, J.; Yao, X.; Gao, J.; et al. SETD3 Downregulation Mediates PTEN Upregulation-Induced Ischemic Neuronal Death Through Suppression of Actin Polymerization and Mitochondrial Function. *Mol. Neurobiol.* **2021**, *15*. [CrossRef]
79. Garza-Rodríguez, M.L.; Oyervides-Muñoz, M.A.; Pérez-Maya, A.A.; Sánchez-Domínguez, C.N.; Berlanga-Garza, A.; Antonio-Macedo, M.; Valdés-Chapa, L.D.; Vidal-Torres, D.; Vidal-Gutiérrez, O.; Pérez-Ibave, D.C.; et al. Analysis of HPV Integrations in Mexican Pre-Tumoral Cervical Lesions Reveal Centromere-Enriched Breakpoints and Abundant Unspecific HPV Regions. *Int. J. Mol. Sci.* **2021**, *22*, 3242. [CrossRef]
80. Gershey, E.L.; Haslett, G.W.; Vidali, G.; Allfrey, V.G. Chemical studies of histone methylation. Evidence for the occurrence of 3-methylhistidine in avian erythrocyte histone fractions. *J. Biol. Chem.* **1969**, *244*, 4871–4877. [CrossRef]
81. Woodcock, C.B.; Yu, D.; Zhang, X.; Cheng, X. Human HemK2/KMT9/N6AMT1 is an active protein methyltransferase, but does not act on DNA in vitro, in the presence of Trm112. *Cell Discov.* **2019**, *5*, 50. [CrossRef]
82. Gao, J.; Wang, B.; Yu, H.; Wu, G.; Wan, C.; Liu, W.; Liao, S.; Cheng, L.; Zhu, Z. Structural insight into HEMK2-TRMT112-mediated glutamine methylation. *Biochem. J.* **2020**, *477*, 3833–3838. [CrossRef] [PubMed]

MDPI
St. Alban-Anlage 66
4052 Basel
Switzerland
Tel. +41 61 683 77 34
Fax +41 61 302 89 18
www.mdpi.com

Life Editorial Office
E-mail: life@mdpi.com
www.mdpi.com/journal/life

www.ingramcontent.com/pod-product-compliance
Lightning Source LLC
LaVergne TN
LVHW070147100526
838202LV00015B/1909